ECOLOGICAL TIME SERIES

edited by

Thomas M. Powell

Department of Intergrative Biology
University of California-Berkeley

John H. Steele

Woods Hole Oceanographic
Institution

CHAPMAN & HALL

I(T)P An International Thomson Publishing Company

New York • Albany • Bonn • Boston • Cincinnati
• Detroit • London • Madrid • Melbourne • Mexico City
• Pacific Grove • Paris • San Francisco • Singapore
• Tokyo • Toronto • Washington

Cover Design: Andrea Meyer, emDASH Inc.

Printed in the United States of America

For more information, contact:

Chapman & Hall
One Penn Plaza
New York, NY 10119

International Thomson Publishing
Berkshire House 168-173
High Holborn
London WC1V 7AA
England

Thomas Nelson Australia
102 Dodds Street
South Merlbourne, 3205
Victoria, Australia

Nelson Canada
1120 Birchmount Road
Scarborough, Ontario
Canada, M1K 5G4

Chapman & Hall
2-6 Boundary Row
London SE1 8HN

International Thomson Editores
Campos Eliseos 385, Piso 7
Col. Polanco
11560 Mexico D.F. Mexico

International Thomson Publishing Gmbh
Königwinterer Strasse 418
53228 Bonn
Germany

International Thomson Publishing Asia
221 Henderson Road
#05-10 Henderson Building
Singapore 0315

International Thomson Publishing-Japan
Hirakawacho-cho Kyowa Building, 3F
1-2-1 Hirakawacho-cho
Chiyoda-ku, 102 Tokyo
Japan

1 2 3 4 5 6 7 8 9 10 XXX 01 00 99 97 96 95

Library of Congress Cataloging-in-Publication Data

Ecological time series / Thomas M. Powell, and John H. Steele. editors
p. cm.
Includes bibliographical references and index.
ISBN 0-412-05191-5(HB) — ISBN 0-412-05201-6(PB)
1. Ecology–Statistical methods. 2. Time-series analysis.
I. Powell, T. M. (Thomas M.) II. Steele, John H.
QH541.15.S72E26 1995
574.5'0151955—dc20 94-16248
CIP

Please send your order for this or any Chapman & Hall book to **Chapman & Hall, 29 West 35th Street, New York, NY 10001, Attn: Customer Service Department.** You may also call our Order Department at 1-212-244-3336 or fax your purchase order to 1-800-248-4724.

For a complete listing of Chapman & Hall's titles, send your requests to **Chapman & Hall, Dept. BC, One Penn Plaza, New York, NY 10119.**

Contents

Preface

This book results from a summer school held at Cornell University in 1992. The participants were graduate students and postdoctoral researchers selected from a broad range of interests and backgrounds in ecological studies. The summer school was the second in a continuing series whose underlying aim—and the aim of this volume—is to bring together the different methods and concepts underpinning terrestrial, freshwater, and marine ecology.

The first volume in the series focused on patch dynamics in these three ecological sectors. Here we have endeavored to complement that volume by extending its comparative approach to the consideration of ecological time series. The types of data and the methods of collection are necessarily very different in these contrasting environments, yet the underlying concept and the technical problems of analysis have much in common.

It proved to be of great interest and value to the summer school participants to see the differences and then work through to an appreciation of the generalizable concepts. We believe that such an approach must have value as well for a much larger audience, and we have structured this volume to provide a comparable reading experience.

Support for the 1992 Cornell summer school was provided by the following organizations:

National Science Foundation, Division of Ocean Science and Environmental Biology (NSF Grant No. OCE-9123451)

Office of Naval Research (ONR Grant No. N00014-92-J-1527)

National Marine Fisheries Service

We are grateful for the cooperation and participation of the Center for Applied Mathematics at Cornell University, where John Guckenhemier and Dolores Pendell provided invaluable assistance in the handling of administrative arrangements and logistics, and of the Cornell Theory Center, one of four NSF National Supercomputer Centers. At Woods Hole Oceanographic Institution, Mary Schumacher coordinated the review and editing of manuscripts. Her attention to both form and content ensured a coherent volume. Finally, we wish to thank the summer school participants, who made our month at Cornell a most stimulating and enjoyable experience.

Thomas M. Powell
Davis, California

John H. Steele
Woods Hole, Massachusetts

February 1994

Contributors

Put O. Ang, Jr.
Fisheries Research Laboratory
Department of Fisheries and Oceans
P.O. Box 550
Halifax, NS B3J 2S7 Canada

Arturo H. Ariño
Department of Ecology
Faculty of Sciences
University of Navarra
E-31080 Pamplona, Spain

Robert A. Armstrong
Program in Atmospheric and Oceanic Sciences
Sayre Hall, P.O. Box CN 710
Princeton University
Princeton, NJ 08544-0710 U.S.A.

Fortunato A. Ascioti
Istituto di Metodologie Avanzate di Analisi Ambientale C.N.R.
Via S. Loja
Zona Industriale-85050
Tito Scalo (Pz) Italy

Patrick J. Bartlein
Department of Geography
University of Oregon
Eugene, OR 97403 U.S.A.

W. Breck Bowden
Department of Natural Resources
University of New Hampshire
Durham, NH 03824 U.S.A.

Fred Brauer
Department of Mathematics
University of Wisconsin
Madison, WI 53706 U.S.A.

Wallace S. Broecker
Lamont-Doherty Geological Observatory of Columbia University
Palisades, NY 10964 U.S.A.

Tormod V. Burkey
Department of Ecology and Evolutionary Biology
Guyot Hall
Princeton University
Princeton, NJ 08544-1003 U.S.A.

Angel Capurro
Universidad de Belgrano
Departamento de Investigaciones
Zabala 1851 piso 12 box 11
1426 Buenos Aires, Argentina

Carlos Castillo-Chavez
Biometrics Unit and Center for Applied Mathematics
341 Warren Hall
Cornell University
Ithaca, NY 14853-8701 U.S.A.

Bernard Cazelles
URBB
INSERM U 263 (National Institute of Health and Medical Research)
Université Paris 7, Tour 73
75 251 Paris cedex 5, France

F. S. Terry Chapin
Department of Integrative Biology
University of California, Berkeley
Berkeley, CA 94720 U.S.A.

James E. Cloern
U.S. Geological Survey, MS 496
345 Middlefield Road
Menlo Park, CA 94025 U.S.A.

Linda A. Deegan
The Ecosystems Center
Marine Biological Laboratory
Woods Hole, MA 02543 U.S.A.

Douglas H. Deutschman
Department of Ecology and Evolutionary Biology
Eno Hall
Princeton University
Princeton, NJ 08544-1003 U.S.A.

Robert R. Dickson
Ministry of Agriculture, Fisheries and Food
Directorate of Fisheries Research, Fisheries Laboratory
Lowestoft NR33 OHT, Suffolk, U.K.

Michael Dodd
Biology Department
The Open University
Walton Hall
Milton Keynes MK7 6 AA, U.K.

Jonathan Dushoff
Department of Ecology and Evolutionary Biology
Princeton University
Princeton, NJ 08544-1003 U.S.A.

Jean-Marc Guarini
University of Western Brittany
Biological Oceanography Laboratory
6, avenue le Gorgeu
29 287 Brest, France

John Guckenheimer
Center for Applied Mathematics
504 ETC Building
Cornell University
Ithaca, NY 14853 U.S.A.

Anne E. Hershey
Department of Biological Sciences
University of Minnesota-Duluth
Duluth, MN 55812 U.S.A.

Patricia Himschoot
Biometrics Unit
322 Warren Hall
Cornell University
Ithaca, NY 14853 U.S.A.

John E. Hobbie
The Ecosystems Center
Marine Biological Laboratory
Woods Hole, MA 02543 U.S.A.

Alan D. Jassby
Division of Environmental Studies
University of California
Davis, CA 95616 U.S.A.

George W. Kipphut
Center for Reservoir Research
Murray State University
Murray, KY 42071 U.S.A.

George W. Kling
Department of Biology
University of Michigan
Ann Arbor, MI 48109-1048 U.S.A.

Simon A. Levin
Department of Ecology and Evolutionary Biology
Eno Hall
Princeton University
Princeton, NJ 08544-1003 U.S.A.

Karin E. Limburg
Institute of Ecosystem Studies
Box AB
Millbrook, NY 12545 U.S.A.

John J. Magnuson
Center for Limnology
University of Wisconsin
680 N. Park Street
Madison, WI 53706 U.S.A.

Mitchell P. McClaran
325 Biological Sciences East Building
School of Renewable Natural Resources
College of Agriculture, University of Arizona
Tucson, AZ 85721 U.S.A.

Michael E. McDonald
Department of Chemical Engineering
University of Minnesota-Duluth
Duluth, MN 55812 U.S.A.

Patricia F. McDowell
Department of Geography
University of Oregon
Eugene, OR 97403 U.S.A.

Frédéric Ménard
Hôpital Saint-Louis
Département de Biostatistique et Informatique Médicale
1, avenue C. Vellefaux
75 475 Paris cedex 10, France

Anthony F. Michaels
Bermuda Biological Station for Research, Inc.
17 Biological Station Lane
Ferry Reach, GE01 Bermuda

Michael C. Miller
Department of Biology
Cincinnati University
Cincinnati, OH 45221 U.S.A.

W. John O'Brien
Department of Systematics and Ecology
University of Kansas
Lawrence, KS 66044 U.S.A.

Tsung-Hung Peng
Environmental Sciences Division
Oak Ridge National Laboratory
P.O. Box 2008
Oak Ridge, TN 37831-6335 U.S.A.

Bruce J. Peterson
The Ecosystems Center
Marine Biological Laboratory
Woods Hole, MA 02543 U.S.A.

Jacqueline Potts
Rothamsted Experimental Station
Harpenden, Herts AL5 2JQ U.K.

Thomas M. Powell
Division of Environmental Studies
University of California, Davis
Davis, CA 95616 U.S.A.

Edward B. Rastetter
The Ecosystems Center
Marine Biological Laboratory
Woods Hole, MA 02543 U.S.A.

Jorge L. Sarmiento
Program in Atmospheric and Oceanic Sciences
Sayre Hall, P.O. Box CN 710
Princeton University
Princeton, NJ 08544-0710 U.S.A.

Mary E. Schumacher
Marine Policy Center
Woods Hole Oceanographic Institution
Woods Hole, MA 02543 U.S.A.

David J. Shafer
School of Ocean and Earth Science and Technology
Division of Biological Oceanography
1000 Pope Road, MSB 632
University of Hawaii
Honolulu, HI 06822 U.S.A.

Gaius R. Shaver
The Ecosystems Center
Marine Biological Laboratory
Woods Hole, MA 02543 U.S.A.

Richard D. Slater
Program in Atmospheric and Oceanic Sciences
Sayre Hall, P.O. Box CN 710
Princeton University
Princeton, NJ 08544-0710 U.S.A.

Andrew R. Solow
Marine Policy Center
Woods Hole Oceanographic Institution
Woods Hole, MA 02543 U.S.A.

John H. Steele
Woods Hole Oceanographic Institution
Woods Hole, MA 02543 U.S.A.

Konstantinos I. Stergiou
National Center for Marine Research
Fisheries Laboratory
Agios Kosmas, Hellinikon
Athens 16604 Greece

Jason D. Stockwell
Department of Zoology
Erindale College
University of Toronto in Mississauga
3359 Mississauga Road North
Mississauga, Ontario L5L 1C6 Canada

Thomas J. Stohlgren
National Park Service
Natural Resource Ecology Laboratory
Colorado State University
Fort Collins, CO 80523 U.S.A.

John R. Vande Castle
LTER Network Office
College of Forest Resources
University of Washington
Seattle, WA 98195 U.S.A.

Elizabeth L. Venrick
Scripps Institution of Oceanography
University of California at San Diego
La Jolla, CA 92093-0227 U.S.A.

Susan C. Warner
Department of Biology
208 Erwin W. Mueller Laboratory
Eberly College of Science
Pennsylvania State University
University Park, PA 16802 U.S.A.

Thompson Webb III
Department of Geological Sciences
Brown University
Providence, RI 02912-3128 U.S.A.

Jianguo Wu
Department of Ecology and Evolutionary Biology
Eno Hall
Princeton University
Princeton, NJ 08544-1003 U.S.A.

ECOLOGICAL TIME SERIES

Introduction

Mary E. Schumacher and John H. Steele

The human sense of the passage of time is among the most intimate perceptions that we have of the external world. Our feeling for rates of change is as closely linked, if not more so, to our own activities as to events beyond our control. We form vivid memories of singular events such as droughts, floods, and hurricanes largely by reference to the stages of our own lives at the time of their occurrence and the daily routines they interrupt; but we are unaware or only dimly aware of processes operating at a much slower pace to produce change that is barely perceptible over the course of a human lifetime. Similar limitations characterize our perception of space. We cannot see beyond our local domain and so must rely on second-hand evidence to assess emerging global-scale concerns.

Not only do these natural limitations color our understanding of our own terrestrial environment; the problem is compounded when we must extend our terrestrial thinking to the processes that occur within the ocean. The human perspective on time and space scales is a key factor that pervades any attempt to compare trends in ocean physics and biology with corresponding changes on land.

This point is acknowledged in the organization and range of topics encompassed by this volume. The seemingly narrow focus of the book is on the analysis and interpretation of time series, but clearly it is necessary to consider such other subjects as population processes, community structure, and, especially, patch dynamics. For this reason, the authors guide us through a broad spectrum of ideas and applications that illuminate the way we view the workings of ecosystems.

In the first part, on analysis and methodology, the chapters frame the basic concepts that are necessary for any environmental study, give an account of modern techniques, and describe some of the important problems that arise in the study of longer time series, particularly in the context of our changing climate.

The second part of the book deals with the wide range of time scales at which we can describe change. Just as our own characteristic scales are very different from those of organisms with lifetimes measured in days, so the time scales of ocean physics are generally much longer than those of corresponding processes in the atmosphere. Thus, the appropriate comparisons of ecological systems on land and in the sea may require us to view the two sectors at very different time scales. The chapters in this section examine temporal variability in the Atlantic and Pacific Oceans, at the coastal boundary, and in the planning of landscape studies, to demonstrate that there is variability at all measured scales for any natural system. The general lesson is that, for practical reasons, we have to be selective and so must start by choosing some conceptual basis.

These critical choices are illuminated in the third part of the book, where results of major investigations in different arenas, and at different time scales, are reviewed. If there is a single conclusion, it is that the world is not, nor has it ever been, near an equilibrium state, whether we are considering human disease or forest disturbance. Consequently, there is no law of diminishing returns as we extend our data and our perspectives on change in the natural environment. This conclusion is exemplified by the diversity reported here: studies of lakes and oceans, of arctic ecosystems, and of human epidemics. Notwithstanding the richness and diversity of these different ecological systems, there are common themes that enhance our understanding of them all. The challenge to the reader is to discern the general processes or "rules" that can serve to unify studies in fields as diverse as ocean climatology and epidemiology.

PART I

Analysis and Methodology

1

Can Ecological Concepts Span the Land and Ocean Domains?

John H. Steele

Ecological studies of marine and terrestrial systems are usually carried out in separate institutions, published in different journals, and funded from disparate sources. Some of the reasons for these distinctions are understandable. There are obvious differences in the technologies (ships versus jeeps) and in the styles of life of the organisms (gravity is a minor problem in the open ocean). But more important are the differing time scales of processes in the two environments, and of the organisms themselves. Air temperature and rainfall are highly variable on a day-to-day basis, whereas the ocean, below the surface layers, changes much more slowly. Conversely, trees have life cycles that are very much longer than those of the primary producers in the ocean, the phytoplankton.

These differences in characteristic time scales have an immediate relevance to our study of time series. What we mean by "long" time sequences can be quite different on land or in the ocean, not merely in terms of the lifetimes of the organisms but also in relation to the temporal processes of the environment in which the plants or animals are living.

In a review of historical patterns in fish stocks, Caddy and Gulland (1983) said: "An examination of landing trends . . . reveals four basic patterns: steady, cyclical, irregular and spasmodic. It is not claimed that the boundaries between these groups are exact." Examples in this book show the problems in constructing categories. In particular, the distinction between irregular and spasmodic may be only a matter of time scale. Caddy and Gulland continue:

> The causes of variation in fish landings fall under two main headings which are not mutually exclusive: fluctuations in the marine environment and variations in fishing intensity.

This chapter is adapted from J. H. Steele, 1991, "Can Ecological Theory Cross the Land-Sea Boundary?" *Journal of Theoretical Biology* 153:426–436, by permission of Academic Press.

For predominantly terrestrial studies, Connell and Sousa (1983) stated:

> Long-term studies reveal a continuum of temporal variability. There is no clear demarcation between assemblages that may exist in an equilibrium state and those that do not. Only a few examples of what might be stable limit cycles were found. There was no evidence of multiple steady states.

Evidence for multiple steady states was considered inapplicable when "the physical environment is different in the different alternative states."

These authors proposed that, rather than the classical idea of stability, "the concept of persistence within stochastically defined bounds is more applicable to ecological systems." Connell and Sousa do not define these bounds, and they summarize the "prevalent view" that one should minimize the effects of physical perturbations so that the real processes affecting populations and communities can be identified.

These comments epitomize the assumptions used in marine and terrestrial studies. The marine view would be that relatively large-scale physical processes are the significant factors in population variability (apart from human harvesting). Community dynamics are assumed to smooth out the effects of variability in the physical environment. Technically, for fish populations, this is done by a formal stock-recruit relation (Rothschild 1986) that assumes density dependence in the larval or juvenile stages where we usually have negligible data.

Terrestrial concepts stress the significance of density dependence through detailed consideration of prey-predator/competitive/community interactions. Environmental perturbations will not be "scrutinized in nearly as much detail as the effects of strictly biological interactions" (Connell and Sousa 1983). The search for explicit evidence for density dependence derives predominantly from the terrestrial literature (e.g., Hassell et al. 1989; den Boer 1991).

These studies illustrate how, in both sectors, land and sea, the interaction between physical and biological processes is considered. But the comparison also demonstrates that there is a significant difference in emphasis, with a focus on internal mechanisms in terrestrial studies and on external physical forcing for marine populations.

Although we are concerned here mainly with time series, the spatial dimension must be included when we want to compare scales of processes on land and in the sea. Once again, there are technical as well as conceptual differences in the way we handle spatial pattern in these two regimes. These are illustrated by the impact of remote sensing. Satellite images reveal complex small-scale features in ocean properties, such as near-surface chlorophyll, that are unobservable from ships, whereas the terrestrial images enable us to see beyond our local landscapes.

Are the patterns in space and their time scales on land and in the sea so fundamentally different that we must maintain the separations of these subdisciplines of ecology? Or can we benefit from the diversity in generalizing about the basic ecological processes? I shall argue that, once the differences between scales are recognized and their implications understood, then we will see the common problems and the general issues that can unite the divergent sectors of ecological research.

Physical Scales

At the longest geological time scales, the earth can be regarded as a single entity subject to changes induced by plate tectonics. At millennial periods of time, we think of the ocean and atmosphere as a closely coupled system. But when we come nearer to the human scales of decades to centuries, we are aware of the differences between the responses of air and sea.

The physics of the ocean and atmosphere have the same basic fluid dynamics (Pedlovsky 1977). This underlying similarity makes it possible to compare particular processes; the differences in density and viscosity mean that these processes have very different scales. Cyclonic systems in the atmosphere have spatial scales around 1000 km and last for about a week. The equivalent eddies in the ocean with diameters near 100 km can exist for months to years. Figure 1-1 compares the physical scales and illustrates the problem in climate modeling, where these

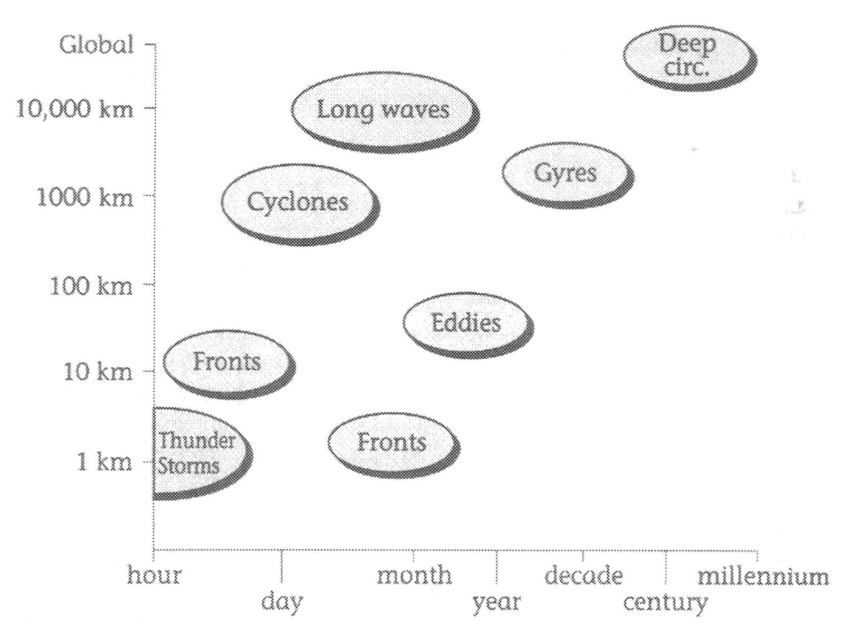

Figure 1-1. Space-time scales for major features of atmospheric (left) and ocean (right) physical systems.

two very different space/time relations have to be coupled. Obviously, the differences illustrated in Figure 1-1 will have an impact on the organisms inhabiting each sector.

But there is a further important distinction between the two media. The variability of each as a function of temporal scales is quite distinct. It is apparent that sea temperature is much more predictable at small scales than air temperature. Conversely, we have come to appreciate that the great heat capacity of the ocean drives climatic variability at longer and larger scales through phenomena such as El Niño.

These differences in variability as functions of time scale apply to other parameters such as currents and wind speed. We can express these relations as power spectra, which relate variance to frequency (Figure 1-2). When the regular and predictable features, such as the seasonal cycle, are removed, the atmospheric variance is nearly constant over a wide range of scales (white noise), whereas variability in the ocean increases continuously with period (red noise). The merging of the two graphs at millennial scales implies the close coupling of long-term climatic change in the two environments (Steele 1985; Steele and Henderson 1994; and references therein).

These great differences in variability (note the log scales in Figure 1-2) must have had a strong influence on the evolving response of organisms in the two habitats. The consequences are suggested by the biological examples added to Figure 1-2. This comparison of physical aspects forms a basis for discussion of the spatial and temporal scales in the biology.

Biological Scales

To compare and contrast the two environments in terms of both their physics and the general biology of their constituent organisms, we shall use the same portrayal in terms of scale. Certain simple but general characteristics can be used, such as the body sizes of the organisms, the periods of their life cycles, and the spatial ambit of their populations. These properties can be depicted simultaneously on spatial and temporal scales. For individuals, different lengths of life can be considered as points on a time scale from days to decades; body sizes, measured as lengths, on a scale from millimeters to meters. For populations, we can have scale defined biologically, in terms of area occupied by reproductive units, or statistically, as the largest spatial scale within which variations in population structure exhibit coherence.

As a starting point, the relation of size of organism to length of life is well known to display regularities for both marine and terrestrial populations (Bonner 1965; Sheldon et al. 1972). Within the general body size-to-life time relationship, it is possible to separate the species into categories with broad trophic status. For pelagic marine populations (Figure 1-3A), the lowest trophic levels, phyto-

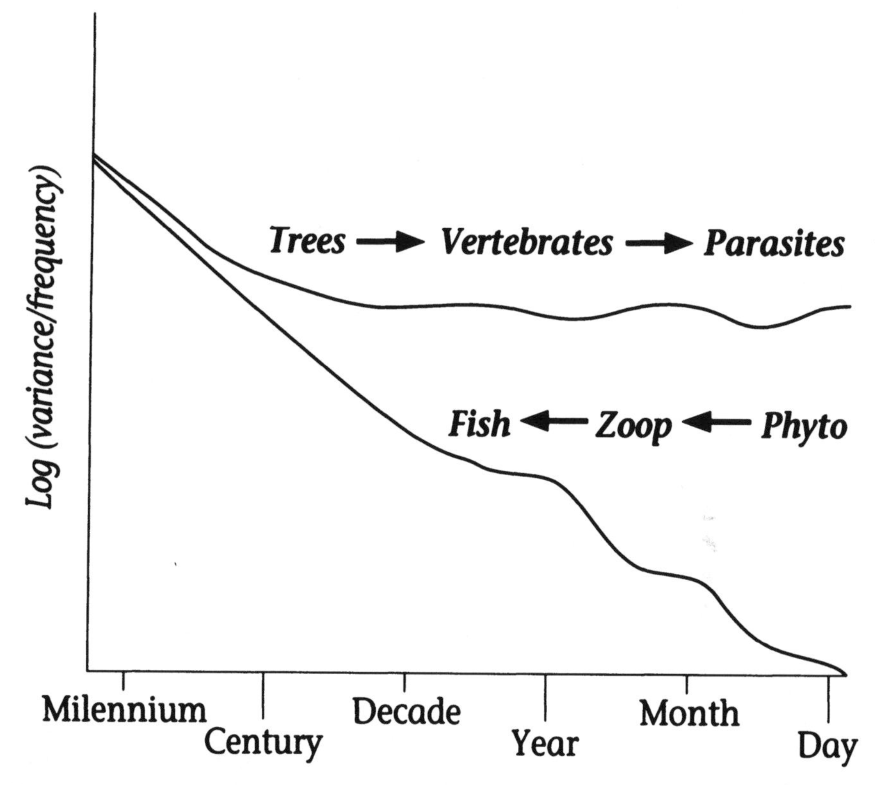

Figure 1-2. Schematic presentations of the power spectra for atmospheric and oceanic temperature (omitting regular cycles) with examples of time scales of particular food chains (adapted from Steele 1985).

plankton and autotrophic bacteria, have the smallest body sizes and the shortest time scales (of the order of days). The top predators are the largest species and have the longest life spans. Thus spatial and temporal scales increase as we move up the food chain.

The terrestrial ranking is more complex and diverse. One selection (Figure 1-3B, from Bonner 1965) shows plants (trees) as the longest-lived; the large mammalian herbivores are next, though small invertebrate herbivores, such as aphids, may be the shortest-lived species. Similarly, invertebrate predators may be short-lived, while vertebrate predators, including man, are long-lived. True predators are generally larger (Cohen et al. 1990), but it is interesting that most ecological theory at present focuses on parasites and parasitoids (Shorrocks 1993). There, a frequent, but not general, pattern can be seen as the inverse of the marine sequence, with the longest-lived species at the base of the food chain (Figure 1-2). Certainly there are short-lived plants—annuals, for example—though these are considerably longer-lived than phytoplankton. Perennial grasses

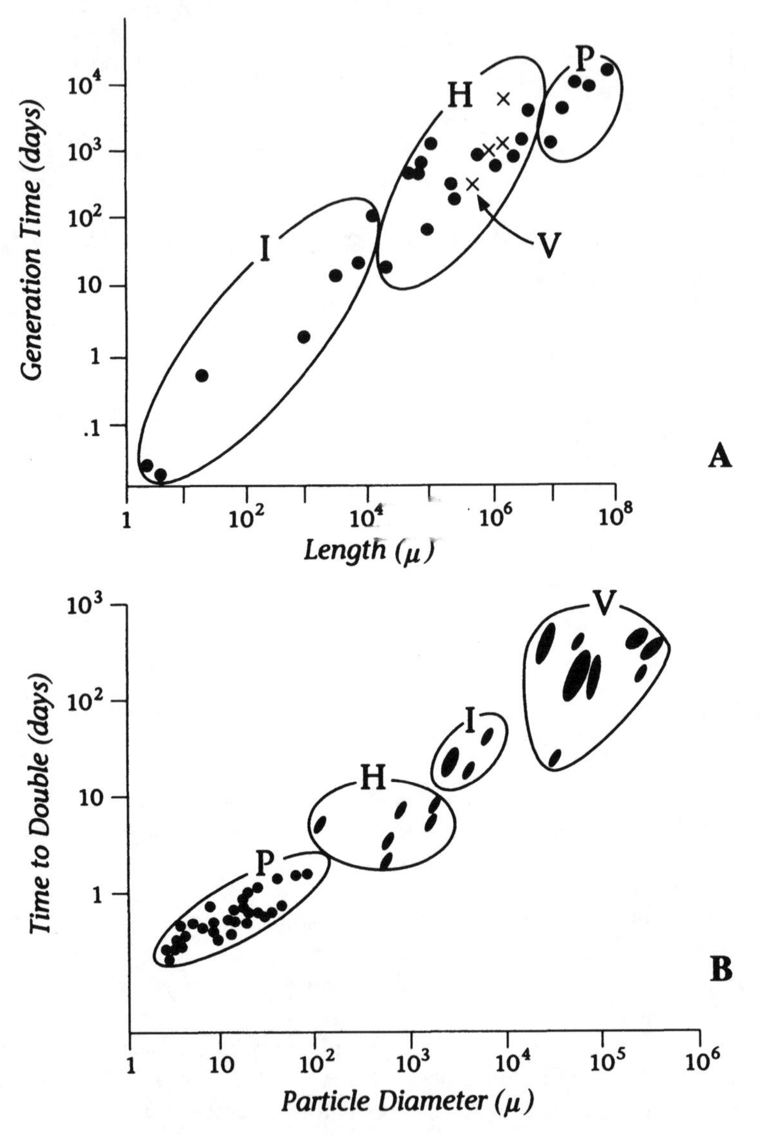

Figure 1-3. Relation of size to growth for plants (P), herbivores (H), invertebrates (I), and vertebrates (V). (A) from Sheldon et al. (1972) for pelagic marine ecosystems; (B) from Bonner (1965) using only the terrestrial species.

occupy an ambiguous position, but if their root systems are the essential component, then they have a multiyear time scale—decades or longer.

For animals, with increasing body size comes an increase in the individual's home range (Southwood 1976). Some general scales for ecological processes are presented in Figure 1-4. In terrestrial environments (Figure 1-4A), processes

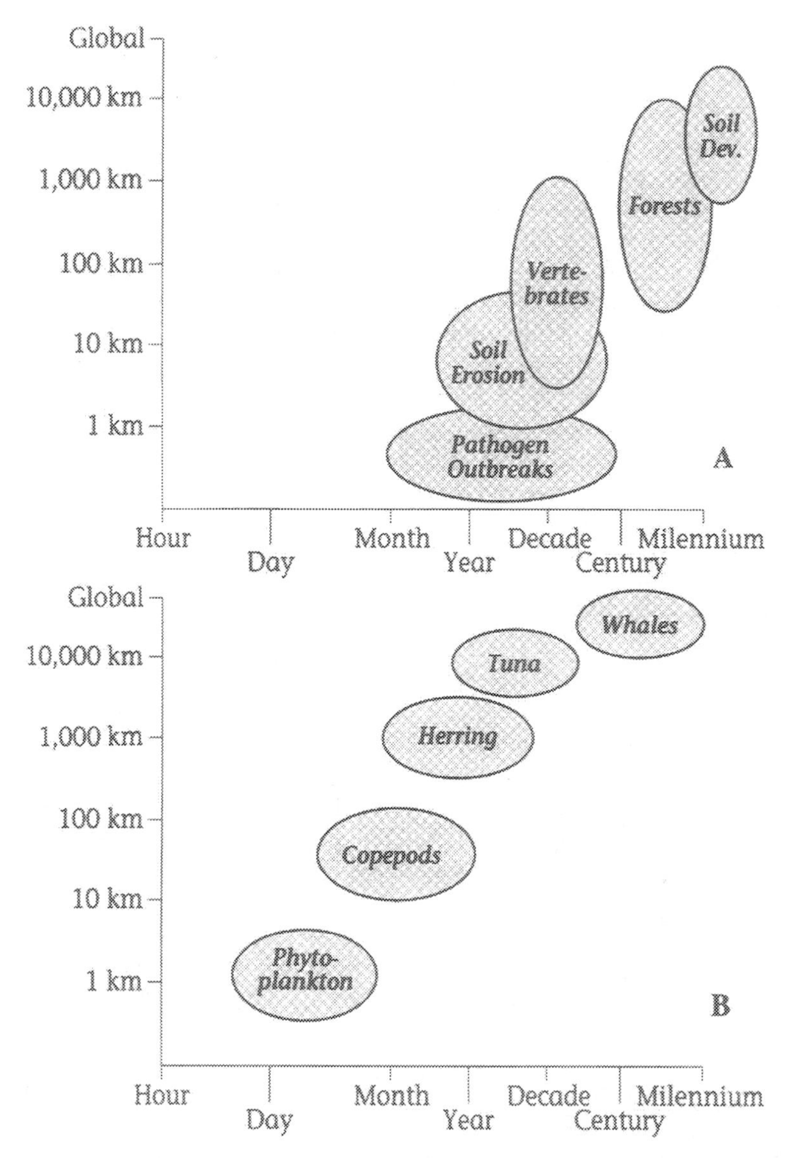

Figure 1-4. Schematic presentation of ecological scales for (A) terrestrial processes (from Delcourt et al. 1983); and (B) marine components (from Steele 1978).

such as soil development and the consequent response of forest or grassland systems take place over centuries to millennia (Delcourt et al. 1983). This is typified by the changing patterns of tree species over North America (Davis 1981), and it is at these time scales that comparisons are made with climate change (McDowell et al., Chapter 15 of this volume).

For marine environments, the scales of food-chain processes were presented in a simplified form by Steele (1978). This presentation demonstrated that the space-time scales of the physics and biology were nearly coincident, and so it is likely that the interactions are closely coupled. Thus herring may depend on specific, highly turbulent areas of the seabed as suitable spawning sites (Sinclair 1987). Their larvae can utilize fronts as sources of enhanced food supply (Richardson et al. 1986), and the juveniles and adults use mesoscale gyres for migration to complete the life cycle (Cushing 1982; Corten 1986). The close relation between physical scales and the corresponding scales for ocean plankton at behavioral, population, and community levels were expressed in a "Stommel diagram" by Haury et al. (1978).

Thus, a close connection between the variability of physical and biological processes in the open ocean is to be expected at time scales comparable to the life cycles of the trophic components. In particular, the characteristic life times of major groups such as zooplankton may be associated with peaks in the physical variance spectrum (Walsh 1977).

Comparing Responses

On the basis of these scale descriptions of physical and ecological processes, it is possible to speculate about the reasons for the differences in ecological theories appropriate to the two environments (Figure 1-5). It has been suggested (Steele 1985; Sinclair 1987) that, in the pelagic marine environment, adaptations have evolved that depend closely on the persistent features in ocean dynamics. But populations must also accept and utilize the variability in the ocean physics, and this can be the basis for the near-universality of a pelagic larval phase (Steele 1985).

It must be noted that there are "aerial plankton" such as aphids and spiderlings that use air currents to disperse. They may read the weather and respond when conditions are suitable. Thus, there can be opportunistic use of the large short-term variability in the atmosphere.

Conversely, for those terrestrial systems that have very much longer time scales than the atmosphere, it is appropriate to regard the purely atmospheric variability as short-term noise. The reasons for these separate evolutionary paths probably depend on the nature of the physical variability as a function of space and time scales. The ocean is less variable or "noisy" than the atmosphere at short time scales, from days to a few years. This difference can be seen in the temperature data from each environment (Steele 1985), but we require a more detailed examination of the physical-biological interactions in each case to understand how these relations actually operate.

One reason for the different ecological interpretations in the oceans versus the land is that comparisons are made at similar time scales. Thus, quasi-periodic

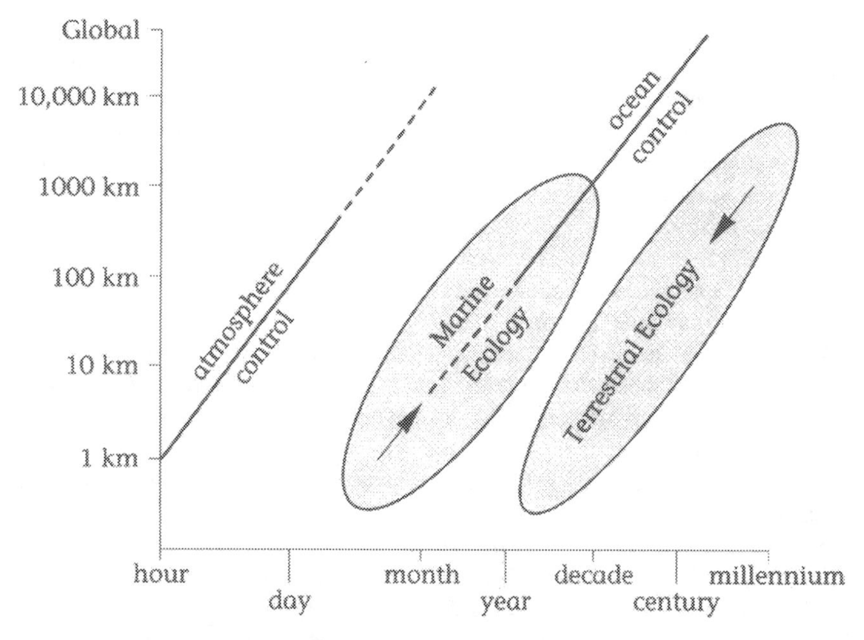

Figure 1-5. Space-time scales for atmospheric and marine physics (lines) and for terrestrial and marine ecosystems (ellipses); derived from Figures 1-1 and 1-4. The arrows indicate the general directions of trophic processes.

events at decade-long time scales in forests are explained in terms of tree-pest interactions (Ludwig et al. 1978). In contrast, similar changes in fish populations are related to possible changes in the physical environment (Southward 1980; Cushing 1982).

At the longest time scales, trends in tree distribution in North America, over centuries to millennia, are viewed as the direct result of climate changes since the last ice age. Thus terrestrial "climate" explanations do exist, but at very different time scales.

There are similar changes in overall distribution of marine organisms at comparably long time scales, and these are used as indices of physical climate change (CLIMAP 1976). This is to be expected, because at time scales longer than centuries, the ocean and atmosphere probably respond as a single system.

Discussion

The space-time diagrams of Figure 1-4 and the spectral analyses of variability are merely descriptions of the scales of physical and biological processes. They do not, by themselves, explain the interactions, but they can suggest speculations about the general nature of such interactions. Moreover, they provide explanations of some of the different approaches ecologists have taken in their studies of marine and terrestrial environments. For example, the concept of ecological succession that has been so pervasive in terrestrial ecology does not seem applicable to the changes in marine systems, where we explain changing community patterns in a physical context.

Similarly, the theories of competitive interactions between species do not transfer easily into studies of the structure of open sea marine ecosystems, perhaps because the pelagic marine environment does not permit detailed observation. It is interesting to note that, for studies of environments, such as coral reefs and rocky shores, the earlier focus on competition among adults for space has been replaced by an emphasis on larval recruitment (Roughgarden et al. 1988; Sale 1982). It is only a partial exaggeration to regard this as a shift from a terrestrial to a marine paradigm.

Conversely, the theories of patch dynamics in marine systems, based on physical processes (Denman and Powell 1984), appear to have little relevance to terrestrial studies (Pickett and White 1985). The marine models use the physical theories of spatially continuous fluid environments with turbulent diffusion as the main passive dispersive mechanism for marine organisms (Steele 1978). The diffusion model has been applied to terrestrial systems, but it can seem inappropriate to structurally and spatially discontinuous landscapes where active animal movement is the critical factor and is highly dependent on these fragmented patterns. Recently, attempts have been made to apply percolation theory to these problems using the analogy of porous media to landscapes (Gardner et al. 1989). At the same time, satellite images of ocean temperature and color have made biological oceanographers aware of the complexities—and often discontinuities—in ocean conditions (Steele 1989). We need to examine possible bases for comparisons of theory and observation. In both systems, patchiness is now recognized as being not merely statistical variability, but probably an essential component of the maintenance of complicated ecological relations within the system. If the reasons are different in each domain, then this should increase our overall understanding and testing of theories.

Another example concerns the generality of conclusions in food web theory (Cohen 1989; Pimm 1982). Cohen has shown that there are surprising uniformities in the proportion of species classed as basal, intermediate, and top (19%, 53%, and 29%, respectively) within 113 documented food webs and has explained this with a topological model. Of this number, nine are from the pelagic ocean realm. Given the quite different ambit and age-size patterns displayed in Figures 1-1 and 1-3, it might seem surprising that there is such agreement. But the basal and top percentages are similar, and the statistical correspondences between theory and observation are poorer for these categories than for the intermediate group. The extension of these topological theories into the aspects of dynamics and size structure may help answer some of these questions (Cohen, pers. comm.) and, again, test the value of such comparisons across environments.

Lastly, there is considerable attention being given to the need for long time series data on terrestrial ecosystems. In the marine sector, fisheries data provide the main source of valuable information on the long-term variability of populations (Cushing 1982). But these data also illustrate the difficulties, if not the impossibility, of deriving "process" conclusions such as the elusive "stock-recruit" relations that are purported to provide the density dependence for marine fish stocks (Rothschild 1986). Similarly, it proves very difficult to demonstrate density dependence in terrestrial populations (Reddingius and Boer 1989). Hassell et al. (1989), in a study of univoltine populations, indicated that data on 10 to 20 generations were necessary. Solow and Steele (1990) demonstrated this result theoretically under simple conditions. If long time series programs are to be successfully combined with process studies, then we need to define the hypotheses being investigated—especially in relation to the relative scales of physical variability and ecological interactions. It is to be expected that there will be scale differences in the two environments (Figure 1-5), so there can be different interpretations of "long term." Will there be useful methods and theories that can apply to the two sectors?

In addition to such purely scientific questions, there are immediate social and economic issues of global and regional significance. These, too, can be described in terms of the space-time representation (Figure 1-5).

Our increasing agricultural, industrial, and demographic capacities to alter terrestrial systems are on large spatial scales but occur much more rapidly than the natural changes. Again, these new changes approach the rates associated with the underlying scales of ocean phenomena. In a mathematical context, any terrestrial model of natural systems would formulate the large-scale atmospheric and ecological processes as fast and slow variables, respectively; whereas, in the marine case, the rates need to be comparable. Now, in looking at the direct and indirect consequences of humanity's effects on the terrestrial environment, we have to consider comparable rates for both physical and biological factors. Our intrusion in the terrestrial world increases the spatial scales of environmental alteration and also increases the rates at which they occur. These anthropogenic changes shift some terrestrial processes toward the marine scales in Figure 1-5, and thus make them close to those climate rates that are determined by ocean processes (Steele 1989). We know, mathematically, that we can expect qualitative changes in the character of the responses, depending on whether the scales match or not. Thus, the ability to generalize, or at least compare, across the sectors is a pressing theoretical issue.

This comparison of marine and terrestrial dynamics has more than theoretical interest. As we utilize marine and terrestrial environments, the consequences, deliberate or accidental, depend on the responses to physical and chemical change.

The imposition of terrestrial standards for marine problems may produce too strict or too lax criteria—or, most likely, quite inappropriate ones. The practical need for general comparative studies is as important and as immediate as the scientific basis.

I have emphasized the separation between marine and terrestrial ecological studies because there are major differences in concepts, in organization, and in funding. The longstanding nature of this dichotomy implies that convergence of these sub-disciplines will not be easy and cannot be rapid. Our present global and regional concerns about interactions among the land, the sea, and the air provide reasons for government and international agencies to assist in bringing these groups together. But the major motive must be scientific interest. Can we define a set of case studies from marine, freshwater, and terrestrial environments that provide the basis for comparisons? What future data are required to enhance the interchange? Above all, can theoretical work illuminate the problems? Will such a synthesis increase our understanding of the components, by focusing our efforts in understanding the different scales of the various changes and how they are related?

These are not new questions, but usually they arise on the periphery of ecological research. If we wish to have global ecosystem dynamics as a critical component of the new earth system studies, then we must make the integration of these components of ecology a central and pressing issue.

Acknowledgments

Many of the speculations presented here are the fruits of discussions with Joel Cohen, Paul Dayton, Si Levin, Tim Kratz, Stuart Pimm, and Bob Ricklefs. The failings are my own. The work was supported by NSF Grant No. OCE-9123451, with partial funding provided under ONR grant N00014-92-J-1527. WHOI Contribution No. 8623.

References

Bonner, J. T. 1965. *Size and Cycle: An Essay on the Structure of Biology*. Princeton University Press, Princeton, NJ.

Caddy, J. F., and J. A. Gulland. 1983. Historical patterns of fish stocks. *Marine Policy* **83**:267–278.

Clark, W. C. 1985. Scales of climate impacts. *Climatic Change* **7**:5–27.

CLIMAP. 1976. The surface of the ice age earth. *Science* **191**:1131–1137.

Cohen, J. E. 1989. "Food Webs and Community Structure." In J. Roughgarden, R. M. May, and S. A. Levin (eds.), *Perspectives in Ecological Theory*. Princeton University Press, Princeton, NJ.

Cohen, J. E., F. Briand, and C. M. Newman. 1990. *Community Food Webs: Data and Theory*. Springer-Verlag, Berlin.

Connell, J. H., and W. P. Sousa. 1983. On the evidence needed to judge ecological stability or persistence. *American Naturalist* **121**:789–824.

Corten, A. 1986. On the causes of recruitment failure of herring in the central and northern North Sea in the years 1972–78. *Journal du Conseil Intérnational pour l'Exploration de la Mer* (ICES) **42**:281-294.

Cushing, D. H. 1982. *Climate and Fisheries*. Academic Press, New York.

Davis, M. B. 1981. "Quaternary History and the Stability of Forest Communities." In West, Shugart, and Botkin (eds.), *Forest Succession, Concepts and Application*. Springer-Verlag, New York.

Delcourt, H. R., P. A. Delcourt, and T. Webb. 1983. Dynamic plant ecology: The spectrum of vegetational change in space and time. *Quarterly Science Review* **1**:153–175.

den Boer, P. J. 1991. Seeing the trees for the woods: Random walks or bounded fluctuations of population size. *Oecologia* **86**:484–491.

Denman, K. L., and T. M. Powell. 1984. Effects of physical processes on planktonic ecosystems in the coastal ocean. *Oceanography and Marine Biology Annual Review* **22**:125–168.

Dickinson, R. E., and A. Henderson-Sellers. 1988. Modelling tropical deforestation. *Quarterly Journal of the Royal Meteorological Society* **114**:439–462.

Eppley, R. W., E. Stewart, R. Abbott, and U. Heyman. 1985. Estimating ocean primary production from chlorophyll. *Journal of Plankton Research* **7**:57–70.

Frost, B. W. 1987. Grazing control of phytoplankton stock in open subarctic Pacific Ocean. *Marine Ecology Program Series* **39**:49–68.

Gardner, R. H., R. V. O'Neill, M. G. Turner, and V. H. Dale. 1989. Quantifying scale dependent effects of animal movements with simple percolation models. *Landscape Ecology* **3**:217–227.

Glantz, M. 1987. Drought in Africa. *Scientific American* **256**:34–40.

Hassell, M. P., J. Latto, and R. M. May. 1989. Seeing the wood for the trees: Detecting density dependence from existing life table studies. *Journal of Animal Ecology* **58**:883–892.

Haury, L. R., J. A. McGowan, and P. H. Wiebe. 1978. "Patterns and Processes in the Time-Space Scales of Plankton Distributions." In J. H. Steele (ed.), *Spatial Patterns in Plankton Communities*. Plenum, New York.

Ludwig, D., D. D. Jones, and C. S. Holling. 1978. Qualitative analysis of insect outbreak systems: The spruce budworm and forest. *Journal of Animal Ecology* **44**:315–332.

NRC. 1986. *Global Change in the Geosphere-Biosphere*. National Research Council, National Academy Press, Washington, D.C.

Pedlovsky, J. 1977. *Geophysical Fluid Dynamics*. Springer-Verlag, New York.

Pickett, S. T. A., and P. S. White. 1985. *The Ecology of Natural Disturbance and Patch Dynamics*. Academic Press, New York.

Pimm, S. L. 1982. *Food Webs*. Chapman & Hall, New York.

Pomeroy, L. R., E. C. Hargrave, and J. J. Alberto. 1988. "The Ecosystem Perspective." In L. R. Pomeroy and J. J. Alberts (eds.), *Concepts of Ecosystem Ecology*. Springer-Verlag, New York.

Reddingius, J., and P. J. Boer. 1989. On the stabilization of animal numbers. Problems of testing: Confrontation with data from the field. *Oecologia* **79**:143–149.

Richardson, K., M. J. Heath, and N. J. Pihl. 1986. Studies of a larval herring patch in the Buchan area. *Dana* **6**:1–10.

Rothschild, B. J. 1986. *Dynamics of Marine Fish Populations*. Harvard University Press, Cambridge, MA.

Roughgarden, J., S. Gaines, and H. Possingham. 1988. Recruitment dynamics in complex life cycles. *Science* **241**:1460–1466.

Running, S. W. 1990. "Estimating Terrestrial Primary Production by Combining Remote Sensing and Ecosystem Simulation." In R. J. Hobbs and H. A. Mooney (eds.), *Remote Sensing of Biosphere Functioning*. Springer-Verlag, New York.

Sale, P. F. 1982. Stock-recruit relationships and regional coexistence in a lottery competitive system: A simulation study. *American Naturalist* **120**:139–159.

Sarmiento, J. L., J. R. Toggwiler, and R. Majjar. 1988. Ocean carbon cycle dynamics and atmospheric CO_2. *Philosophical Transactions of the Royal Society* **A:325**:3–21.

Sheldon, R. W., A. Prakash, and W. Sutcliffe. 1972. The size distribution of particles in the ocean. *Limnology and Oceanography* **17**:327–340.

Shorrocks, B. 1993. Trends in the *Journal of Animal Ecology:* 1932–92. *Journal of Animal Ecology* **62**:599–605.

Sinclair, M. 1987. *Marine Populations*. University of Washington Press, Seattle, WA.

Solow, A., and J. Steele. 1990. On sample size, statistical power, and the detection of density dependence. *Journal of Animal Ecology* **59**:1073–1076.

Southward, A. J. 1980. The western English Channel—an inconstant ecosystem. *Nature* **285**:361–366.

Southwood, T. R. E. 1976. "Bionomic Strategies and Population Parameters." In R. M. May (ed.), *Theoretical Ecology*. Saunders, Philadelphia.

———. 1977. Habitat, the templet for ecological strategies. *Journal of Animal Ecology* **46**:337–365.

Steele, J. H. 1978. "Some Comments on Plankton Patches." In J. H. Steele (ed.), *Spatial Pattern in Plankton Communities*. Plenum, New York.

———. 1985. Comparison of marine and terrestrial ecological systems. *Nature* **313**:355–358.

———. 1989. "Discussion: Scale and Coupling in Ecological Systems." In J. Roughgarden, R. M. May, and S. A. Levin (eds.), *Perspectives in Ecological Theory*. Princeton University Press, Princeton, NJ.

Steele, J. H., and E. W. Henderson. 1994. Coupling between physical and biological scales. *Philosophical Transactions of the Royal Society of London* B **343**:5–9.

Stommel, H. 1963. Varieties of oceanographic experience. *Science* **139**:572–576.

Walsh, J. J. 1977. "A Biological Sketchbook for an Eastern Boundary Current." In E. D. Goldberg, I. N. McCave, J. J. O'Brien, and J. H. Steele (eds.), *The Sea,* Vol. 6. Wiley Interscience, New York.

———. 1988. *On the Nature of Continental Shelves.* Academic Press, New York.

2

Fitting Population Models to Time Series Data

Andrew R. Solow

Like many scientists, population ecologists exhibit a kind of schizophrenia when it comes to time series. On one hand, the development of mathematical models representing the dynamics of one or more populations has a long and glorious history (e.g., Kingsland 1985). On the other hand, when confronted with actual population time series data, most ecologists fall back on familiar empirical models (e.g., ARMA models) that have little or no connection to the underlying ecological processes. In the few exceptions in which actual population models are used (e.g., Hassell et al. 1976), model fitting tends to be *ad hoc*. The purpose of this chapter is to outline how formal statistical methods can be used to fit a population model to data and, at the same time, to point out that this approach may be problematic.

First, however, a word is in order about the advantages of fitting population (as opposed to empirical) models to population data. Of course, an outstanding advantage arises when the population model is correct. In fact, no model is exactly correct, and both population models and empirical models are best viewed as approximations to reality. The question then becomes: which type of model is a better approximation? To the extent that population models capture the main features of population dynamics, they are probably better approximations than empirical models. If this is so, it is more useful, in terms of interpretation, to estimate biological parameters like growth rate and carrying capacity than empirical parameters like autoregressive and moving average coefficients.

On the statistical side, because nonlinear biological models admit a greater variety of behaviors than low-order linear models, they tend to provide a more parsimonious description of the data. A good nonbiological example of this is simple one-dimensional climate models, which provide as good a fit to data using three or four interpretable physical parameters as autoregressive models with 10 or 12 parameters. Loosely speaking, for a fixed amount of data, statistical inference is more powerful the more parsimonious the model. Finally, the advantage of formal, as opposed to *ad hoc*, model fitting is that all the machinery of formal statistical inference (e.g., confidence intervals, hypothesis tests, etc.) becomes applicable. Similarly, the same battery of formal and informal methods for checking the adequacy of an empirical model could and should be used when fitting a population model.

The Basic Model

Let y_t be population density (possibly after transformation) in period t. The basic model I will consider in this chapter is:

$$\begin{aligned} y_t &= F(y_{t-1};\alpha) \qquad t = 1, 2, \ldots \\ &= F^t(y_0;\alpha) \end{aligned} \tag{1}$$

where F is a function specified up to the value of the parameter α (which may be vector valued) and F^t is the t-fold convolution of F with itself. The extension to higher-order models (i.e., those in which y_t depends on $y_{t-1}, y_{t-2}, \ldots, y_{t-p}$) is straightforward. Throughout this chapter, I will use the logistic or quadratic map:

$$y_t = \alpha y_{t-1}(1 - y_{t-1}) \tag{2}$$

as an example. The qualitative dynamics of the quadratic map as a function of α are described in Devaney (1989).

Although models like (1) can exhibit complex dynamics, they are purely deterministic and, if population density follows this model *exactly*, then it is possible to find the true value of α (i.e., to fit the data *perfectly*). For example, in the case of the quadratic map, α can be found exactly from as few as two observations. In practice, it is never possible to fit the data perfectly. One explanation for this is that the assumed form of F is incorrect. A second explanation is that the data contain a stochastic component or error. This chapter focuses on the role of stochastic error.

One implication of stochastic error is that the ecologist is off the hook if his model does not fit the data well. A second implication is that α cannot be found exactly, but must be estimated. It is perhaps insufficiently appreciated that the properties of a particular estimation method depend on the way in which stochastic error enters the basic model. In the next section, I consider three different ways in which stochastic error can be introduced into the basic model and discuss the implications for estimating α.

Estimating α Under Three Error Models

One way in which a stochastic component can be introduced into the basic model is:

$$Y_t = F(Y_{t-1};\alpha) + \epsilon_t \tag{3}$$

where ϵ_t is a random variable. I will assume here and below that ϵ_t is normally distributed with mean 0 and *known* variance σ^2. In practice, σ^2 is unknown and must be treated as a nuisance parameter to be estimated. I will also assume that the error process is serially independent. Again, the extension to the case of serially dependent errors is (at least, in principle) straightforward.

In (3), density is denoted by Y_t (instead of y_t) to reflect the fact that it is a random variable. Under this model, the stochastic error in each period is incorporated into the density in subsequent periods. I will refer to this as process error. The effect of process error on dynamical models has been studied by a number of authors, including Schaffer et al. (1986).

I will consider next maximum likelihood estimation of α. Suppose that observations $\mathbf{y} = (y_1, y_2, \ldots, y_n)$ are available from (3). The likelihood function is defined as:

$$L_1(\alpha) = g(\mathbf{y};\alpha)$$

where $g(\mathbf{y};\alpha)$ is the joint probability density function (pdf) of $\mathbf{Y} = (Y_1, Y_2, \ldots, Y_n)$. This joint pdf can be decomposed into the product of conditional pdfs:

$$g(\mathbf{y};\alpha) = g_1(y_1;\alpha)\, g_{2|1}(y_2 \mid y_1;\alpha); \ldots g_{n|1,2,\ldots,n-1}(Y_n|Y_1, Y_2, \ldots, Y_{n-1};\alpha) \tag{4}$$

where, for example, $g_{2|1}(y_2 \mid y_1;\alpha)$ is the conditional pdf of Y_2 given $Y_1 = y_1$. Consider the last term on the right-hand side of (4). By the Markovian structure of the basic model and the independence of the error process, this term reduces to:

$$g_n|_{1,2,\ldots,n-1}(y_n|y_1, y_2, \ldots, y_{n-1};\alpha) = g_{n|n-1}(y_n \mid y_{n-1};\alpha).$$

It follows that:

$$L_1(\alpha) = g_1(y_1;\alpha) \prod_{t=2}^{n} g_t|_{t-1}(y_t|y_{t-1};\alpha). \tag{5}$$

For the process error model, the conditional distribution of Y_t given $Y_{t-1} = y_{t-1}$ is Gaussian with mean $F(y_{t-1};\alpha)$ and variance σ^2. After some algebra, it follows that, apart from irrelevant constants:

$$\log L_1(\alpha) = \log g_1(y_1;\alpha) - \sum_{t=2}^{n} (y_t - F(y_{t-1};\alpha))^2. \tag{6}$$

A standard approach is to condition on $Y_1 = y_1$, so that the first term on the right-hand side of (6) is a constant, and the maximum likelihood estimator of α corresponds to the ordinary least squares estimator.

As an example, consider the quadratic map (2). In this case, because α enters F linearly, its least squares estimator has a particularly simple form:

$$\alpha_1 = \frac{\sum_{t=2}^{n} y_t y_{t-1} (1 - y_{t-1})}{\sum_{t=2}^{n} (y_{t-1} (1 - y_{t-1}))^2}. \tag{7}$$

That ordinary least squares estimation—the statistical properties of which (and computational algorithms for which) are well understood—is appropriate under (3) is a strong result. It is, however, a two-edged sword. The use of ordinary least squares is implicitly connected to the process error model with Gaussian errors of constant variance. If this model is inappropriate, then ordinary least squares estimation may be badly biased.

A second way in which a stochastic component can enter the basic model is through observation error. That is, while the true dynamics of the population are governed by (1), the observations have the form:

$$X_t = y_t + \epsilon_t. \tag{8}$$

I will call this the observation error model. The effect of observation error on fitting dynamical models has been considered by Berliner (1988). Observation error, which will always arise when density is estimated by sampling, is fundamentally different from process error. To begin with, observation error is not incorporated into the dynamics of the population. Moreover, under (8), the observations themselves do not follow the basic model. That is:

$$X_t \neq F(X_{t-1}; \alpha).$$

Under the observation error model, X_t has a normal distribution with mean $F(y_{t-1}; \alpha) = F^t(y_0; \alpha)$ and variance σ^2. It follows that, apart from irrelevant constants, the log-likelihood function is:

$$L_2(x_0, \alpha) = \sum_{t=1}^{n} (X_t - F^t(x_0; \alpha))^2 \tag{9}$$

where now x_0 is an unknown nuisance parameter. The maximum likelihood estimate of α can be found by maximizing (9). In most cases, this must be done numerically. Some care must be taken in doing so because the likelihood function may be extremely complicated with many local maxima. For example, for the quadratic map, for fixed x_0, F^t is a polynomial of degree $2^t - 1$ in α.

It is well known that least squares estimation is biased in the presence of observation error (e.g., Fuller 1987). The severity of this bias depends on F and σ^2. For the quadratic map, Solow (1994) showed that this bias can be severe even when the signal-to-noise ratio in the observations is large.

A third way in which a stochastic component can enter the basic model is through the parameter α itself. For example, suppose that:

$$Y_t = F(Y_{t-1}; \alpha_t) \tag{10}$$

where

$$\alpha_t = \alpha + \epsilon_t. \tag{11}$$

That is, α_t is a normal random variable with mean α and variance σ^2. I will refer to this as parameter error. The likelihood function under parameter error has the same form as (5). However, unless α enters F linearly, the conditional distribution of Y_t given $Y_{t-1} = y_{t-1}$ is non-normal. Provided α enters F linearly—that is, provided:

$$F(y;\alpha) = \alpha F(y) \tag{12}$$

conditional on $Y_{t-1} = y_{t-1}$—the parameter error model can be rewritten as:

$$\begin{aligned} Y_t &= \alpha F(Y_{t-1}) + \epsilon_t F(y_{t-1}) \\ &= \alpha F(y_{t-1}) + \beta_t \end{aligned} \tag{13}$$

where β_t is an error with mean 0 and:

$$\mathrm{Var}(\beta_t) = (\sigma(F(y_{t-1}))^2. \tag{14}$$

That is, provided (12) holds, the parameter error model is equivalent to the process error model, with the important difference that the errors do not have constant variance. In that case, maximum likelihood estimation corresponds to *weighted* least squares (e.g., Silvey 1970). It is easy to show that, in this case, the ordinary least squares estimator of α is unbiased, but not fully efficient. More important, statistical inference (i.e., a confidence interval or an hypothesis test) about α based on ordinary least squares will be incorrect. In the case of the quadratic map, the weighted least squares estimator of α is given by:

$$\alpha_3 = \frac{1}{n-1} \sum_{t=2}^{n} \frac{y_t}{y_{t-1}(1 - y_{t-1})}. \tag{15}$$

To summarize, it is possible to imagine a stochastic component entering the basic model in at least three ways: through the process itself, through the observations, and through the parameter α. Of course, it is also possible to imagine these errors operating simultaneously. Even when attention is restricted to one source of error at a time, the form of the maximum likelihood estimator of α depends on the type of error. To put it in more negative terms, an estimator of α that is appropriate under one type of error can behave badly under another type of error.

It is important to point out that the incorporation of either process error or parameter error into a well-behaved deterministic model can cause serious problems. For example, if the deterministic model has an invariant region (i.e., a region to which population fluctuations are confined), then process or parameter error will lead excursions outside this region, resulting in rapid extinction. As an illustration, try simulating values from the quadratic map with process error.

One way to avoid this problem is to change the form of F. Another is to "fix" the error model. An example of the latter is given in Schaffer et al. (1986), where the variance of the process error is assumed to shrink near the boundary of the invariant region. If this kind of state-dependent error structure is to be taken seriously, then it, too, has implications for model fitting. Specifically, observations near the boundary, having smaller variance, tend to represent the purely deterministic dynamics of the population. For this reason, it may be possible to use such observations to estimate α with great precision. As before, this form of state-dependent error implies that it should be possible to fit the data almost exactly near the boundary.

Returning to the general problem, from a statistical point of view, the appropriate approach is to consider the general model:

$$\begin{aligned} Y_t &= F(Y_{t-1};\alpha_t) + \epsilon_t \\ \alpha_t &= \alpha + \epsilon'_t \\ X_t &= Y_t + \epsilon''_t \end{aligned}$$

where ϵ_t, ϵ'_t, and ϵ''_t are process, parameter, and observation errors. In addition to fitting this model, it is possible in this setting to test for the presence of different error types. Unfortunately, fitting such a model, particularly when the unrealistic assumption of serially independent error processes with known variances is relaxed, is no simple matter. Moreover, it would be nearly impossible to estimate the model parameters with any degree of precision when the number of observations is as small as it usually is for population time series.

What Is an Ecologist to Do?

At a general level, the time may have come for ecologists to think carefully about what they hope to get out of their models. One view is that, as with models in other disciplines, ecological models are best viewed as a tool for gaining qualitative insight about population dynamics. Under this view, if the results of model experiments are robust to changes in model specification and the introduction of stochastic error, then they may be accepted with some confidence. Unfortunately, starting from this rather innocuous view, it is easy to end up believing that the model is correct. If the results that flow from the model are robust, then there is probably not too much harm in this. But this chapter suggests that such robustness may be too much to expect.

If the decision is made to pursue model fitting, then the main practical conclusion of this chapter is that the same kind of thought that goes into specifying the deterministic component F needs to go into specifying the stochastic component. Misspecification of either component can lead to erroneous conclusions. Unfortunately, this problem arises well short of formal model fitting. Even the interpretation of a plot of Y_t against Y_{t-1} (or X_t against X_{t-1}) depends on assumptions about the stochastic component.

Incidentally, the same kind of problem arises with empirical models. For example, if Y_t follows an AR(1) model and $X_t = Y_t + \epsilon_t$ (i.e., Y_t observed with error), then X_t follows an ARMA(1,1) model. Failure to recognize observation error would lead either to concluding that the population follows the wrong model or, if an AR(1) model is fit to the observed data, to biased estimation of the true autoregressive parameter.

A second practical conclusion is that the behavior of the basic model can be quite different from that of the basic model with process or parameter error. Specifically, many simple forms of F are *fragile* in the face of these kinds of errors. This suggests that attention should be focused on developing simple, robust forms of F.

Acknowledgments

This work was supported by NSF Grant No. OCE-9123451 and by ONR Grant No. N00014-92-J-1527. WHOI Contribution No. 8653.

References

Berliner, L. M. 1988. Likelihood and Bayesian prediction of chaotic systems. *Journal of the American Statistical Association* **86**:938–952.

Devaney, R. L. 1989. *Introduction to Chaotic Dynamical Systems*. Benjamin-Cummings, Menlo Park, CA.

Fuller, W. 1987. *Measurement Error Models*. Wiley and Sons, New York.

Hassell, M. P., J. H. Lawton, and R. M. May. 1976. Patterns of dynamical behavior in single species models. *Journal of Animal Ecology* **45**:471–486.

Kingsland, S. E. 1985. *Modelling Nature*. University of Chicago Press, Chicago.

Schaffer, W. M., S. Elnner, and M. Kot. 1986. Effects of noise on some dynamical models in ecology. *Journal of Mathematical Biology* **24**:479–523.

Silvey, S. D. 1970. *Statistical Inference*. Chapman & Hall, London.

Solow, A. R. 1994. Observation error and least squares estimation in population models. *Ecology*. (Submitted.)

3

Estimate of Interhemispheric Ocean Carbon Transport Based on ΣCO_2 and Nutrient Distribution

Tsung-Hung Peng and Wallace S. Broecker

The partial pressure of CO_2 in the atmosphere (pCO_2) has been rising ever since the industrial revolution in the early 1800s. The release of CO_2 from fossil fuel combustion is the main source of the extra amount of CO_2 in the atmosphere. Other sources, such as deforestation and soil disturbances from agricultural practice, may have also contributed to the rise. Although the fossil fuel CO_2 release is relatively well documented [Marland and Boden (1991) give an average of 5.4 gigatons (Gt) of carbon released per year for the 1980s], the release from land use is uncertain (Post et al. 1990). Direct observation of the increase in atmospheric CO_2 partial pressure in the past few decades indicates that approximately 55% to 57% of the amount of carbon released from fossil fuel has remained in the atmosphere (Keeling et al. 1989a). To balance the carbon budget, the remaining anthropogenic CO_2 molecules must have entered the ocean or terrestrial biosphere. The amount of carbon taken up by the ocean is estimated to be in the range between 26% and 35% of fossil fuel CO_2 (Peng 1986). These estimates are made by various ocean models calibrated with the distribution of geochemical tracers in the ocean. An average for oceanic uptake in the 1980s is estimated to be about 2.1 gigatons of carbon per year (Gt C/year). Depending on the magnitude of the amount of carbon released from the land biosphere, we need to search for other carbon sinks (so-called missing sinks) and to quantify their sizes in order to balance the carbon budget.

The conclusion from a search of missing carbon sinks over the last decade points to the organic matter reservoir in the terrestrial biosphere as only one possible sink that is big enough to receive and store extra carbon flux (Broecker and Peng 1993). As a consequence of increased abundance of CO_2 in the atmosphere and an additional amount of fixed nitrogen generated from internal combustion engines such as automobiles, the rate of photosynthesis may have been increased and more carbon may have been stored in tree trunks and soil humus. Hence, while the terrestrial biosphere reservoir is being trimmed around the edges by deforestation processes, it is becoming more lush in the interior (i.e., global greening) by the fertilization processes. Assuming an average deforestation in the 1980s to be about 1.6 Gt C/year, a balanced global carbon budget would require a global greening to produce a carbon sink flux of about 2.0 Gt C/year in the terrestrial biosphere.

The spatial patterns and temporal variations of atmospheric CO_2 partial pressure have been used to derive the geographical distribution of CO_2 sources and sinks (Tans et al. 1990). During the last decade, atmospheric pCO_2 at high latitudes in the Northern Hemisphere was, on average, 3 μatm higher than that over the high latitudes in the Southern Hemisphere. By contrast, Tans et al. (1990) predict a pole-to-pole meridional pCO_2 gradient of 5.7 to 7.3 μatm. Their predictions are based on an atmospheric general circulation model with geographical distribution of fossil fuel combustion, seasonal terrestrial CO_2 exchange, and tropical deforestation, together with an oceanic sink acting strongly in the Southern Hemisphere. Models with verified atmospheric mixing show that the currently observed north-to-south gradient allows a cross-equatorial transport of CO_2 of an amount approximately equal to that required to produce the observed annual increase of atmospheric CO_2 inventory in the Southern Hemisphere. This model prediction means that the total uptake of CO_2 by the combined land biosphere and sea surface in the Southern Hemisphere must be close to zero. This result is considered by Tans et al. (1990) as evidence that the uptake of fossil fuel CO_2 by the ocean is considerably smaller than that calculated from ocean models calibrated with the tracer distribution. If this is correct, then the amount of excess CO_2 accounted for in the ocean and atmosphere reservoirs is only about one-half of the total CO_2 release through fossil fuel use and land use. Because the only other reservoir of significance is the terrestrial biosphere, this imbalance between production and inventory leads to the conclusion that the growth rate of terrestrial biota has been enhanced, presumably, by CO_2 fertilization. This enhancement must be large enough to produce increases in mass of the forest and soil humus reservoirs that more than compensate for losses by land use. Furthermore, to account for the low magnitude of the observed interhemispheric gradient, this enhancement must take place in the North Temperate Zone. Tans et al. (1990) suggested that such terrestrial sinks could be as large as 2.0 to 3.4 Gt of carbon per year.

The unexpected lower pCO_2 gradient observed in the atmosphere could also be explained, however, by a quite different mechanism that was proposed by Keeling et al. (1989b), Heimann and Keeling (1986), and Keeling and Heimann (1986). This mechanism involves a reverse atmospheric CO_2 gradient prior to the industrial revolution. The release of fossil fuel CO_2 into the Northern Hemisphere by the industrialized countries slowly balances such a natural south-to-north gradient; eventually it reaches a point where the gradient begins to tilt to the other direction. This hypothesis is supported with the observed evolution of the atmospheric CO_2 difference between Mauna Loa, Hawaii, and McMurdo,

Antarctica. The difference has grown from a value near zero in the early 1960s to about 3 μatm in the last decade. This interpretation eliminates the compelling need to call for a sizeable terrestrial carbon sink to take up excessive carbon budget in the Northern Hemisphere.

Keeling and Heimann (1986) proposed a preindustrial, south-to-north interhemispheric transport of about 1 Gt of carbon through the atmosphere. This northward transport of carbon would have to have been compensated by a southward transport of carbon through the ocean. To evaluate the proposed mechanism for carbon transport through the atmosphere, we could examine the interhemispheric transport of carbon through the ocean. The objective of this chapter is to undertake such evaluation from the oceanographic perspective. Although some conclusions have been reported elsewhere (Broecker and Peng 1992), in this chapter we stress the details of analysis of the observed oceanographic data and how the interhemispheric transport of oceanic carbon fluxes are calculated from these results.

Methodology

The most likely candidate for interhemispheric CO_2 transport is the Atlantic Ocean conveyor circulation (Broecker 1991), because the lower limb of the conveyor carries ~20 Sverdrups (Sv) of deep water to the Southern Hemisphere. This is also called North Atlantic Deep Water (NADW), which was exposed to the Northern Hemisphere atmosphere before sinking into the deep sea. If the northward-flowing return water in the upper Atlantic carries less dissolved carbon than the southward-flowing deep water, then the conveyor could have acted as an interhemispheric carbon pump. If this pump transported a gigaton of carbon per year toward the Southern Hemisphere ocean before the industrial revolution, then the southward-flowing NADW would have to carry ~ 130 μmol/kg more total dissolved inorganic carbon (ΣCO_2) than the northward-flowing upper water.

Hence, the evaluation of the potential for NADW to carry excess ΣCO_2 in the preindustrial time is the central issue. Because the transfer of CO_2 across the sea-air interface is the main factor in determining the amount of CO_2 that has entered the sea water, we need to develop a method to estimate the magnitude of such CO_2 for the surface water. Assuming that the CO_2 exchange process does not take place, the expected total dissolved inorganic carbon content (ΣCO_2^{EXP}) could be computed by using the following equation (Broecker and Peng 1992):

$$\Sigma CO_2^{EXP} = 1900\frac{S}{35.00} + \frac{1}{2}\left(ALK + NO_3 - 2310\frac{S}{35.00}\right) + 127PO_4 \qquad (1)$$

where S, ALK, NO_3, and PO_4 are the observed salinity, alkalinity, and NO_3 and PO_4 concentrations, respectively, for the water sample of interest. We choose a reference salinity to be 35.00%, a reference ΣCO_2 concentration to be 1900 μmol/kg, and a reference alkalinity to be 2310 μeq/kg. For biological cycling, the carbon-to-phosphorus ratio in marine organic matter is taken to be 127 (Peng and Broecker 1987). As we are interested only in ΣCO_2 change caused by gas exchange across the air-sea interface, other changes resulting from salt variation, organic material cycling, and $CaCO_3$ formation and dissolution have to be eliminated. Hence, equation (1) is normalized first to a salinity of 35% to remove dilution or enrichment of carbon concentration introduced by changes in freshwater input. The second term of the equation corrects for the alkalinity changes caused by the formation and dissolution of $CaCO_3$ and the change in nitrate concentration caused by the nutrient use. The last term is the correction for the cycling of organic material.

The observed S, ALK, NO_3, and PO_4 values used for computing ΣCO_2^{EXP} are taken from measurements made as part of the geochemical expeditions during the GEOSECS Atlantic, 1972–73 (Bainbridge 1981); GEOSECS Pacific, 1973–74 (Broecker et al. 1982); GEOSECS Indian, 1978 (Weiss et al. 1983); TTO North Atlantic Study, 1981 (1986); Hudson Cruise, 1982 (1984); Tropical Atlantic Study, 1982–83 (1986); and SAVE South Atlantic, 1987–89 (1989).

The difference between the observed ΣCO_2 and the expected ΣCO_2,

$$\Delta\Sigma CO_2 = \Sigma CO_2^{OBS} - \Sigma CO_2^{EXP}, \qquad (2)$$

is a measure of the amount of carbon either gained or lost by any given seawater sample through exchange of CO_2 with the atmosphere. The observed ΣCO_2 values are taken from the same sources as those just mentioned for ΣCO_2^{EXP} computations in equation (1).

Results and Discussion

The computed $\Delta\Sigma CO_2$ values plotted against water temperatures for surface waters in the Southern Hemisphere ocean are shown in Figure 3-1A. The GEOSECS results are displayed in the upper panel; the TTO and SAVE results are shown in the lower panel. The main feature in common is that both tropical and polar waters have similar low values of $\Delta\Sigma CO_2$, whereas temperate waters have more positive values. This feature is in contrast with that shown in Figure 3-1B, where the $\Delta\Sigma CO_2$ values for surface waters from the Northern Hemisphere portion of the Atlantic Ocean are plotted against water temperature. A distinctive difference is that the $\Delta\Sigma CO_2$ values increase steadily with decreasing temperature in the North Atlantic. Included in Figure 3-1B for low-temperature regions (high-latitude ocean) are data of measurements made both during winter months (Hudson in 1982) and during summer months (TTO North Atlantic Study in 1981). The most positive $\Delta\Sigma CO_2$ values in the world ocean are found in the

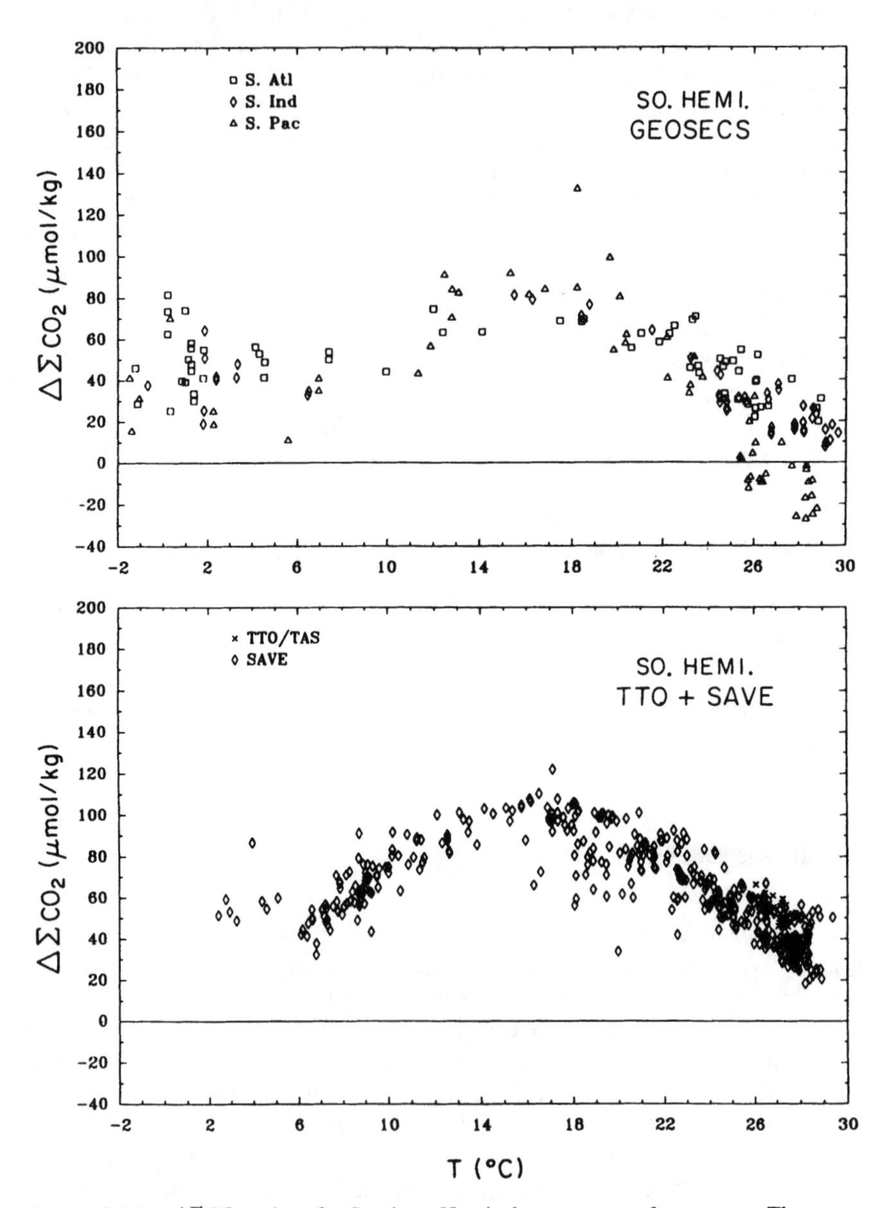

Figure 3-1A. $\Delta\Sigma CO_2$ values for Southern Hemisphere ocean surface waters. The upper panel is for GEOSECS in three oceans. The lower panel is for the TTO Tropical Atlantic Study (in 1983) and SAVE (in 1988) in the Atlantic Ocean.

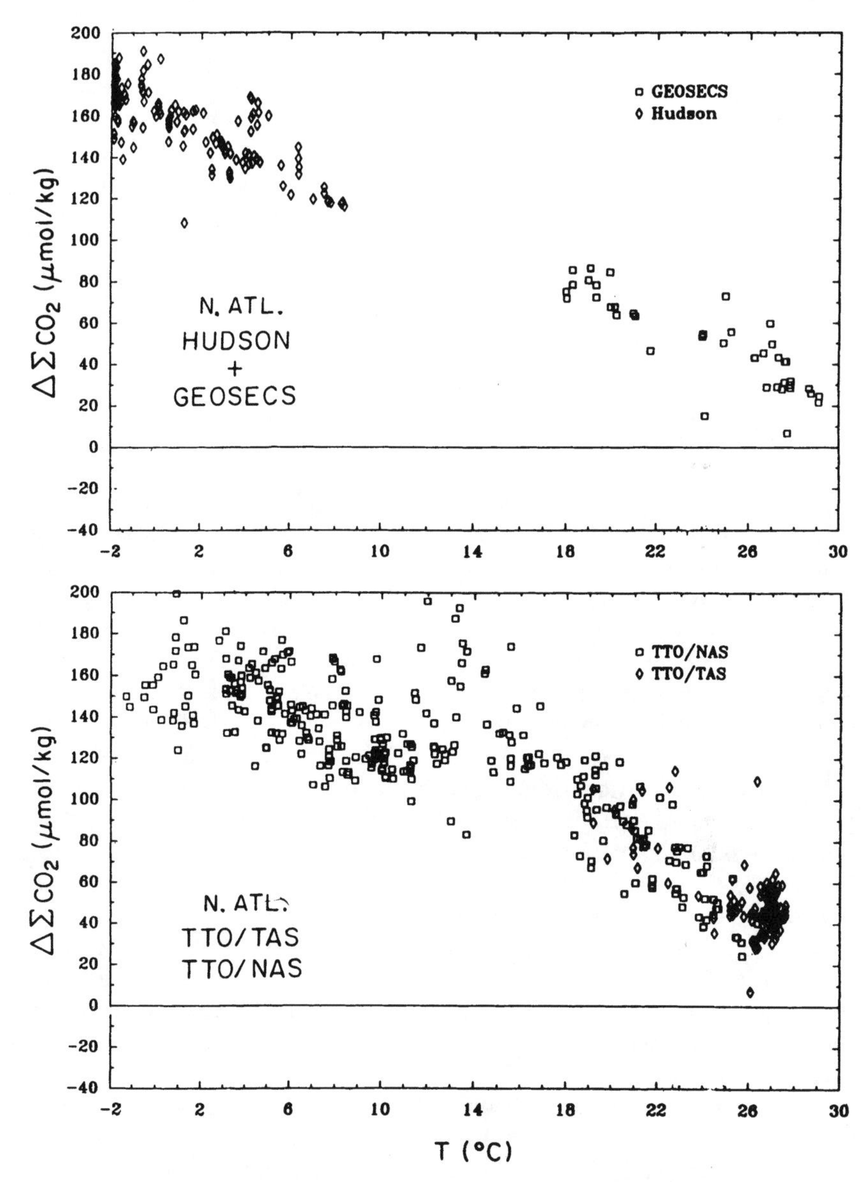

Figure 3-1B. $\Delta\Sigma CO_2$ values for North Atlantic surface waters. Results from the GEOSECS (in summer, 1972) and the Hudson-82 (in winter, 1982) expeditions are shown in the upper panel. In the lower panel are shown the results from the TTO North Atlantic Study (in 1981) and the TTO Tropical Atlantic Study expeditions (in 1983).

northern North Atlantic Ocean. This conclusion is also supported by measurements made in the Northern Hemisphere portion of the Pacific Ocean. While fewer data are available for the northern Pacific, the general trend shows that the maximum $\Delta\Sigma CO_2$ values occur in the temperate region, which is similar to that shown in Figure 3-1A for the Southern Hemisphere.

A possible interfering factor in the geographical distribution of $\Delta\Sigma CO_2$ is the excess of ΣCO_2 in surface water resulting from the uptake of CO_2 from the atmosphere over the last century because of the rising atmospheric CO_2 content caused by fossil fuel combustion. The extent of build-up of excess ΣCO_2 depends on the temperature of the water and the year when measurements were made. The relationship between the pCO_2 rise and the ΣCO_2 increase is controlled by the thermodynamic buffer factor *R*, defined by the following equation:

$$R = \frac{\delta pCO_2/pCO_2}{\delta\Sigma CO_2/\Sigma CO_2} \tag{3}$$

where δpCO_2 and $\delta\Sigma CO_2$ are changes in pCO_2 and ΣCO_2, respectively. The value of *R* ranges from about 9 for tropical waters to about 14 for polar waters. Thus, for the same length of time exposed to the same atmospheric pCO_2 increase, the extent of excess dissolved ΣCO_2 build-up is larger for warm than for cold surface waters. Assuming an equilibrium between atmosphere and surface water at the time the ocean surveys were made, the increases in ΣCO_2 expected in the Atlantic Ocean are shown in Figure 3-2. In these calculations, the preindustrial atmospheric pCO_2 is taken to be 280 μatm. The atmospheric pCO_2 levels during the period of ocean survey are taken from Keeling et al. (1989a). As expected, the warm water in the low latitudes has higher excess ΣCO_2 build-up than cold water in the high latitudes. In general, equilibrium excess ΣCO_2 values range from 30 to 60 μmol/kg. Results from the SAVE cruises have the highest values because these samples were taken most recently and thus allowed the longest time of exposure to the highest atmospheric pCO_2 level.

However, these values do not represent the actual build-up because of the finite rate of CO_2 exchange between atmosphere and sea surface. For any given surface water, the expected excess ΣCO_2 content should be less than that for the assumed chemical equilibrium between ocean and atmosphere. Tracer-calibrated ocean-atmosphere models for carbon cycle show that the resistance to fossil fuel CO_2 uptake posed by vertical mixing processes is about seven times larger than that posed by the air-sea gas exchange process. Hence, on the average, the surface water excess ΣCO_2 value is ~85% of the equilibrium increase. To remove the effect of fossil fuel CO_2 uptake on the estimates of $\Delta\Sigma CO_2$, 85% of the equilibrium excess ΣCO_2 increase has to be subtracted from the observed ΣCO_2 data. Values of $\Delta\Sigma CO_2$ after such corrections for the Atlantic Ocean are shown in the right panels of Figure 3-3. The uncorrected $\Delta\Sigma CO_2$ values are also shown in the same figure (left panels) for comparison. Although the corrected $\Delta\Sigma CO_2$

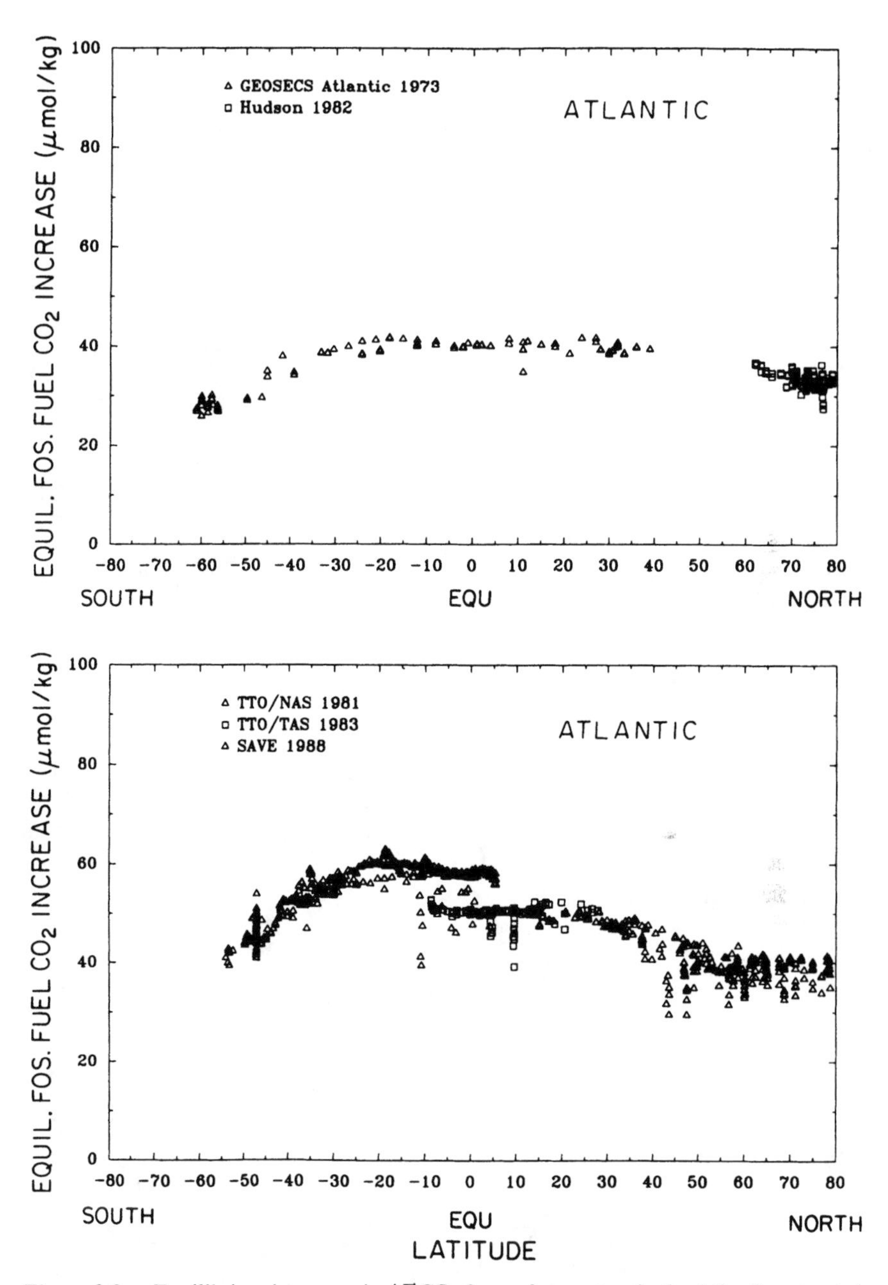

Figure 3-2. Equilibrium increases in $\Delta\Sigma CO_2$ for surface waters in the Atlantic expected from the increase in atmospheric CO_2 content from a preindustrial value of 280 μatm to the CO_2 content at the time the water sample was obtained. The magnitude of the increase is a function of water temperature.

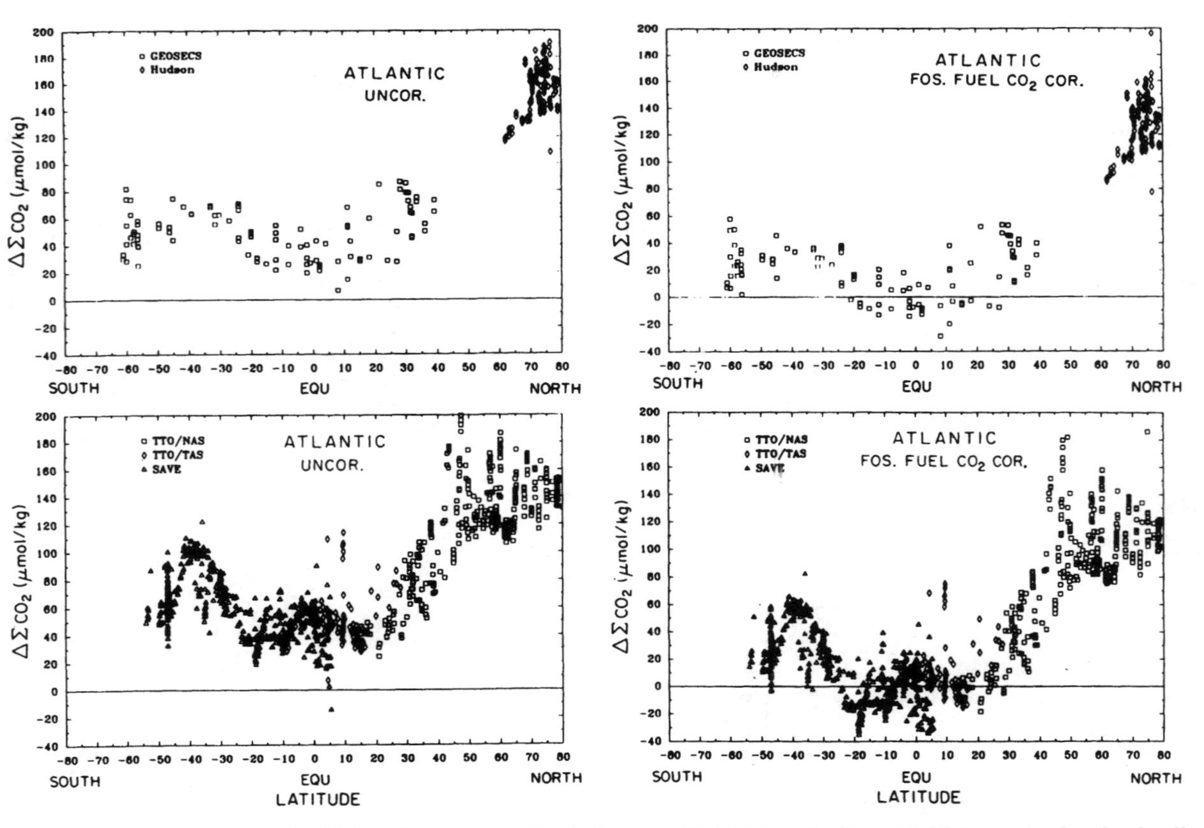

Figure 3-3. Comparison of $\Delta\Sigma CO_2$ values in the Atlantic Ocean with (right) and without (left) the correction for fossil fuel CO_2 uptake. Results from GEOSECS and Hudson-82 are shown in the upper panels. In the lower panels are shown results from TTO/NAS, TTO/TAS, and SAVE.

values are lower than the uncorrected ones, this correction does not alter the pattern of $\Delta\Sigma CO_2$ distribution.

The time of sample collection during the ocean geochemical surveys may also play a role in the $\Delta\Sigma CO_2$ distribution pattern. However, where overlap of these surveys occurs, the differences of estimated $\Delta\Sigma CO_2$ corrected for fossil fuel uptake between the expeditions are small (see Table 3-1). Also, at high latitudes, no large winter-summer difference is seen. Hence, the major differences in $\Delta\Sigma CO_2$ are not caused by the seasonal variations in the observed data.

To understand the reason why waters in the northern Atlantic have high $\Delta\Sigma CO_2$ values, we need to explore a crucial factor in sea-air CO_2 exchange: the expected CO_2 partial pressure for surface waters (pCO_2^{EXP}), which is defined as the CO_2 partial pressure that would exist during preindustrial times if there were no transfer of CO_2 between the ocean and the atmosphere. In essence, it is the pCO_2 that would exist if the water had the expected ΣCO_2 concentration (ΣCO_2^{EXP}). Values of pCO_2^{EXP} can be computed,

$$pCO_2^{EXP} = f(T,S,ALK,\Sigma CO_2^{EXP}),$$

by using Takahashi's carbonate chemistry equations (Peng et al. 1987). Differences in pCO_2^{EXP} for various latitudinal zones in the global ocean would give an indication of potential ocean sinks and sources for preindustrial carbon cycle. Because the choice of coefficients in the ΣCO_2^{EXP} equation was arbitrary (in order to bring the mean $\Delta\Sigma CO_2$ value for global ocean close to zero), the mean pCO_2^{EXP} value does not need to equal that for the 1850 atmosphere. Indeed, the average value of calculated pCO_2^{EXP} is somewhat lower than 280 μatm. However, as the regional difference in pCO_2^{EXP} is the main concern, this small mismatch

Table 3-1. Comparison of the fossil fuel CO_2 corrected $\Delta\Sigma CO_2$ and PCO_2^{EXP} results obtained on three different expeditions in the North Atlantic averaged by latitude belt. (As can be seen, the differences between the expeditions are small. Also, no large winter-summer difference is seen at high latitudes.)

Latitude belt	GEOSECS Summer			TTO-NAS Summer			Hudson Winter		
°N	No. Stns.	$\Delta\Sigma CO_2$* μmol/kg	pCO_2^{EXP} μatm	No. Stns.	$\Delta\Sigma CO_2$* μmol/kg	pCO_2^{EXP} μatm	No. Stns.	$\Delta\Sigma CO_2$* μmol/kg	pCO_2^{EXP} μatm
80–70	0	—	—	11	110	115	41	133	143
70–60	0	—	—	15	94	153	8	102	177
60–50	0	—	—	40	97	157	0	—	—
50–40	0	—	—	23	101	167	0	—	—
40–30	6	33	237	34	48	233	0	—	—
30–20	6	27	260	17	12	265	0	—	—

*Fossil fuel corrected.

is of no consequence. It should not be confused with the observed surface water pCO_2 values from the global geochemical survey programs, which are affected by the build-up of industrial CO_2 in the atmosphere. The pCO_2^{EXP} value simply represents the CO_2 partial pressure for surface waters expected during the preindustrial period if no CO_2 transport through the atmosphere occurred.

Values of pCO_2^{EXP} for Southern Hemisphere surface waters are shown in Figure 3-4A. As can be seen, tropical and polar waters have high pCO_2^{EXP} values. However, as shown in Figure 3-4B, the pCO_2^{EXP} values are lower in the polar region of the North Atlantic Ocean than in any other regions. Therefore, the high-latitude region in the North Atlantic Ocean was a potential carbon sink in the preindustrial time. The CO_2 gas from other parts of the ocean tends to move through the atmosphere into the northern Atlantic region. As a result, the $\Delta\Sigma CO_2$ values for the northern North Atlantic are expected to be higher than those for other parts of the ocean. The high values of pCO_2^{EXP} in tropical waters reflect the potential carbon source in the equatorial ocean. This is consistent with the low $\Delta\Sigma CO_2$ values shown in Figures 3-1A and 1B. However, the important point to be made is not what happened in the tropics, because the CO_2 lost to the tropical atmosphere that created deficiencies in ΣCO_2 would not lead to an interhemispheric gradient. Rather, it is the difference in $\Delta\Sigma CO_2$ between the northern North Atlantic and the Antarctic that matters. As shown in Figure 3-4A, relatively high pCO_2^{EXP} values in the Antarctic result in low $\Delta\Sigma CO_2$ values, which are in contrast with the highest $\Delta\Sigma CO_2$ values in the northern North Atlantic.

Why do waters in the northern North Atlantic have low pCO_2^{EXP} values? The answer lies in the temperature and nutrient concentrations of the northern North Atlantic waters. Taking water temperature and PO_4 content as variables, the pCO_2^{EXP} values are calculated for a water sample with a salinity of 35%, a ΣCO_2 concentration of 1900 μmol/kg, and an alkalinity of 2310 μeq/kg [i.e., coefficients used in equation (1)]. The formation and destruction of marine organic matter by photosynthesis and respiration, respectively, affect the ΣCO_2 concentration, and hence change pCO_2 values. A C/P ratio of 127 and an N/P ratio of 16 is used for computing such biological effects. Contours of pCO_2^{EXP} values as a function of water temperature and PO_4 content are shown in Figure 3-5. As can be seen, for waters with the same PO_4 concentration, the pCO_2^{EXP} values increase with rising temperature. For waters of the same temperature, the figure shows that the higher the nutrient content, the higher the pCO_2^{EXP} values. Thus, regardless of low nutrient content, the pCO_2^{EXP} values for warm tropical and temperate surface waters are high because of high water temperature. For cold Antarctic surface waters, however, the negative low-temperature effect is cancelled out by the positive high-nutrient effect. The Antarctic surface waters have on average 1.6 μmol/kg of PO_4 concentration, yielding just the amount of excess respiration ΣCO_2 concentration to compensate for the lowering of pCO_2^{EXP} caused by the low temperature. Hence, the pCO_2^{EXP} values are also high for

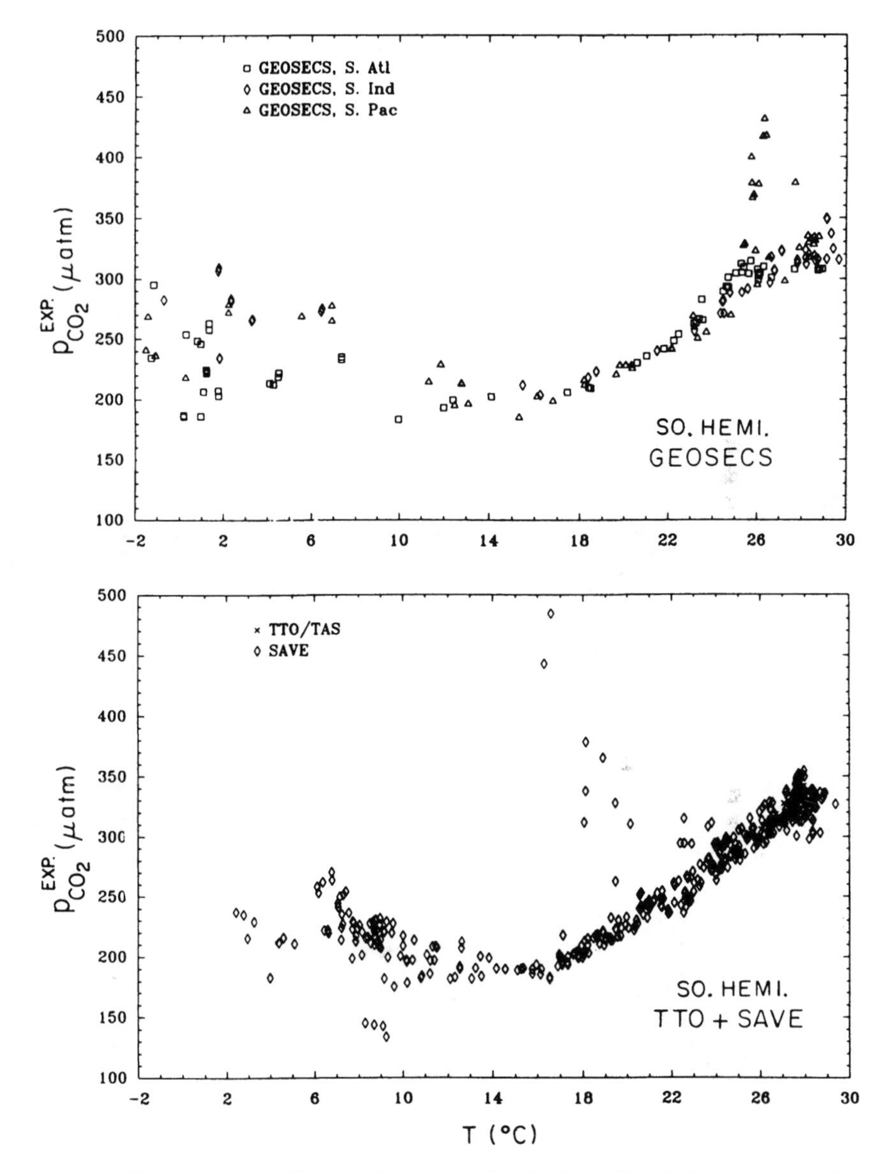

Figure 3-4A. Expected CO_2 partial pressures for Southern Hemisphere ocean surface waters. GEOSECS results are shown in the upper panel; TTO and SAVE results are shown in the lower panel.

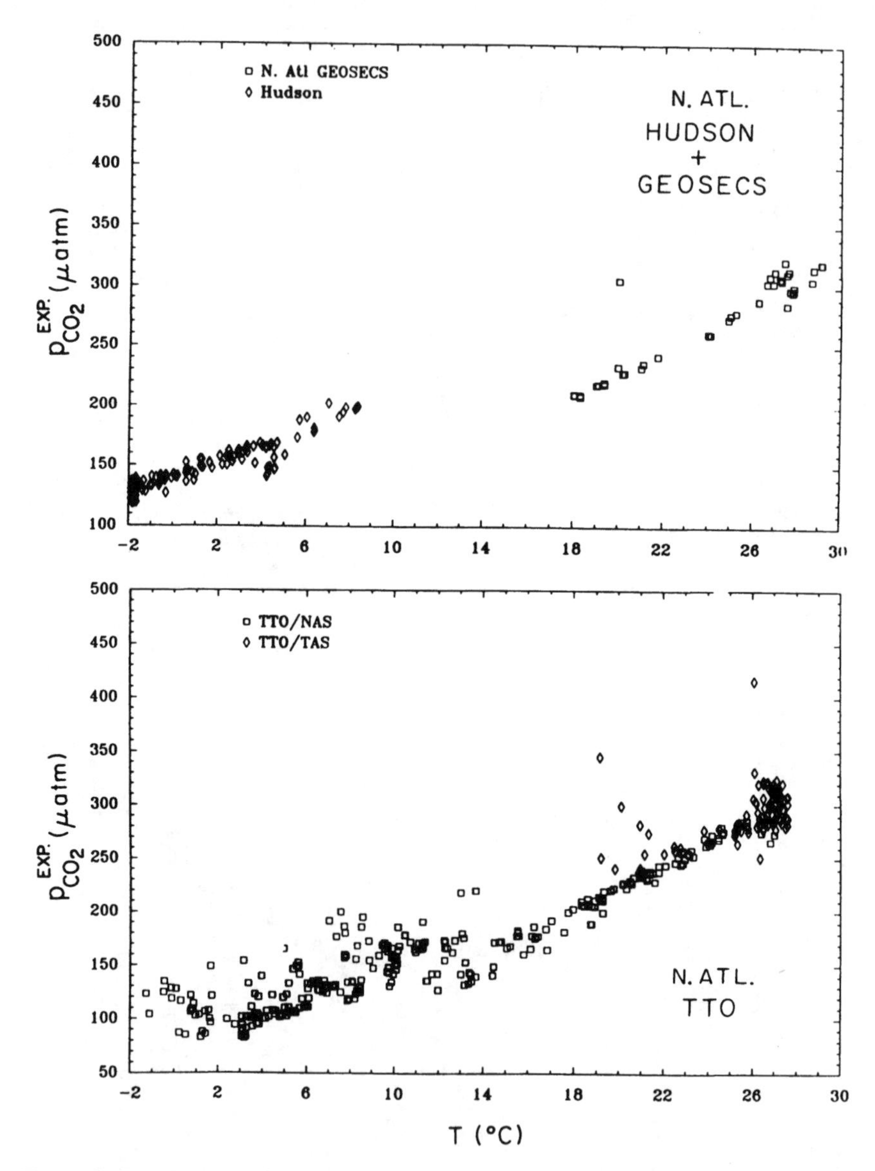

Figure 3-4B. Expected CO_2 partial pressures for Northern Hemisphere Atlantic surface waters. GEOSECS and Hudson-82 results are shown in the upper panel; TTO results are shown in the lower panel.

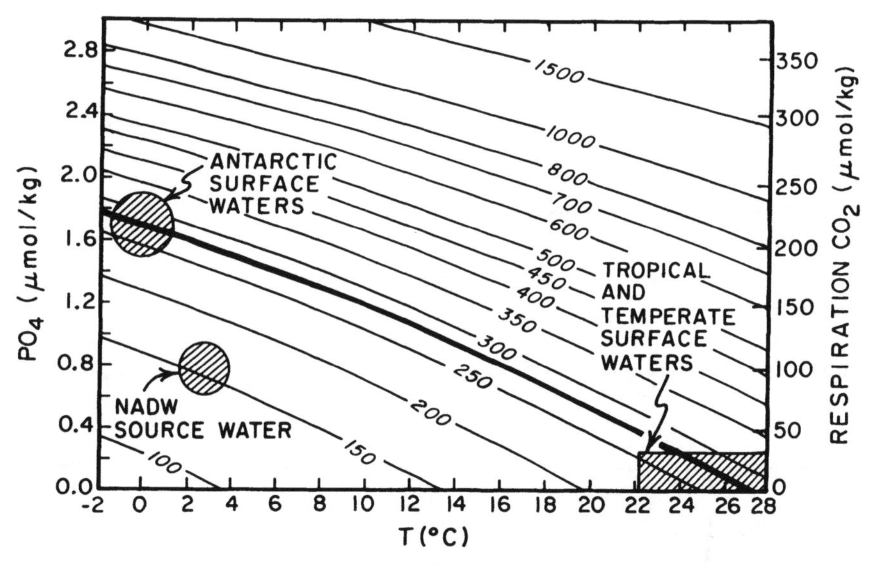

Figure 3-5. Contours of expected CO_2 partial pressures (in μatm) for a water sample with a salinity of 35%, a ΣCO_2 concentration of 1900 μmol/kg, and an alkalinity of 2310 μeq/kg as a function of water temperature and phosphate content. A C/P ratio of 127 and an N/P ratio of 16 are assumed for the marine organic matter formed by photosynthesis and destroyed by respiration. The heavy contour is that representing the preindustrial atmosphere (280 μatm). Note that in the real ocean, freshwater-induced salinity and $CaCO_3$-induced alkalinity differences create additional texture in the surface water CO_2 partial pressure distribution. This diagram shows that the high nutrient content compensates for the low temperature of Antarctic surface waters. By contrast, because of their lower nutrient content, surface waters in the northern Atlantic tend to have much lower CO_2 partial pressures.

Antarctic surface water, in spite of its low temperature. By contrast, surface waters in the northern North Atlantic have only ~0.8 μmol/kg of PO_4. The amount of excess respiration ΣCO_2 associated with this PO_4 is too small to offset the effect of lower temperature. The difference of 0.8 μmol/kg PO_4 concentration between the northern North Atlantic and the Antarctic surface waters corresponds to ~100 μmol/kg of ΣCO_2 concentration (0.8 × 127). With a buffer factor of 14 for cold waters, this 5% decrease in ΣCO_2 yields a 70% reduction in pCO_2 values. This explains why northern North Atlantic surface waters have much lower pCO_2^{EXP} values than Antarctic surface waters.

The low nutrient content in northern North Atlantic surface waters, and hence in the NADW, is the consequence of global conveyor circulation. Thus, the interaction between biological cycling and water transport that dictates low nutrient contents in Atlantic deep waters relative to other deep waters in the world ocean also creates the tendency for transport of CO_2 through the atmosphere to the surface of the northern North Atlantic. At a steady state in the preindustrial time, these transports must be matched by compensating transport through the ocean. In the next section, estimates of the magnitude of this matching ocean transport flux will be made.

Interhemispheric Transport Flux

A map showing the geographical distribution of the corrected $\Delta\Sigma CO_2$ values for fossil fuel CO_2 for surface waters in the world ocean is given in Figure 3-6. A distinctive feature is the high $\Delta\Sigma CO_2$ value (100 μmol/kg) in the northern North Atlantic Ocean and the low $\Delta\Sigma CO_2$ value (25 μmol/kg) in the high-latitude

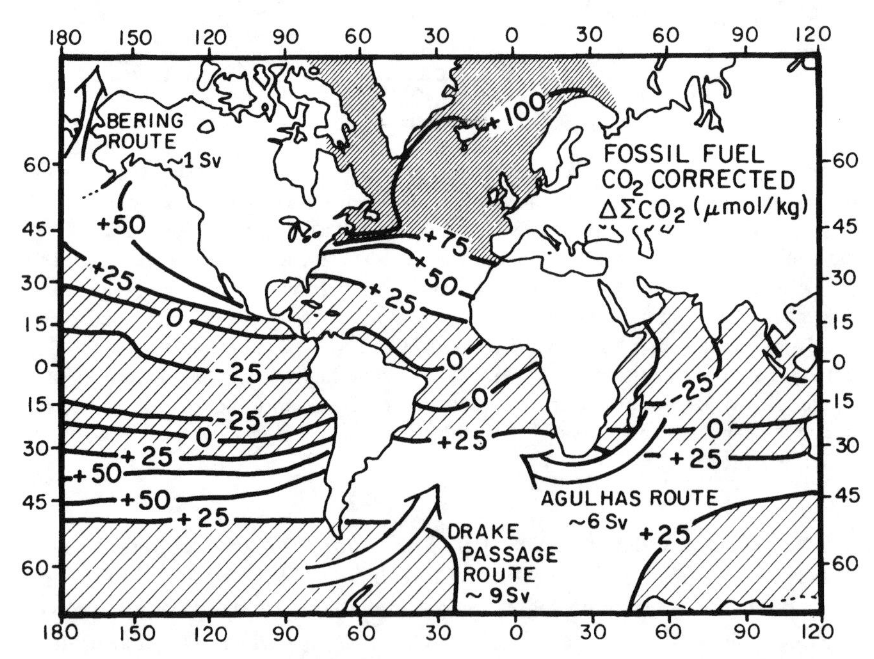

Figure 3-6. Map showing the geographical distribution of corrected $\Delta\Sigma CO_2$ values for fossil fuel CO_2 for surface waters based on measurements of salinity, PO_4, NO_3, ΣCO_2, and alkalinity made during the GEOSECS and TTO expeditions. In the temperate North Atlantic, where geographic overlap exists between these two expeditions, the data sets are concordant. It should be noted that winter $\Delta\Sigma CO_2$ values computed from measurements on winter surface waters in the Atlantic north of Iceland during the Hudson 1982 expedition yield values similar to the summer values obtained during the TTO expedition. The return routes for surface water balancing export from the Atlantic via the conveyor's lower limb (i.e., NADW) are shown by the arrows, with rough estimates of the magnitude of their contributions in Sverdrups ($10^6/m^3/s$). A total of 16 Sv is shown. Northward-flowing deep water contributes the remaining 4 Sv of return flow.

Southern Ocean. Although the lowest $\Delta\Sigma CO_2$ values are found in tropical regions of the Pacific and Indian Oceans, these low values are not important because, as discussed before, they would not give rise to an interhemispheric CO_2 gradient. The northern North Atlantic surface waters with high $\Delta\Sigma CO_2$ values are the main sources of the NADW. After the formation of NADW, this deep water flows the length of the Atlantic into the Antarctic, where it joins the circumpolar current and eventually upwells to the surface. To the extent that this upwelling occurs in the Antarctic, the excess CO_2 carried by the southward-flowing deep water is discharged into the Southern Hemisphere atmosphere. The resulting higher atmospheric pCO_2 in the Southern Hemisphere tends to cause a northward transport of CO_2 through the atmosphere, and thus completes the cycle.

The export of NADW from the Atlantic Ocean in the deep is balanced by an import from the south along the surface, at the intermediate depths, and along the bottom. As shown in Figure 3-6, the major surface return flows are thought to be about 9 Sv from Drake Passage around the tip of South America via the circumpolar current, 6 Sv or so around the tip of Africa via the Agulhas retroflection (Gordon 1986), and about 1 Sv via the Bering Strait (Overland and Roach 1987). These surface waters all have $\Delta\Sigma CO_2$ values similar to those for Antarctic surface waters. The Antarctic Intermediate Water (AAIW) and Antarctic Bottom Water (AABW) are the major deep return flows of about 4 Sv. The contrast between the $\Delta\Sigma CO_2$ contents of the AAIW and AABW contributions to this return flow, and the $\Delta\Sigma CO_2$ content in the NADW produced in the northern Atlantic, is shown by the $\Delta\Sigma CO_2$ versus depth plot (Figure 3-7) for a GEOSECS station from 12° S in the western Atlantic. These $\Delta\Sigma CO_2$ values are not corrected for fossil fuel CO_2 uptake, because the water below the 500-m depth has not yet incorporated a significant amount of this anthropogenic CO_2. As can be seen, the southward-flowing NADW clearly carries an excess of ΣCO_2 over the northward-flowing waters of overlying AAIW and underlying AABW.

The contrast in $\Delta\Sigma CO_2$ values exemplified above could be extended to the full range of waters in the deep Atlantic. To identify waters, we use a parameter PO_4*, which is defined as:

$$PO_4^* = PO_4 + \frac{O_2}{175} - 1.95$$

where 175 is the average molar ratio of oxygen consumed to PO_4 released in the process of respiration for organic particulate matter in the deep sea, and 1.95 is an arbitrarily chosen constant (Broecker et al. 1991a). Thus, PO_4* is a quasi-conservative property; a decrease in the O_2 contribution to PO_4* value as a consequence of respiration should be balanced by an increase in PO_4. The northern component water (pure NADW) has a PO_4* value of 0.73 ± 0.03 μmol/kg, whereas the pure southern component water has a value of 1.6 ± 0.05 μmol/kg. Waters produced by mixing between these two endmember waters will have

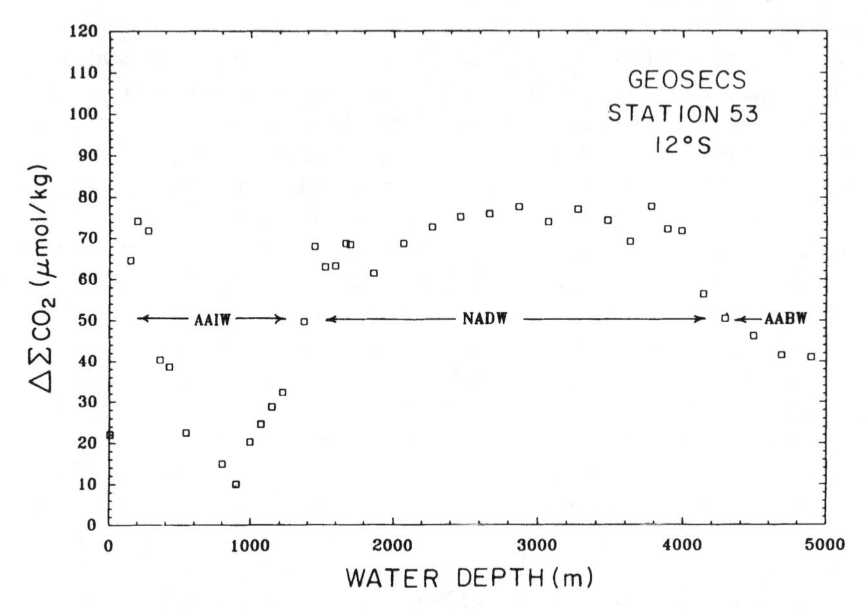

Figure 3-7. $\Delta\Sigma CO_2$ versus water depth at GEOSECS station 53 in the western Atlantic (12° S). The southward-flowing North Atlantic Deep Water (NADW) has higher values than the northward-flowing, overlying Antarctic Intermediate Water (AAIW) and underlying Antarctic Bottom Water (AABW) of southern origin.

a PO_4* value between these two limits. Hence, the PO_4* value for a given deep water in the Atlantic Ocean provides a measure of the relative contribution of the waters formed in the northern North Atlantic (northern component) and the waters formed in the Antarctic (southern component). Figure 3-8 shows plots of $\Delta\Sigma CO_2$ values against PO_4* for all waters deeper than 1000 m in the Atlantic Ocean between 40° N and 40° S. Two linear trends are visible, one for intermediate waters with depths between 1000 and 2000 m, and the other for waters deeper than 2000 m. Clearly, deep waters of northern origin have higher $\Delta\Sigma CO_2$ values than those of southern origin. The $\Delta\Sigma CO_2$ values for pure NADW range from 80 to 115 μmol/kg. These values are consistent with the corrected values for fossil fuel CO_2 for northern North Atlantic surface waters. The value for southern component water, 20 μmol/kg, also agrees well with the corrected values for fossil fuel CO_2 for Antarctic surface waters. Hence, a difference of about 80 μmol/kg is expected between the northern North Atlantic- and Antarctic-derived waters.

We conclude that the tendency for ocean transport of CO_2 from the Northern to the Southern Hemisphere in preindustrial times is surely there. On the basis of current meter measurements (Dickson et al. 1990), geostrophic flow calculations (Bryden and Hall 1980), and radiocarbon measurements (Broecker et al. 1991b),

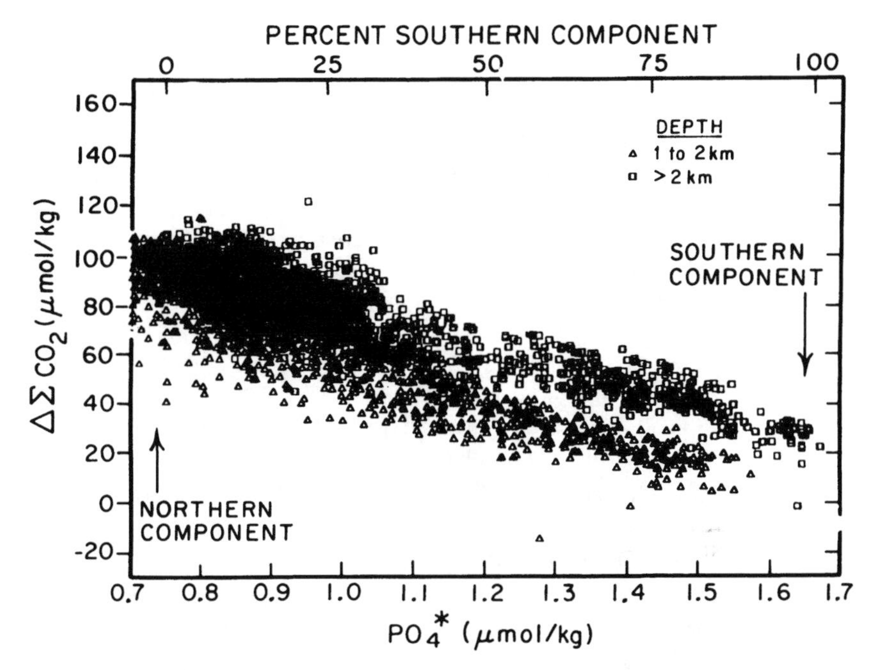

Figure 3-8. $\Delta\Sigma CO_2$ versus PO_4* for intermediate waters (triangles) and deep waters (squares) in the Atlantic Ocean (40° N to 40° S). PO_4* is a measure of the contribution of Antarctic-derived waters to any given water parcel. The northern component (pure NADW) is formed in the northern North Atlantic; the southern component is formed in the Antarctic. The PO_4* and $\Delta\Sigma CO_2$ values are calculated from measurements made as part of the GEOSECS, TTO/NAS, TTO/TAS, and SAVE expeditions in the Atlantic Ocean.

the Atlantic conveyor transport is estimated to be about 20 Sv. With the difference of 80 μmol/kg in $\Delta\Sigma CO_2$ between southward-flowing NADW and the northward-flowing return waters, the transport of 0.6 Gt of carbon per year from the Northern Hemisphere to the Southern Hemisphere could have been possible in preindustrial times. This flux can be compared with that of 0.26 Gt per year estimated by Brewer et al. (1989) for the net southward transport across a section at 25° N in the Atlantic. Because their estimate is the difference in carbon fluxes between upper and lower waters, and their upper water estimate was not corrected for fossil fuel CO_2 build-up, their carbon transport flux must be considered as only a lower limit. Assuming the current total atmospheric inventory of CO_2 in the Southern Hemisphere increases by 1.5 Gt of carbon per year (one half of global increase) under the observed north-south atmospheric gradient of 3 μatm, the estimated interhemispheric carbon flux of 0.6 Gt per year would require a south-to-north gradient of 1.2 μatm in preindustrial times.

Acknowledgments

We thank T. Takahashi for providing carbon chemistry and nutrient datasets for the GEOSECS, TTO, and SAVE cruises. R. B. Cook and L. W. Cooper gave internal reviews. Research was sponsored by the Carbon Dioxide Research Program, Environmental Sciences Division, Office of Health and Environmental Research, U.S. Department of Energy, under contract DE-AC05-84OR21400 with Martin Marietta Energy Systems, Inc. Publication No. 4213, Environmental Sciences Division, Oak Ridge National Laboratory.

References

Bainbridge, A. E. 1981. GEOSECS Atlantic Expedition, Hydrographic Data Vol. 1. National Science Foundation, Washington, D.C.

Brewer, P. G., C. Goyet, and D. Dyrssen. 1989. Carbon dioxide transport by ocean currents at 26° N latitude in the Atlantic Ocean. *Science* **246**:477–479.

Broecker, W. S. 1991. The great ocean conveyor. *Oceanography* **4**:79–89.

Broecker, W. S., and T.-H. Peng. 1992. Interhemispheric transport of carbon dioxide by ocean circulation. *Nature* **356**:587–589.

———, 1993. *Greenhouse Puzzles*. Eldigio Press, Lamont-Doherty Earth Observatory, Palisades, NY.

Broecker, W. S., D. W. Spencer, and H. Craig. 1982. GEOSECS Pacific Expedition, Hydrographic Data Vol. 3. National Science Foundation, Washington, D.C.

Broecker, W. S., S. Blanton, W. M. Smethie, Jr., and G. Ostlund. 1991a. Radiocarbon decay and oxygen utilization in the deep Atlantic Ocean. *Global Biogeochemical Cycles* **5**:87–117.

Broecker, W. S., A. Virgilio, and T.-H. Peng. 1991b. Radiocarbon age of water in the deep Atlantic revisted. *Geophysical Research Letters* **18**:1–3.

Bryden, H. L., and M. M. Hall. 1980. Heat transport by currents across 25° N latitude in the Atlantic Ocean. *Science* **207**:884–886.

Dickson, R. R., E. M. Gmitrowicz, and A. J. Watson. 1990. Deep water renewal in the northern North Atlantic. *Nature* **344**:848–850.

Gordon, A. L. 1986. Interocean exchange of thermocline water. *Journal of Geophysical Research* **91**:5037–5046.

Heimann, M., and C. D. Keeling. 1986. Meridional eddy diffusion model of the transport of atmospheric carbon dioxide. 1. Seasonal carbon cycle over the tropical Pacific Ocean. *Journal of Geophysical Research* **91**:7765–7781.

Hudson Cruise 82-001, Vol. 1. 1984. Physical Chemical Data Report, Scripps Institution of Oceanography, San Diego.

Keeling, C. D., and M. Heimann. 1986. Meridional eddy diffusion model of the transport of atmospheric carbon dioxide. 2. Mean annual carbon cycle. *Journal of Geophysical Research* **91**:7782–7796.

Keeling, C. D., R. B. Bacastow, A. F. Carter, S. C. Piper, T. P. Whorf, M. Heimann, W. G. Mook, and H. Roeloffzen. 1989a. A three-dimensional model of atmospheric CO_2 transport based on observed winds: 1. Analysis of observational data. *Geophysical Monographs* 55, American Geophysical Union, Washington, D.C., pp. 165–231.

Keeling, C. D., S. D. Piper, and M. Heimann. 1989b. A three-dimensional model of atmospheric CO_2 transport based on observed winds: 4. Mean annual gradients and interannual variations. In D. H. Peterson (ed.), Aspects of climate variability in the Pacific and the Western Americas. *Geophysical Monographs* 55, American Geophysical Union, Washington, D.C., pp. 305–363.

Marland, G., and T. A. Boden. 1991. "CO_2 Emissions—Modern Record." In T. A. Boden, R. J. Sepanski, and F. W. Stoss (eds.), *Trends '91: A Compendium of Data on Global Change*. ORNL/CDIAC-46, Carbon Dioxide Information Analysis Center, Oak Ridge National Laboratory, Oak Ridge, TN, pp. 386–507.

Overland, J. E., and A. T. Roach. 1987. Northward flow in the Bering and Chukchi Seas. *Journal of Geophysical Research* **92**:7097–7105.

Peng, T.-H. 1986. Uptake of anthropogenic CO_2 by lateral transport models of the ocean based on the distribution of bomb-produced ^{14}C. *Radiocarbon* **28**:363–375.

Peng, T.-H., and W. S. Broecker. 1987. C/P ratios in marine detritus. *Global Biogeochemical Cycles* **1**:155–161.

Peng, T.-H., T. Takahashi, W. S. Broecker, and J. Olafsson. 1987. Seasonal variability of carbon dioxide, nutrients and oxygen in the northern North Atlantic surface water: Observations and a model. *Tellus* **39B**:439–458.

Post, W. M., T.-H. Peng, W. R. Emanuel, A. W. King, V. H. Dale, and D. L. DeAngelis. 1990. The global carbon cycle. *American Scientist* **78**:310–326.

SAVE (South Atlantic Ventilation Experiment), 1988–1989. Preliminary Shipboard Chemical Physical Data Report (Legs 1–5), Scripps Institution of Oceanography, San Diego.

Tans, P. P., I. Y. Fung, and T. Takahashi. 1990. Observational constraints on the global atmospheric CO_2 budget. *Science* **247**:1431–1438.

TTO/NAS (Transient Tracers in the Ocean, North Atlantic Study). 1986. Shipboard Physical Chemical Data Report, Scripps Institution of Oceanography, San Diego.

TTO/TAS (Transient Tracers in the Ocean, Tropical Atlantic Study). 1986. Shipboard Physical Chemical Data Report, Scripps Institution of Oceanography, San Diego.

Weiss, R. F., W. S. Broecker, H. Craig, and D. Spencer. 1983. GEOSECS Indian Ocean Expedition, Hydrographic Data Vol. 5. National Science Foundation, Washington, D.C.

4

Integration of Spatial Analysis in Long-Term Ecological Studies

John R. Vande Castle

Ecological research often involves investigations covering small field plots to areas the size of watersheds (Kareiva and Andersen 1988). Remote sensing and geographic information system (GIS) technology has allowed comparison of field measurements with satellite-based measurements and expanded site-specific research to include landscape and regional processes. More recent interest in landscape ecology has provided models to investigate ecological patterns and processes (Turner and Ruscher 1988). These models provide tools for analyzing ecological data in a spatial framework based on the organizational structure of the data.

Long-term ecological studies are generally focused on temporal changes of ecological processes. Although the spatial area of the research is important, the focus remains linked to the temporal domain of the study. Data from long-term studies is seldom integrated into the structure used in spatial analysis. Ecological research is geared toward temporal or spatial phenomena, but the spatial and temporal components are seldom considered at the same time. The research and analysis of spatial and temporal data generally proceed in parallel. Consequently, many of the analytic tools are applicable to either temporal or spatial data, not to both. Although the tools used for temporal or spatial analysis might be different, however, the actual methods are often identical.

The concepts of scale and location, in space and time, are key points for integration of spatial and temporal information in comparative research. The organizational structure for both spatial and temporal information is similar, but again is often not integrated. The importance of spatial size and resolution of landscape data is conceptually no different from the time length and sampling interval of a long-term data set. The organizing framework used for spatial analysis is often not considered in temporal studies, yet it provides a powerful tool where it can be applied to temporal data.

Spatial Organization of Remote Sensing Data

In spatial analysis, GIS technology provides a structured framework for organizing and processing data. Integration of data from many diverse sources is accomplished in a GIS, as all data are referenced by their location. This capability to integrate data from many sources is particularly powerful for data integration and modeling. An important point to remember about any GIS is that many of the internal analysis tools used for processing data are no different from more conventional processing techniques. The modeling tools contained within a GIS are, for the most part, standard techniques for statistical description and processing, including such techniques for correlations and for principal component analysis, filter techniques, and spectral analysis. Many of the routines contained within GIS packages are useful in themselves for visualization purposes. Routines for integration of data, filtering, and modeling can be incorporated into many types of data, as long as they can be referenced at least as a matrix of rows and columns.

Data obtained from satellite-based remote-sensing platforms are often arranged as grid cells of a raster data file. Researchers less familiar with data generated by remote-sensing technology may consider the images produced to be pictures rather than true data. In reality, these pictures are nothing more than the graphical representation of a massive spreadsheet of data. Each element is a data point, or "cell value," occupying (in spreadsheet terms) a row and column position. Satellite data in general are nothing more than data values of sensor measurements produced one cell or one column at a time, generating a row of data. Any single row or column of data would represent a transect of measurements. In simplified terms, a GIS manages the data it contains as a "digital map" or GIS layer. It is the rows and columns of the data that provide the structural framework of the image. In a GIS, these rows and columns of data are translated into location information, of latitude (rows) and longitude (columns). The pattern generated by these rows and columns of data form the actual satellite images.

Remote-sensing data represent one of the few measurements that can be used for ecological intercomparisons. An example is spatial comparison of vegetation indices (NDVI), a generic measure of vegetation cover, which can be calculated from satellite data (Price 1987). Figure 4-1 shows the results from an NDVI analysis of a full Landsat Thematic Mapper (TM) scene. The image covers an area of 185 km $\times$ 175 km over central New Mexico. The city of Albuquerque is in the upper left corner of the image, vegetation along the Rio Grande river from north to south is a clear pattern in the image, and the Sevilleta National Wildlife Refuge is in the lower left corner. The image is referenced within the GIS in Universal Transverse Mercator coordinates with a resolution of 25 m. The actual data values of the NDVI calculation range from -0.0 to $+1.0$, and the data can be stored within the GIS as floating point numbers from the NDVI calculation.

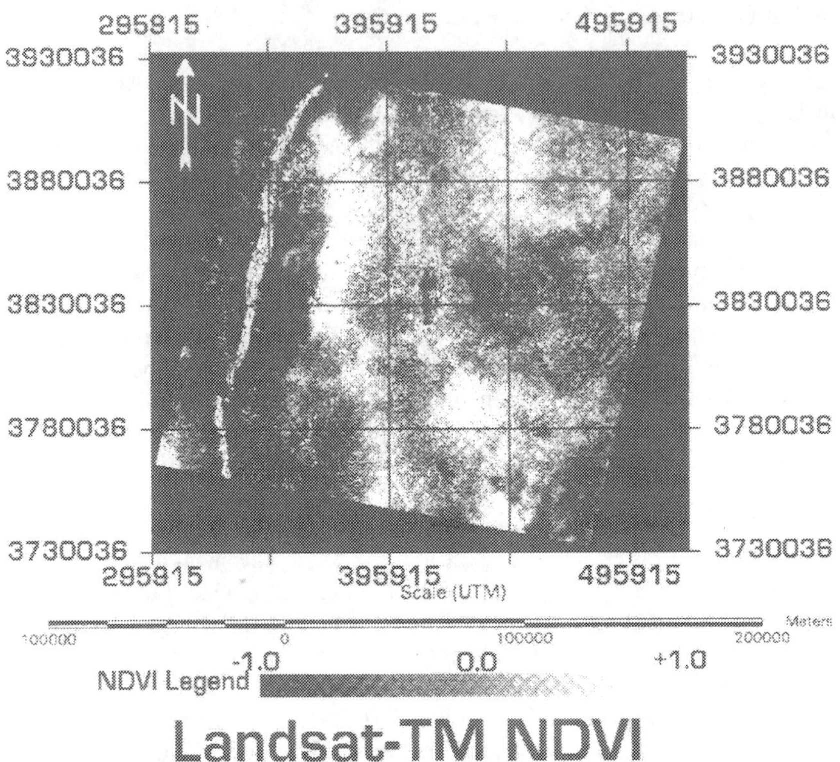

Figure 4-1. Normalized difference vegetation index (NDVI) data calculated from a full Landsat Thematic Mapper (TM) satellite image over New Mexico and referenced to a Universal Transverse Mercator (UTM) projection. Low vegetation index values, indicating a lack of living vegetation, appear dark in the image; bright areas indicate land covered with active vegetation.

The spatial organization of the GIS data is more clearly seen in Figure 4-2, where the NDVI data of a small section (0.3 × 1.0 km) of Figure 4-1 is enlarged. Spatial statistics of many forms can be generated within the GIS systems, but the graphical output of the GIS is able to show the general patterns formed by the individual measurements of the data.

Temporal Organization of Time Series Data

Many methods have been used for analysis of time series data. Basic statistics as well as more complex techniques, such as Fourier transforms (Blackman and

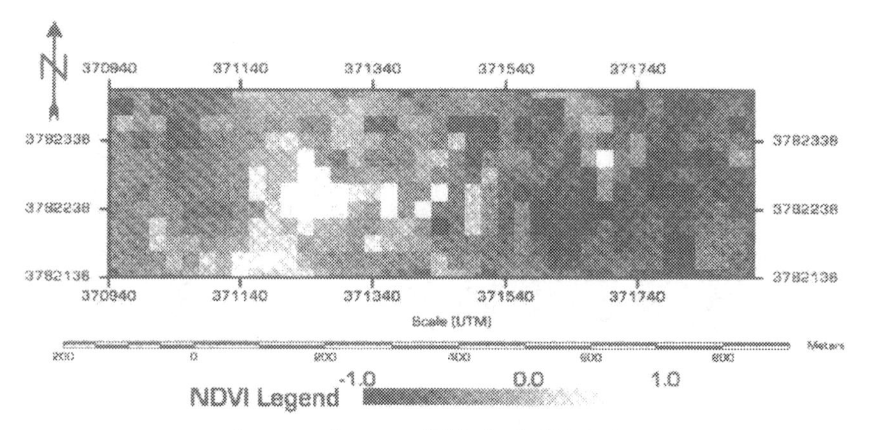

Figure 4-2. NDVI data of a 1.0-km × 0.3-km area subset from Figure 4-1. These data become a 40 × 12 element data array. Each element, or square, represents the average NDVI value for a 25-m area of the ground.

Tukey 1958), are often used to describe patterns in a data series. Detection of patterns becomes more difficult within larger data sets. However, it is possible to apply the organization used for GIS analysis to conventional time series data.

The data in Figure 4-3 represent monthly averages of daily subsurface temperature measurements in Lake Michigan (Vande Castle 1985). The strong annual cycle is clearly observed, but the length of the time series and the large amount of information hide details such as monthly variation. Annual averages are often used for intercomparison of long-term data, but this summary of the data represents a significant loss of detail for comparison. Structural organization based on the previous GIS data examples permits more complete comparisons.

The organization of time series data will depend to a large degree on the

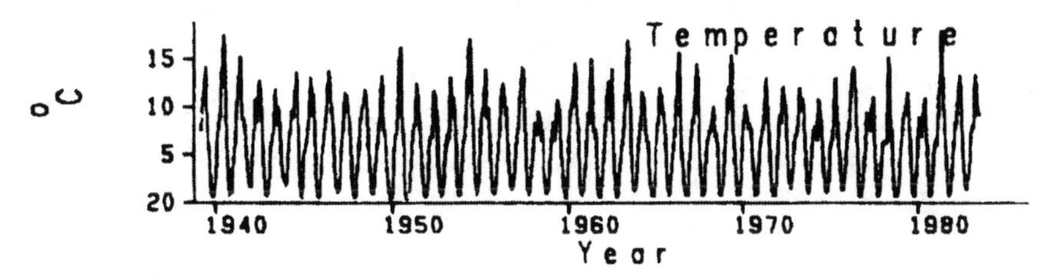

Figure 4-3. Long-term time series data of Lake Michigan subsurface water temperature plotted on a conventional plot of temperature versus monthly average data (from Vande Castle 1985).

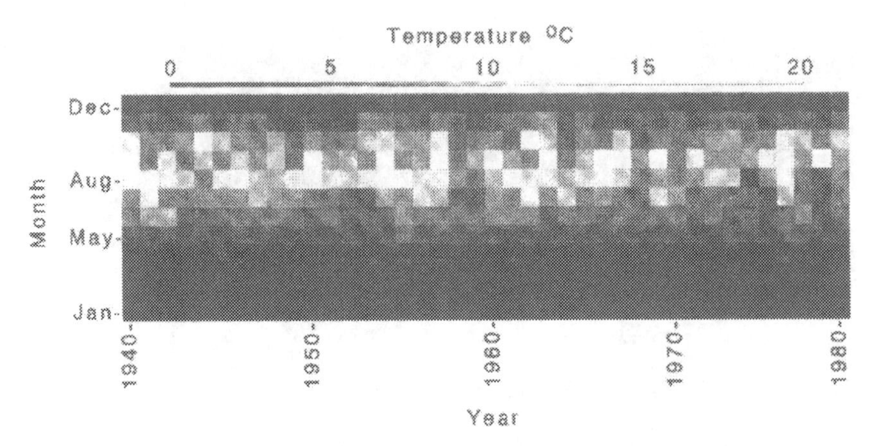

Figure 4-4. The same temperature time series data as for Figure 4-3, here organized into rows representing years and columns representing months. Each element, or square, represents one month of data.

type of data. However, for time series data based on annual measurements, an organizational structure based on the annual cycle provides a simplified approach for analysis of temporal patterns of the data. The data in Figure 4-3 are individual observations broken into months of the year. These same data are replotted in Figure 4-4, where the temporal domain is expanded as an image, similar to the spatial information presented in Figure 4-2. Although the data are seen more clearly in color output, the grey-scale images here are able to show a large amount of detail. An important feature of this type of data transformation is the small amount of data (40 × 12) compared to remote-sensing datasets, which are commonly two orders of magnitude larger.

The temporal data of Figure 4-4 are able to show clearly the overall annual cycle as the water temperature increases during the summer and cools again late in the year. General features such as the warmest temperatures in August and September are evident, but variation between years is a dominant feature of the data. Other data plotted in a similar format can show dominant features such as a lack of annual or interannual variation, or a gradual shift in the data indicating a directional trend. The monthly observations have been used here as a simple approach to restructuring the analysis. This structure can be expanded to other resolutions such as annual displays of weekly, daily, or even hourly observations. Shorter-term investigations can use the day rather than the year as the organizational unit with grids of days versus hours of the day.

Conclusion

The organizing structure used in GIS analysis provides a framework for analyzing other forms of data. Organization of temporal data based on annual cycles is just one approach for investigating trends and patterns within the data. The parallel approaches used in temporal and spatial analysis are beginning to converge as research tools based on similar principles become more integrated. This type of analysis is in essence the same process used for generation of three-dimensional perspectives of data, and this is an important point. Tools used for data processing and visualization are based on common methods and are converging in their general applicability. Capabilities, or at least the tools of GIB analyses, are not restricted to spatial analysis, and, as information systems and visualization systems emerge, the capability and ease of processing temporal information in a structured framework will become more common.

References

Blackman, R. B., and J. W. Tukey. 1958. *The Measurement of Power Spectra*. Dover, New York.

Kareiva, P., and M. Andersen. 1988. "Spatial Aspects of Species Interactions: The Wedding of Models and Experiments." In A. Hastings (ed.), *Community Ecology*. Springer-Verlag, New York.

Price, J. C. 1987. Calibration of satellite radiometers and the comparison of vegetation indices. *Remote Sensing of the Environment* **21**:15–27.

Turner, M., and C. Ruscher. 1988. Changes in landscape pattern in Georgia, USA. *Landscape Ecology* **1(4)**:241–251.

Vande Castle, J. R. 1985. A Study of Long-term Changes in the Phytoplankton Community and the Seasonal Change of Alkaline Phosphatase Activity in Lake Michigan. Ph.D. dissertation, University of Wisconsin, Milwaukee.

5

Dynamical Systems Theory for Ecologists: A Brief Overview

John Guckenheimer

This chapter provides a brief overview of computational tools based upon modern dynamical systems theory that help ecologists (and other scientists) with the analysis of models and time series data. Little more is intended than to give pointers to more extensive and detailed accounts of the methods described. The subject of dynamical systems theory, nonlinear dynamics, or chaos theory is a large edifice; there is no way that a few pages can suffice to give a reasonable survey of its contents.

For the purposes of this chapter, *dynamical systems* are mappings or systems of ordinary differential equations defined on a finite-dimensional Euclidean space. For example, a mapping $F(h,p)$ of the plane to itself can serve as a model of an interacting host and predator with discrete generations. The map F "updates" current host and predator populations from generation n to generation $n + 1$. As has been appreciated for a long time (Beddington et al. 1975), the sequence of predicted populations from such a model (the model *trajectories*) can be very complicated, sometimes displaying what is now called chaos. Dynamical systems theory has emphasized the study of qualitative properties of trajectories and how the underlying *phase space* of the system is filled by trajectories. Using the mathematics of *transversality* (a part of differential topology), the theory seeks to classify *generic* properties that are found in "typical" systems. Building rigorous mathematical foundations under these fleeting images is a substantial undertaking (Guckenheimer and Holmes 1983; Shub 1987). The reader is warned that this mathematical theory requires care in its use and that there are many pitfalls that easily lead to its misapplication.

We shall discuss only deterministic systems, though we shall touch upon the issue of "noise" when we discuss data. To simplify the discussion, we shall henceforth discuss continuous time systems rather than make statements that apply uniformly to both discrete time and continuous time systems. Analogous theory (usually a bit simpler) is applied for discrete time systems, but the terminology gets confused a bit when both discrete and continuous time systems are discussed simultaneously.

Phase Portraits

The fundamental problem of a dynamical model is the determination of trajectories from their initial conditions. Generally, the only way to compute this information is by numerical integration. This yields the trajectory bit by bit, accumulating long-time information from the concatenation of short-time information provided directly by the model. Frequently, "simulation" is interpreted as the process of formulating realistic models and building the means to compute trajectories. This allows us to answer "what if" questions about the system behavior beginning from specified initial conditions. This type of simulation often does not provide enough information, particularly in circumstances in which there is a loose connection between the model and "reality." In such situations, we are interested in questions about many trajectories and how these fit together. For example, we might want to know which initial conditions lead to extinction of a particular species in a population model. This is hard to determine just by computing trajectories, because each trajectory computation requires a large number of arithmetic operations. Therefore, one would like to have qualitative information about how the phase space of a system is divided into sets of trajectories with similar behavior. The resulting picture is called (colloquially) the phase portrait of the system.

One of the remarkable features of dynamical systems theory is that there are "universal" geometric patterns that are found in the phase portraits of "generic" systems. These patterns form a taxonomy that allows us to describe the phase portrait of the system concisely and to answer questions about the system behavior quickly, without computing trajectories. A few terms regarding this taxonomy are needed for our discussion. First, there is a division of trajectories as wandering and nonwandering. The wandering trajectories are "transient," in that initial conditions close enough to the wandering trajectory never return to a small enough region around this trajectory. The asymptotic or long-term behavior of the system is found in the set of nonwandering trajectories. Second, there is a topological classification of types of trajectories into equilibrium points (the trajectory is a single point), periodic orbits (the trajectory returns exactly to its initial point at a later time), and the more complicated categories of quasi-periodic and chaotic orbits. We shall focus attention upon equilibrium points and periodic orbits to make our discussion simpler and more concrete.

Near an equilibrium point or periodic orbit, a dynamical system can be "linearized." This means that we compute the matrix of first derivatives of the system and compare the behavior of nonlinear and linear systems in the vicinity of these points. The linear systems have "eigenvalues" that are exponential rates of growth and decay for solutions to the linear system. If all of these eigenvalues yield nonzero rates of growth and decay, then the trajectory is called hyperbolic. The geometries of linear and nonlinear systems are similar to one another near hyperbolic equilibria and periodic orbits. In particular, there are stable and unstable manifolds tangent to the directions of decaying and growing eigenvalues that consist of those trajectories that approach the equilibrium point or periodic orbit as time tends to $\pm\infty$. When there are only decaying eigenvalues, the equilibrium point or periodic orbit is called an attractor, and its stable manifold forms a region around the orbit called its basin of attraction.

Heuristically, a basic task is to divide the phase space into basins of attraction for its attractors, some of which may be more complicated than individual equilibrium points or periodic orbits. One way to specify these sets is to determine the boundaries that divide them. The stable manifolds of nonattracting equilibrium points and periodic orbits frequently form parts of these boundaries, so they are objects that one would like to compute. Much of the classical theory of nonlinear dynamical systems was developed for continuous time systems whose phase space is the plane. These planar vector fields have neither quasi-periodic nor chaotic trajectories, so their phase portraits are much simpler than the general case. For generic planar systems, the attractors are equilibrium points and periodic orbits, and the boundaries of the domains of attraction are formed by unstable periodic orbits, the stable manifolds of equilibrium points and trajectories that are unbounded in forwarded time.

We now come to the task of computing information about phase portraits. Several basic mathematical algorithms that can be used in the analysis of dynamical systems are listed below. We assume that the system is defined by an (autonomous) set of differential equations of the form $\dot{x} = f(x)$ where x is an n-dimensional vector and f: $R^n \rightarrow R^n$ is a vector-valued function of x.

(1) *Numerical integration algorithms*. Numerical integration algorithms compute approximate trajectories to continuous time vector fields. The simplest method is the Euler method, in which the approximation $x(t + h) \approx x(t) + h\dot{x}(t)$ for small values of h is used to compute an approximation of $x(t + h)$ from $x(t)$. By iteratively using this approximation, approximations to $x(t + nh)$ can be computed for increasing n. Improvements to the Euler method for many purposes have been constructed, and there is an extensive mathematical literature on numerical integration (Henrici 1962; Hairer and Wanner 1991) with dozens of distinct algorithms.

(2) *Root finding*. Equilibrium points are zeros of f. The location of an equilibrium point can be computed by algorithms like Newton's method that seek to solve the system of equations $f = 0$. Newton's method is itself a (noninvertible) discrete dynamical system defined by $F(x) = x - (Df)^{-1}(f(x))$. It has the property that at zeros of f, $F(x) = x$ and $DF = 0$. This implies that all of the zeros of f are strongly attracting fixed

points of F. Nonetheless, many initial points for Newton's method may not converge to zeros of f. There are polynomial root-finding algorithms that are guaranteed to converge to zeros of f, but they tend to be much more complex than Newton's method and liable to numerical errors. The practical task of finding roots of systems of equations is still a difficult problem for which there is no single algorithm of choice.

(3) *Eigenvalue and eigenvector solvers*. Once an equilibrium point has been located, one would like to determine its stability properties and its stable and unstable manifolds. The linearization of the vector field at the equilibrium point can be computed with finite difference methods, symbolically from explicit formulas for Df, or from more sophisticated automatic differentiation methods (Griewank et al. 1992). With the matrix derivative Df at an equilibrium point, the QR algorithm (with shifts) has become a reliable, preferred method for computing the eigenvalues and eigenvectors of Df. The eigenvectors determine the tangent spaces to the stable and unstable manifolds at the equilibrium point. Although the linear stable and unstable manifolds are readily computed in this fashion, computing large regions on stable and unstable manifolds of dimension larger than one is a substantial problem that has yet to be solved satisfactorily.

With these three types of algorithms, a great deal can be determined about the phase portrait of a dynamical system. One would also like to apply root-finding methods to locate periodic orbits. This can be done by computing return maps for a cross-section to a flow. Return maps are discrete maps that describe the next point of intersection with a hypersurface for initial points that lie on that surface. Limitations on the accuracy of numerical integration algorithms diminish the reliability of root finding for return maps. It is even more difficult to obtain accurate values for the derivative of a return map to use in Newton's method, so that algorithms that locate periodic orbits with iterative methods have not found widespread acceptance, except in the context of continuation described below.

We illustrate these ideas with a model for the dynamics of spruce budworm outbreaks drawn form the work of Ludwig et al. (1978). The equations describing this model are

$$\dot{B} = r_B\left(1 - \frac{b}{KS}\frac{T_E^2+E^2}{E^2}\right) - \beta\frac{B^2}{(\alpha S)^2+B^2}$$

$$\dot{S} = r_s S\left(1 - \frac{SK_E}{EK_S}\right)$$

$$\dot{E} = r_E E\left(1 - \frac{E}{KE}\right) - P\frac{B}{S}\frac{E^2}{T_E^2+E^2} .$$

The variables in this model represent budworm density (B), the average size of trees (S), and a measure of the health of trees and condition of their foliage (E). The parameters of the model have biological interpretations (see Ludwig et al. 1978 for details), and estimates for the value of these parameters were derived from data. Table 5-1 presents the parameter values listed by Ludwig et al. (1978) on which they based their simulations, except that here we have divided the values of B and S by 10,000, which results in a proportional decrease in the values of the parameters K_S and β.

Figures 5-1 and 5-2 show two phase portraits from this model projected into the (E,B) plane. All of the parameters have the values listed in Table 5-1 except r_B and T_E. The parameter r_B represents the exponential growth rate (per year) of the budworm larvae and has the value 0.33 in the displayed phase portraits. The factors that include T_E represent the effect of stress due to predation on foliage production, but they were not systematically examined in the analysis conducted on the model. Therefore, we have chosen to emphasize the effect of this parameter upon the dynamics of the model. In Figure 5-1, $T_E = 0.2$, and in Figure 5-2, $T_E = 0.12$. Figure 5-1A shows a system with four equilibrium states, one along the line $R = 0$ representing extinction of the budworm. Close to this is a stable equilibrium with a low budworm population. Figure 5-1B shows more detail of the region near these equilibria. The stable equilibrium is marked by a triangle. The second stable equilibrium has a pair of complex eigenvalues with negative real part, implying that the approach to this equilibrium is oscillatory. The last equilibrium is a saddle point, and its stable manifold divides the regions of trajectories that approach the two stable equilibria. The trajectories shown in the figure lie along the stable and unstable manifolds of the saddles. In Figure 5-2A, the high budworm equilibrium has become unstable and is now "surrounded" by a periodic orbit that is a stable limit cycle. Figure 5-2B shows the region near the limit cycle. Note that the unstable manifold of one of the saddles comes close to the equilibrium "inside" the limit cycle and then winds out toward it.

Bifurcations

Dynamical systems that are formulated as models for physical or biological systems have parameters. Parameters are coefficients in the definition of the

Table 5-1. Parameter values, adjusted, from Ludwig et al. (1978)

r_B	1.52
K	355
β	4.32
α	1.11
r_s	0.095
K_s	2.544
K_E	1. 0
r_E	0.92
P	0.00195

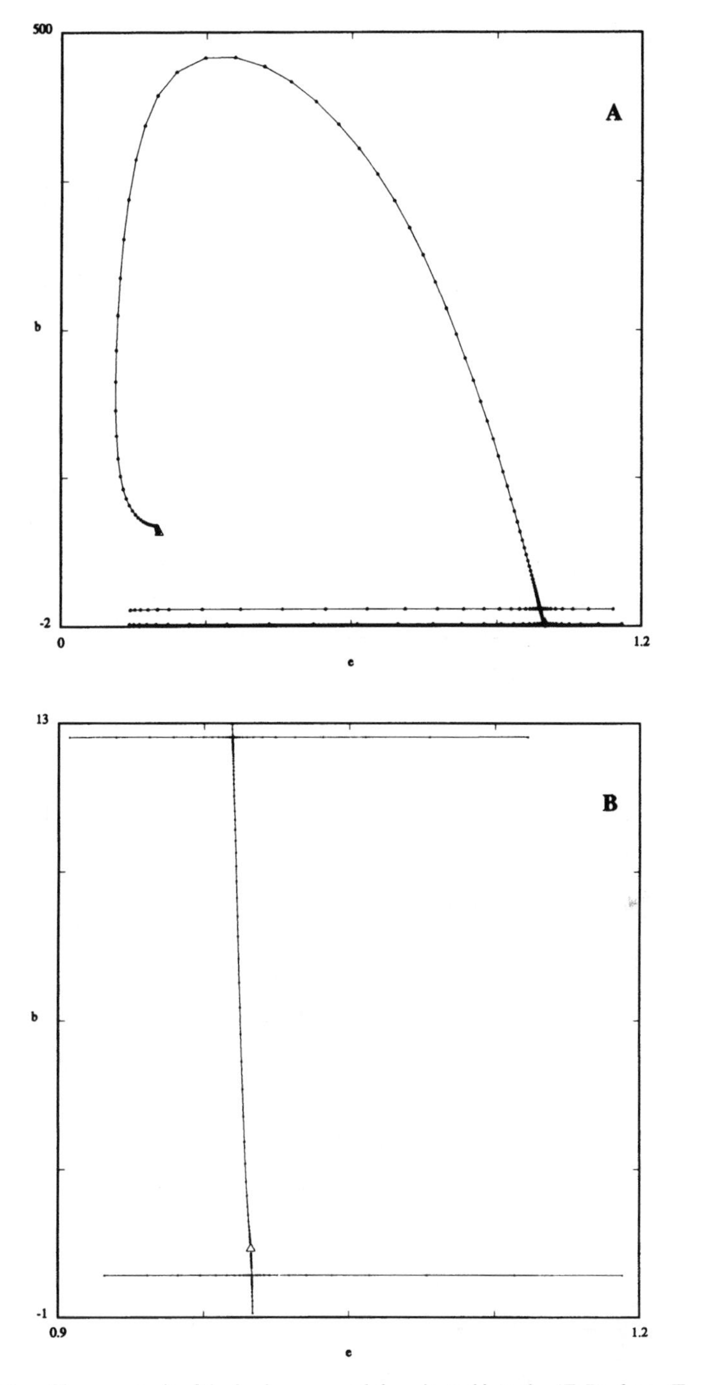

Figure 5-1. Phase portrait of the budworm model projected into the (E,B) plane. Trajectories computed to lie along the stable and unstable manifolds of saddle points are displayed. The computed points are displayed as dots and connected by segments. Attracting equilibria are shown as triangles. (B) shows more detail of the lower right corner of the diagram.

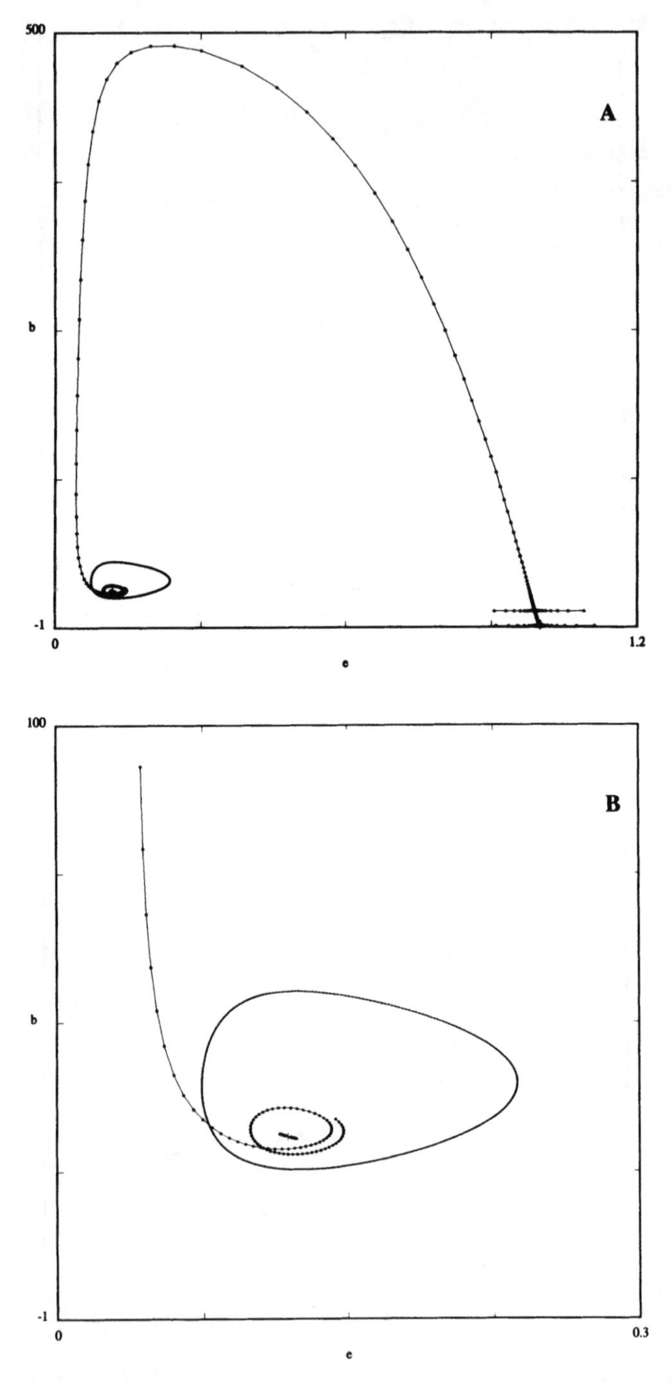

Figure 5-2. Phase portrait of the budworm model projected into the (E,B) plane. Trajectories computed to lie along the stable and unstable manifolds of saddle points and a stable limit cycle are displayed. (B) shows more detail of a neighborhood of the limit cycle. The unstable manifold of the saddle spirals out toward the limit cycle when the portion shown is extended.

vector field that are determined by measurements or estimated by other methods, but they are not logically fixed nor are they dynamical variables that change in time. For example, a birth rate in a population model is a parameter if the birth rate is regarded as unchanging. As parameters are varied in a dynamical system model, the phase portrait of the system may undergo qualitative changes. These parameter values and the corresponding changes are called bifurcations, and their study is called bifurcation theory. The study of bifurcation in a model can help locate those parameter regions in which the model displays a desired behavior. An understanding of generic patterns of bifurcation can also help to guide numerical studies of a system, indicating regions of the parameter space in which qualitative changes in a phase portrait occur rapidly as parameters are varied, and lending confidence to the interpretation of phenomena that occur slowly on long time scales. As the number of parameters that are varied simultaneously in generic families increases, one encounters more degenerate and complex bifurcations. The codimension of a bifurcation is the number of parameters that must be varied simultaneously to encounter a bifurcation of that type in a generic family. Some examples are given below. The classification of bifurcations by codimension is a complicated and intricate task that is still a very active area of research. It is unlikely that there will ever be a satisfying classification that extends beyond codimension two or three, but the theoretical results are still very useful in numerical studies.

The simplest bifurcation take place at equilibria, and they are called local bifurcations. At local bifurcations, the number of equilibrium points can change, or the stability properties of a persistent equilibrium may change. As examples, we describe the generic codimension one bifurcations of equilibria. There are two ways in which an equilibrium point can fail to be hyperbolic: it may have either a zero eigenvalue or a pair of pure imaginary eigenvalues. Each of these cases yields a codimension one bifurcation. The saddle-node bifurcation is associated to an equilibrium point at which there is a simple zero eigenvalue. A pair of equilibria coalesce with one another at a saddle-node bifurcation. On one side of the bifurcation in the parameter space, there are two equilibria, and on the other side there is none. A simple family in which a saddle-node bifurcation occurs is the one-dimensional vector field $\dot{x} = \lambda - x^2$. At a Hopf bifurcation of an equilibrium, there is a single pair of pure imaginary eigenvalues. A simple two-dimensional family with a Hopf bifurcation is

$$\dot{x} = \lambda x - \omega y + (ax - by)(x^2 + y^2)$$
$$\dot{y} = \omega x + \lambda y + (ay + bx)(x^2 + y^2).$$

As one passes through a Hopf bifurcation, a family of periodic orbits emerges from (or collapses onto) the equilibrium point. Systematic accounts of these codimension one bifurcations may be found in Guckenheimer and Holmes (1983).

The analysis of local bifurcations begins with the computation of normal forms. The normal forms are prototypical systems, like the ones defined above with saddle-node and Hopf bifurcations, with the bifurcations of the given type. Given a family with a bifurcation of that type, there are coordinate transformations that reduce the given system to the normal form (or to an approximation of the normal form). The computational analysis of local bifurcations usually begins with an attempt to compute the coefficients that appear in the normal form after this coordinate transformation. Some coefficients determine important qualitative information about the bifurcation. For example, the coefficient a in the Hopf normal form plays a role in determining the stability and direction of branching for the adjacent periodic orbits.

Global bifurcations are bifurcations that involve more than just changes in the immediate vicinity of an equilibrium or periodic orbit. One type of global codimension one bifurcation involves homoclinic orbits: trajectories that are asymptotic to an equilibrium point in both forward and backward time. For codimension one bifurcations of planar vector fields, homoclinic orbits occur at saddle points, and they limit families of periodic orbits whose period becomes infinite. In higher-dimensional vector fields, homoclinic orbits can be more complex. For example, in three-dimensional vector fields, there are homoclinic orbits that are surrounded by chaotic trajectories [the Silnikov bifurcations discussed in Guckenheimer and Holmes (1983)].

In multiparameter families of vector fields, the changes that take place near a bifurcation of codimension larger than one can be quite complicated. The Takens-Bogdanov bifurcation (Guckenheimer and Holmes 1983) is an illustrative example. Takens-Bogdanov bifurcation involves an equilibrium that has a zero eigenvalue of multiplicity two. In the vicinity of the Takens-Bogdanov bifurcation in parameter space, there are curves along which saddle-node, Hopf, and homoclinic bifurcations take place. The unfolding of the Takens-Bogdanov bifurcations describes the geometry in both phase space and parameter space for the qualitative changes that occur as parameters are varied near the Takens-Bogdanov bifurcation.

There are computational tools that aid in the analysis of bifurcations. One of the key algorithmic concepts is the idea of a continuation method. If $G(x) = y$ is a system of n equations in $n + k$ unknowns, then the implicit function theorem states that the set of solutions is a smooth submanifold of dimension k near points where the matrix of derivatives DG has an $n \times n$ nonsingular submatrix. If $k = 1$, then the implicit function theorem also gives a vector field, one of whose trajectories consists of solutions to the equations. A continuation algorithm combines a root-finding algorithm with a numerical integration algorithm to solve the system of equations. A segment of an approximate trajectory is computed by numerical integration and then refined by use of the root-finding algorithm. Convergence properties of the root finding can be used to help control step sizes for the numerical integration.

Continuation methods can be used to compute a variety of data concerning bifurcations. For example, if one uses continuation to track an equilibrium point with a varying parameter, then it is possible to monitor the eigenvalues of the equilibrium in order to detect the location of Hopf and saddle-node bifurcations. Larger systems of equations that incorporate explicit criteria in terms of derivatives for the existence of bifurcations can also be formulated. Using these larger systems, curves of Hopf and saddle-node bifurcations can be computed in two-parameter families. This strategy can be pursued further to formulate explicit systems of equations that determine points of higher codimension bifurcations. Continuation can be used to track bifurcations of codimension k and monitor criteria that determine bifurcations of codimension $k + 1$ while varying $k + 1$ parameters. When a bifurcation of codimension $k + 1$ is detected, then k can be augmented by 1 and the process begun anew. Once bifurcations of codimension k have been located, information about their unfoldings can be used to map the adjacent parameter regions.

Figures 5-3 and 5-4 show two parameter space pictures that illustrate this process for the budworm model of Ludwig et al. (1978). Figure 5-3 was produced by starting at a high budworm equilibrium point and following this equilibrium

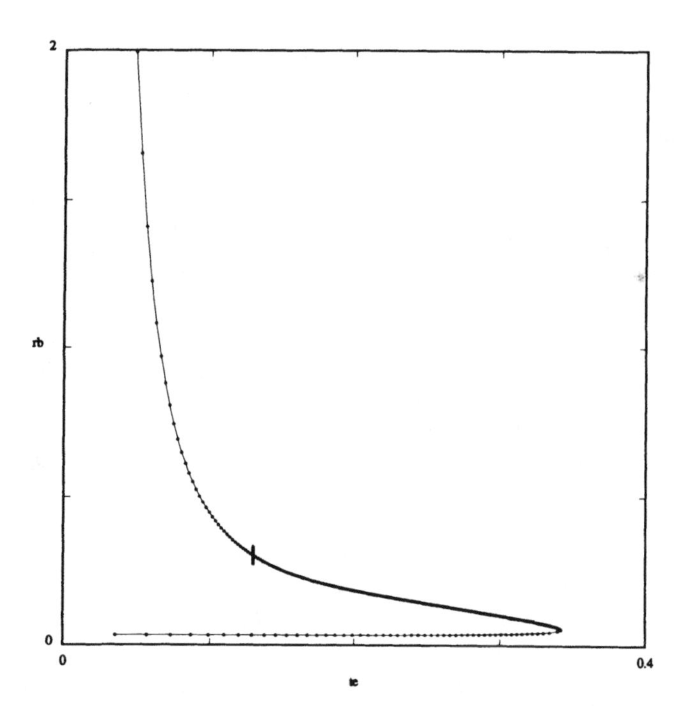

Figure 5-3. The curve of Hopf bifurcations in the (T_E, R_B) plane. A point of Hopf bifurcation was initially located by continuing a curve of equilibria for varying R_B along the short vertical segment in the figure. The lower branch of the Hopf bifurcation curve corresponds to equilibria with the "unbiological" $E < 0$.

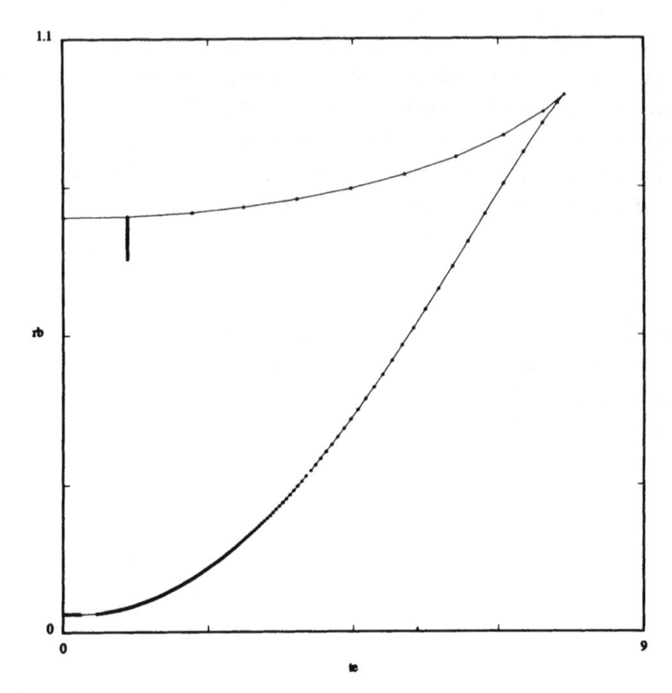

Figure 5-4. The curve of saddle-node bifurcations in the (T_E, R_B) plane.

as the parameter R_B was varied. The vertical segment in the figure represents the range of values of T_E that were examined. The eigenvalues at the equilibrium point were monitored as the equilibria were computed. A complex pair of these eigenvalues was observed to cross the imaginary axis. Making R_B a second active parameter and using an initial point near where these eigenvalues had small real magnitude, we used a continuation method and an algorithm designed to find Hopf bifurcation points to trace the curve of Hopf bifurcations in the (R_B,T_E) plane. For parameters below this curve, the high budworm equilibrium is unstable; above the curve the high budworm equilibrium is stable. Figure 5-4 shows data on saddle-node bifurcations in the (R_B,T_E) plane. Once again, R_B was used as the continuation parameter for tracking an initial curve of equilibria. There is a closed region in the plane enclosed by the saddle-node curve. Inside this curve, there are three equilibria with positive budworm populations. Outside the curve there is one. The curve has a cusp at a value of R_B close to 1. At this point, all three of these equilibria coalesce.

The state of the art in implementing algorithms that compute parameter space maps is not satisfactory and needs further improvement. There are only a small number of programs that have been written to compute local bifurcations directly, and these do not go far into the hierarchy of bifurcations of increasing bifurcation. For global bifurcations, even less has been done. The accuracy of numerical integration algorithms limits the robustness of algorithms that are designed to compute aspects of global bifurcations. For example, the application of continuation methods to tracking periodic orbits by computing fixed points of return maps requires that one be able to evaluate the derivatives of a return map. If the return map itself can be computed with only moderate accuracy, then finite difference methods will do an even poorer job of computing these derivatives, and the root-finding methods are likely to give erratic performance. The development of bifurcation diagrams for specific vector fields requires a good understanding of theoretical results and their implications for numerical computations. Computing these diagrams is currently a highly interactive process, as one tests whether computations give results that are consistent with the theory and refines the calculations when they are not.

Data Analysis

In addition to the simulation tools described above, there are methods that have been developed for the analysis of time series data that are based upon the study of chaotic systems. These methods are based on the assumption that the process generating the data is nonlinear and that traditional statistical methods based upon fitting the data to a stochastic linear system are inappropriate. This assumption is bolstered by the fact that there are very simple discrete dynamical systems (for example, the "tent map" of the unit interval $f(x) = 1 - 2|x - 1/2|$) whose trajectories appear to be completely "random" when viewed with these traditional techniques. Most trajectories for the tent map have autocorrelation functions that are delta functions and power spectra that are constant. These are the hallmarks of "white noise," despite the fact that the trajectory comes from a completely deterministic system.

The nonlinear time series methods view the clustering of points in a multidimensional space and their divergence in time. In deterministic, "dissipative" systems, most trajectories tend toward attracting sets that are "small" subsets of the phase space. On the other hand, "random" systems are expected to have trajectories that explore open regions of the phase space. One goal of the time series methods is to describe the size of the portion of phase space visited by a trajectory. Probabilistic reasoning is used to give a mathematical foundation for the procedures that give technical meaning to the concept of size in this discussion. The outcome of the analysis is a concept of dimension that can be used as the statistic that measures the size of the attractor.

One assumes that the observed trajectory is "typical" and that its asymptotic behavior is described by a measure or probability distribution that vanishes off the attractor. In addition, one assumes that the attractor is "ergodic," so that the time spent near different parts of the attractor by typical trajectories is the same for most trajectories in the attractor's basin. With these hypotheses, one constructs an approximation to the measure on the attractor by distributing masses of equal weight at each point of the observed trajectory. The next step is the definition of the dimension of a measure. The observation that the volume of a ball of radius r in d-dimensional space is proportional to r^d is generalized in the following way. If x is a point of the attractor and $r > 0$, then a typical trajectory spends a portion of its time in the ball of radius r centered at x that is equal to the measure $m_x(r)$ of this ball, using the attractor's measure. Letting r tend to zero, the dimension of the measure at x is d if $m_x(r)$ tends to zero as does r_d. Probability theory implies that this dimension will be the same for almost all points x of the attractor, so the typical value is called the dimension of the measure.

There are two additional concepts that provide statistics about the complexity of an attractor: entropy and Lyapunov exponents. Entropy, like the dimension defined above, is a property of measures. We shall not give a complete definition, but shall restrict this discussion to the entropy of a partition **A**. If an attractor is partitioned into n pieces $A_1, A_2, \ldots A_k$ and F is the "time 1" map obtained by moving points one unit of time along their trajectories, then the entropy gives us a (weighted) average exponential rate of decrease in the sizes of sets of the form $A_{i_0} \cap F^{-1}(A_{i_1}) \cap \ldots \cap F^{-n}(A_{i_m})$. Heuristically, if $m(A_j)$ is the probability that a typical point lands in A_j after m units of time and there is probabilistic independence in where points land after several units of time, then a typical set of the form $A_{i_0} \cap F^{-1}(A_{i_1}) \cap \ldots \cap F^{-n}(A_{i_m})$. is likely to have the set A_j appear approximately $m(A_j)/(n + 1)$ times among the A_{i_1}. If one assumes that this is valid, then

$$m(A_{i_0} \cap F^{-1}(A_{i_1}) \cap \ldots \cap F^{-n}(A_{i_m})) \approx \Pi m(A_j)^{m(Aj)/n}$$

Taking the limit of the natural logarithm of this last expression gives the formula for the entropy of the partition:

$$\mathrm{h}(\mathbf{A}) = \lim \frac{1}{n} \sum m(A_j)\ln(m(A_j))$$

(assuming that the limit indeed exists).

The Lyapunov exponents are defined as the exponential rates by which the derivative along a flow stretches trajectories. The map F^n that flows points n units of time along their trajectories has a derivative that maps the unit sphere in a tangent space at a point x into an ellipsoid in the tangent space at the point $F^n(x)$. As $n \to \infty$, the lengths of the axes of this ellipsoid grow at definite exponential rates for "typical points," and these rates are the Lyapunov exponents of the trajectory. Measuring the exponents is rather tricky, because the strongest stretching tends to make the matrices become almost singular. Considerable care is required to obtain accurate calculations of the Lyapunov exponents that are not the largest ones in the system.

This discussion of statistics associated with the complexity of an attractor has been abstract, so let us try to describe succinctly what one can do to estimate the statistics from observed or computed data. There are alternate strategies for the computations that we describe, and a substantial literature has emerged concerning the pros and cons of different variants. Assume that one is given N observations in a discrete time series $x_i = x(i\Delta t)$ of vectors that are assumed to be measurements at regular time intervals of a typical trajectory lying on an attractor. To compute the fractal dimension of the attractor, interpoint distances can be used to estimate the scaling of the volume of balls centered at a point of the attractor (say, one of the observed points). In a sorted list of the interpoint distances, the kth smallest distance is an estimate for the radius of a ball with volume k/N. Relying upon the remark that the volume of the ball of radius r is proportional to r^d, the asymptotic slope of a log-log plot of the function relating volume to distance is the dimension of the attractor. There are many problems in making good statistical estimates for this dimension (Guckenheimer 1984), but the technique can be used to assess whether a seemingly random time series could have been produced by a deterministic system with a low-dimensional chaotic attractor.

Similar ideas can be used to estimate the entropy of a system from an observed time series. Here the goal is to assess the extent to which volumes become mixed by flowing forward in time. The spread of sets in time can be estimated by combining successive vectors $x_i, x_{i+1}, \ldots, x_{i+m-1}$ into vectors in mn-dimensional space and looking at scaling properties as m increases. If we use a metric on mn-dimensional space that is the maximum of the coordinates, then the volume of balls of fixed radius decreases exponentially in a system with positive entropy. By using measurements of interpoint distances, this decrease can be estimated by fixing a radius r and constructing the function that gives the volume of a typical ball of radius r in mn space as a function of n.

There are two fundamental difficulties with the use of algorithms based upon these descriptions. First, we seldom have the means of measuring all of the phase space variable of a system. In most cases, we have a single time series u_i. There is a means of overcoming this problem based upon the assumption that the time series comes from a "generic" measurement of the system. A variant of the Whitney embedding theorem due to Takens (1980) implies that the m vectors $u_i, u_{i+1}, \ldots, u_{i+m-1}$ will yield an embedding of the attractor of a system if m is at least twice the dimension of the attractor. The second difficulty has to do with the accuracy of the calculations. The expected time to realize a specified interpoint distance in the time series increases exponentially with the dimension of the attractor, because the volumes of balls decrease at an exponential rate with the dimension d. This means that vast amounts of data are needed to cope with high-dimensional attractors. There is no apparent cure for this obstacle, and a reasonable statistical theory for estimating the dimension of an attractor from short time series (say, less than a thousand points) has yet to be created. Note that even if the data are "exact" and free of "noise," one is trying to fit slopes to a function that cannot be expected to be smooth (Guckenheimer 1984). These time series methods give a useful tool for searching for patterns in a dataset, but they are hardly a panacea for automatically discerning the interesting aspects of a times series.

Conclusion

This chapter has provided a brief indication of the types of computation that can be done with dynamical models and with time series data to ascertain some of their qualitative properties. These methods and tools are based upon insights that come from dynamical systems theory, and they have been implemented in a number of software packages of varying degrees of sophistication and user friendliness. We hope that the functionality and power of these computational tools will continue to increase rapidly over the next few years, and that experience gained on case studies will lead to a greater understanding of the intrinsic limits of our ability to analyze dynamical models and time series data.

References

Back, A., J. Guckenheimer, M. Myers, F. Wicklin, and P. Worfolk, 1992. Computer assisted exploration of dynamical systems. *Notices of the American Mathematical Society* **39(4)**:303–309.

Beddington, J., C. Free, and J. Lawton. 1975. Dynamic complexity in predator-prey models framed in difference equations. *Nature* **255**:58–60.

Doedel, E. 1986. *AUTO: Software for Continuation and Bifurcation Problems in Ordinary Differential Equations*. California Institute of Technology Press, Pasadena.

Ermentrout, B. 1990. *Phaseplane, Version 3.0*. Brooks-Cole.

Griewank, A., D. Juedes, J. Srinivasa, and C. Tyner. 1992. *Adol-C* (preprint). Argonne National Laboratory, Oak Ridge, TN.

Guckenheimer, J. 1984. Dimension estimates for attractors. *Contemporary Mathematics* **28**:357–367.

Guckenheimer, J., and P. Holmes. 1983. *Nonlinear Oscillations, Dynamical Systems, and Bifurcations of Vector Fields*. Springer-Verlag, New York.

Hairer, E., and G. Wanner. 1991. *Solving Ordinary Differential Equations II*. Springer-Verlag, New York.

Henrici, P. 1962. *Discrete Variable Methods in Ordinary Differential Equations*. Wiley, New York.

Hubbard, J., and B. West. 1990. *MacMath*. Springer-Verlag, Berlin.

Khibnik, A., Y. Kuznetsov, V. Levitin, and E. Nikolaev. 1992. *LOCBIF, Interactive LOCal BIFurcation Analyzer* (preprint).

Kocak, H. 1989. *Differential and Difference Equations through Computer Experiments*. Springer-Verlag, Berlin.

Ludwig, D., D. Jones, and C. Holling. 1978. Qualitative analysis of insect outbreak systems: The spruce budworm and forest. *Journal of Animal Ecology* **47**:315–332.

Parker, T., and L. Chua. 1989. *Practical Numerical Algorithms for Chaotic Systems*. Springer-Verlag, New York.

Shub, M. 1987. *Global Stability of Dynamical Systems*. Springer-Verlag, New York.

Takens, F. 1980. "Detecting Strange Attractors in Turbulence." In D. A. Rand and L. S. Young (eds.), *Dynamical Systems and Turbulence*, Lecture Notes in Mathematics 898, pp. 366–391.

Yorke, J. 1989. *Dynamics: A Program for IBM PC Clones*. University of Maryland.

6

The Natural History of Time Series

Robert R. Dickson

In a volume that describes the collection, analysis, and interpretation of the key time series by which environmental variation and its effects are understood—all of them screened, calibrated, and corrected with considerable thought and care—it is perhaps appropriate to remind ourselves that the business of achieving a homogeneous series is neither easy nor obvious. When collected over decades, time series are almost bound to be affected by subtle changes in the site and method, which are important to environmental change only insofar as they obscure it; and even in the case of shorter series, records may still suffer from instrumental errors and malfunctions of greater or lesser subtlety.

The reason we cannot simply screen these errors out is usually twofold: either the errors and assumptions build so insidiously that we do not know they are there; or, if we do spot them, we are uncertain as to whether an unexpected result looks unusual because it is wrong or because a variable has merely changed outside the range of our past experience.

The best we can do is to pass on the sense that errors are possible on a wide range of fronts and to illustrate this point with a sufficiently broad selection of cases to show how. In doing so, we concentrate and therefore exaggerate the problems; but if it seems quirky to use a "natural history" analogue to unify a rather disparate set of cases, the important point is that all the examples described are real, and, although these specific examples may not recur, their generic equivalents almost certainly will.

Birth Rate, Mortality Rate, and Longevity

For time series, as with any population, it is possible to point to individuals of great age: the hydrographic monitoring of fixed routes and stations in the Baltic and North seas begun by Otto Pettersson and Gustav Ekman in 1890, some of which can be traced through to the present day in, for example, the Norwegian Sections Program; the nearly 100-year working of the Faroe-Shetland Channel Section by the Scots (begun in 1902), set across the principal inflow of heat and salt to the Norwegian Sea; the 60-year series on the planktonic ecosystem and its changes stemming from the Continuous Plankton Recorder Survey of the North Atlantic (begun in 1931); the almost equally venerable dataset on the ecosystem of the eastern Pacific provided by the CalCOFI program (begun in 1949); the 40-year hydrographic record of the Panulirus station (Station S) off Bermuda; and many shorter, post-war records at a wide scattering of sites.

Here, however, we must be concerned more with the general, and less with these venerable exceptions, important though they are. And here the pattern of birth rate, mortality rate, and death rate of time series is far from reassuring. It is a sobering fact, captured by Duarte et al. (1992), that although there has been an exponential increase in the initiation of new time series-monitoring programs in European marine stations in recent decades (Figure 6-1), there has also been an exponential increase in their termination, so that "long-term monitoring programs are, paradoxically, among the shortest projects in marine sciences: many are initiated but few survive a decade" (Duarte et al. 1992). Put differently, policy makers are rather easily startled into initiating actions—doubtless driven by worries about ecosystem change and the role of humans in provoking it—but the time scales of policy, funding, and even of scientific "fashion" are not normally imbued with much stamina. "Consequently," as Duarte et al. (1992)

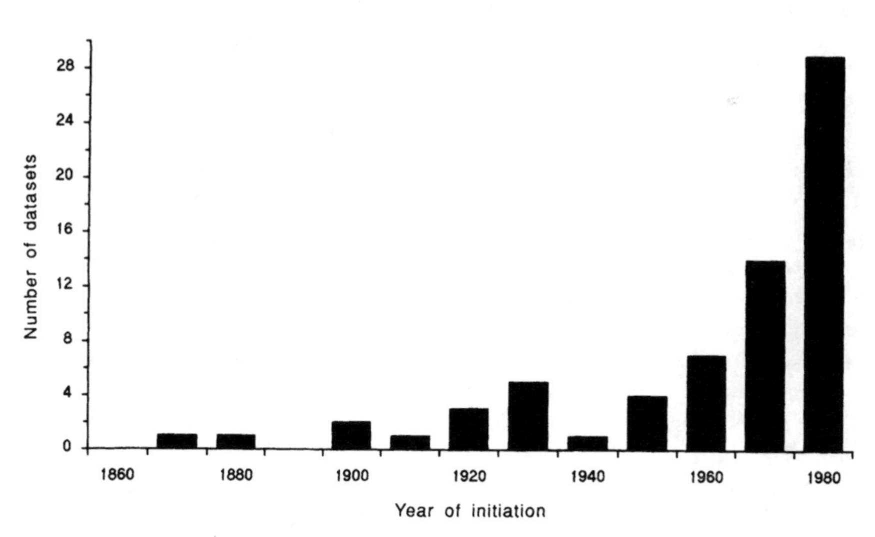

Figure 6-1. Rate of initiation of long-term monitoring programs by European marine stations. (Reprinted with permission from C.M. Duarte, J. Cebrian, and N. Marba, Uncertainty of detecting sea change, *Nature* 356:190. Copyright © 1992 by Macmillan Magazines Limited.)

point out, "the continuation of long-term monitoring programs is often heavily dependent on the personal effort and dedication of individual scientists." It all begins to sound much more like the efforts of the Wildfowl Trust to save the Ne-Ne goose than a healthy and self-sustaining stock of time series, viable because they are needed.

The same concerns apply in space as well as in time. In the global synthesis of hydrographic data to 1000 m compiled by Levitus (1988), we find that the rapid expansion of coverage between the 1920s and 1930s was not maintained (Figure 6-2). The same Pacific-Atlantic bias was maintained into the 1970s and, with temporary exceptions such as that provided by the World Ocean Circulation Experiment, it is likely that much the same underlying bias exists today.

In ocean science there is one question more basic than whether we can partition environmental and ecosystem change into its natural (climatic) and anthropogenic components. It is whether we can recognize that change has even taken place. The pattern of stalled coverage and short-lived series described above is the best possible rationale for the "global" and "permanent" design features of the Global Ocean Observing System (GOOS).

Growth and Maturity

When we analyze the post-war bibliography generated by the 62-year time series of the Continuous Plankton Recorder Survey of the North Atlantic and classify the major papers according to type, we find that the elaboration of our understanding grows steadily with time (Figure 6-3). Initially, these papers provide description. Then the samples become adequate to contribute to the systematics of the plankton; then successively to comment on variability at increasing time scales, from seasonal change through seasonal dynamics to interannual change and lastly to ecosystem change—involving a complex of variability across decades and trophic levels. Thus, over four and one-half decades, the value of the time series *per year* has grown in terms of the breadth of understanding that we can mine from it.

To some this point is obvious: you cannot know about interannual change until records are more than one year long! But it is nonetheless worth emphasizing to funding agencies that adding another year to a 60-year record may be inherently more valuable than starting anew elsewhere; and it is no surprise to learn, in view of the semi-permanent aims of the Global Ocean Observing System, that Figure 6-3 is abstracted from the initial justification for that project—*The Case for GOOS* (1992).

Whether the fluctuating component of a record is the signal to be understood or the noise to be overcome, an increasing utility with increasing record length is a feature of many time series; and in the case of the long, slow shifts associated with climate change, the period over which a time series develops its usefulness or interpretability may be many decades.

Douglas (1992) provides an excellent example of a maturing series from sea-

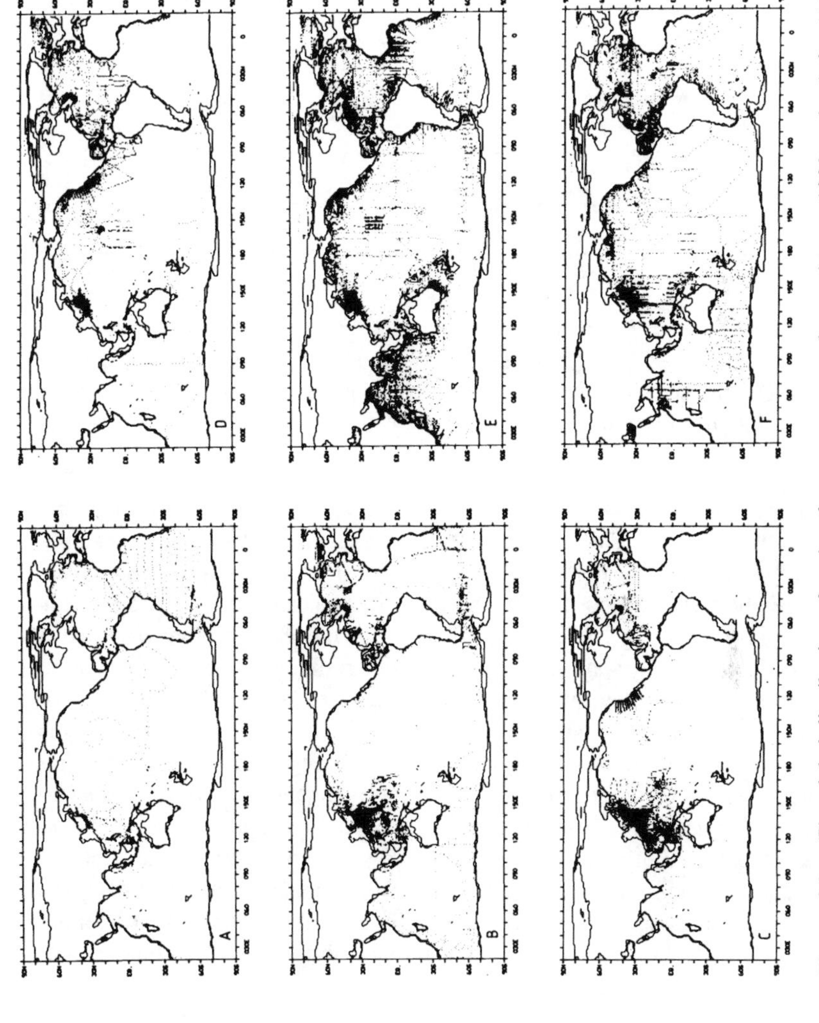

Figure 6-2. The global distribution of station data temperature observations at 1000 m depth for (A) 1920–29, (B) 1930–39, (C) 1940–49, (D) 1950–54, (E) 1960–64, and (F) 1970–74. (From Levitus 1988.)

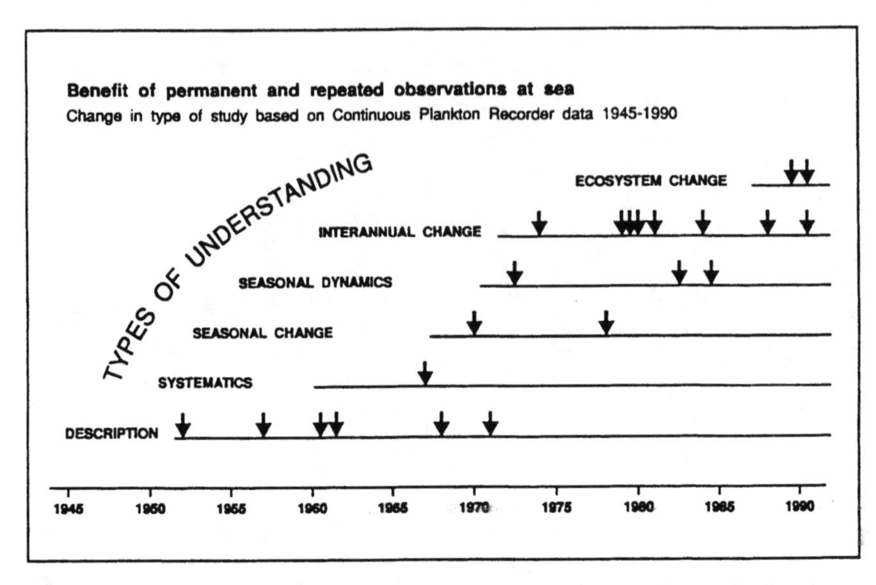

Figure 6-3. The progressive elaboration of our understanding as a time series lengthens, illustrated by the date and type of key papers from the post-war bibliography of the Continuous Plankton Recorder Survey of the North Atlantic. (From *The Case For GOOS* 1992)

level records, and one in which there is a most practical reason for wishing to define the age at first maturity. The problem is this: since greenhouse warming scenarios commonly forecast an acceleration of sea-level rise (i.e., a deviation from a purely linear rise), how long a record do you need before you can tell if such an effect is under way? By analyzing sea-level records of a wide range of durations (from 10 to 140 years) from 10 geographical groups and by determining the rate of acceleration (positive or negative) in each, Douglas shows (Figure 6-4) that a wide range of acceleration estimates is possible in records shorter than 40 years (because of the effects of interdecadal variability), but that for records longer than 50 years accelerations are constant and practically zero. The one, dramatic figure thus makes two clear and relevant points: (1) it is unlikely that a useful index of sea-level acceleration can be obtained in records shorter than 50 years unless the interannual and interdecadal variations of sea level can be understood, modeled, and removed; and (2) there is no evidence for any acceleration in the past 100 years that is either statistically significant or in any way comparable to the accelerations that are predicted to accompany greenhouse warming.

Health and Disease

All instrumental records are subject to errors of varying subtlety due to instrument malfunction. There is usually little trouble detecting and correcting these while

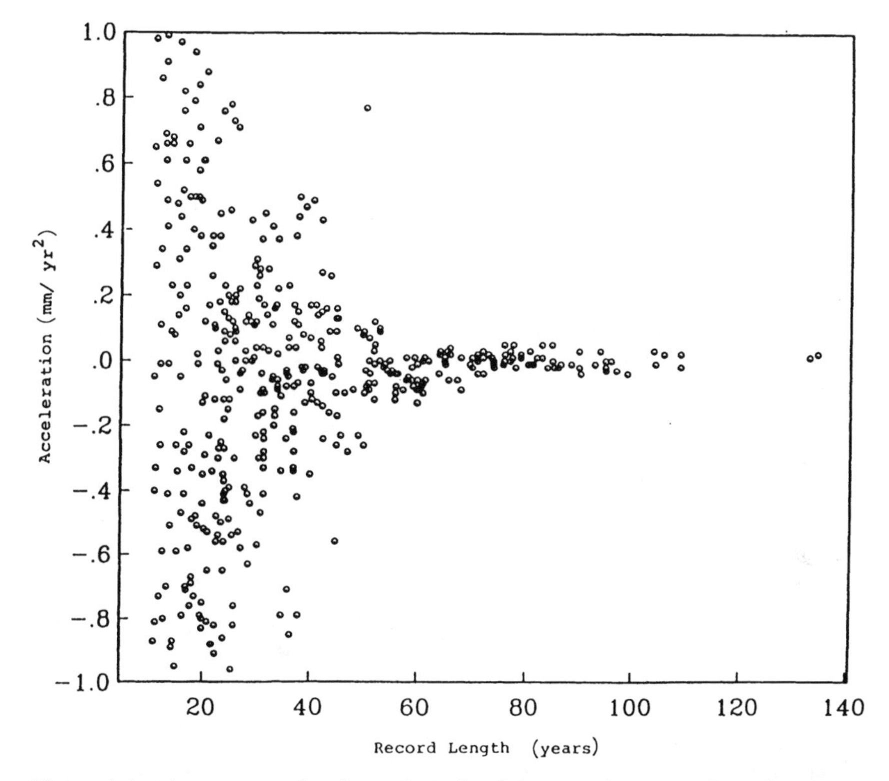

Figure 6-4. Apparent accelerations of sea level for records greater than 10 years in length. Low frequency variations of sea level heavily corrupt the computation of an acceleration parameter for records less than about 50 years in length. (Reprinted with permission from B.C. Douglas, 1992, Global sea level acceleration, *Journal of Geophysical Research* 97:12699–12706.)

the instrument is accessible; calibration, maintenance, and intercomparison can be carried out as the series develops. With autonomous instruments, however, which operate remotely, we may have only the eventual record to assess whether the time series is healthy or whether it is flawed in some way.

There may be little point in giving examples of the ways in which an instrument may perturb a time series, as these are likely to be too instrument- or case-specific to make worthwhile general points. Reports of instrument malfunction are sufficiently rare in the scientific literature, however, to suggest that such examples might have at least some general value in generating an appropriate degree of skepticism about instrumental records.

The four examples that follow are taken from current meter records because these have appropriate characteristics for our purpose: they are inaccessible and cannot normally be interrogated while recording; they are simple in operation, recording a small number of variables (usually current speed, direction, and temperature) on magnetic tape at fixed, preset intervals (usually 10 or 60 minutes); and the moorings from which the instruments are suspended are also simple. The instruments that record current speed by means of a *horizontal propeller* are held away from the mooring wire on some type of metal A-frame, whereas the class of instruments that use *S-rotors* are suspended in wire from a gimballed rod (see Figure 6-5). In each case the essence is that the instrument should be freely swivelling so that it can be directed into the flow by means of its tail fin and turn as the tide turns. The flow direction is measured by referring the alignment of the instrument case to an internal compass, which is located at the bottom of the case, at the end farther from the rotor. In the records that follow, the main tidal signal is semi-diurnal (12.42 hours), but there will be evidence also of the fortnightly spring-neap cycle in tidal current strength.

With a few notes to provide essential background information, these records are first presented and later explained (see Appendix I at the end of this chapter) to allow readers to exercise their own powers of deductive reasoning. The point of this approach is that (with the exception of experience) the reader will have roughly the same basis for assessing error as the data originator. To reduce the reader's disadvantage, the reader is forewarned that there is a problem and that in each case the problem originates with the mechanics of the *mooring* rather than with the electronics of the current meter.

Case 1 presents 33-hour contemporaneous time series of current speed and direction for two instruments from the Irish Sea (Figure 6-6A,B). The instruments are, as normal, changing their direction every 12.5 hours with the tide, yet one record is healthy and one is sick. Which is sick, and what is it suffering from?

Case 2 appears at first sight to concern a problem similar to that in Case 1. Here Figure 6-7 compares 9-day (i.e., ~ 1300 × 10-minute values) records from two instruments that are closely spaced on the same mooring wire—an A-frame propeller instrument (solid line) and an in-wire S-rotor instrument (dotted line), set 9 m and 4 m above the seabed in 25-m water depth in the Irish Sea. Speeds and directions compare well at both the beginning and the end of the record, but in the middle the correspondence in direction is lost. The extra sensor (pressure) shows that this occurs during neap tides, when the in-wire instrument continues to perform well. Though the reader will not be able to deduce what is going wrong with the other instrument, it should nevertheless be possible to deduce why this case is *not* identical to Case 1.

Case 3 also compares two records from instruments of different type set close together on the same instrument wire, the A-frame instrument held off-wire at 12-m depth with the in-wire instrument 1 m lower (Figure 6-8). The specific mooring design is that shown in Figure 6-5. Again, the records from both current meters are plotted together, this time in the form of five successive 25-hour panels (150 × 10-minute values) of current speed, and with directions shown

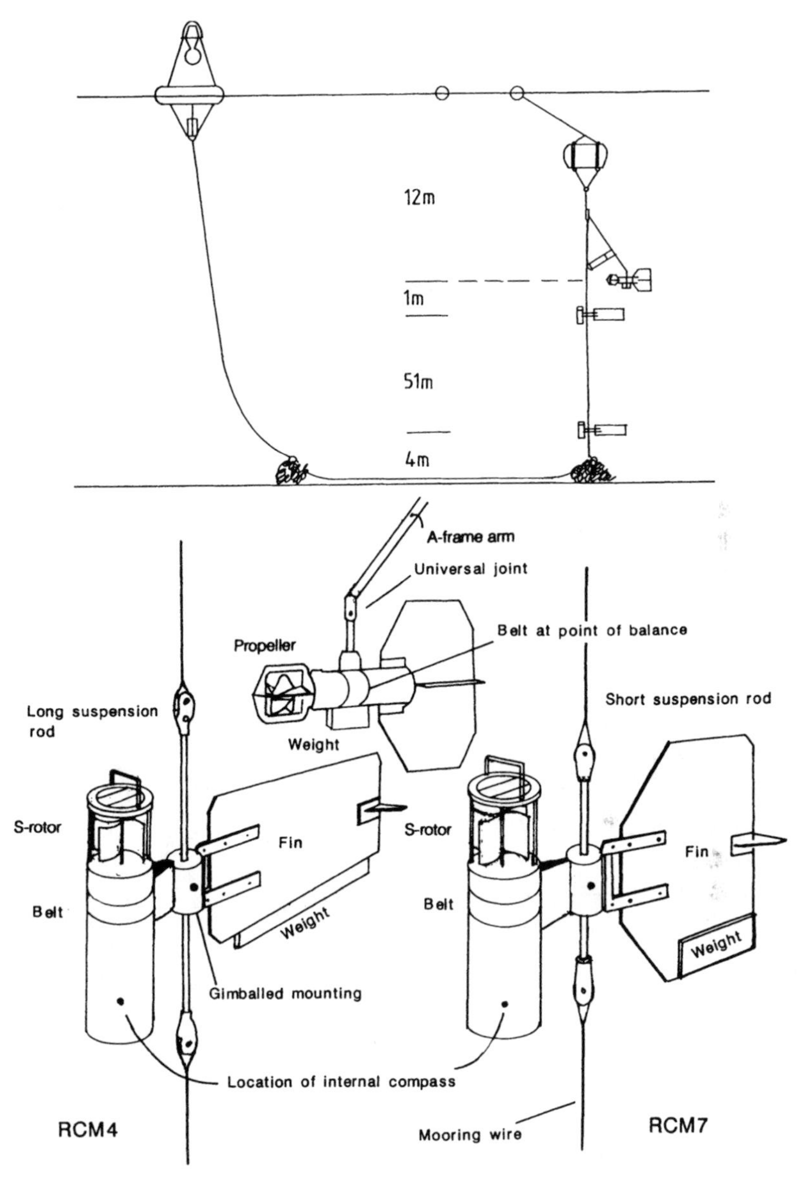

Figure 6-5. Typical shelf-seas types of current meter mooring, showing instrument wire suspended from subsurface float (to decouple the instruments from the direct effects of wave action). The current meters are of two main types: a horizontally mounted propeller-driven instrument, held off the wire by an A-frame, suspended at its port of balance and swivelling freely by means of a universal joint; and an in-wire instrument suspended from a gimballed rod that may be of longer or shorter type. The essential features of these instruments are shown diagrammatically.

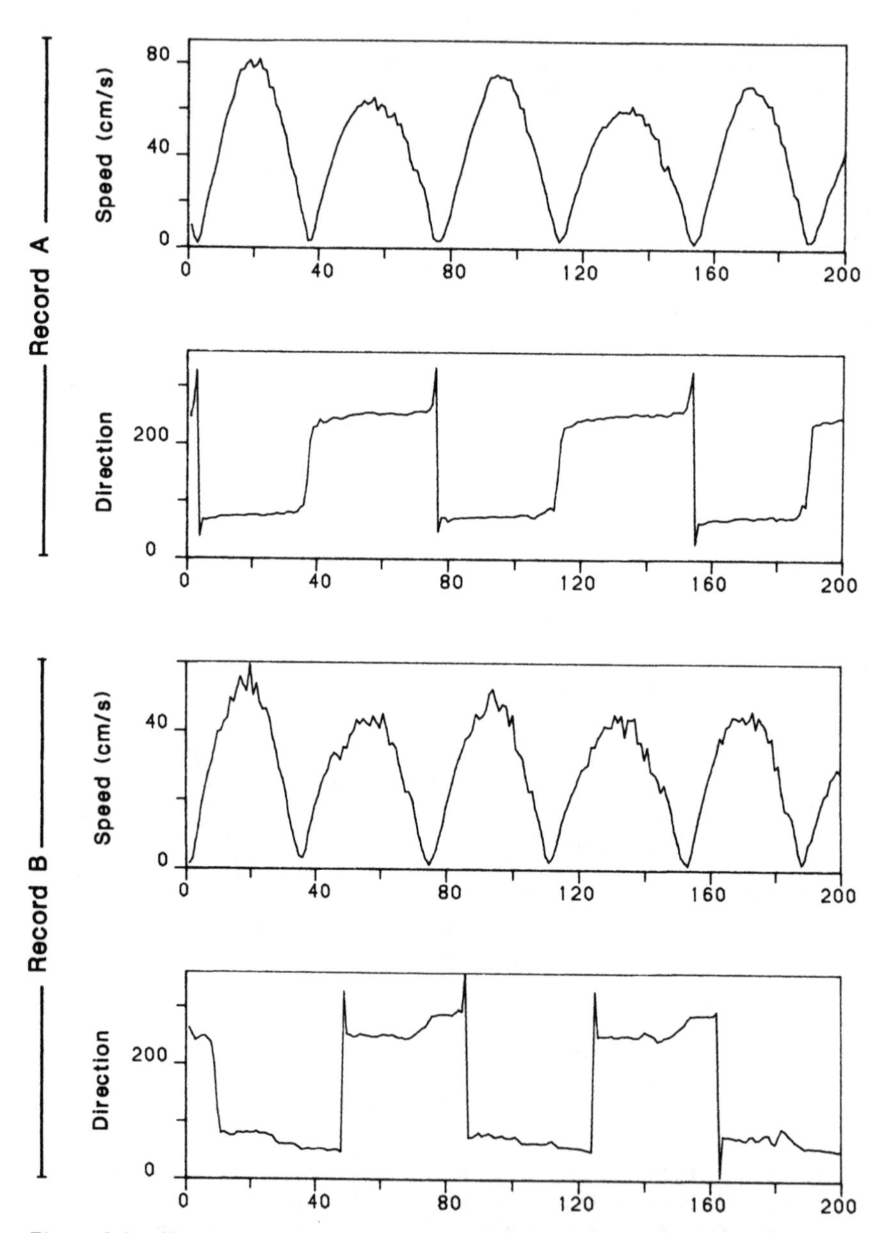

Figure 6-6. Simultaneous 33-hour time series of speed and direction (recorded every 10 minutes) for two current meters on the same mooring in the Irish Sea. One record is healthy, one is sick. Which is which?

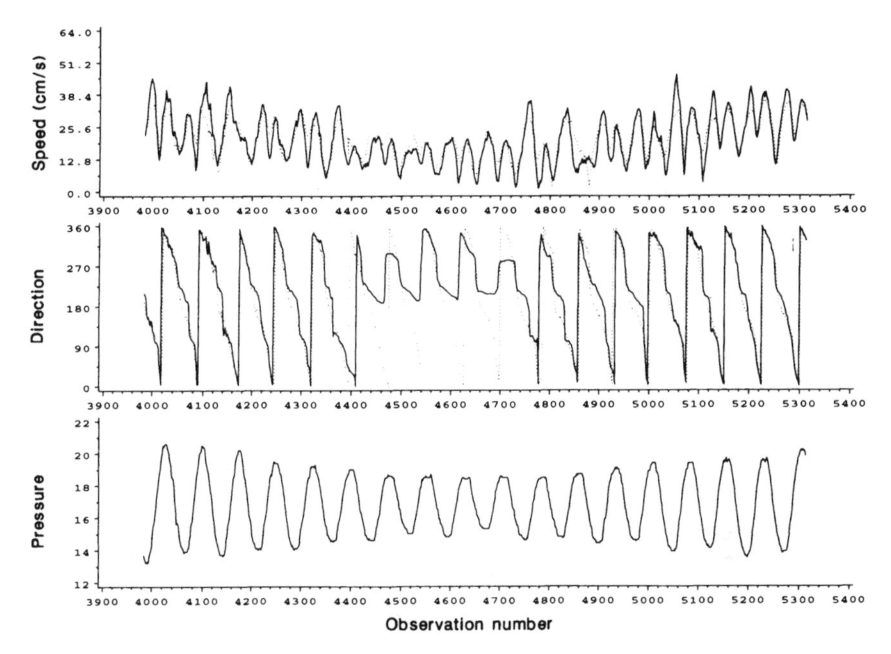

Figure 6-7. Comparison of 9-day records of current speed and direction from an A-frame, propeller instrument (solid line) and an in-wire, S-rotor instrument (dotted line), set close together on the same instrument wire in the Irish Sea. A representative pressure reading (third panel) indicates the change from spring to neap tides over this period. Speeds and directions compare reasonably well at springs, but the correspondence in direction is lost at neaps. For explanation, see text.

for the bottom panel only. The in-wire, S-rotor instrument is shown by the dashed line throughout.

The salient features of these records are as follows. To start with (in panel A), the record is entirely normal; the S-rotor is overrunning at high speeds compared with the propeller, but this is a well-known characteristic of S-rotors. In panels B and C, however, this overrunning first reduces, then disappears; and by panel D the S-rotor is *underrunning* the propeller. Something is going wrong, and, by panel E, whatever it is proves terminal, with speeds reducing on both instruments but with the in-wire instrument (significantly) dying first. Directions cease to change at the same time. What on earth is happening to this mooring?

Case 4 compares scatter plots of the north versus the east current component for two in-wire current meters set only 1.5 m apart on the same Irish Sea mooring (Figure 6-9). They are remarkably different. For the RCM4 type, the plot is reasonably linear (as it should be), but the plot is banana shaped for the RCM7, with an increasing deviation from the straight line at high current speeds. There are two main pieces to this particular puzzle. First, the problem occurs or at

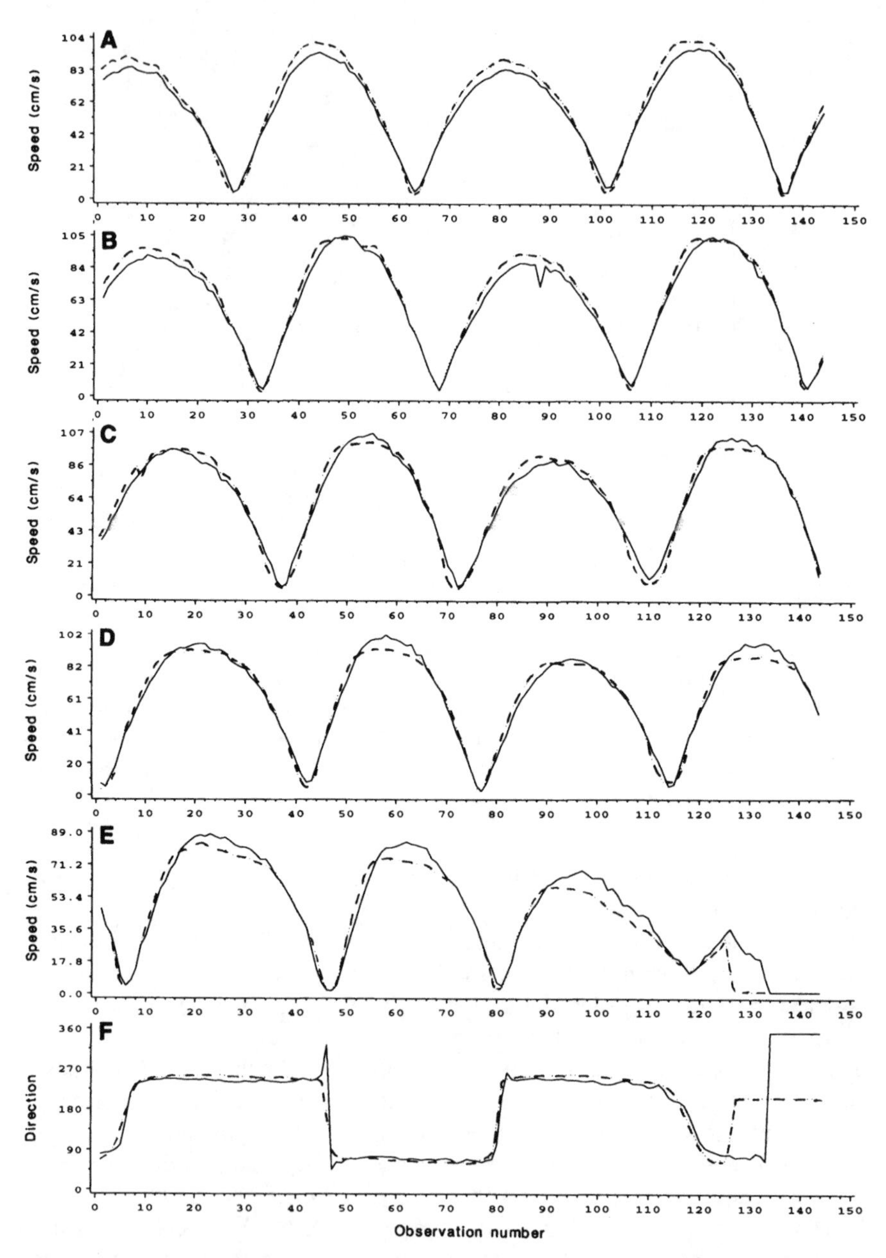

Figure 6-8. Comparison of current speeds from an A-frame-mounted (solid line) and an in-wire (dashed line) instrument, set close together on the same instrument wire over five successive 25-hour periods (panels A through E). The corresponding time series of direction are shown for the last of these periods only (panel F). For explanation, see text.

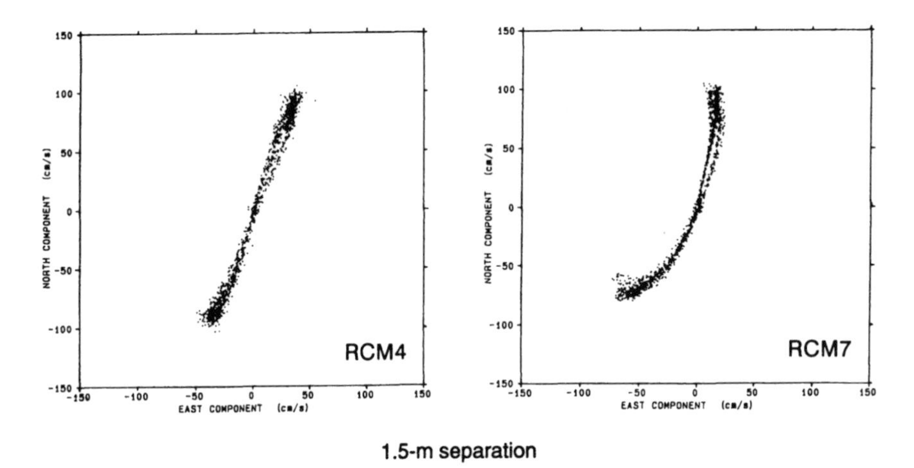

Figure 6-9. Scatter plots of north versus east current component for two in-wire current meters of different type, set 1.5 m apart on the same Irish Sea mooring and over the same 7.8-day period. The RCM7 instrument, characterized by a shortened suspension rod, shows an increasing deviation from the straight line at high current speeds. For explanation, see text.

least worsens at high tidal current speeds. Second, it occurs on the RCM7 instrument, whose main difference from the RCM4 is a shortened suspension rod (see Figure 6-5). What is the problem and what should we do about it?

For our interpretations (yours may be better!), see Appendix I.

Extreme and Unacceptable Behavior

Because errors and spikes affect all data, those charged with handling and reducing large data sets frequently need to employ techniques for the automatic discarding of unacceptable values. While this works well for many purposes—such as combing a Celsius SST-deck for remaining Fahrenheit temperatures—the characteristic "red" spectra of many oceanographic time series (Wunsch 1992) imply a continuing dilemma in deciding whether a given extreme data point is in error or merely unprecedented in our (relatively short-term) experience.

In the past, data were sometimes discarded according to arbitrary rules based on "unusualness"—e.g., fluctuations of more than 3 standard deviations about the mean, or some such definition—but these procedures were seldom well founded. Thus, when we now attempt to find evidence of the century-long maximum in surface salinity that is thought to have passed through the northern North Atlantic during the 1930s, we find the following tantalizing report for Cape Farewell, south Greenland: that "apart from 1933 *when the salinity observations were thought to be in error,* the highest salinity recorded by the U.S. Coast

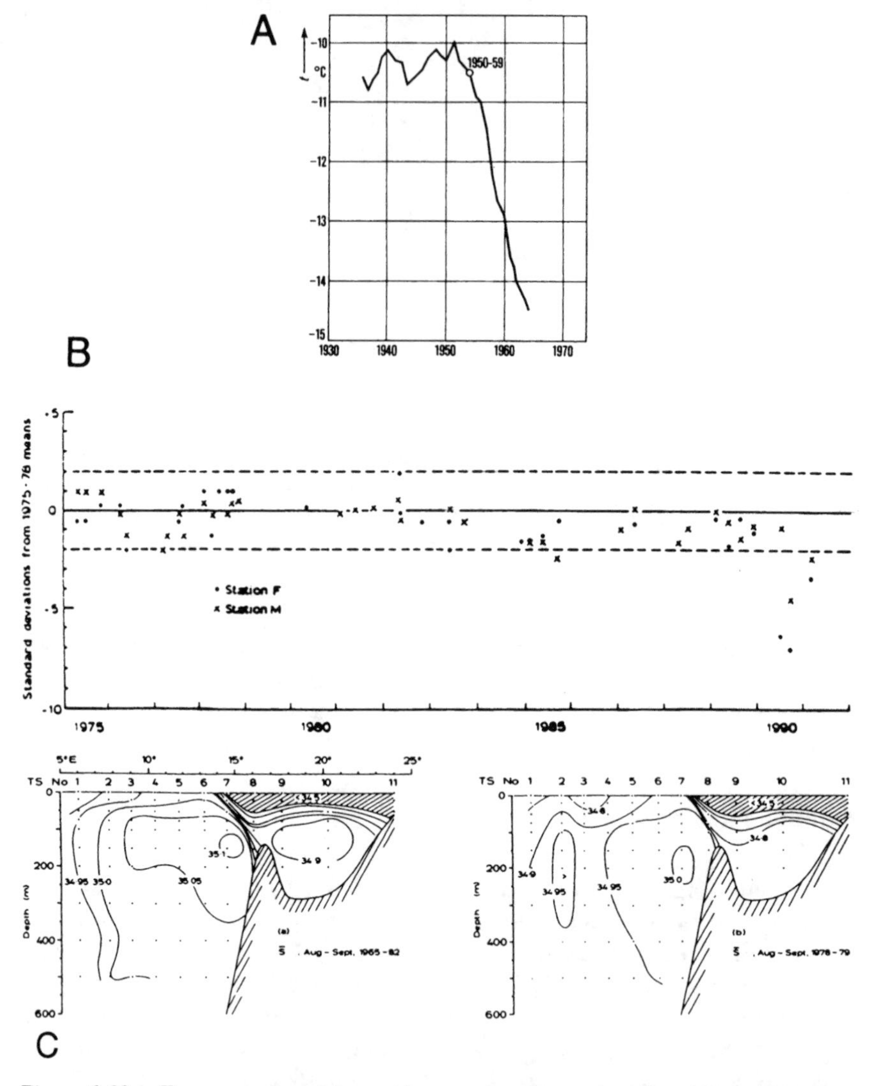

Figure 6-10. Four examples of the rapid onset of extreme anomalies: (A) rapid cooling by 4° C in the running 10-year means of air temperature at Franz Josef Land (from Scherhag 1970); (B) freshening of the Labrador Sea Water layer in the Rockall Trough by more than 7 standard deviations in 1990 (Part of Figure 3b from D.S. Ellett 1993; transit times to the NE Atlantic of Labrador Sea water signals from ICES *CM 1993*/C:25); (C) dwindling of the extent of Atlantic water (salinity > 35) in the Norwegian Atlantic current passing West Spitzbergen in 1978–79, compared with its long-term mean distribution for 1965–82 (from Dickson et al. 1988b); and (D) increase by > 4 standard deviations in the abundance of the Atlantic forms *Thalassiothrix longissima* and *Metridia lucens* in the plankton of the North Sea in 1988–89 (from Dickson 1991).

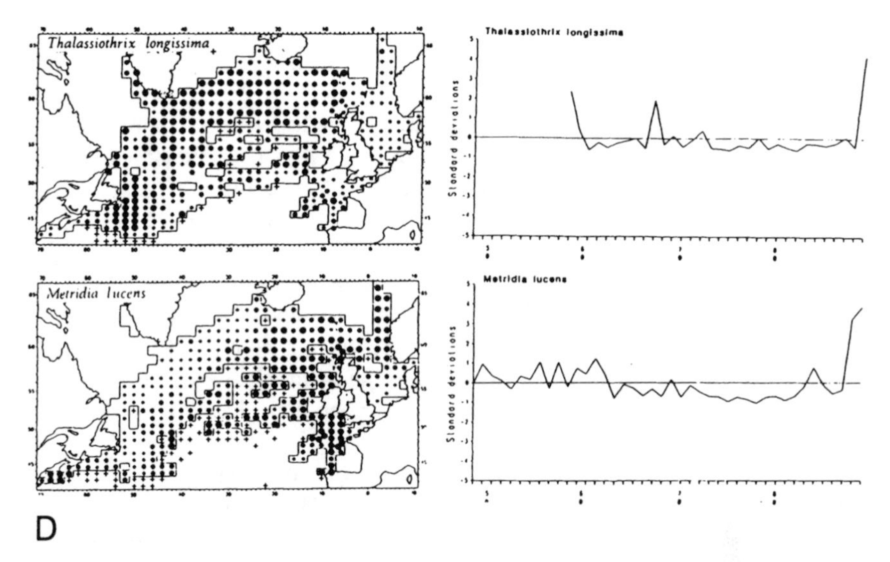

Figure 6-10. (*Continued*).

Guard in any of their 22 observations along this section between 1928 and 1959 was a value of 35.07% observed in both August 1931 and July 1934" (Harvey 1962, p. 17).

The scientific literature now contains so many examples of sudden wild extrema that are nonetheless believed to be correct that we are much more cautious than we would once have been in discarding the unusual or the unexpected. Figure 6-10 will make the point, showing:

- the cliff-like decline by 4° C in *decade-mean* air temperatures at Franz Josef Land (Scherhag 1970), but at a time when the polar airflow was maximal over European arctic and subarctic seas;
- the precipitous but nonetheless expected decline, by more than 7 standard deviations, in the salinity of the Labrador Sea Water layer in the Rockall Trough in 1990 (Ellett 1993), heralding the arrival in the East Atlantic of conditions resulting from renewed deep convection at OWS BRAVO after 1986;
- the almost total lack of "Atlantic water" (conventionally defined as water $\geq$ 35 psu) in the Norwegian *Atlantic* Current off West Spitzbergen in 1978–79 as the Great Salinity Anomaly signal (Dickson et al. 1988b) passed through; and
- the steep increase in the abundance of the planktonic forms *Thalassiothrix* and *Metridia* in the North Sea in the late 1980s, but at a time when other

"Atlantic" components of the plankton were also showing unprecedented abundance there (Dickson 1991, from CPR records).

These are all (real) anomalies of such scale as to defy any automatic filtering process based on probability; but as gross extrema, it would be surprising indeed if other components of the ecosystem or physical environment were not undergoing equally radical changes at the same time. For example, the factors that brought a record abundance of Atlantic plankton to the North Sea in 1989–90 (Figure 6-10D) also appear to have caused an unprecedented warming in the 120-year record of surface temperature at the Torungen Lightvessel in the Skagerrak (Figure 6-11, from Johan Blindheim, MRI, Bergen, pers. comm.). The "Atlantic" origin of this change is shown by the fact that the warming event is evident only in winter-spring, when the surface water of the open Atlantic is warm compared with that of the Skagerrak.

The examples that best argue the case for forbearance in dealing with extrema, however, are often best recognized in retrospect. The reports of the Swedish Deep Sea Expedition aboard the *Albatross* in 1947–48 record the fact that, during passage of the Red Sea, bottom temperatures of around 25° C were recorded at one station at 1930 m depth, much warmer than normal for Red Sea bottom water (Bruneau et al. 1953). Although by no means as extreme as the subsequent T and S observations by *Atlantis II* and *Discovery* (Figure 6-12A, from Swallow

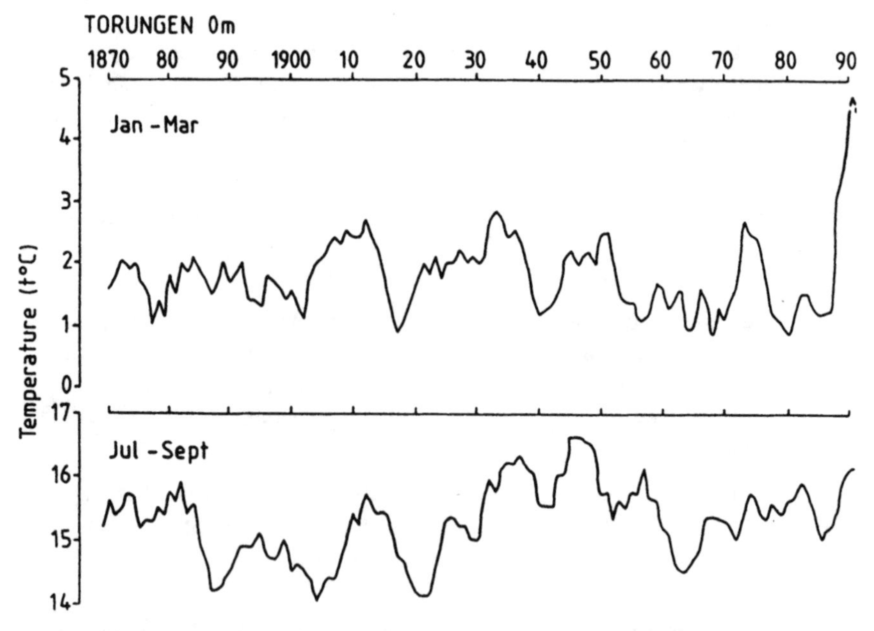

Figure 6-11. Time series of sea surface temperature for the Torungen Lightvessel in the Skagerrak in January–March and July–September, 1870–1991. (Kindly provided by Johan Blindheim, MRI, Bergen, pers. comm.)

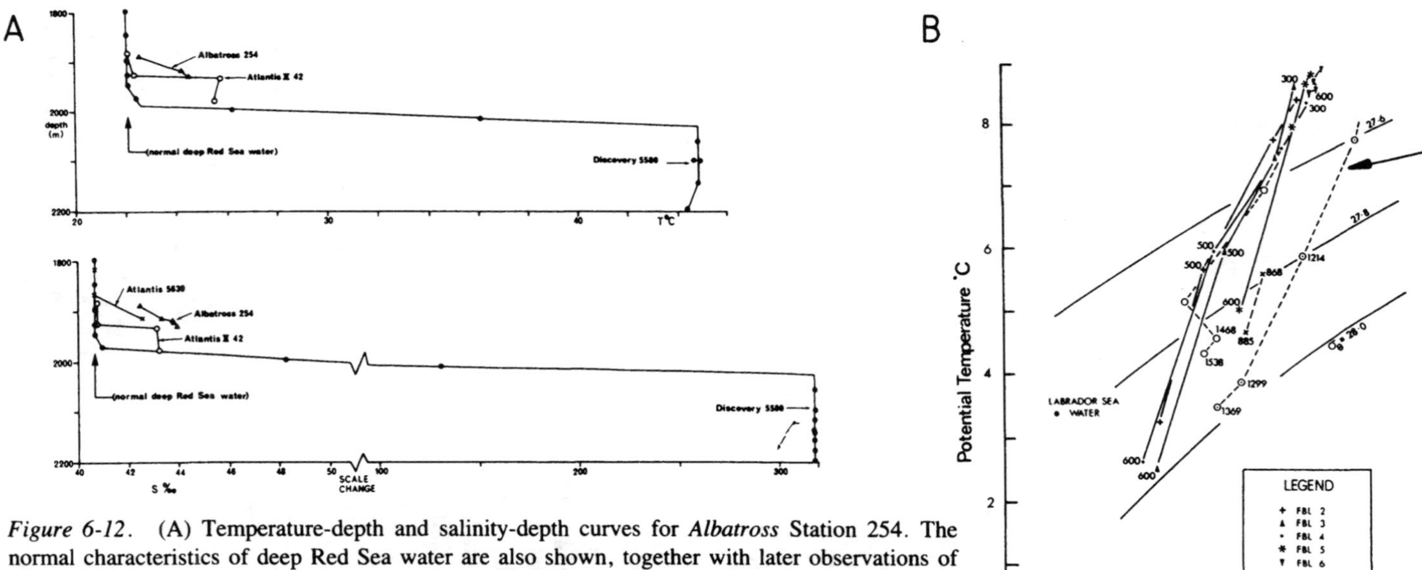

Figure 6-12. (A) Temperature-depth and salinity-depth curves for *Albatross* Station 254. The normal characteristics of deep Red Sea water are also shown, together with later observations of abnormal bottom water conditions by *Atlantis*, *Atlantis II*, and *Discovery*. (Reprinted by permission from J.C. Swallow, 1969, "History of the Exploration of the Hot Brine Area of the Red Sea: DISCOVERY Account," In E.T. Degens and D.A. Ross (eds.), *Hot Brines and Recent Heavy Metal Deposits in the Red Sea*. Springer-Verlag, New York, pp. 3–9.) (B) Potential temperature-salinity diagrams for stations in the northern Rockall Trough. The anomalous profile by USCGS *Evergreen* in 1965 is arrowed (from Ellett and Roberts 1973).

1969), these *Albatross* observations were nonetheless grossly abnormal according to the established wisdom of the day. They were, in fact, the first evidence the world had for the existence of hot brines in the floor of the Red Sea, and as such features were then unsuspected, it is to the expedition's great credit that the observation was retained, albeit without comment.

A more recent and more local example concerns the unusually dense water that the USCGS *Evergreen* found on the floor of the northern Rockall Trough in November 1965 (Figure 6-12B, from Ellett and Roberts 1973), which is now regarded as among our first and best evidence of an intermittent dense overflow of Norwegian Sea deep water across the Wyville-Thompson Ridge (Ellett and Roberts 1973). Although these strange observations were omitted from the plotted salinity section (*Evergreen* Stations 17-C to 29-C; Elder 1970, his Figure 29), they were sensibly retained in the tabulated data (Station 26-C, p. 105).

Thus, we have become less prone to discard extrema, and we have begun to realize that such extreme outliers can be of extraordinary importance in demonstrating how the ocean-atmosphere system works. These extrema are frequently the only signals that are large enough to be traced unambiguously through an oceanographic record that is both gappy and of variable quality in space and time. And so it will be by following the progress of the Great Salinity Anomaly (Dickson et al. 1988b) and the Labrador Sea Water Anomaly (just described) that we obtain our first direct confirmation of spreading rates in the North Atlantic at surface and subsurface depths.

At the same time, however, we have probably become more skeptical in our use of proxy data during rare and extreme events, simply because events may be *so extreme* that the proxy no longer applies. Thus, for example, although the 10° C isotherm position at 200 m depth in the North Atlantic is normally so well embedded in the Ocean Polar Front as to define its location, Dickson et al. (1984) have shown that, in the case of extreme events, this association with the OPF may break down. During the earlier Great Salinity Anomaly event in 1910–14, the 10° C isotherm at 200 m lay 250 nautical miles east of the main frontal gradient.

Heredity

It is a natural assumption that the accuracy and other characteristics of a new sensor or method will apply to a time series from the date of its introduction. However, there are ways in which the new system (and the new time series) can be made to inherit the faults of the old until long after the changeover has taken place. These inherited characteristics can be very difficult to detect, because they are often the result of deliberate attempts to smooth over such transitional changes (i.e., to make them disappear), and our instincts are to prefer and believe seamless transition to steplike change when new sensors are introduced.

Parker (1990) documents a good example in his analysis of the "Effects of Changing the Exposure of Thermometers at Land Stations"—part of the background documentation that the Intergovernmental Panel on Climate Change (IPCC) relied on to determine the realism of past evidence of climate change. What Parker shows is that ever since observations at land stations became widespread in the mid-nineteenth century, there have been substantial and systematic changes in the exposure of thermometers, with sizeable impacts on perceived climatic trends.

In India, the general practice since the 1870s had been to house thermometer cages under thatched shelters. Tests later showed that this method caused overheating (by 0.3° C–0.5° C in the monthly mean during the day and 0.2° C–0.7° C at night) and, as a result, the Stevenson Screen was introduced in the 1920s. The error was perpetuated, however, by the fact that *reporting stations adjusted their new data to the earlier exposure*; as a result, the relative warmth of land stations after 1890, which is supposed to be due to this bias, continued until at least 1935 (see Figure 6-13, from Parker 1990 but using land data provided by P.D. Jones. Note that the curve is unreliable before the 1880s because there were fewer than 10 locations for comparison). A modern analogue might be the

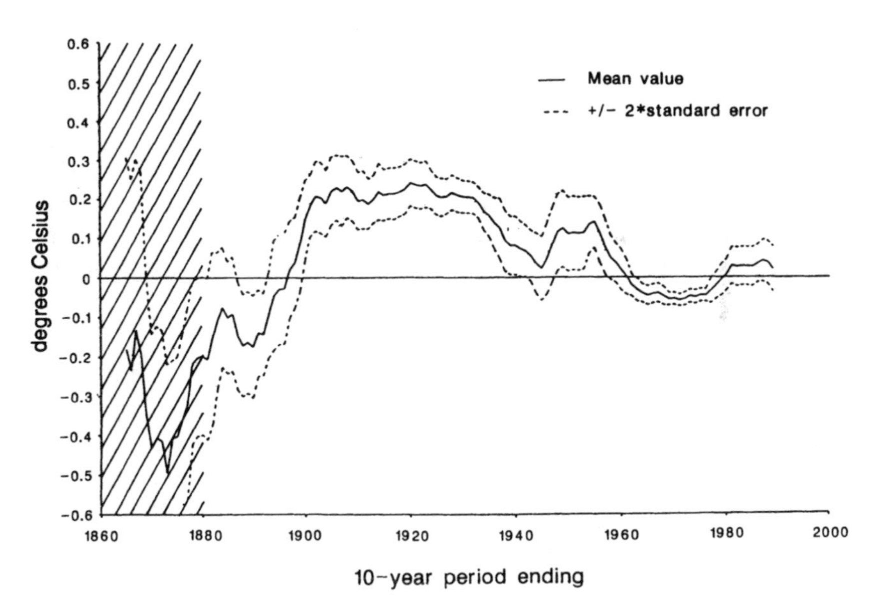

Figure 6-13. Ten-year running mean differences between air temperatures at coastal or island sites in the tropics (20° N–20° S), and sea-surface temperatures nearby. The excess warmth of land status between 1900 and 1940 is initially due to the effect of the type of shelter in which the thermometers were exposed, but after the 1920s, when this method was abandoned, the continued overwarming is due to the fact that data were adjusted to be consistent with earlier records. (From Parker 1990.)

continued use of the same algorithm for the reduction of satellite data during a period of temporary or permanent change in the conditions for which the algorithm was designed.

In addition to spreading an error in time, the effect of procedural errors or inconsistencies in either observation or analysis can also become geographically widespread through spatial averaging, and if the initial problem is large its effects can be spread globally. In Figure 6-14, for example, the NOAA-NESDIS satellite-based analysis of snow cover extent for the Northern Hemisphere shows a rapid increase from the late 1960s to the early 1970s, discussed by Kukla and Kukla (1974), which is now accepted to be largely due to inconsistent mapping of part of the Eurasian Sector, including the Tibetan Plateau (Wiesnet et al. 1987).

The area of science in which "heredity" continues to be most commonly encountered, however, concerns the special case of the self-fulfilling prophecy in which incorrect assumptions control experiment design, which in turn confirms the assumptions. This problem can affect the whole gamut of observation from single readings to time series. One spectacular example will demonstrate its workings.

Until recently, long incubations were the approved methodology for measuring primary production in the oligotrophic ocean using the ^{14}C method. It was argued that the time of incubation should be proportional to algal growth rates; and that since these were "well known" to be slow in the tropics and subtropics, with doubling times of days, incubations were proportionately long (up to 48 hours). When measured after these long incubations, low productivity rates were confirmed. However, when Gieskes et al. (1979) shortened the incubation period to 2 hours, the phytoplankton growth rates were, in fact, shown to be high, with generation times of hours, not days, and primary production rates were *5–15 times higher* than by the established method, and with a high associated activity

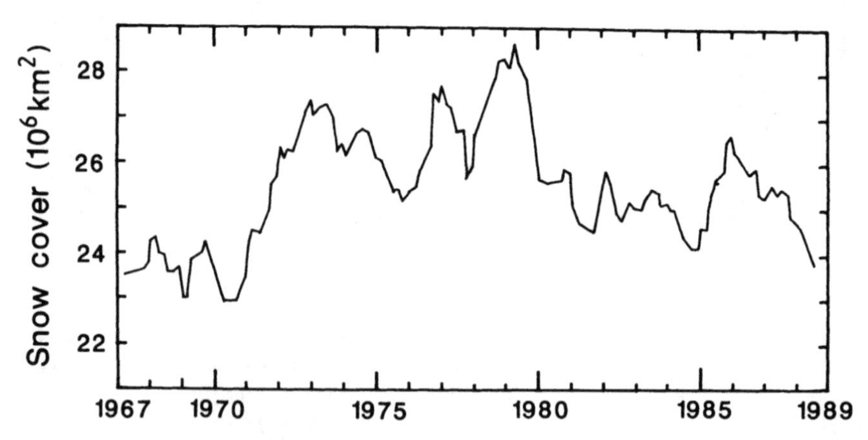

Figure 6-14. NOAA-NESDIS Northern Hemisphere snow cover extent, 12-month running means, 1967–88. (From Barry 1990, by permission.)

by grazers. It now seems clear that during long incubation the photochemical destruction of pigments and the development of zooplankton grazing were jointly responsible for reducing the algal biomass and thus for perpetuating the illusion of low production rates in oligotrophic waters.

Camouflage and Mimicry

Not all parameters are as they seem. During the early phase of the North Pacific Experiment (NORPAX) in the 1970s, there was some low key debate as to the relative merits of using sea-level pressure anomaly (SLP DM, DM signifying departure from the mean) or 700 mb height anomaly (700 DM) to describe changes in the pressure field over the North Pacific and to relate these changes to other variables. Generally, it was argued that the upper level field offered a useful simplification of events at ground level, and 700 DM was more widely used.

Skepticism prompted Figure 6-15, in which 27-year time series of both fields have been correlated for each 5° grid intersection in the North Pacific and surrounding coasts, with the correlation coefficients plotted and contoured. Although there is little relationship over the continents, the correlation jumps to > 0.8 immediately upon crossing the coast, and over much of the central North Pacific the correlation exceeds 0.9, with a peak in excess of 0.99 in mid-ocean.

Although the pressure field can be expected to be more barotropic over the ocean, such a closeness of fit over 27 years surely points to a simpler explanation—that the two "parameters" are in fact the same and that, remote from land and the upper air soundings of land stations, the 700 DM field is a simple conversion of the observed SLP observations; i.e., SLP DM = 700 DM.

In fact, occasions where one parameter masquerades as another are not uncom-

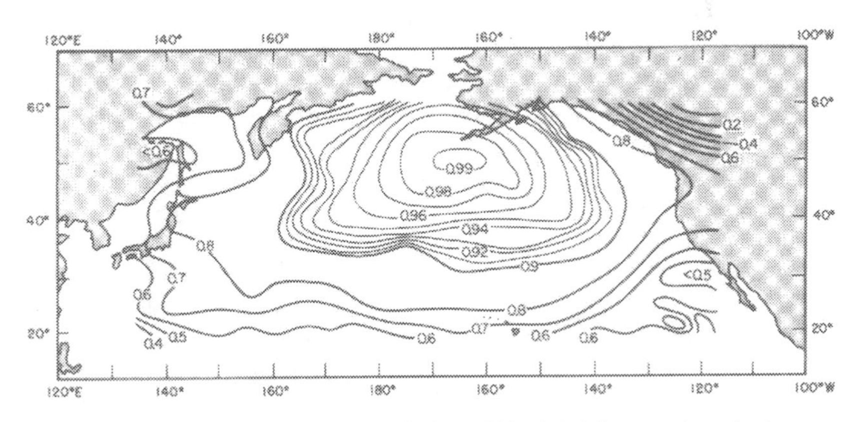

Figure 6-15. Correlation coefficients of winter 700 mb height anomaly and winter sea-level pressure anomaly over the North Pacific and adjacent land areas, 1948–74.

mon, and this may be particularly misleading when different derivatives of the same thing give the impression of a precision that they do not possess. In calculating U.K. fishing effort, for example, "hours fishing" *sounds* like a more precise number than "days' absence," but given the fact that its reporting is optional rather than mandatory and includes time spent searching rather than just fishing, the former is at best as good as the latter.

The Individual and the Population

To the extent that long time series are costly in terms of the sustained effort required, they tend also to be sparsely distributed in space and time. Their interpretation, in other words, begs questions of representativeness; representativeness immediately involves scale, and scale entails purpose.

Although no more true than in other branches of science, with time series we are often forced into pragmatism, because if no confirmatory series exist, we usually cannot go back and establish them. Some of our pragmatic assumptions, of course, are later confirmed, and we may even forget—through long familiarity—that assumptions were ever required. Johan Hjort (1943) manages beautifully to recapture one such chain of assumption—which now forms the accepted basis of all fisheries management—when he refers to "the hypothesis, which at the beginning had seemed so daring . . . that the statistics of the catches of the fishermen or research vessels may be considered as representative of the existing stock"; that "samples collected from a series of years disclosed the important fact of fluctuations in the stock from one year to another"; and that "from the closer understanding of these changes arose the conviction that such changes were governed . . . by the influence of the environment."

One example should reinforce the point that, whether or not we are able to answer them, questions of representativeness and scale are implied by all time series. We use the example of the depth ($\approx$ time) variation of properties through a seabed sediment core, as this is one example where we can, merely by subdividing a core or taking new ones, establish new "time series" over any scale we wish.

When we subdivide even a single box core into more than 100 separate subsamples, we find from the ^{210}Pb profile that the biological mixing rate can vary by orders of magnitude in subcores only a few centimeters apart (Smith and Schafer 1984). Plainly, a representative stratigraphic series cannot be achieved merely by reducing the distance between samples. Any subcore will share elements of similarity due to regional similarities in sedimentation and bioturbation rate, yet will also contain elements that are stratigraphically localized and unique—for example, the change in sediment properties, sediment mixing, and contaminant invasion due to a single recent burrow.

The general point is that, whether we are dealing with true time series or their stratigraphic equivalents, they are all likely to be the result or synthesis of variability over a wide range of time scales. The significance of this is that the spatial coherence of change (which is our only real measure of "representativeness") is usually directly related to the periodicity of the fluctuation.

This space-time relationship has been elegantly illustrated for zooplankton biomass variations by Haury et al. (1978) [see also McGowan (1989, 1990)], so that (as in Figure 6-16) short-term variations are generally on small spatial scales, whereas long-lived fluctuations are spatially extensive. Similar space-time relationships have been shown by Holling (1992) to apply to the forest ecosystem, carnivore mammals, and birds.

Thus, questions of the spatial representativeness of a time series are wrongly posed. Any time series will be coherent over a range of space scales according to the mix of its component periodicities. As the CPR Survey clearly shows for the post-war trends in zooplankton biomass over the eastern Atlantic (Figure 6-17), the longest period trends are correlated in all areas covered by the survey, while the highest frequency fluctuations are not.

We discover, therefore, that the real question is not one of representativeness

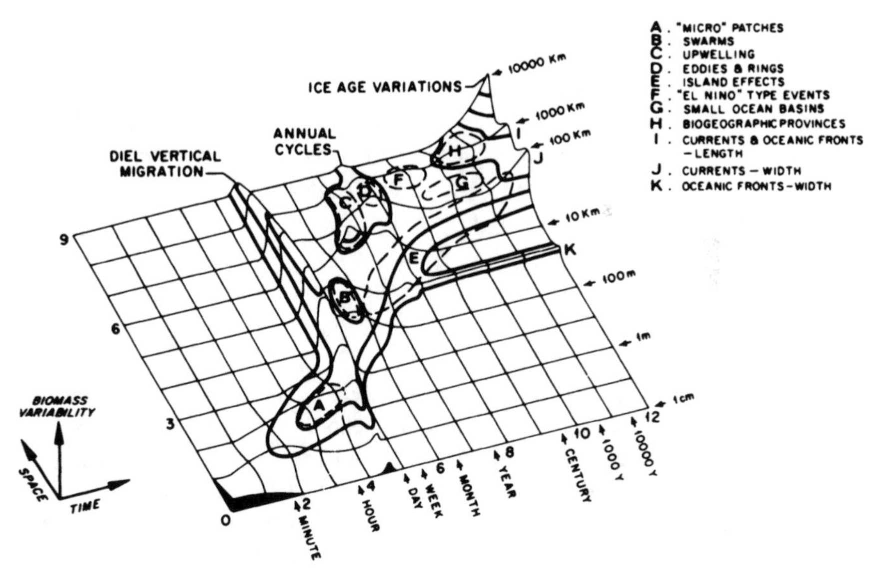

Figure 6-16. A space-time diagram of zooplankton biomass variations. Small spatial changes are short lived, whereas larger spatial changes are long lasting. (Reprinted with permission from L.R. Haury, J.A. McGowan, and P.H. Wiebe, 1978, "Patterns and Processes in the Time-Space Scales of Plankton Distribution," in J.H. Steele (ed.), *Spatial Patterns in Plankton Communities*. Plenum, New York.)

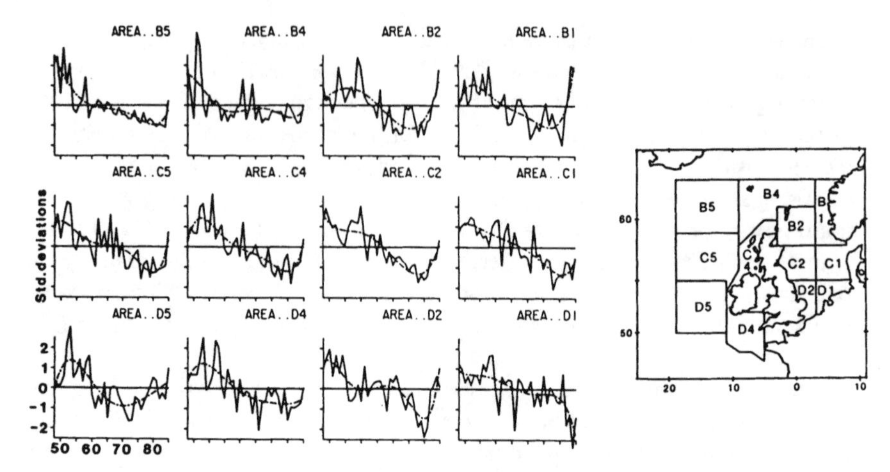

Figure 6-17. Post-war trends in zooplankton abundance in 12 CPR Standard Areas around the British Isles (see chart). (Reprinted from R.R. Dickson, P.M. Kelly, J.M. Colebrook, W.S. Wooster, and D.H. Cushing, 1988, North winds and production in the eastern North Atlantic, *Journal of Plankton Research* 10:151–169, by permission of Oxford University Press.)

but of how these various scales of coherence interact. As Levin (1988) sums it up: "predictability is inextricably intertwined with variability, and with the temporal and spatial scales of interest. Thus, a central challenge in ecological theory must be an elaboration of the understanding of how scales relate, how systems behave on multiple scales and how the measurements and dynamics of particular phenomena vary across scales."

Evolution

It is appropriate that the final section of this paper should deal with problems analogous to those of evolution, because as technology advances and changes are introduced, something very similar to evolutionary pressure can act on a time series as it lengthens. Of course, the intention is always the worthy one of *improved accuracy*, but whether the changes are introduced through successive improvements to the observing instruments or to the analytical methods on which the time series depend, the net result may nevertheless be detrimental, feeding through to effect trends in the data as well as the intended gains in data quality and handling. Such cases can hardly be described under our "ill-health" analogue, because the improvements are real enough; but, like conventional evolutionary pressures, they can result in the loss of time series—or at least of their usefulness—unless care is taken to remove the effects of our interference from the changes we are trying to observe.

One example will explain the problem and give some idea of its potential effects. Since 1951, the upper air climatology over Finland has been routinely monitored using radiosondes equipped with thermometers and hygrometers. Over the years, as summarized in Table 6-1, the height of the surface inversion layer appeared to decrease dramatically, with its mean annual height at Ilmala-Jokioinen and Sodankylä reducing by 24% and 28%, respectively, between the 1950s and 1980s (Huovila and Tuominen 1990). These authors demonstrate convincingly that this change was by no means real, but was the result of a progressive reduction in the thermometer lag error arising from the introduction of a succession of improved radiosonde types from the Vaisala RS11 of the 1950s through RS12, RS15, and RS18 to the RS80 of the 1980s. Progressively, the 11.5-second lag coefficient of the bimetal sensor in RS11 was reduced to 2.5 seconds in the capacitive sensor of the RS80, and the inversion layer was recorded earlier in the ascent as a result. A similar error affected the hygrometers, especially in the upper troposphere and stratosphere, where the relative humidity is very low. The hair hygrometer of the RS11 was unable to reach such low values during the radiosonde ascent, because its lag coefficient at −20° C was 400 seconds compared with 30 seconds in the RS80, and, as a result, the mean relative humidity at the 300 hPa level in North Finland in 1984–88 was 31% lower in winter and 13% lower in summer than it was in 1951–55! These effects of an evolving technology are of course lent a more than local significance by

Table 6-1. Mean height (meters) of surface inversion at night in 1951–55 (A), 1963–69 (B), and 1982–88 (C). P = mean percentage of inversion nights. (From Huovila and Tuominen 1990.)

	Ilmala-Jokioinen				Sodankylä			
Month	A	B	C	P	A	B	C	P
I	277	305	290	30	406	404	329	54
II	334	326	302	36	397	380	354	52
III	321	273	222	46	355	316	260	54
IV	312	235	205	48	287	259	181	57
V	278	242	196	63	268	205	192	57
VI	265	238	193	70	318	240	156	54
VII	275	223	188	68	292	254	184	58
VIII	256	224	193	64	308	274	190	57
IX	273	219	188	52	259	253	185	49
X	208	206	182	36	324	269	245	35
XI	217	202	187	23	389	315	306	41
XII	238	241	214	26	365	356	308	54
Year	**275.5**	**241.4**	**209.1**	**46.4**	**333.4**	**291.8**	**240.7**	**51.2**

two remarks of Huovila and Tuominen. First (page 7): "The influence of the lag error of radiosonde thermometers and hygrometers on measured vertical temperature and humidity profiles has long been recognized. . . . However, no practical methods or guidelines for taking into account the lag errors have come into operational use." The second remark, from their abstract, is more salutary: "These examples demonstrate the very significant influence of the lag error on some parts of the old radiosonde data. . . . Although this study only refers to the Vaisala radiosonde and Finland, similar conclusions can without doubt be drawn concerning old radiosonde data recorded by other radiosonde types in other countries."

To this we might well add: " . . . and in many other data series, subject to progressive upgrades of instruments and methods over time!"

Appendix I: Current Meter Malfunctions

As mentioned earlier, all four of these problems arose through defects in the mechanics of the mooring, not primarily with the current meter.

Case 1. Here the instrument in Figure 6-6B is clearly suffering from a stiff suspension. Although the current direction should change shortly after each slack water period as flood turns to ebb or vice versa, in fact the instrument does not register a change in direction until long after slack water has passed and the speed has built to around half a knot (~ 25 cm/s). The current meter would work well enough if the A-frame itself were prevented from turning, because the instrument would still be able to rotate from its universal joint. Thus, it is the universal joint itself that is stiff, preventing the current meter from swinging freely and easily from its A-frame.

Case 2. At first sight, this appears to be a variant of the stiff suspension problem of Case 1. The A-frame instrument (solid line) matches the good performance of the in-wire current meter (dotted line) until the strong tides of springs give way to the weaker tides of neaps. As soon as the last strong flow of springs occurs (~ observation 4375), the instrument ceases to swing smoothly with the tide, and directional movement seems restricted until strong flows resume around observation 4750. It might appear then that either the A-frame or the current meter is fouled in some way, so that only the stronger spring tides are capable of pushing the whole fouled assembly around with the tide. The reader cannot be expected to deduce the real reason.[1] But the reader may notice that the above

[1]The metal belt that suspends the barrel of the instrument horizontal at its point of balance has slipped, the current meter has become unbalanced and hangs nose-down or tail-down as a result, and the tilt has allowed the compass needle to foul the roof of the chamber in which it floats. Except at slack water (see "steps" A-B, C-D in Figure 6-18), the *spring* tides act strongly enough on the fin to raise the instrument toward the horizontal so that the compass frees off and "catches up" with where it ought to be. At neaps, the period of the dragging compass needle extends from one slack water to the next without ever "catching up."

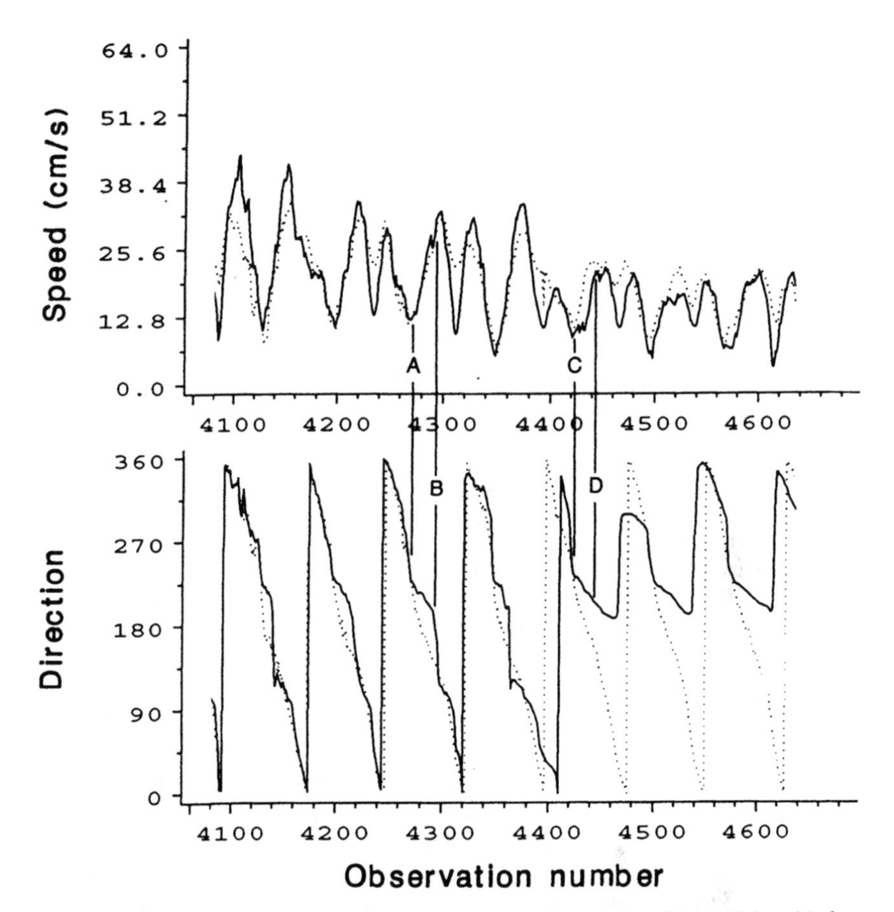

Figure 6-18. Expansion of a segment of the records compared in Figure 6-7, with four events labeled. At time A during spring tides, the compass card fouls its chamber only during slack water, as the unbalanced instrument tilts, but is freed off and allowed to "catch up" during the subsequent period of peak speeds (time B). During neaps the compass again fouls (time C), but peak tidal speeds are insufficiently strong (time D) to level the instrument and free off the compass.

explanation fails to explain why the current meter swings more completely (if a little jerkily) during the weaker parts of the spring tidal cycle than during the (stronger) peaks of the neap tides.

Case 3. "What on earth?" is right! The mooring is sinking. A leak has developed in the subsurface float (see mooring diagram, Figure 6-5), and, as the water leaks in, the instruments are gently lowered to the seabed, where speeds go to zero and directions cease to change. The S-rotor instrument feels these effects first and worst because it is in wire and works less and less efficiently as the

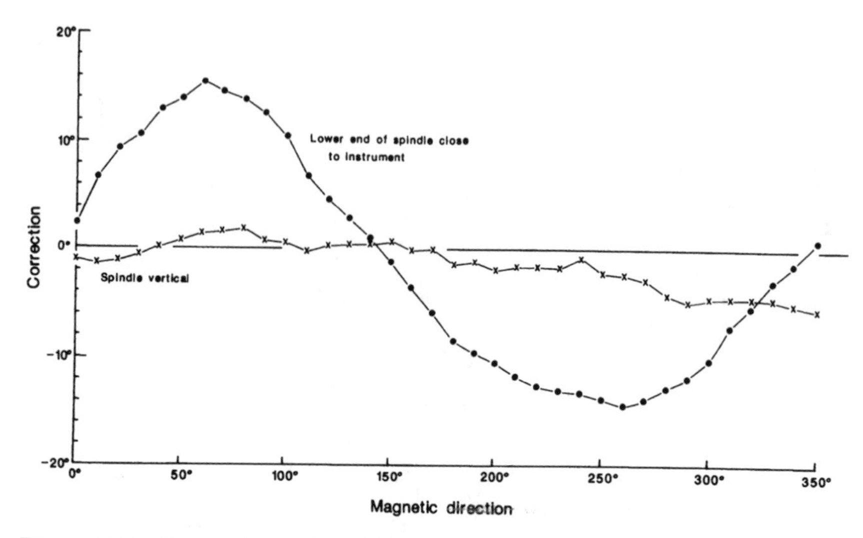

Figure 6-19. Curve of compass calibration error for *the same* RCM7 instrument with the short suspension rod and standard shackle in place, but with the spindle vertical (x-x-x), and with the spindle at maximum tilt, so that its lower end lies close to the base of the instrument (●-●-●).

wire leans over. Hanging from its A-frame, the propeller instrument is isolated from the effects of the sagging wire and works effectively for much longer.

Case 4. The problem is, we believe, caused by the shortening of the suspension rod. As the mooring leans over at high flow speeds, the suspension rod tilts relative to the gimballed instrument, which remains upright. This is no problem as long as the rod is the long type (Figure 6-5), but, when the shortened rod tilts, the large mass of metal (the end of the instrument wire plus metal thimble plus ¾-inch shackle) is brought too close to the instrument's internal compass, which, as mentioned earlier, is located at the bottom of the instrument case. This explanation is confirmed when we calibrate *the same RCM7 instrument* on the compass table using the short suspension rod set vertically and at maximum tilt. The compass correction increases from $\pm 3°$ to $\pm 14°$ on tilting the spindle (Figure 6-19). Action? The new short rods were replaced by the old-type longer rods on all RCM7s, and the problem ceased.

References

Barry, R. G., 1990. "Observational Evidence of Changes in Global Snow and Ice Cover." In D. E. Parker (ed.), *Observed Climate Variations and Change: Contributions in Support of Section 7 of the 1990 IPCC Scientific Assessments*.

Bruneau, L., N. G. Jerlov, and F. F. Koczy. 1953. Physical and chemical methods. *Report of the Swedish Deep Sea Expedition III, Physics and Chemistry* **4:**101–112 (with two appendices).

The Case for GOOS. 1992. (Anonymous.) GOOS Publ. Serv. 1. IOC, Paris.

Dickson, R. R. 1991. Monitoring the health of the ocean: Defining the role of the Continuous Plankton Recorder in global ecosystem studies. IOC/INF–869.

Dickson, R. R., P. M. Kelly, J. M. Colebrook, W. S. Wooster, and D. H. Cushing. 1988a. North winds and production in the eastern North Atlantic. *Journal of Plankton Research* **10:**151–169.

Dickson, R. R., S.-A. Malmberg, S. R. Jones, and A. J. Lee. 1984. An investigation of the earlier great salinity anomaly of 1910–14 in waters west of the British Isles. ICES C.M. 1984/Gen:4 (mimeo).

Dickson, R. R., J. Meincke, S.-A. Malmberg, and A. J. Lee. 1988b. The "Great Salinity Anomaly" in the northern North Atlantic, 1968–1982. *Programs in Oceanography* **20:**103–151.

Douglas, B. C., 1992. Global sea level acceleration. *Journal of Geophysical Research* **97:**12699–12706.

Duarte, C. M., J. Cebrian, and N. Marba. 1992. Uncertainty of detecting sea change. *Nature* 356:190.

Elder, R. B. 1970. Oceanographic observations between Iceland and Scotland, July–November 1965. *U.S. Coast Guard Oceanographic Report* 28.

Ellett, D. J. 1993. Transit times to the NE Atlantic of Labrador Sea water signals. ICES *CM 1993*/C:25.

Ellett, D. J., and D. G. Roberts. 1973. The overflow of Norwegian Sea Deep Water across the Wyville-Thompson Ridge. *Deep-Sea Research* **20:**819–835.

Gieskes, W. W. C., G. W. Kraay, and M. A. Baars. 1979. Current ^{14}C methods for measuring primary production: Gross underestimates in oceanic waters. *Netherlands Journal of Sea Research* **13(1):**58–78.

Harvey, J. G. 1962. Hydrographic conditions in Greenland waters during August 1960. *Annals of Biology* (Copenhagen) **17:**14–17.

Haury, L. R., J. A. McGowan, and P. H. Wiebe. 1978. "Patterns and Processes in the Time-Space Scales of Plankton Distribution." In J. H. Steele (ed.), *Spatial Patterns in Plankton Communities*. Plenum, New York, pp. 277–327.

Hjort, J. 1943. Preface. *Annals of Biology* (Copenhagen) **1:**5–6.

Holling, C. S. 1992. Cross-scale morphology, geometry and dynamics of ecosystems. *Ecological Monographs* **62(4):**447–502.

Huovila, S., and A. Tuominen. 1990. On the influence of radiosonde lag-error on upper-air climatological data in Finland. 1951–88. *Finnish Meteorological Institute Meteorological Publication* 14.

Kukla, G. J., and H. J. Kukla. 1974. Increased surface albedo in the Northern Hemisphere. *Science* **183:**709–714.

Levin, S. A. 1988. Ecology in theory and application: Applied mathematical ecology. *Proceedings*, Trieste, 1986, Biomathematics 19. Springer-Verlag, Heidelberg.

Levitus, S. 1988. Decadal and pentadal distributions of hydrographic stations at 1000 m depth for the world ocean. *Programs in Oceanography* **20:**83–101.

McGowan, J. A. 1989. Pelagic ecology and Pacific climate. In D. H. Peterson (ed.), Aspects of Climate Variability in the Pacific and the Western Americas. AGU *Geophysical Monographs* **55:**141–150.

———. 1990. Climate and change in oceanic ecosystems: The value of time-series data. *Trends in Ecology and Evolution* **5(9):**293–299.

Parker, D. E. 1990. "Effects of Changing Exposure of Thermometers at Land Stations." In D. E. Parker (ed.), *Observed Climate Variations and Change: Contributions in Support of Section 7 of the 1990 IPCC Scientific Assessments*.

Scherhag, R. 1970. *Die gegenwartige Abkuhlung der Arktis. Beil z. Berliner Wetterkt.*. 105/70 so 31/70.

Smith, J. N., and C. T. Schafer. 1984. Bioturbation processes in continental slope and rise sediments delineated by Pb-210, microfossil and textural indicators. *Journal of Marine Research* **42:**1117–1145.

Swallow, J. C. 1969. "History of the Exploration of the Hot Brine Area of the Red Sea: DISCOVERY Account." In E. T. Degens and D. A. Ross (eds.), *Hot Brines and Recent Heavy Metal Deposits in the Red Sea*. Springer-Verlag, New York, pp. 3–9.

Wiesnet, D. R., C. F. Ropelewski, G. J. Kukla, and D. A. Robinson. 1987. "A Discussion of the Accuracy of NOAA Satellite-Derived Global Seasonal Snow Cover Measurements." In B. B. E. Goodison, R. G. Barry, and J. Dozier (eds.), *Large-Scale Effects of Seasonal Snow Cover*. IAHS Publication 166. IAHS Press, Wallingford, U.K., pp. 15–25.

Wunsch, C. 1992. Decade to century changes in the ocean circulation. *Oceanography* **5(2):**99–106.

7

Detecting Periodicity in Quantitative versus Semi-Quantitative Time Series

Jason D. Stockwell, Tormod V. Burkey, Bernard Cazelles, Frédéric Ménard, Fortunato A. Ascioti, Patricia Himschoot, and Jianguo Wu

Ordered categorical (or *semi-quantitative*) data are frequently encountered in ecology (e.g., Steen et al. 1990; Ménard et al. 1993). Researchers often resort to semi-quantitative measures (to describe abundance patterns, age or stage structures, environmental factors, etc.) to reduce processing time and/or because of financial constraints, while retaining an acceptable level of accuracy.

Time series analysis is a powerful tool for detecting underlying patterns and mechanisms in time-ordered data and is being employed increasingly in ecology (Platt and Denman 1975; Shugart 1978; Weber et al. 1986; Rose and Leggett 1990). Nearly all "traditional" methods of time series analysis require that data be from a continuous state space (from here on referred to as *quantitative data*). On the other hand, semi-quantitative data are from a finite set and cannot be analyzed using traditional methods. There are several techniques that deal with semi-quantitative time series, but these are primarily concerned with assessing short-term dependence between observations (e.g., Markov chain models). Few techniques exist for detecting periodicities. There is a body of theory for binary time series analysis (Kedem 1980), but techniques to explore data with more than two categories are less well developed.

Here we briefly review a technique for dealing with time-ordered, semi-quantitative data—the contingency periodogram as developed by Legendre and Legendre (1979) and as further explored by Legendre et al. (1981).We explore the potential change in ability to detect periodicities when the form of a dataset changes from quantitative to semi-quantitative (five and two categories), and how these changes may affect interpretation of results.

Our approach is to analyze four datasets. Two of these are simulated models, one of a quasi-periodic population, the other of a truly chaotic population. These models reflect the growing interest in chaos in ecology. The other two datasets consist of a horizontal temperature profile from Lake Ontario, and of recorded abundance of a zooplankter at a Mediterranean reference station. The first three datasets are quantitative, and thus allow comparison between traditional time series methods and the contingency periodogram (after categorizing). The last dataset was collected as semi-quantitative data and thus can be analyzed only by means of the contingency periodogram.

Methods

Traditional Time Series Analysis

When time series are sufficiently long, and data are quantitative, time series analysis in the traditional sense (e.g., Chatfield 1989; Diggle 1990) can be a powerful tool. The reader is urged to examine the abundant literature on this subject for a more complete understanding of these techniques.

The estimated spectrum of a time series partitions the variance of the series into its component frequencies (time^{-1}) or wavenumbers (length^{-1}). Peaks in the spectrum indicate what scales are contributing the most variance to the series. In this manner, periodicities, if present, are detected. The estimated spectrum is calculated by smoothing the raw periodogram. For our analyses, we have calculated the raw periodograms of the datasets in their quantitative form, *but we have not smoothed them*. This facilitates direct comparison with the contingency periodogram of the datasets in their semi-quantitative forms.

The Contingency Periodogram

The contingency periodogram (Legendre et al. 1981) is one method that can potentially identify periodicities in short, semi-quantitative time series with two or more categories. As with other time series methods, the data must be stationary (constant mean and variance) and detrended.

The contingency periodogram is based on contingency tables. To illustrate this method, we give an example taken directly from Legendre et al. (1981). Table 7-1 shows a time series of $N=16$ observations. Each observation can be one of three states (categories). The series is divided into nonoverlapping subseries of period T, which can range from 2 to $N/2$. In the example, T ranges from 2 to 8. A contingency table is then constructed for each period T. The rows of the contingency table correspond to the states of the data. The columns are the sequential observations in period T. The values in the contingency table are the number of observations in a given state S of the process (row) made at one observation of the period T (column). Table 7-2 shows the contingency table for $T=5$ in the example time series. The value of 3 in Row Three (state 1), Column Two (observation 2) signifies that there are a total of three state 1's located in the second position when the series is divided into nonoverlapping subseries of $T=5$ (positions 2, 7, and 12 in Table 7-1).

Table 7-1. Example time series of 16 observations that can each be one of three states (from Legendre et al. 1981).

States	1	2	3	4	5	6	7	8	9	10	11	12	13	14	15	16	Totals
3				X	X				X					X	X		5
2			X			X		X		X			X				5
1	X	X					X				X	X				X	6

Table 7-2. Contingency table for the period T = 5, for the example time series in Table 7-1 (from Legendre et al. 1981).

	Observation axis (X)					
States	1	2	3	4	5	Total
3	0	0	0	3	2	5
2	1	0	3	0	1	5
1	3	3	0	0	0	6
Total	4	3	3	3	3	16

For each of the T contingency tables, the entropies H (uncertainties) are estimated following information theory:

$$H(S) = -\sum_{i=1}^{s} \frac{N_i}{N} \log \frac{N_i}{N} \tag{1}$$

$$H(X) = -\sum_{j=1}^{T} \frac{N_j}{N} \log \frac{N_j}{N} \tag{2}$$

$$H(SX) = -\sum_{i=1}^{s} \sum_{j=1}^{T} \frac{N_{ij}}{n} \log \frac{N_{ij}}{N} \tag{3}$$

where $H(S)$ is the uncertainty of the states, $H(X)$ is the uncertainty of the observation axis, $H(SX)$ is the uncertainty of the entire contingency table, N_{ij} is the value in row i and column j of the contingency table, N_i and N_j are the totals of row i and column j, respectively, and s is the number of states. The contingency statistic is:

$$H(S \cap X) = H(S) + H(X) - H(SX). \tag{4}$$

Simply put, $H(S \cap X)$ is a measure of how much uncertainty in the states (categories) is explained by a given period T.

If the rows and columns are independent, the quantity $2N^*H(S \cap X)$ is asymptotically distributed as χ^2 with $(s - 1)(T - 1)$ degrees of freedom when $\log_e$ is used (Kullback 1959). However, as Legendre et al. (1981) point out, each value of the contingency periodogram (i.e., $H(S \cap X)$ for each period T) is not independent, and therefore one cannot test each value for significance. Instead, if a value is below its critical value, then it is definitely too small to be different from zero.

If the value exceeds its critical value, it still may not be different from zero, "but their ecological implications may be considered" (Legendre et al. 1981). Figure 7-1 shows the contingency periodogram for the example series in Table 7-1. A peak is evident at the period $T = 5$.

The number of states (categories) and their definitions are clearly important. However, the contingency periodogram does not take into account the ordering of the states, and, as a result, one may be less certain of identifying actual periodicities in semi-quantitative data sets (Legendre et al. 1981). Despite this potential problem, results form Legendre et al. (1981) and from this study indicate that the contingency periodogram is a useful method for detecting periodicities in semi-quantitative time series. Legendre et al. (1981) propose a method for dividing rank-ordered variables into states when there are no available ecological or biological criteria. In our analyses, we have chosen to use five and two categories for all of our examples. Five was chosen to match the original number of categories in one of our datasets (salp abundance), and two was chosen arbitrarily as some value less than five to generate decreasing resolution in the data. In the salp example, we use ecological criteria for establishing the partition

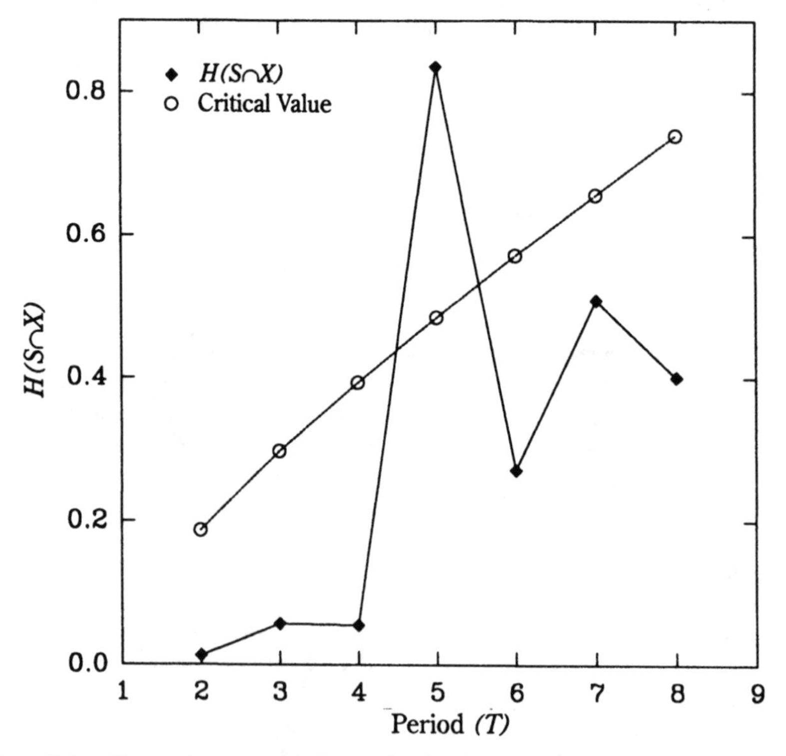

Figure 7-1. The contingency periodogram for the time series in Table 7-1 (from Legendre et al. 1981). Critical values are $\chi^2_{(s-1)(T-1)}/2N$ where $s = 3$ and $N = 16$.

values of the binary series. Otherwise, series are partitioned to give near-equal numbers of data points in each category, or are partitioned based on an approximate $\log_2$ scale. In each example, we identify how data are partitioned.

Also note that this method allows missing values, which are usually the rule in long-term ecological studies. For a more rigorous development of the contingency periodogram, see Legendre et al. (1981).

Analyses

Example 1: A Seasonal Model of Population Dynamics

We generated a time series from a population dynamics model developed by Burkey and Stenseth (in press). The model population is territorial during summer months and nonterritorial during winter months. The population does not reproduce during the winter, which introduces seasonal and strong yearly patterns. We chose parameters to generate a quasi-periodic series. For details of the model, see Burkey and Stenseth (in press).

Figure 7-2 shows a 100-year (stationary) time series of the model as (A) quantitative data, (B) quantitative data partitioned into five ordered categories, and (C) quantitative data partitioned into an ordered binary series. Samples are taken 10 times per year, giving a total of $N = 1000$ data points. The partition values are 11, 25, 45, and 80 for the five categories. This represents an approximate doubling of the category ranges and also provides a nearly equal number of points in each category (although there are fewer points in the last category). In the binary case, the 0's consist of the first two classes (≤ 25) in the five-category case, and the 1's consist of the last three classes (> 25) in the five-category case.

The raw periodogram (Figure 7-3) reveals the strong annual pattern (10 time steps), as well as the semiannual and seasonal patterns, despite the quasi-periodic nature of the model. The presence of a number of smaller peaks at other frequencies indicates harmonics or other possible dynamics, but these cannot be considered significant, as the large peaks are several orders of magnitude greater.

Figures 7-4A and B show the contingency periodograms for both semi-quantitative series. Annual and semi-annual periodicities ($T = 10$ and $T = 5$, respectively, plus harmonics) are detected in both cases. However, a quarterly cycle that exceeds its critical value in the five-category series (Figure 7-4A) does not exceed its critical value in the binary series (Figure 7-4B). This is expected: as the resolution of the data decreases, the ability to detect finer-scale periodicities also decreases. Thus, it would appear that five categories is a sufficient number to detect periodicities down to a seasonal level, while (presumably) also reducing costs associated with collecting and processing quantitative measures.

It is worth noting that, when examining the contingency periodogram for values that exceed their critical values, we are performing multiple tests. Typically

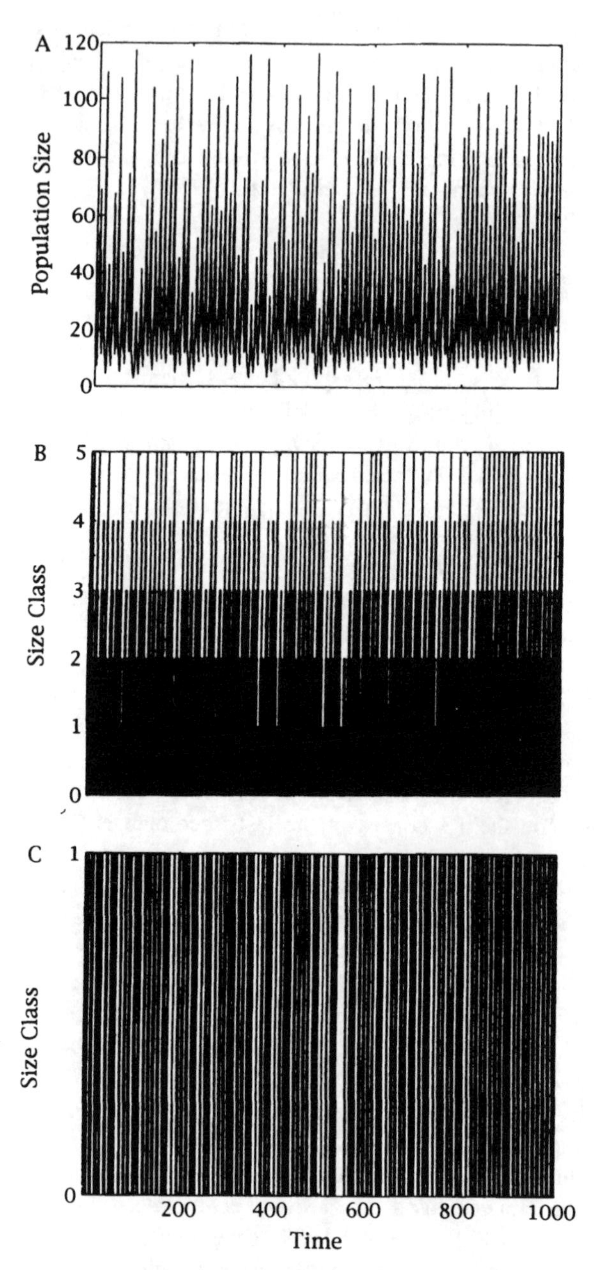

Figure 7-2. Time series generated from a seasonal model of population dynamics (Burkey and Stenseth, in press) as: (A) quantitative data, (B) quantitative data partitioned into five groups, and (C) quantitative data partitioned into two groups. Parameters of the model were chosen to generate a quasi-periodic series. Ten time units equals one year.

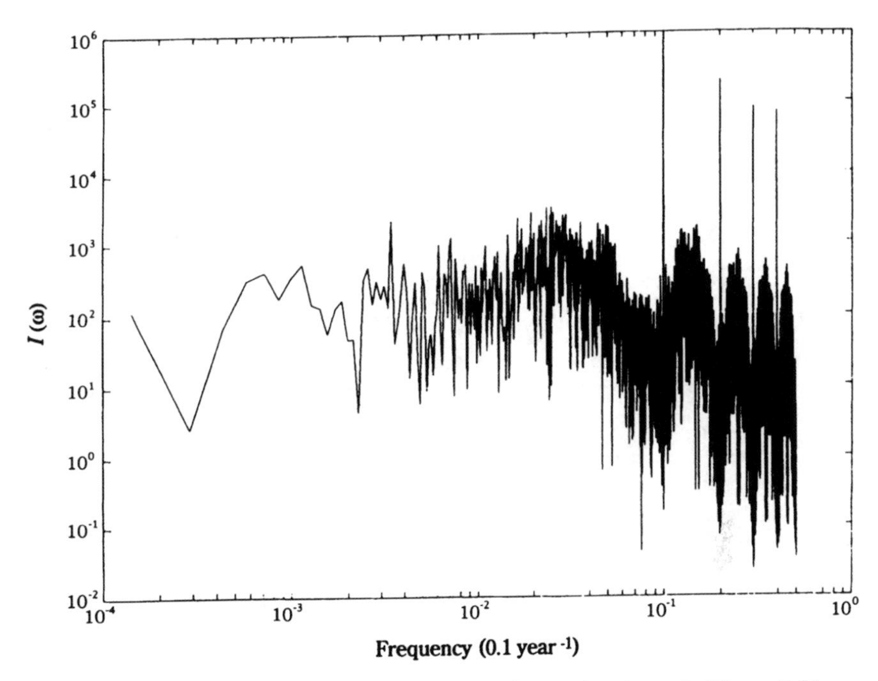

Figure 7-3. Raw periodogram of quantitative time series shown in Figure 7-2A.

one should apply a Bonferroni correction by dividing the α-value (0.05) by the number of tests (in this case, T = 2 to 100, or 99). However, a Bonferroni approach will not work when the number of comparisons (T) gets large (possibly at $T \geq 10$; A. Solow, pers. comm.). It may be possible to develop a portmanteau test for global significance based on randomization. If any of the contingency periodogram values exceeds this portmanteau critical value, then independence would be rejected.

Example 2: A "Super" Chaotic Model

A chaotic signal represents extreme deterministic, nonlinear behavior: it is aperiodic and gives a flat spectrum. Thus, it is unpredictable on a long-term basis (May 1974; Parker and Chua 1987).

A chaotic time series was generated using the simple logistic equation:

$$X(t+1) = abe^{-bX(t)/m}. \tag{5}$$

This model was studied by Vandermeer (1982a,b) and resembles the population fluctuations of rare species. For example, for an insect population X growing on plant seeds, the model gives the number of insects at generation t + 1 given the number of insects at previous generation t under the following assumptions: (1)

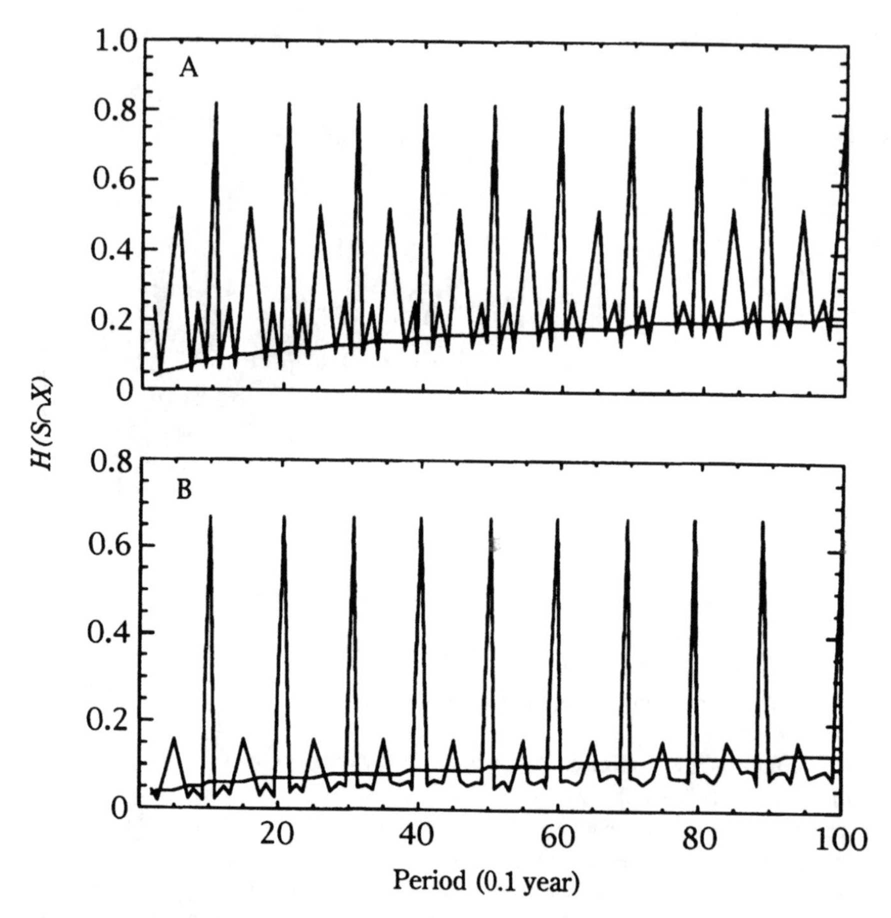

Figure 7-4. Contingency periodograms of semi-quantitative time series shown in Figures 7-2B and 7-2C, respectively: (A) five categories and (B) two categories.

any individual insect lays an average of b eggs at random, (2) only one larva can survive on a single seed with survivorship rate a, and (3) the total number of available seeds is m.

Vandermeer (1982a,b) showed that for very large values of the product ab (i.e., large growth rate), the model enters a particular region of dynamical behavior called "super" chaos. This behavior is characterized by a stable linear relationship between the numbers of years of low numbers in a population n and the amplitude of the population burst X_b. Thus, there exists the possibility of predicting n from X_b (or, in Vandermeer's words, to "resolve" the chaotic behavior).

A time series of $N = 1000$ points (Figure 7-5) was generated by iterating this "super" chaotic version of equation (5). Figure 7-5 also shows the population X of the time series partitioned into semi-quantitative measures (five and two

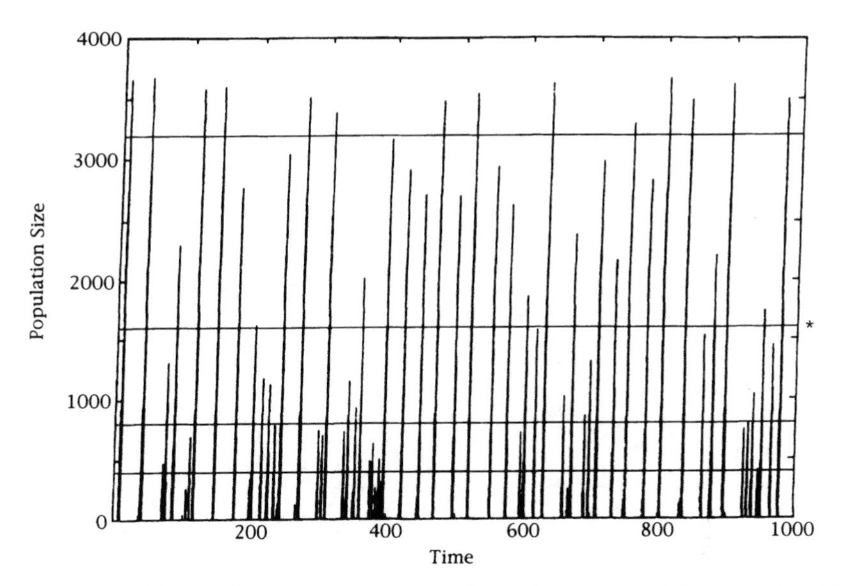

Figure 7-5. Time series generated from the "super" chaotic model in equation (5) (Vandermeer 1982a,b). Solid lines indicate partitioning for five categories. Asterisk indicates partitioning for two categories.

categories). The categories were classified using a geometric progression with base 2 for the five-category case. The binary series was partitioned at the value separating the third and fourth categories in the five-category series.

Despite the "resolution" property shown by Vandermeer (1982a,b), the raw periodogram of this time series (not shown) appears to be random (i.e., a flat periodogram), as expected. The contingency periodogram of the five-category series (Figure 7-6A) reveals no values different from zero, exemplifying the aperiodic nature of the signal. A similar result is found in the binary case (Figure 7-6B). Thus, the analyses for the quantitative form and both semi-quantitative forms of this "super" chaotic time series give similar results: no detectable periodicities.

Example 3: Application to Limnological Data

Figure 7-7 shows temperature for a horizontal transect at 30-m depth from Lake Ontario in November 1991. The data were collected with a conductivity-temperature-depth (CTD) probe mounted with other instruments on an Endeco V-fin (Sprules et al. 1992). The sampling interval between data points is 1 second, or approximately 2.1 m. If we assume temporal homogeneity, then we can investigate the periodicities of the spatial structure (in one dimension) of the data.

The data were detrended with a linear regression, and the raw periodogram

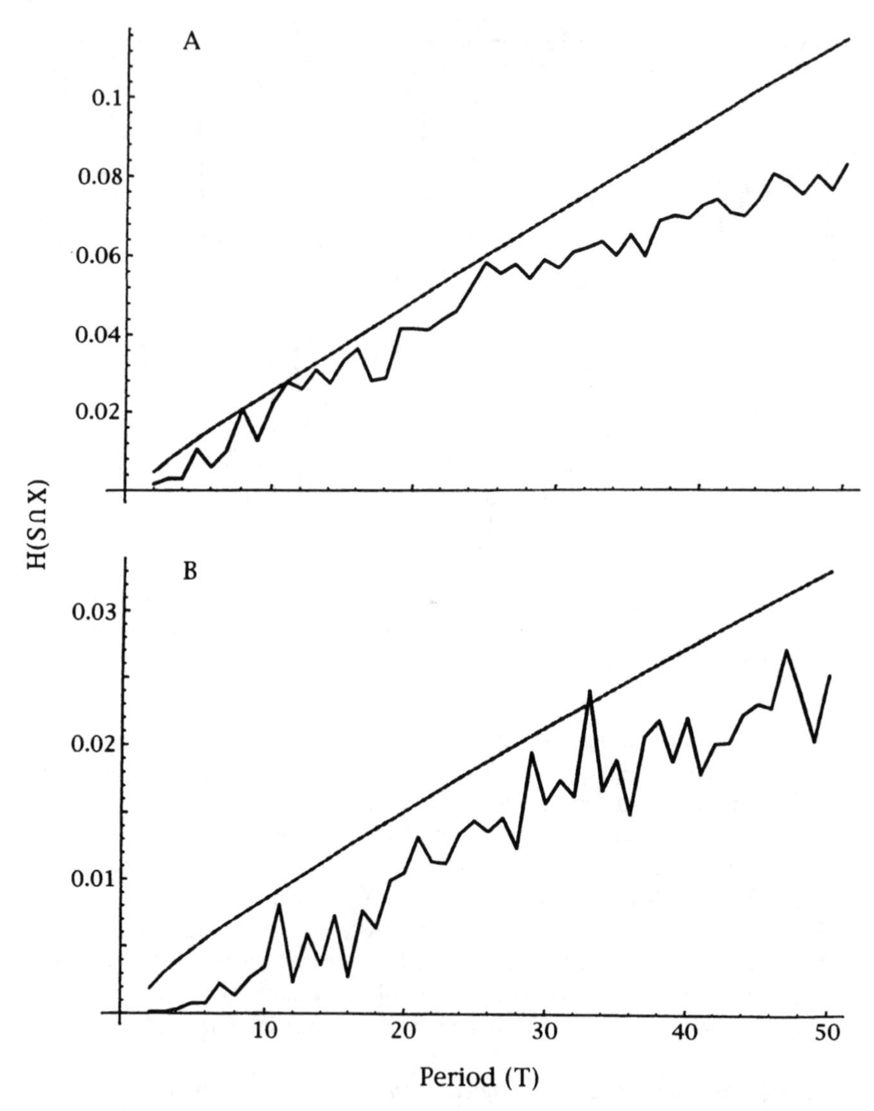

Figure 7-6. Contingency periodograms of (A) five-category and (B) two-category time series generated from the "super" chaotic model of equation (5) shown in Figure 7-5.

was calculated for the residuals. The residuals were then partitioned into semi-quantitative series (five and two categories). The categories were determined by ensuring a roughly equal number of data points in each category for both semi-quantitative series.

The raw periodogram (Figure 7-8) shows a dominating, long-term trend in the temperature, evident at the lowest wavenumbers (largest spatial scales). The

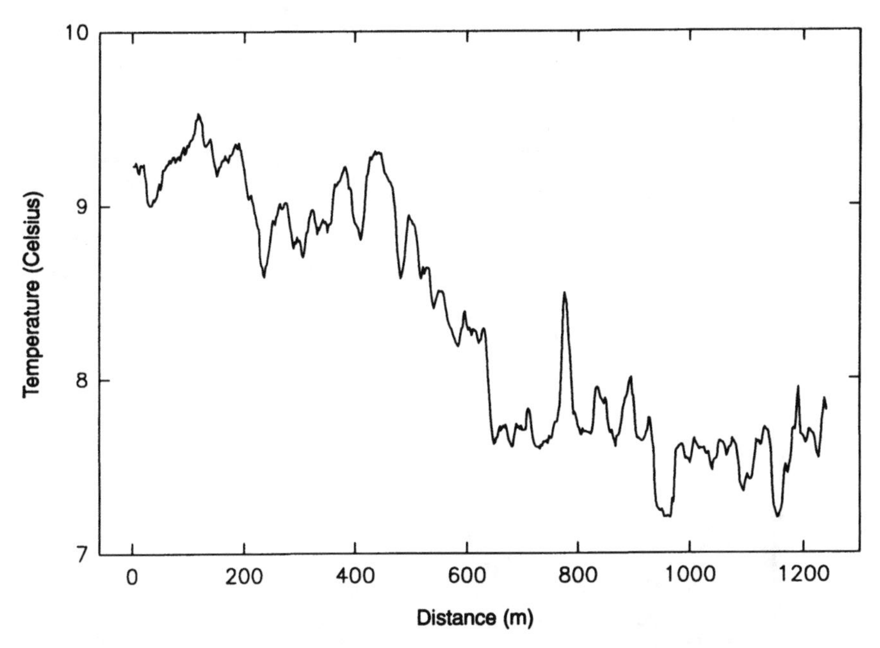

Figure 7-7. Horizontal temperature profile at 30-m depth in Lake Ontario, November 1991.

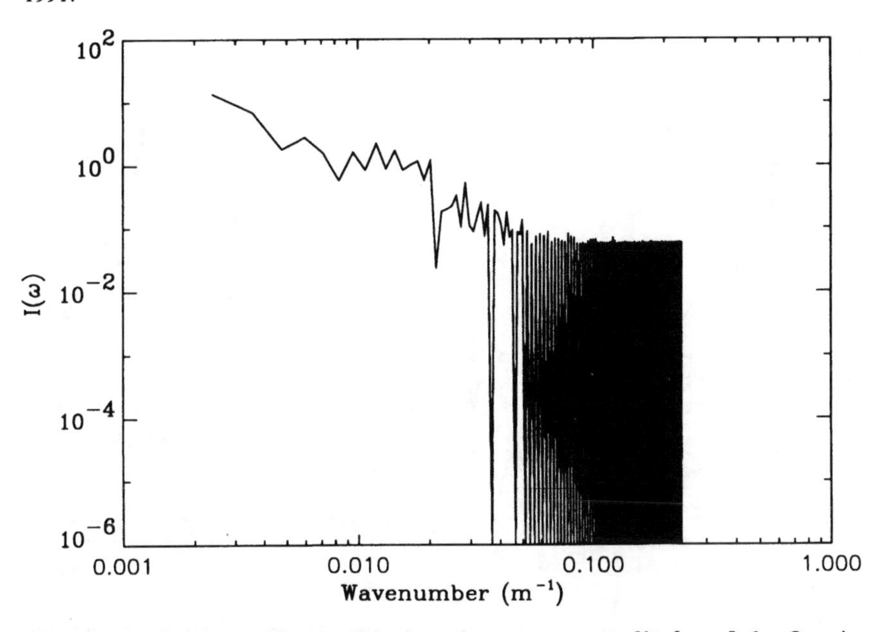

Figure 7-8. Raw periodogram of horizontal temperature profile from Lake Ontario shown in Figure 7-7.

contingency periodogram for five categories (Figure 7-9A) identifies a minor peak at approximately 294 m. The contingency periodogram for the binary series (Figure 7-9B) shows a very strong peak centered around 357 m, also corresponding to the long-term trend in the quantitative data.

The long-term trend is detected in each of the semi-quantitative series, although

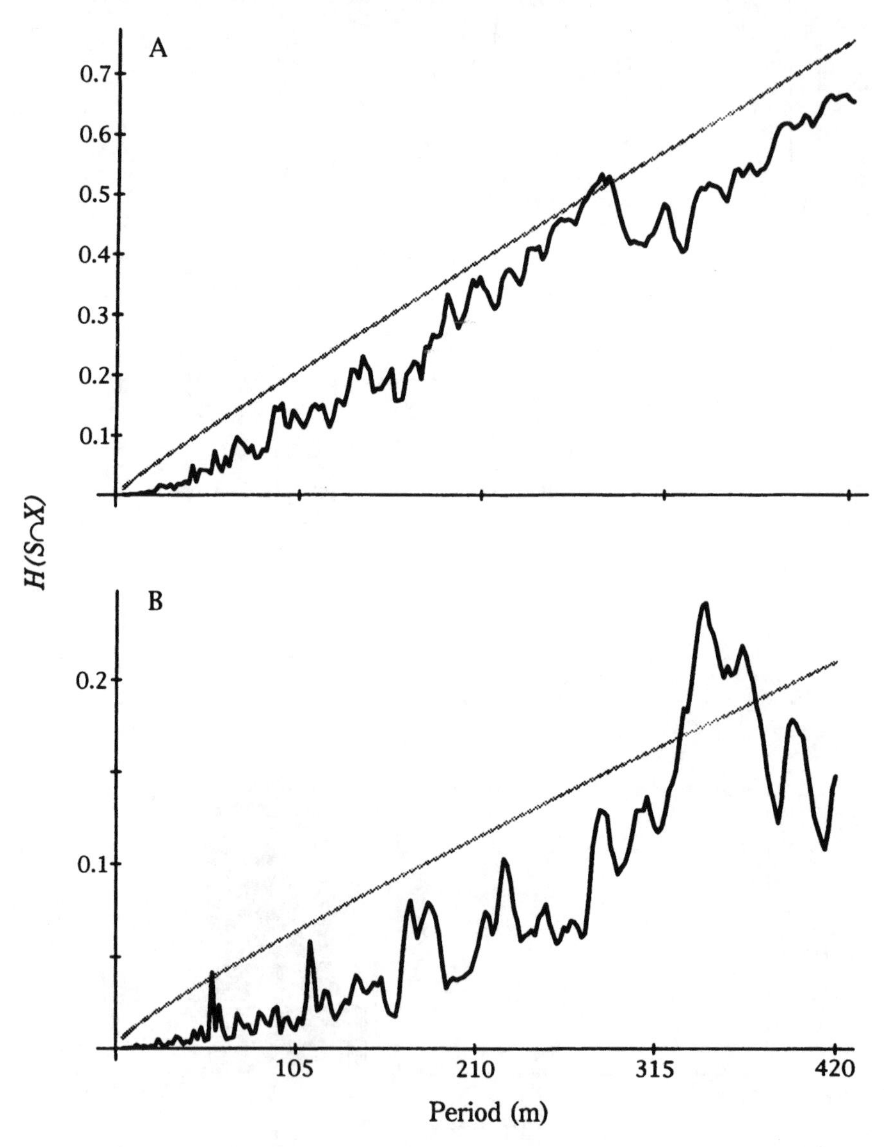

Figure 7-9. Contingency periodograms for the horizontal temperature profile from Lake Ontario as (A) five categories and (B) two categories.

the peak is not very strong in the five-category case. A shift from five to two categories shifts the peak of the contingency periodogram from 294 to 357 m. This raises an important question: how does categorizing affect results and their subsequent interpretation? More important, do the number of categories and the partition values between categories properly "arrange" the given data to give results that are meaningful to the questions being asked? Basically, this technique consists of matching the scale of the sampling with the scale of the question (as shown in the following example).

Example 4: Application to Marine Biology

The salp (*Thalia democratica*) is a gelatinous zooplanktonic filter feeder. Its reproductive strategy is such that a few individuals may give rise to quick population bursts (blooms) in time and space. This generally occurs in the spring. Samples were collected weekly from 1967 to 1990 at a fixed station (Bay of Villefranche-sur-Mer, western Mediterranean) with vertical net hauls. Abundance classes were approximately based upon a geometric progression with base 4.3 (Table 7-3), as determined by Frontier (1969). Missing values are coded as -1. Frontier (1969) showed that this semi-quantitative determination is well suited for describing the spatial and temporal variations of zooplanktonic species at classical regional scales in oceanographic studies.

The semi-quantitative data capture the salient features of the series: strong seasonality with periods of few or no salp, high year-to-year variability with respect to intensity and duration of blooms, and the presence of missing values (Figure 7-10). Note that the sampling design is not well suited for precise determination of the size of the population: salp are distributed in swarms, and vertical sampling at a fixed station precludes the evaluation of spatial fluctuations. *We are interested here in the possible occurrence of blooms, which can be detected by this semi-quantitative measure.*

To reduce the salp series to a binary series, we use biological criteria. We consider that fluctuations between classes 1 and 2 in the original series are attributable to sample fluctuations. Therefore, those two classes form the first category. The second category consists of classes 3 to 5 in the original series, which correspond to a higher concentration of salp (the possible occurrence of a bloom).

Table 7-3. Abundance classes (categories) for the salp time series in Figure 7-10.

Number of salp per haul	Abundance class
0	1
1 to 3	2
4 to 17	3
18 to 80	4
> 80	5

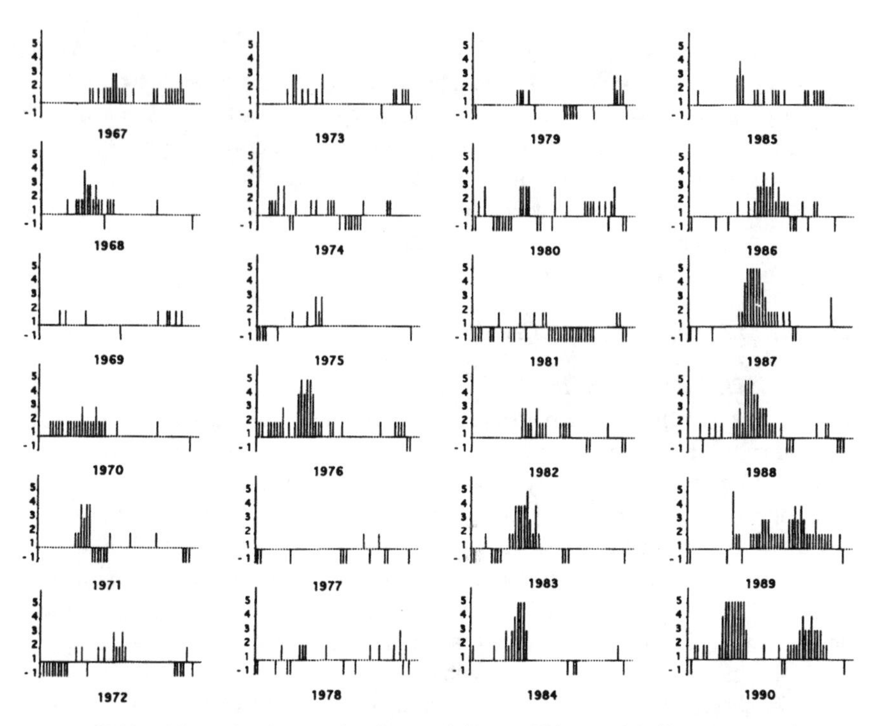

Figure 7-10. The salp time series from 1967 to 1990 at a Mediterranean reference station. Missing values are coded as −1. (Reprinted from F. Ménard, S. Dallot, and G. Thomas, 1993, A stochastic model for ordered categorical time series. Application to planktonic abundance data, *Ecological Modelling* 66:101–112. By permission of Elsevier Science Publishers.)

In the five-category case, a very weak periodicity exists at 52 weeks (Figure 7-11A), signifying a possible yearly cycle. In the two-category case, a strong peak occurs at week 52, as well as later harmonics (Figure 7-11B). A slight peak is also evident at week 26, indicating a possible semi-annual cycle in salp blooms.

It seems obvious that the abundance of salp at time t depends on the abundance at previous times. Ménard et al. (1993) have shown, using a periodic Markov model, that significant information is contained in the abundance of the previous week. Considering this, an alternative approach might be to base the statistic $H(S \cap X)$ on transition probabilities (e.g., the probability that salp abundance changes from category i to category j over one unit of time). However, the advantage of the approach by Legendre et al. is that it does not require the dependence pattern of the series to be explicitly modeled. Its asymptotic distribution will be χ^2 as long as the series can be considered ergodic and stationary under the null hypothesis of no periodicity.

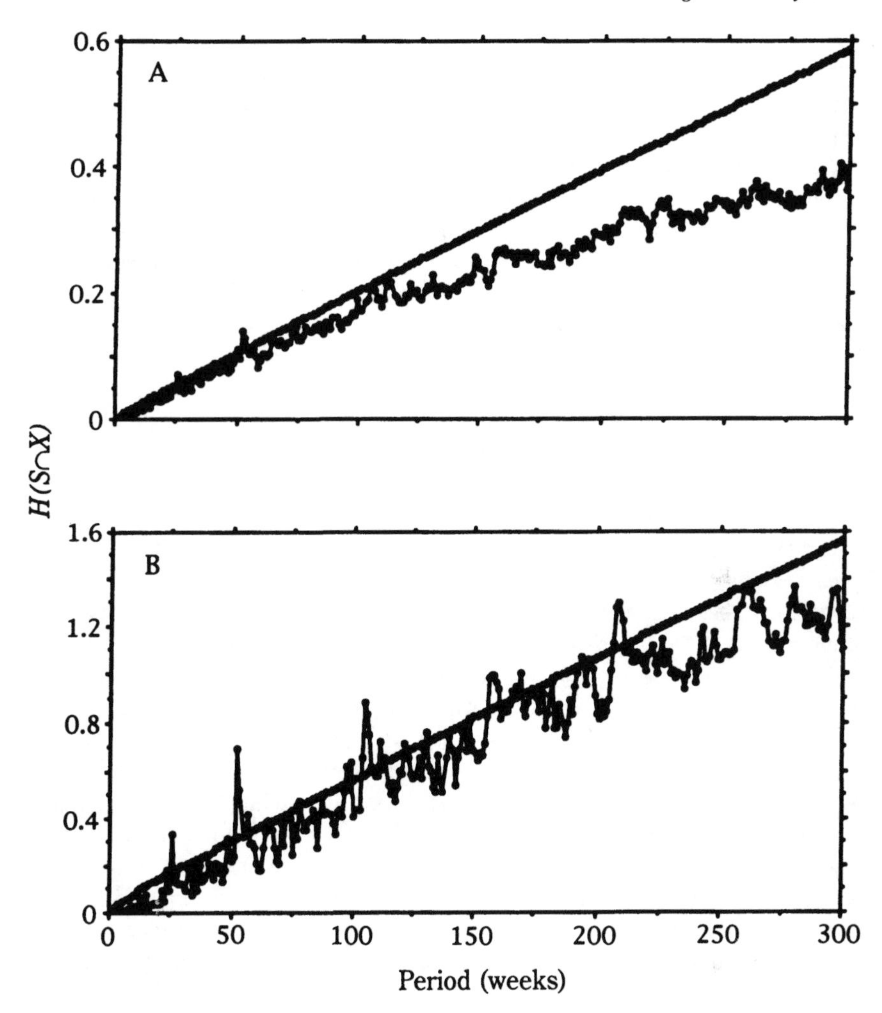

Figure 7-11. The contingency periodograms for the salp time series as (A) five categories (original series) and (B) two categories.

Conclusion

The use of time series analysis in ecology has been increasing in recent years. Its application to ecological data is appropriate and useful when sufficiently long, quantitative datasets are available. In many instances, however, semi-quantitative time series are the norm, and few methods have been proposed to analyze these types of time series for periodicities.

Here we have reviewed the use of one such method, the contingency periodogram, to investigate the ability to detect periodicities when time series are reduced from quantitative to semi-quantitative data. In general, the results are quite consistent: periodicities (or lack of periodicities) are preserved despite decreases in data resolution (Examples 1, 2, 3). For the salp series (Figure 7-10), the contingency periodogram identified a yearly periodicity in blooms in both the original (five-category) and the binary series (Figure 7-11). The $H(S \cap X)$ values in this example appear suspect, likely because of the strong autocorrelation within the dataset, and so have been adjusted by accounting for first-order dependence. We suggest caution when applying the contingency periodogram to field data. Appropriate checks must be made to ensure that short-term serial dependence or long-term trends are not obscuring or biasing results.

As the resolution in a dataset decreases, the ability to detect finer-scale structure also decreases (Example 1). However, as the contingency periodogram does not take into account the ordering of categories (e.g., a change from category 1 to category 2 is considered the same as a change from category 1 to category 4), there may be a loss of information (in terms of contingency periodogram analysis) when the number of groups is *increased*. This appears to be the case for the temperature data (Figure 7-7), where the contingency periodogram shows a much stronger peak in the binary form than in the five-category form.

There is also a discrepancy in the location of the peak in the contingency periodogram between five versus two categories for the temperature data (Figure 7-9). The researcher must decide how much discrepancy can be tolerated without losing the ability (or at least the potential) to answer the question(s) being asked. For instance, in the seasonal model (Figure 7-2), if we are interested in dynamics occurring at the temporal scale of years, then sampling with two categories appears to be sufficient (and more economical). If we are interested in seasonal dynamics, however, then an increase in the number of categories (and an increase in costs) is more appropriate. If we are interested in looking at weekly or monthly dynamics, then some number of categories ≥ 5 (all the way up to quantitative measures) is more appropriate. At the same time, sampling frequency should increase as well, to examine shorter time-scale fluctuations. In the salp analysis (Example 4), the categorization of the data was sufficient to help answer the question we were asking. When the number and partition values of categories do not properly reflect the questions being addressed, the effects can be deleterious. Further research on the effects of differing partition values is needed.

Acknowledgments

The authors would like to extend a warm thank-you to Tom Powell and John Steele for an interesting, informative, and fun-filled workshop. Very special thanks are also given to Dolores Pendell, for "always being there" during the workshop, and to Mary Schumacher, for her assistance with the final product after the workshop. Andy Solow provided guidance during the early development of this chapter and constructive comments on later drafts. W.G. Sprules also provided helpful comments.

References

Burkey, T. V., and N. C. Stenseth. In press. Population dynamics of a territorial species in a patchy and seasonal environment. *Oikos*.

Chatfield, C. 1989. *The Analysis of Time Series: An Introduction*. Chapman & Hall, New York.

Diggle, P. J. 1990. *Time Series: A Biostatistical Introduction*. Oxford Science Publications, New York.

Frontier, S. 1969. Sur une méthode d'analyse faunistique rapide du zooplancton. *Journal of Experimental Marine Biology and Ecology* **3:**18–26.

Kedem, B. 1980. *Binary Time Series*. Lecture Notes in Pure and Applied Mathematics, Vol. 52. Marcel Dekker, New York.

Kullback, S. 1959. *Information Theory and Statistics*. Wiley and Sons, New York.

Legendre, L., M. Fréchette, and P. Legendre. 1981. The contingency periodogram: A method for identifying rhythms in series of nonmetric ecological data. *Journal of Ecology* **69:**965–979.

Legendre, L., and P. Legendre. 1979. *Ecologie Numérique. 2. La Structure des Données Écologiques*. Masson, Paris.

May, R. M. 1974. Biological populations with non-overlapping generations: Stable points, stable cycles, and chaos. *Science* **186:**645–647.

Ménard F., S. Dallot, and G. Thomas. 1993. A stochastic model for ordered categorical time series. Application to planktonic abundance data. *Ecological Modelling* **66:**101–112.

Parker, T. S., and L. O. Chua. 1987. Chaos: A tutorial for engineers. *IEEE Proceedings* 75(8).

Platt, T., and K. L. Denman. 1975. Spectral analysis in ecology. *Annual Review of Ecological Systems* **6:**189–210.

Rose, G. A., and W. C. Leggett. 1990. The importance of scale to predator-prey spatial correlations: An example of Atlantic fishes. *Ecology* **71:**33–43.

Shugart, H. H. (ed.). 1978. *Time Series and Ecological Processes*. SIAM, Philadelphia.

Sprules, W. G., B. Bergström, H. Cyr, B. R. Hargreaves, S. S. Kiliam, H. J. MacIsaac, K. Matsushita, R. S. Stemberger, and R. Williams. 1992. Non-video optical instruments for studying zooplankton distribution and abundance. *Archiv für Hydrobiologie Beiheft* **36:**45–58.

Steen, H., N. G. Yoccoz, and R. A. Ims. 1990. Predators and small rodent cycles: An analysis of a 79-year time series of small rodent population fluctuations. *Oikos* **59:**115–120.

Vandermeer, J. 1982a. To be rare is to be chaotic. *Ecology* **63:**117–168.

———. 1982b. On the resolution of chaos in population models. *Theoretical Population Biology* **22:**17–27.

Weber, L. H., S. Z. El-Sayed, and I. Hampton. 1986. The variance spectra of phytoplankton, krill, and water temperature in the Antarctic Ocean south of Africa. *Deep-Sea Research* **33:**1327–1343.

PART II

Comparisons of Scales

8

Physical and Biological Scales of Variability in Lakes, Estuaries, and the Coastal Ocean

Thomas M. Powell

What is meant by "scale"? Why is it an important concept in ecology? Why does scale play a particularly prominent role when one speaks of "coupling"? Can one make nontrivial, general statements, if phrased in terms of scale alone, that apply to several seemingly different systems? Can other generalizations that highlight "scale" and "coupling" help us understand why some ecological systems differ so greatly from others? I shall address aspects of these questions here, with particular emphasis on results from lakes, estuaries, and the coastal ocean. I begin with some elementary notions, then review how scale considerations enter the coupling between physical and biological systems in the size scale regime between 100 m and 100 km—a regime of great ecological interest and the scale of most lakes, estuaries, and the coastal ocean. Other papers in this volume (i.e., Roughgarden et al. 1989, see footnote at the bottom of this page) focus on, for example, terrestrial systems and the general question of how large, complex systems are structured; I end, then, by advancing some speculations about these and other areas I have neglected. Finally, some publications on theory in biological oceanography (and related disciplines) have recently appeared: one is nearly a tutorial (Platt et al. 1981) and the other assays the use and prospects for ecosystem theory (Ulanowicz and Platt 1985). I therefore largely avoid the topics that these authors have addressed so well.

Elementary Considerations

Intuitively, one speaks of the spatial scale of a problem as the distance one must travel before some quantity of interest changes significantly. Two adjacent parcels

---

of water separated by 1 mm are likely to have very similar concentrations of chlorophyll, for example, or similar concentrations of a single phytoplankton species (within the limits of sampling error, of course). Conversely, water parcels separated by 1 km might be expected to differ greatly in these quantities. An observer would understand that the relevant spatial scale was greater than 1 mm and less than 1 km. This example raises two points. First, when one speaks of scale one is considering variability. A spatial scale cannot be assigned to a process, or quantity, that is uniform in space. The terms "scale" and "scale of variability" are synonymous. Second, the spatial scale may depend upon the varying quantity, or process, under consideration. For example, the chlorophyll concentration in each of the two parcels of water separated by 1 km could be identical, whereas the phytoplankton species content could differ substantially. The "biological" scale (determined by the varying phytoplankton concentration field) can, in principle, differ from another "biological" scale, as determined by chlorophyll variation.

The intuitive discussion about spatial scales applies equally to time, time scales, and temporal variability. The time scale, then, is that period over which one waits to see a significant change in some quantity of interest.

The entire discussion can be formalized. Let $q(x)$ be any quantity that varies with spatial distance x. Let $\langle q \rangle_{RMS} = \langle (q - \langle q \rangle)^2 \rangle^{1/2}$ and $\langle dq/dx \rangle_{RMS} = \langle (dq/dx)^2 \rangle^{1/2}$. The brackets ($\langle \ \rangle$) indicate a spatial average. Then one defines L_q as the spatial scale over which $q(x)$ varies significantly, or, simply, the spatial scale

$$L_q = [\langle dq/dx \rangle_{RMS}]^{-1} \langle q \rangle_{RMS}. \tag{1}$$

Similarly, defining $\overline{q}_{RMS} = [\overline{(q - \overline{q})^2}]^{1/2}$ and $\overline{dq/dt}_{RMS} = [\overline{(dq/dt)^2}]^{1/2}$, where the overbar indicates a time average, one defines T_q as the time scale over which $q(t)$ varies significantly, or, simply, the time scale:

$$T_q = [\overline{dq/dt}_{RMS}]^{-1} \overline{q}_{RMS}. \tag{2}$$

Note the connection to variance (i.e., variability) and quantities like the variance spectrum. Other similar definitions are, of course, possible.

We often speak loosely of the spatial scale as the area (or length or volume) in space that we are considering, and of the time scale as the duration of our attention to a phenomenon. We can be found admonishing our colleagues to remember that ecological interactions may differ if we consider larger (or smaller) space or time scales (I translate this roughly to mean that if the observer had looked at a larger area, or for a longer time, different results might have been obtained). The mere size of the area under consideration is rarely the quantity of direct interest, however. We are usually much more concerned with knowing the largest size scale that one must observe to capture "most" of the variance. Though it is often true that at the very largest space and time scales more of new variance enters a system, we are often restricted to smaller intervals in space and time for unavoidable logistic and practical reasons. Accordingly, dominant processes may act over considerably smaller time or space scales than the full duration, or distances, over which we observe. A satisfactory formalism to quantify the variability—variance—as a function of spatial or temporal scale is necessary in order to determine what the relevant scales are in a problem. Time series analysis (and its analogue in space), including spectral techniques, is one approach (Williamson 1975; Shugart 1978; Chatfield 1984). Other approaches, for example, multivariate statistics such as principal components or empirical orthogonal functions (Williamson 1972; Denman and Mackas 1978; Chatfield 1980; Gauch 1982), are also useful and may not be as demanding (or as powerful) as the time series techniques. We shall not be surprised, then, to find that model calculations of variance spectra are those that may tell us most about the scale questions (and are those most easily compared to actual data concerned with these questions).

Why are considerations of scale important in ecology, especially in theoretical ecology? Because they offer the prospect of simplification—how simplification of our models might arise and, equally important, when and why one might not expect such simplification.

The small-scale physical environment in aquatic systems differs from that at large scales. Life "in the small" is dominated by viscous effects (Purcell 1977; Vogel 1981; Berg 1983); the Coriolis "force" is the prominent actor at large scales (Pedlosky 1979; Gill 1982). A formalism to utilize these facts—dimensional analysis—plays a major role in determining the forces that dominate physical phenomena at given space and time scales. Further, and most important, there is often a pronounced separation of scales—that is, scales where one effect dominates may not overlap the range of scales where another effect holds sway. One may disregard certain effects within certain time and space scales, thus achieving a simplification. Finally, though separation of different effects at differing scales is the "zeroth order" model, some coupling between scales is always present. For example, a fluctuation in velocity, that is, variance—mechanical energy in this case—enters most physical systems at large scale. Ultimately, its impact is dissipated to heat at small scale via viscosity (see Figure 8-1). Nonetheless, the coupling is sufficiently weak that one may neglect viscous effects when considering phenomena over ocean basin scales. In summary, the physical environment in the sea is divided into separate spatial and temporal regimes; though coupled to one another, the regimes maintain their integrity because the coupling is weak enough. Finally, the coupling between the separate regimes must be via nonlinear terms in the governing equations of motion.

A similar separation describes the ecological structure in the sea (and in lakes and estuaries). Figure 8-2A shows a measure of the spatial scales of variability for phytoplankton community composition, zooplankton community composition, and zooplankton biomass. Note the separation between dominant scales for each quantity. Note also the differences between the alongshore and cross-shelf scales; this anisotropy is evidence for the control the physical environment exerts

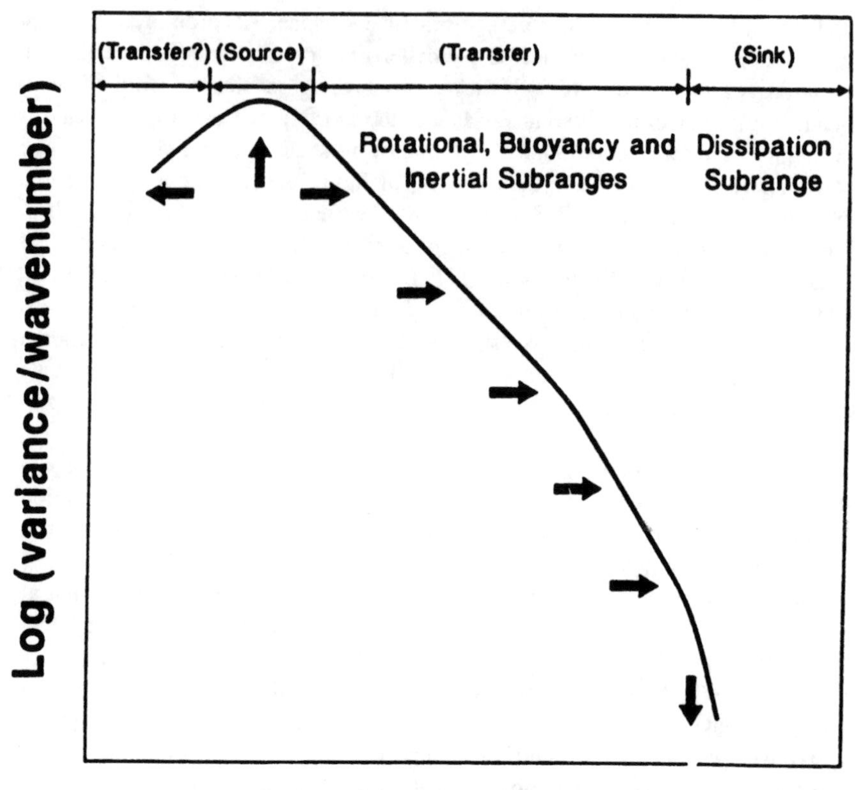

Figure 8-1. Spectrum of a scalar tracer versus (size scale)$^{-1}$ = wavenumber = 2π/ wavelength. "Energy," or variance, enters at large scales, cascades from large scale to small, and is lost through dissipation at small scales. Rotational, buoyancy, or inertial forces dominate subranges of the cascade/transfer region. (After Mackas et al. 1985.)

on the spatial distribution of these planktonic organisms. Figure 8-2B shows estimates of particle size and doubling times for phytoplankton, zooplankton, invertebrate carnivores or omnivores, and fish communities. With the provisos noted above, these are taken as surrogates for the scales of variability. Note again the separation of spatial and temporal regimes. We anticipate that these separations point to simplifications that ecological theorists can advantageously incorporate, following the practice of modelers of the physical environment.

Physical-Biological Coupling

Figure 8-3 shows a prototype "box model" for the budget of any quantity q (which might be phytoplankton biomass, the mass of a nutrient such as a nitrate,

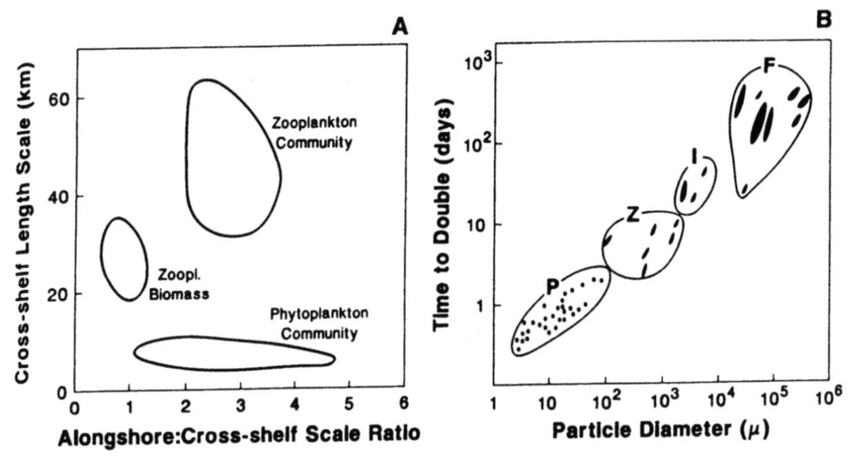

Figure 8-2. (A) 75% confidence regions for the scales of variability of zooplankton biomass, zooplankton community composition, and phytoplankton community composition along the coast of British Columbia. (After Mackas 1984.) (B) Doubling times versus particle size distributions for phytoplankton (P), zooplankton (Z), invertebrate carnivores or omnivores (I), and fish (F). (After Sheldon et al. 1972; and Steele, 1978, "Some Comments on Plankton Patches," in J. H. Steele, (ed.), *Spatial Patterns in Plankton Communities,* Plenum Press, New York. By permission of Plenum Press.)

or the density of a particular species of zooplankton). The governing equations in Steele's (1974) model are of this form, for example. The rate of change of q obeys a balance equation of the form

$$dq/dt = F_I - F_O + P - C. \tag{3}$$

F_I and F_O are input and output fluxes of q, respectively; P and C are the rates of internal production and consumption of q within the "box"—any volume of water. In aquatic systems dominated by plankton, the input and output fluxes, F_I and F_O, are usually the result of physical transport process (though biological transport via swimming, for example, can enter); often, but not always, they can be an order of magnitude or more greater than the *in situ* rates, P − C. Accordingly, the first task is to express the coupling of the physical environment to the ecological structures in these aquatic systems. The general question of physical-biological coupling has been reviewed by Denman and Powell (1984) and Mackas et al. (1985). Other references can be found in these two pieces. Several useful generalizations emerged from these works. First, the dominant temporal scales seemed to be set by biological processes, but the characteristic spatial scales were largely determined by the physics. Second, a "linear, first-order" hypothesis is that if a concentration of "energy" (variance) were to be found at a certain spatial scale in a physical phenomenon—say, a front—then one

F_I = input flux
F_O = output flux
P = internal production
C = internal consumption

Figure 8-3. "Box model" for a quantity q, which can change by inputs (F_I), outputs (F_O), internal production (P), and consumption (C). The budget for q can be expressed as $dq/dt = F_I - F_O + P - C$.

might expect to find a concentration of variance at the same scale in biologically interesting quantities. (A biological oceanographer told me that he made it a point to be on board ship, taking data, when two of his physical oceanographer colleagues were on cruises. "If those two have found something interesting," he said, "there's bound to be something interesting for me, too.") This second hypothesis says nothing about the coupling of one scale to another—only about the source regions for biological variance.

The coupling of ecological phenomena occurring at different scales is an area that is particularly ripe for theoretical investigation. For logistic reasons, experimenters can rarely cover a sufficiently large area in a short enough time to truly measure synoptically over several scale ranges. Have things changed because of the changes in observed scale, or have they "merely" changed *in situ* with time? We are stuck in our parochial little ponds—even on the ocean! Perhaps the only way to investigate these questions now is with theoretical calculations, until satellite remote sensing can provide a greater number of more useful biological indicators. Because some, perhaps most, of the coupling between scales is accomplished by physical processes, we will consider these models in detail.

100-Meter–100-Kilometer Scales

Early patch models in the sea—plankton growing with constant growth rates and diffusing with constant diffusivity, called KISS models after an appealing grouping of the first letters of the names of early investigators (Kierstead, Slobodkin, and Skellam)—are reviewed by Okubo (1980, 1984). Models for the power spectrum of spatial fluctuations in phytoplankton biomass that assume a constant growth rate in isotropic turbulence (Denman and Platt 1976; Denman et al. 1977) led to predictions that agree with a large number of measurements in lakes, estuaries, and the ocean (Platt 1978). Some years later Denman (1983) realized that as long as the growth rate was only a function of time, not space (including a constant function, of course), then the growth rate could be transformed out of equations for the spectral distribution. Unhappily, then, these models proclaimed that the magnitude of growth rates of phytoplankton, *if spatially homogeneous,* played no role in determining patchiness; phytoplankton acted only as passive tracers, like dye. With a Lagrangian particle statistics representation of the large-scale, turbulent velocity field, Bennett and Denman (1985) modeled phytoplankton concentrations in a spatially varying random growth rate field $r(x,t)$. Results were obtained for two cases: $r(x,t)$ held fixed in space, and $r(x,t)$ advanced with the flow. In each case the initial spectral distribution was peaked around the characteristic spatial scale set by the initial phytoplankton concentration: i.e., it was "patchy." The spectrum "decayed," however, to a noisy spectrum proportional to k^{-1} (k is 2π/"wavelength," i.e., 2π/size scale). (See Figure 8-4.) Three results are noteworthy. First, the model provided a pathway for variance to show up at small scales once initiated at large scales—a variance cascade to smaller scales as in the physical turbulent transfer. Second, a growth rate field that varied spatially, but in a random sense, was insufficient to maintain an initially patchy distribution of phytoplankton. Third, the few synoptic measurements at these large scales (e.g., Gower et al. 1980; Denman and Abbott 1988) show a spectral distribution closer to k^{-2}, not k^{-1}. The present intense exploration of satellite records may help resolve this discrepancy.

The Bennett-Denman calculation presents a much more realistic description of the essentially two-dimensional turbulence field at the larger scales of interest than do studies by previous workers (i.e., isotropic turbulence occurs only at small scales, < 10 m). Further, the dominant features of the model can be phrased largely in terms of only the temporal and spatial scales of the processes at work. This model also has the potential to incorporate more complex biological dynamics than previous efforts, i.e., calculations that involve two coupled fields—a nutrient and a phytoplankton, for example, or a phytoplankton species and a zooplankton predator. The biological dynamics need not be restricted to linear functions. It is not clear that calculations incorporating such complications will lead to results similar to the Bennett-Denman calculations, i.e., "running down" to noise. For example, we expect that a predator-prey model of the form

$$\frac{dp}{dt} = rp(1 - \frac{p}{K}) - A(\frac{p^2}{\Gamma + p^2})h = F(p,h) \tag{4}$$

$$\frac{dh}{dt} = B(\frac{p^2}{\Gamma + p^2})h - dh = G(p,h), \tag{5}$$

with logistic growth of the prey p and Holling type III functional (and numerical) response to the predator h, will have three types of solutions: smooth approach

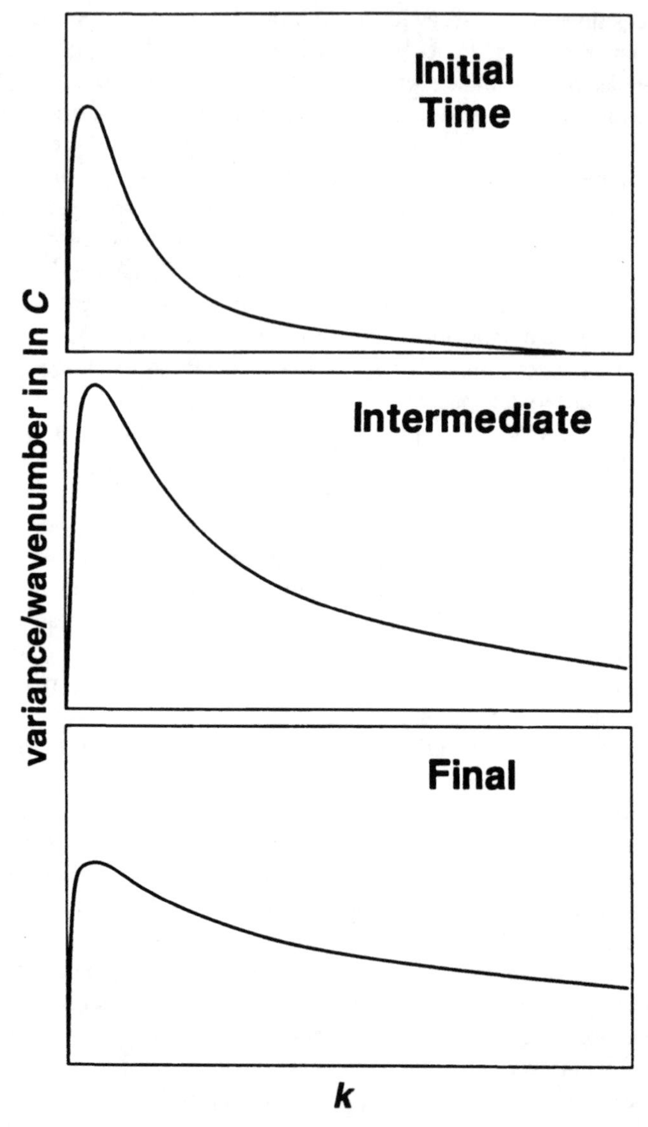

Figure 8-4. Spectrum of (transformed) concentration of phytoplankton versus k [(size scale)$^{-1}$ = 2π/wavelength] for initial, intermediate, and final times. Note the transfer of variance from large to small scales. (After Bennett and Denman 1985.)

to equilibrium; damped, oscillatory approach to equilibrium; and limit cycle activity (see, e.g., Roughgarden 1979 for an analysis with a similar, but not identical, Holling type II model). The *time scale* of the oscillatory behavior depends on the parameters of the model and could be shorter than, of the same order as, or longer than the physical time scale of relaxation to equilibrium.

Accordingly, if the oscillatory time scale is short, one might see large bursts of phytoplankton concentration early in the process of "physical decay," when large spatial scales dominate. Such activity arises from the internal biological dynamics of the system; it provides a mechanism of adding variance to the system at large scales, before the system has had a chance to "run down." Bennett and Denman (1988) have already shown how phytoplankton patchiness can arise from the addition of an annual cycle to a simple two-dimensional turbulence model. The calculated spectral distribution could be more like the observed k^{-2}, not k^{-1}. A simple, one-dimensional simulation model with Lotka-Volterra dynamics and random injections of variance (Steele and Henderson 1977) captured some of this flavor, with calculated spectral shapes between k^{-2} and k^{-3}.

Calculations of the Bennett-Denman type with more complex biological dynamics are possible in principle, but would be extremely tedious. The models would have to allow several or all of the parameters (r, K, A, B, Γ, and d) to behave stochastically. [It should be easier to let only r, A, B, and d vary stochastically; equations (4) and (5) are linear in those quantities.] Another heuristic approach would be to relax some of the physical realism and parameterize the "turbulence" with constant, but unequal, Fickian diffusivities, D_1 and D_2. Two-dimensional reaction-diffusion equations like (6) and (7) would result:

$$\frac{\partial p}{\partial t} = D_1 \nabla^2 p + F(p,h) \tag{6}$$

$$\frac{\partial h}{\partial t} = D_2 \nabla^2 h + G(p,h). \tag{7}$$

The advantage of such a "mimic" of reality is tractability. It would be extremely interesting to see what Bennett-Denman-type results would be obtained for the parameter groups that lead to diffusive instability. Though the turbulence mechanism in the reaction-diffusion calculations differs substantially from the Lagrangian statistics approach, perhaps enough of the "turbulence" acts in Fickian fashion that some smeared, nonuniform "steady" states might emerge (at least transiently). This could be another source of large-scale variance.

The reaction-diffusion calculations also promise to help researchers develop some intuition about the much more difficult statistical endeavors, like those of Bennett and Denman. We know a great deal about one-dimensional reaction-diffusion systems, and investigation continues to be very intense. We know less about two-dimensional systems, but some signal studies are emerging (Mimura et al. 1988). If the pace of discovery for two-dimensional systems continues as rapidly as it has for one-dimensional systems, then we shall soon have a number of results to use for reference. Comparison studies between the reaction-diffusion formulations and statistical calculations (in the style of Bennett and Denman) could be extremely valuable.

Two observations are pertinent. First, equations (6) and (7) contain several parameters that must be estimated. Ecologists are often too ready to apologize for the lack of solid information about "constants" that appear in models. Compared to many other biological disciplines, ecologists can have more confidence about the magnitude of such numbers. This is not the case in developmental biology, for example, where coefficients enter models (quite useful, distinguished, and sophisticated models; see, e.g., Murray and Oster 1984) for which no information whatsoever exists. Estimates of all the constants in equations (6) and (7), certainly to within an order of magnitude, could be made for a phytoplankton-zooplankton predator-prey interaction. Even the earlier dynamical models in the sea (e.g., Steele 1974) tracked around actual systems (the North Sea) to a surprising degree.

Second, there is little field evidence that reaction-diffusion models explain the development of spatial patterns, particularly in plankton systems. There may be a reason for this. Virtually all of the field investigations consider only biomass, not the distribution of individual species (or functional groups of species, perhaps). The subtle distributional patterns of individual species relative to one another could be missed by the coarser measures of mere biomass (e.g., fluorescence in aquatic systems). There is some preliminary, unpublished support for this view. In a large lake—Lake Tahoe, bordering California and Nevada—Peter Richerson and I collected samples along 25-km transects and, with the aid of a small squad of phytoplankton counters, enumerated the samples down to the species. In one representative transect the numerically dominant diatom species, *Cyclotella stelligera* and *Fragilaria crotonensis,* showed little pattern beyond randomness—their spatial autocorrelations are well described as a Markov process with a very short fall-off distance. Conversely, several of the less numerous chlorophyte species show much more large-scale structure and have significant cross-correlations with other species at large-scale separations (see Figure 8-5). That the dominant species have little spatial pattern beyond that determined by the "physics" confirms our earlier analyses of chlorophyll in this large lake (Richerson et al. 1978; Abbott et al. 1982). We have no ready explanation as to why the rarer species, significant contributors to lacustrine planktonic diversity, do seem to show some large-scaled pattern, while the dominants do not. At least one model of phytoplankton competing for dissolved nutrients (Powell and Richerson 1985) predicts that the reaction-diffusion mechanism could lead to patterns at precisely these size scales in this large lake. These remarks underscore an important conclusion: the spatial scale of community composition may differ from that of biomass variability, a result not yet entirely understood (see also Figure 8-2A). A solely empirical approach to this crucial question would involve a monumental effort to count planktonic species from net and bottle samples, an extremely tedious and time-consuming task because of our lack of automatic counting devices. An understanding of biology at the level of interacting species is critical to ecology, and we need to understand how species partition space (if indeed they do). Careful theoretical guidance is in order before we can go sailing off to count literally millions of plankters, consuming millions of dollars.

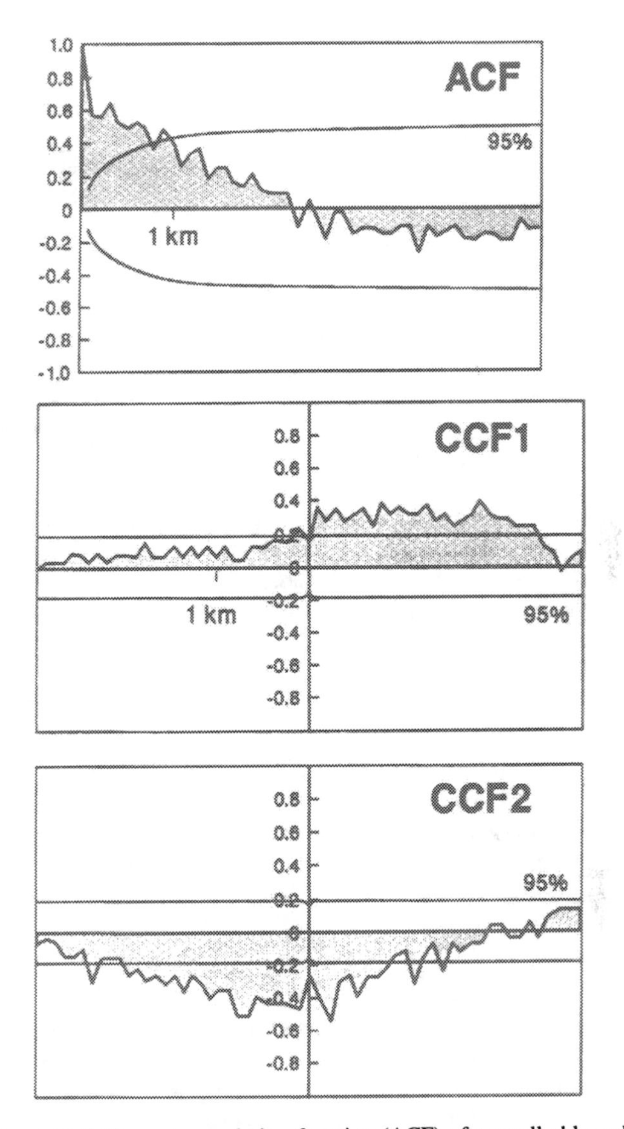

Figure 8-5. (ACF) The autocorrelation function (ACF) of a small chlorophyte (phytoplankton) species versus (lag) distance. Note the large-scale distribution—the ACF falls to e^{-1} of its zero-lag value at approximately 1 km. Similar e^{-1} lengths for the dominant species are no more than 100 m. 95% confidence limits are shown. (CCF1) The cross-correlation function (CCF) of the small chlorophyte with a less numerous diatom, *Fragilaria pinnata,* versus distance. Note the significant positive correlation of the two species abundances when displaced by 1–2 km (large scale). (CCF2) The cross-correlation function of the first chlorophyte with another relatively rare chlorophyte. Note the significant negative correlation between the two species abundances, again at large scale.

Smaller (< 100 m) and Larger (> 100 km) Scales

The nature of the patchy marine environment at very small scales (the scale of phytoplankton—100 m and smaller) has propelled a number of investigators to ask if phytoplankton can maintain themselves on the average level of, for example, nitrogen in the sea (McCarthy and Goldman 1979; Jackson 1980; Lehman and Scavia 1982). The question turns on the spatial and temporal scales of, first, physical variability in the ocean and, second, uptake and assimilation by algae. Everyone agrees that the average level of nitrogen is far too low to maintain the observed growth, and that there are bursts of high-nutrient water at sea. Are these bursts long-lived enough for the algae to take them up and utilize them for growth? The question remains unresolved, but at least one calculation (Jackson 1985) can account for all measurements, both in very oligotrophic waters, like the mid-ocean, and in very eutrophic environments, like the laboratory vessels where some of these experiments were carried out. Theory contributed to this endeavor by calling into question the initial explanations of McCarthy and Goldman, forcing them and others to seek new explanations for the observed phenomena. Jackson's initial calculations were authoritative enough to cause researchers, especially J. Goldman, to seek other small-scale concentrating mechanisms, focusing on the role played by bacterial communities on small particles. The rapid and stimulating interplay between theoretical calculation and field measurements is particularly noteworthy, because the effect of small-scale variance and the nature of its presumed transfer to larger scales have extraordinary consequences for the entire large-scale marine ecosystem.

Similar remarks can be made about the small-scale details of zooplankton feeding that depend critically on the small-scale, low Reynolds number hydrodynamics—"the physics" (Koehl and Strickler 1981). The details of this feeding mechanism have not yet been incorporated into models for zooplankton foraging (e.g., Lehman 1976; Lam and Frost 1976). Such a calculation would quantify how the small-scale behavior of individuals controls large-scale phenomena, because zooplankton grazers are believed to control phytoplankton standing crop in large parts of the sea and lakes. If this connection could be made, perhaps we would have a more solid basis on which to judge the alternative hypothesis that more control is exerted by previously unsuspected microzooplankton grazers, even bacteria, than by the traditional calanoid copepod zooplankton.

At very large scales—whole-ocean basin and global—one of the "linear, first-order" generalizations of Denman and Powell (1984) is seriously violated. In the coupled physical-biological systems, the dominant time scales are no longer controlled by biological processes. Interannual variability of weather patterns and even long climatic variations are prominent. Figures 8-6A and 8-6B from Clark (1985) and Steele (1984) present the space-time regimes of dominant processes on a diagram similar to the one in Figure 8-2. We have excellent evidence that those large-scale events exert considerable influence on ecological

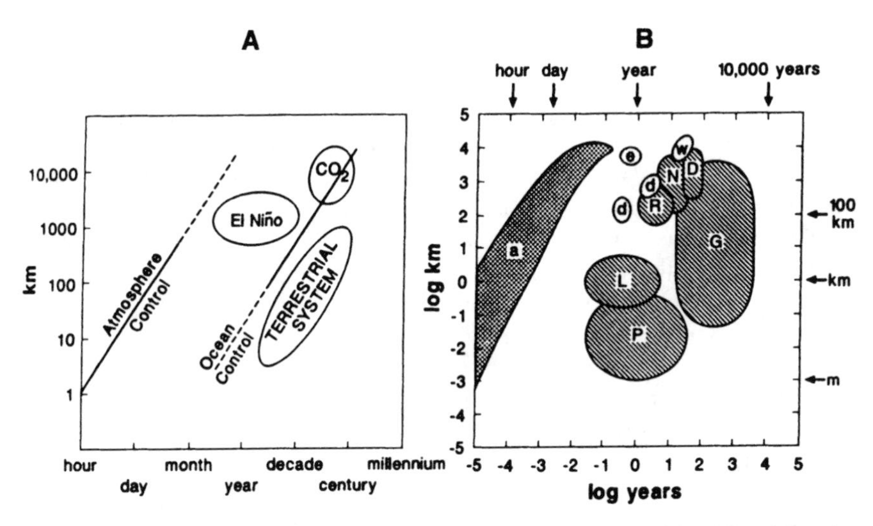

Figure 8-6. (A) Space-time regimes (see Figure 8-2B). Events with large spatial scales generally have associated long time scales. The atmosphere exerts control over time scales of weeks or less. The ocean exerts control over time scales of years or more. (After Steele 1984.) (B) Space-time regimes, continued, including man's impact. (a) atmospheric phenomena; (e) El Niño; (d) drought; (w) global warming trend; (P) ecological processes at the population level; (G) ecological processes over larger geographical extents, i.e., at the landscape or regional level; (L) local farming activities; (R) regional farming activities; (N) national industrial activity; (D) changing global patterns of political and demographic activities. (After Clark 1985.)

processes. The impact of the 1983 El Niño on biological processes in the ocean has been reviewed by Barber and Chavez (1983, 1986). We have increasing evidence that these large-scale, long-term events are also coupled to shorter-term, small-scale variability. Strub et al. (1985) documented the impact of the El Niño phenomenon over the previous 20 years on a small, subalpine lake in northern California; Scavia et al. (1986) showed the impact of interannual variability on processes in Lake Michigan. The mechanisms generating these longer-term events are poorly known, if at all, and uncovering the details of their coupling to smaller-scale, shorter-duration processes promises to be an area of intense activity in the decades ahead. Clearly, theoretical activity will be focused on long-duration, large-scale regimes because relevant measurement programs have such very long time horizons. We simply cannot afford to wait for the data to pour in to determine what the future may bring for our planet.

Oceanic-Terrestrial Coupling

Figure 8-6 shows how oceanic events affect terrestrial processes at the large scale and over the long term. This coupling is important at smaller scales and over shorter times as well. A geographic zone where one might anticipate the effects of such coupling to be prominent is at the land-sea interface—at subtidal and intertidal environments. The empirical situation is confused. Dayton and Tegner (1984) have documented the effects of El Niño events on subtidal kelp forests in southern California, but Paine (1986) sees new effects on the benthic community at Tatoosh Island, which he has studied for decades. Further, "intermediate disturbance" workers (Sousa 1984) have documented the importance of temporal and spatial scale in the intertidal zone. Additional coupled biological-physical models that expand on Jackson and Strathmann's (1981) effort would be a welcome adjunct to some much-needed data collection on the actual physics of "exposure," that is, the physical mechanisms of energy, momentum, and mass transfer in this complex environment. On a larger scale, the barnacle work of Roughgarden, Gaines, and collaborators (e.g., Roughgarden et al. 1987) has signaled the need for a reevaluation of the importance of transport in recruitment processes. Space and time-scale considerations play a critical role in their studies and speculations; moreover, simple models coupled to detailed experimental studies are a hallmark of this group's influential work. Their concerns are mirrored in the work of Botsford and collaborators on the Dungeness crab (*Cancer magister*); again, critical timing of transport of larval stages is thought to be controlled by wind-driven transport in the upper mixed layer along the entire west coast of North America (Johnson et al. 1986). These authors argue that cyclic variations in wind stress lead to the cyclical nature of this process (Johnson et al. 1986). An interesting aspect of Botsford's work is the application of a theoretical model to dismiss a competing hypothesis, namely, that the cycles can be explained solely by the density-dependence in the predator-prey dynamics within larval crab populations. These studies point to a coupling of events at different scales in addition to aquatic-terrestrial coupling.

Terrestrial Systems and Trophic Levels: Conclusion

It is now useful to note some aspects of terrestrial and marine systems as separate entities. For example, there are Steele's (1985) model studies, which have pointed out that the two systems should differ at the longest scales because of the very different climates each encounters. And we can compare the two different systems at shorter time and smaller space scales. Kareiva (1982, 1984, 1986) has made effective use of models that apply equally to plankton systems in his insect studies. A more formal comparison between models of dispersing, interacting populations of insects and dispersing, interacting planktonic populations also bears consideration. One important difference between the models concerns dispersal: insect dispersal depends on the population level (Okubo 1980); plankton are at the mercy of water movements, which results in less behavioral control over their "dispersal."

Interesting comparisons could also be made between some of the plankton patchiness work and the recent studies of Fahrig and collaborators (e.g., Fahrig 1988; Fahrig and Paloheimo 1987). In these, the arrangement of patches in the insect communities might be akin to the arrangement that the physical environment "performs" on plankton. Because dispersal distances scale both the insect and plankton problems, some comparison could be made at the same (and/or different) *dynamical* scales for both systems.

Perhaps even more intriguing are the recent supercomputer analyses of Levin and Buttel (1988). Following a patch model of competing intertidal organisms (Levin and Paine 1974), Levin and Buttel showed that the population variance *decreased* as size scale increased. This behavior is profoundly different from that seen in aquatic communities, where there is much more variance at larger scale (see Figures 8-1 and 8-4). One might expect that in intertidal communities, as one begins to encompass increasingly different habitats, the population variance must increase at some size scale. Is there a scale where a minimum occurs? The details of the transfer of variance between scales in the Levin-Buttel model may be instructive. Lake/pond habitats may also provide interesting comparisons. They are subject to climatic fluctuations (i.e., atmospheric time scales) to some degree, and yet they exhibit many of the aspects of variability seen in marine systems (Richerson et al. 1978).

I have avoided discussing coupling between trophic levels in deference to other authors (i.e., Roughgarden et al. 1989, see the footnote on the first page of this chapter), who have, no doubt, forgotten more about this topic than I may ever know. One subject, however, has attracted attention from freshwater ecologists. Following the ideas of Paine (1966), model investigations by Carpenter and Kitchell (1984) and later field investigations (Carpenter et al. 1987) have demonstrated that higher vertebrate predators can exert considerable control over primary productivity at the lowest autotrophic level in small lake systems—top-down control. Other model studies (Carpenter 1987) suggest that through the "top-down" mechanism significant variance will appear in primary productivity at the time scale associated with the generation time of the dominant predator—a reasonable, if unsupported, assertion. The top-down suggestion does jostle the conventional wisdom on this issue—the large body of knowledge that has demonstrated the control of primary production by environmental fluctuations (e.g., Harris 1986; Harris and Timbree 1986). Note, however, that the time scales of environmental fluctuation are surely linked to the time scales of predator abundance—perhaps through large-scale die-offs or migrations and introductions. Both "top-down" and environmental control must be at work at all lakes, though the degree to which each dominates will surely differ from lake to lake. The time scales of interest are long in this case, and further theoretical analyses are certain to sharpen and restrict the applicability of the top-down hypothesis before enough data will be available to decide the issue.

I conclude by reminding the reader of another kind of coupling—the productive coupling between theory and field measurements and experimentation in all of the areas I have surveyed. The importance of theoretical suggestion (and refutation) cannot be overestimated, especially as we increasingly focus on large-scale, long-term processes for which measurement programs may be so vast and time consuming that we can investigate some questions only theoretically. It is difficult to conceive of further progress on the questions I have raised without a substantial infusion of theoretical guidance.

Acknowledgments

Andrew Bennett, Ken Denman, John Steele, and especially Dave Mackas gave me extensive, thoughtful, and good-humored comments. My reliance on their work best expresses my gratitude. Bennett and Denman pointed out an error in the Gower et al. (1980) study: accordingly, one should compare spectral densities to a k^{-2} form, not k^{-3} as in the original article. John Steele made two figures available; he also composed a masterful and helpful summary (Chapter 12 in Roughgarden et al. 1989; see the footnote on the first page of this chapter) from a wide-ranging discussion in which the strong views of individual conference participants continually threatened to overwhelm the search for consensus. Lenore Fahrig made her work available in advance of publication, as did Bennett and Denman. Conversations with Peter Kareiva, Bob Paine, and Wayne Sousa were especially useful. During the 15 years that I have investigated these and closely related questions, the work of Si Levin, Akira Okubo, Trevor Platt, and John Steele has inspired and challenged me; it is a pleasure to acknowledge them here.

This work was supported by the National Science Foundation through grant OCE-8613749 to Project SUPER.

References

Abbott, M. R., T. M. Powell, and P. J. Richerson. 1982. The relations of environmental variability to the spatial patterns of phytoplankton in Lake Tahoe. *Journal of Plankton Research* **4:**927–41.

Barber, R. T., and F. P. Chavez. 1983. Biological consequences of El Niño. *Science* **222:**1203–10.

———. 1986. Ocean variability in relation to living resources during the 1982–1983 El Niño. *Nature* **319:**279–85.

Bennett, A. F., and K. L. Denman. 1985. Phytoplankton patchiness: Inferences from particle statistics. *Journal of Marine Research*. **43:**307–35.

———. 1988. Large-scale patchiness due to an annual cycle. *Transcripts of the American Geophysical Union (EOS)* **68:**1696. Abstract only, and unpublished manuscript.

Berg, H. 1983. *Random Motion in Biology*. Princeton University Press, Princeton, NJ.

Carpenter, S. R. 1987. Transmission of variance through lake food webs. Unpublished manuscript for workshop on complex interactions in lake communities. Department of Biological Sciences, University of Notre Dame (March).

Carpenter, S. R., and J. F. Kitchell. 1984. Plankton community structure and limnetic primary production. *American Naturalist* **124:**157–72.

Carpenter, S. R., J. F. Kitchell, J. R. Hodgson, P. A. Cochran, J. J. Elser, D. M. Lodge, D. Kretchmer, and X. He. 1987. Regulation of lake primary productivity by food-web structure. *Ecology* **68:**1863–76.

Chatfield, C. 1980. *Introduction to Multivariate Analysis*. Chapman & Hall, New York.

———. 1984. *The Analysis of Time Series*. Third edition. Chapman & Hall, New York.

Clark, W. C. 1985. Scales of climate impacts. *Climatic Change* **7:**5–27.

Dayton, P. K., and M. J. Tegner. 1984. Catastrophic storms, El Niño, and patch stability in a Southern California kelp forest. *Science* **224:**283–85.

Denman, K. L. 1983. "Predictability of the Marine Planktonic Ecosystems." In G. Holloway and B. J. West (eds.), *The Predictability of Fluid Motions*. American Institute of Physics, Conference Proceedings No. 106, New York, pp. 601–602.

Denman, K. L., and M. R. Abbott. 1988. Time evolution of surface chlorophyll patterns from cross-spectrum analysis of satellite color changes. [Unpublished manuscript.]

Denman, K. L., and D. L. Mackas. 1978. Collection and analysis of underway data. In J. H. Steele (ed.), *Spatial Pattern in Plankton Communities*. Plenum, New York, pp. 85–109.

Denman, K. L., and T. Platt. 1976. The variance spectrum of phytoplankton in a turbulent ocean. *Journal of Marine Research*. **34:**593–601.

Denman, K. L., and T. M. Powell. 1984. Effects of physical processes on planktonic ecosystems in the coastal ocean. *Oceanography and Marine Biology Annual Review* **22:**125–68.

Denman, K. L., A. Okubo, and T. Platt. 1977. The chlorophyll fluctuation spectrum in the sea. *Limnology and Oceanography* **22:**1033–38.

Fahrig, L. 1988. A general model of populations in patchy habitats. *Journal of Applied Mathematics* **27(1):**53–66.

Fahrig, L., and J. E. Paloheimo. 1987. Determinants of local population abundance in patchy habitats. Unpublished manuscript.

Gauch, H. G. 1982. *Multivariate Analysis in Community Ecology*. Cambridge University Press, London, England.

Gill, A. E. 1982. *Atmosphere-Ocean Dynamics*. Academic Press, New York.

Gower, J. F. R., K. L. Denman, and R. J. Holyer. 1980. Phytoplankton patchiness indicates the fluctuations spectrum of mesoscale oceanic structure. *Nature* **288:**157–59.

Harris, G. P. 1986. *Phytoplankton Ecology: Structure, Function, and Fluctuation*. Chapman & Hall, London.

Harris, G. P., and A. M. Trimbee. 1986. Phytoplankton population dynamics of a small reservoir: Physical/biological coupling and the time scales of community change. *Journal of Plankton Research* **8:**1011–25.

Jackson, G. A. 1980. Phytoplankton growth and zooplankton grazing in oligotrophic oceans. *Nature* **284:**439–41.

———. 1985. Pulses, aggregates, and diffusion: Life in the small. Forty-eighth annual meeting of the American Society for Limnology and Oceanography, Minneapolis. Abstract only.

Jackson, G. A., and R. R. Strathmann. 1981. Larval mortality from offshore mixing as a link between precompetent and competent periods of development. *American Naturalist* **118:**16–26.

Johnson, D. F., L. W. Botsford, R. D. Methot, Jr., and T. C. Wainwright. 1986. Wind stress and cycles in Dungness crab (*Cancer magister*) catch off California, Oregon, and Washington. *Canadian Journal of Fisheries and Aquatic Science* **43:**833–845.

Kareiva, P. 1982. Experimental and mathematical analysis of herbivore movement: Quantifying the influence of plant spacing and quality on foraging discrimination. *Ecological Monographs* **52:**261–82.

———. 1984. "Predator-prey Dynamics in Spatially Structured Populations: Manipulating Dispersal in a Coccinellid-aphid Interaction." In S. A. Levin and T. G. Hallam (eds.), *Mathematical Ecology*. Lecture Notes in Biomathematics, vol. 54. Springer-Verlag, Berlin.

———. 1986. "Patchiness, Dispersal and Species Interactions: Consequences for Communities of Herbivorous Insects." In J. Diamond and T. J. Case (eds.), *Community Ecology*. Harper and Row, New York, pp. 192–206.

Koehl, M. A. R., and J. R. Strickler. 1981. Copepod feeding currents: Food capture at low Reynolds number. *Limnology and Oceanography* **26:**1062–73.

Lam, R. K., and B. W. Frost. 1976. Model of copepod filtering response to changes in size and concentration of food. *Limnology and Oceanography* **21:**490–500.

Lehman, J. T. 1976. The filter feeder as an optimal forager, and the predicted shapes of feeding curves. *Limnology and Oceanography* **21:**501–16.

Lehman, J. T., and D. Scavia. 1982. Microscale patchiness of nutrients in plankton communities. *Science* **216:**729–30.

Levin, S. A., and L. Buttel. 1988. Measures of patchiness in ecological systems. (Unpublished manuscript.)

Levin, S. A., and R. T. Paine. 1974. Disturbance, patch formation, and community structure. *Proceedings of the National Academy of Sciences USA* **71:**2744–47.

Mackas, D. L. 1984. Spatial autocorrelation of plankton community composition in a continental shelf ecosystem. *Limnology and Oceanography* **29:**451–71.

Mackas, D. L., K. L. Denman, and M. R. Abbott. 1985. Plankton patchiness: Biology in the physical vernacular. *Bulletin of Marine Science* **37:**652–74.

McCarthy, J. J., and J. C. Goldman. 1979. Nitrogenous nutrition of marine phytoplankton in nutrient-depleted waters. *Science* **203:**670–72.

Mimura, M., Y. Kan-on, and Y. Nichiura. 1988. "Oscillations in Segregation of Competing Populations." In T. G. Hallam, L. J. Gross, and S. A. Levin (eds.), *1986 Proceedings of the Trieste Research Conference on Mathematical Ecology*. World Scientific Publishing, Singapore, pp. 711–17.

Murray, J. D., and G. F. Oster. 1984. Generation of biological pattern and form. *IMA Journal of Mathematics Applied in Medicine and Biology* **1**:51–75.

Okubo, A. 1980. *Diffusion and Ecological Models*. Springer-Verlag, New York.

———. 1984. "Critical Patch Size for Plankton and Patchiness." In S. A. Levin and T. G. Hallam, eds., *Mathematical Ecology*. Lecture Notes in Biomathematics, vol. 54. Springer-Verlag, Berlin, pp. 456–77.

Paine, R. T. 1966. Food web complexity and species diversity. *American Naturalist* **100**:65–75.

———. 1986. Benthic community—water column coupling during the 1982–83 El Niño: Are community changes at high latitudes attributable to cause or coincidence? *Limnology and Oceanography* **31**:351–60.

Pedlosky, J. 1979. *Geophysical Fluid Dynamics*. Springer-Verlag, New York.

Platt, T. 1978. "Spectral Analysis of Spatial Structure in Phytoplankton Populations." In J. H. Steele (ed.), *Spatial Patterns in Plankton Communities*. Plenum Press, New York, pp. 73–84.

Platt, T., K. H. Mann, and R. E. Ulanowicz (eds.) 1981. *Mathematical Models in Biological Oceanography*. UNESCO Press, France.

Powell, T., and P. J. Richerson. 1985. Temporal variation, spatial heterogeneity, and competition for resources in plankton systems: A theoretical model. *American Naturalist* **125**:431–64.

Purcell, E. M. 1977. Life at low Reynolds numbers. *American Journal of Physics* **45**:3–11.

Richerson, P. J., T. M. Powell, M. R. Leigh-Abbott, and J. A. Coil. 1978. "Spatial Heterogeneity in Closed Basins." In J. H. Steele (ed), *Spatial Patterns in Plankton Communities*. Plenum Press, New York, pp. 239–76.

Roughgarden, J. 1979. *Theory of Population Genetics and Evolutionary Ecology: An Introduction*. Macmillan, New York.

Roughgarden, J., S. Gaines, and S. Pacala. 1987. "Supply Side Ecology: The Role of Physical Transport Processes." In P. Giller and J. Gee (eds.), *Organization of Communities: Past and Present*. Proceedings of the British Ecological Society Symposium, Aberystwyth, Wales, April 1986. Blackwell Scientific, London.

Scavia, D., G. Fahenstiel, M. Evans, D. Jude, and J. Lehman. 1986. Influence of salmonid predation and weather on long-term water quality in Lake Michigan. *Canadian Journal of Fisheries and Aquatic Science* **43**:435–43.

Sheldon, R., A. Prakash, and W. Sutcliffe. 1972. The size distribution of particles in the ocean. *Limnology and Oceanography* **17**:327–40.

Shugart, H. H. (ed.), 1978. *Time Series and Ecological Processes*. *SIAM*, Philadelphia.

Sousa, W. P. 1984. The role of disturbance in natural communities. *Annual Review of Ecological Systems* **15**:353–91.

Steele, J. H. 1974. *The Structure of Marine Ecosystems*. Harvard University Press, Cambridge, MA.

———. 1978. "Some Comments on Plankton Patches." In J. H. Steele (ed.), *Spatial Patterns in Plankton Communities*. Plenum Press, New York, pp. 1–20.

———. 1984. Long-range plans. Woods Hole Oceanographic Institution, 1984 Annual Meeting, Woods Hole, MA.

———. 1985. A comparison of terrestrial and marine ecological systems. *Nature* **313**:355–58.

Steele, J. H., and E. W. Henderson. 1977. "Plankton Patches in the Northern North Sea." In J. H. Steele (ed.), *Fisheries Mathematics*. Academic Press, New York, pp. 1–19.

Strub, P. T., T. M. Powell, and C. R. Goldman. 1985. Climatic forcing: Effects of El Niño on a small, temperate lake. *Science* **227**:55–57.

Ulanowicz, R. E., and T. Platt (eds.). 1985. Ecosystem theory for biological oceanography. *Canadian Journal of Fisheries and Aquatic Science* **213**:260.

Vogel, S. 1981. *Life in Moving Fluids*. Princeton University Press, Princeton, NJ.

Williamson, M. 1972. The relation of principal component analysis to the analysis of variance. *International Journal of Mathematical Education in Science and Technology* **3**:35–42.

———. 1975. The biological interpretation of time series analysis. *Bulletin of the Institute of Mathematics and its Applications*. **11**:67–69.

9

Year-to-Year Fluctuation of the Spring Phytoplankton Bloom in South San Francisco Bay: An Example of Ecological Variability at the Land-Sea Interface

James E. Cloern and Alan D. Jassby

Estuaries are transitional ecosystems at the interface of the terrestrial and marine realms. Their unique physiographic position gives rise to large spatial variability, and to dynamic temporal variability resulting, in part, from a variety of forces and fluxes at the oceanic and terrestrial boundaries. River flow, in particular, is an important mechanism for delivering watershed-derived materials such as fresh water, sediments, and nutrients; each of these quantities in turn directly influences the physical structure and biological communities of estuaries. With this setting in mind, we consider here the general proposition that estuarine variability at the yearly time scale can be caused by annual fluctuations in river flow. We use a "long-term" (15-year) time series of phytoplankton biomass variability in South San Francisco Bay (SSFB), a lagoon-type estuary in which phytoplankton primary production is the largest source of organic carbon (Jassby et al. 1993).

Previous analysis has suggested that much of the interannual variability in phytoplankton dynamics of SSFB might result from fluctuations in river discharge (Cloern 1991a). This hypothesis was supported by a correlation between the magnitude of the spring phytoplankton bloom and antecedent river flow observed for the years 1980–1987. Our intent here is to explore this suggested hydrologic-biological connection further, with two specific objectives in mind. First, we reconsider the hypothesis that year-to-year fluctuations in phytoplankton biomass are associated with annual fluctuations in river flow, using an extended dataset that includes measurements throughout the period 1978–1992. Second, we illustrate the use of principal component analysis as a technique to identify patterns of spatial-temporal variability in ecological datasets—in this case, time series of observations at fixed sampling locations positioned along a longitudinal transect.

Description of the Dataset

Near-surface chlorophyll *a* concentration and salinity have been mapped along the 40-km axial transect of SSFB since 1978. Sampling frequency has varied among years and seasons, but has usually included at least one profile per month, except for an eight-month hiatus in 1981. Chlorophyll *a* was measured at fixed locations, spaced approximately 3 km apart (Figure 9-1), with a fluorometer that was calibrated each date with discrete measures of chlorophyll *a* concentration. Salinity was measured either with a flow-cell induction salinometer or a CTD. Sampling methods are explained in detail in annual data reports (e.g., Wienke et al. 1992).

During the period January 1978 through June 1992, 4838 measurements of chlorophyll fluorescence and salinity were made along the SSFB transect. Chlorophyll *a* concentration ranged from about 1 to 70 mg/m^3 and reflected the recurrent annual cycle of high phytoplankton biomass during the spring months and generally low biomass during the other seasons (Figure 9-2). Although the SSFB spring bloom has been the subject of considerable research interest (e.g., Cloern 1982; Powell et al. 1986; Cloern 1991a,b), we have not characterized well the year-to-year fluctuation in the spatial pattern or the temporal evolution of phytoplankton biomass during spring. For example, Figure 9-2 shows that the timing, duration, and spatial extent of the spring bloom all vary from year to

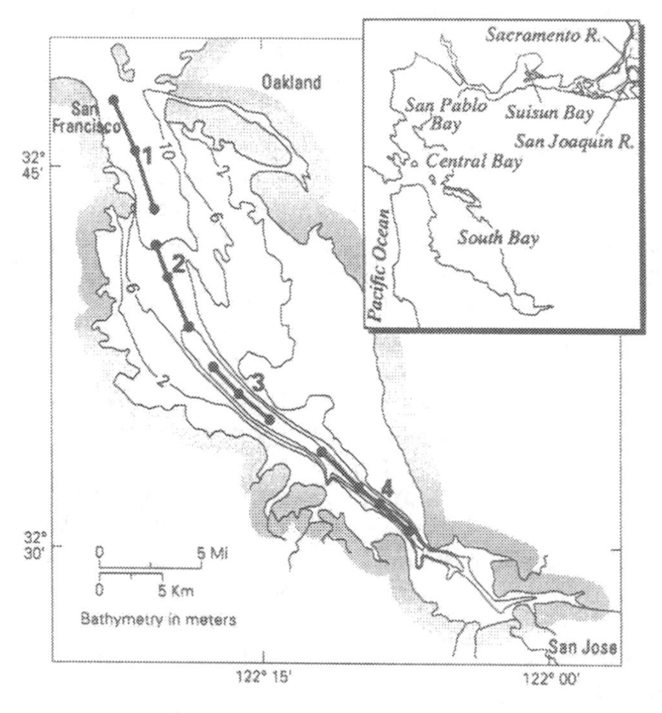

Figure 9-1. Map of South San Francisco Bay showing sampling locations (filled circles) for chlorophyll *a* and salinity along the channel. Measurements were grouped into four "stations" between the seaward estuary (station 1) and landward estuary (station 4).

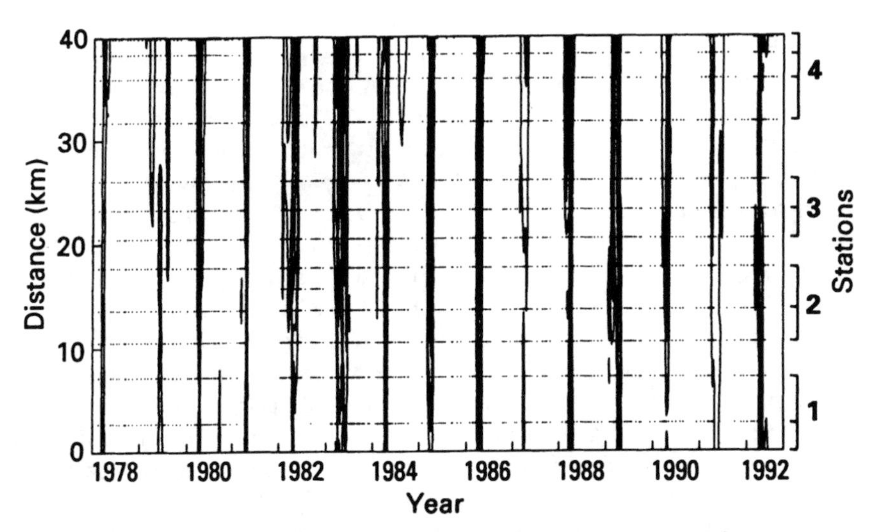

Figure 9-2. Contour plot of near-surface chlorophyll *a*, showing spatial-temporal fluctuations of phytoplankton biomass along the SSFB transect from 1978 to 1992. Locations of the 4838 measurements are shown with the grid of points. The contour interval is every 5 mg/m^3 chlorophyll *a*. Resolution of each contour line is not possible at this scale, but the figure illustrates the complex patterns of phytoplankton patchiness in space and time, including year-to-year fluctuation of spring bloom dynamics.

year. Episodes of high biomass began as early as February in some years and persisted well into May during other years; in some years the spring bloom was most intense at the landward reach of the estuary, whereas during other years it appeared to be a spatially coherent event. Given this patchy distribution of phytoplankton biomass in time and space, are there dominant patterns of variability that can be extracted from this detailed set of observations; and, if so, are these related to fluctuations in river discharge, as suggested previously?

Principal Component Analysis

Jassby and Powell (1990) present some guidelines for the application of principal component analysis (PCA) as a technique to reduce the dimensionality of ecological data sets comprising time series measurements across a spatial grid. We follow these guidelines here and pattern our analysis after the specific application of PCA to identify underlying mechanisms of interannual variability in the phytoplankton productivity of Castle Lake (Jassby et al. 1990) and Lake Tahoe (Jassby et al. 1992). PCA is a multivariate technique for exploratory analysis that can be used to simplify the description of a set of interrelated variables. The method yields a transformed set of new variables (principal components) that are linear combinations of the original variables. Under ideal conditions, most of the variability within the original dataset is captured by only a few principal components, each of which is constrained to be orthogonal, so that each represents a unique mode of variability. One particularly useful result of PCA is a scalar time series of "amplitudes" (or "scores") that measure the time-varying strength of each mode. We will use this result of PCA to explore year-to-year fluctuations in phytoplankton dynamics of SSFB, emphasizing variability during the spring bloom. A thorough description of PCA can be found in texts such as Jolliffe (1986) or Chatfield and Collins (1980). Afifi and Clark (1990) give an elementary and concise description of PCA with examples and some practical advice.

Preliminary steps are required to organize the original measurements into a data matrix appropriate for PCA. As our interest here is interannual variability of the spring bloom, we included measurements only from the first six months of each year. We first aggregated the individual chlorophyll measurements (Figure 9-2) into four "stations" and three "seasons." The station averages were taken as the arithmetic mean chlorophyll *a* concentration measured along subtransects (Figure 9-1) for each date. This step was done (1) to smooth the short-term variability associated with tidal advection during the course of sampling (Powell et al. 1989; Cloern et al. 1989), and (2) to separate the data into the four spatial regimes that have been identified previously from analysis of chlorophyll patchiness along the SSFB transect (Powell et al. 1986). The station averages for each date were then binned into "season" averages for each year, where the seasons were prescribed as 60-day periods corresponding approximately to January–February (JF = days 1–60), March–April (MA = days 61–120), and May–June (MJ = days 121–180). Seasonal averages were calculated by trapezoidal quadrature of the measurements made within each 60-day period.

These preliminary steps yielded a 15×12 matrix of mean chlorophyll *a* concentrations, where the 15 rows ("cases") correspond to individual years from 1978 to 1992, and the columns ("variables") correspond to the 12 station-seasons (e.g., JF1 = mean chlorophyll *a* concentration at station 1 for the first 60 days of each year; MA2 = mean chlorophyll *a* concentration at station 2 for days 61–120). This scheme for binning the original data is appropriate for pursuing questions about interannual fluctuation of the coarse patterns of seasonal and spatial variability around the spring bloom ("coarse," here, means spatial variability at the scale of about 10 km and temporal variability at the scale of months). Following Jassby et al. (1990), we performed PCA on the covariance matrix, and we present results of rotated solutions after varimax rotation of the significant principal components (PCs).

Results of PCA

The first principal component (PC1) extracted by PCA is that linear combination of the original variables having the largest possible variance; the second PC is the unique linear combination of original variables that is both orthogonal to PC1 and has the second-highest variance, etc. The variance associated with each PC is simply the corresponding eigenvalue of the covariance matrix, and the first step of PCA interpretation is to examine the proportion of the total variance in the original dataset that is attributable to each unrotated PC. Figure 9-3 shows this result for the 15 × 12 matrix of yearly variability in mean chlorophyll *a* concentration of SSFB binned by station-season. The first PC accounts for 53% of the total variance, PC2 accounts for an additional 34% of the variance, and PC3 accounts for 7% of the total variance. Several different criteria have been proposed to select the number of PCs to retain for rotation and further analysis. We used the "Rule N" criterion (Priesendorfer 1988), which indicates that the first two PCs are significant; together they account for 87% of the total variance.

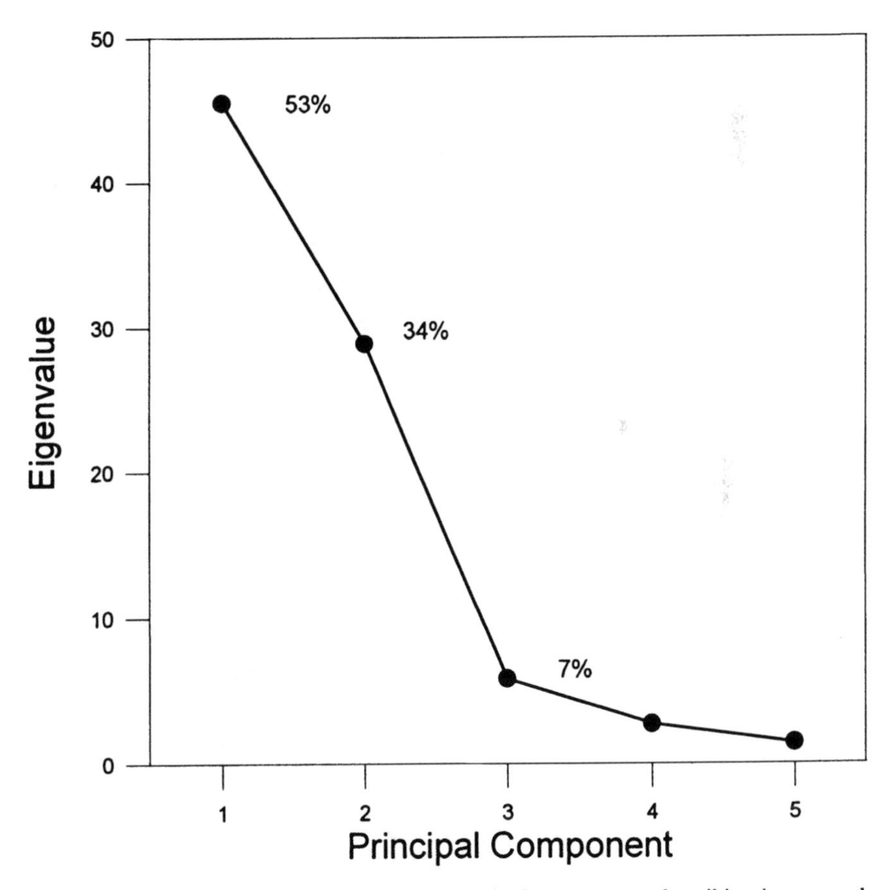

Figure 9-3. Eigenvalues for the first five principal components describing interannual variability of chlorophyll *a* in SSFB. Percentages show the proportion of the total variance that is accounted for by each of the first three PCs.

Associated with each of the original variables are "component coefficients" or "loadings," which measure the covariance between that variable and each PC; these are useful for interpreting the modes of variability associated with each PC. Because we performed PCA on the covariance matrix, the mean value for each bin is irrelevant, and these component coefficients represent the underlying pattern of *anomalies* in the distribution of chlorophyll. Following Jassby et al. (1990), we used "eigenmaps" to display the spatial-temporal patterns of variability among the coefficients. For PC1 (Figure 9-4A), the coefficients were largest for March–April, near-zero for January–February, and small for May–June. In addition, there was a spatial pattern to the distribution of PC1 coefficients, which progressively increased with station number (i.e., largest coefficients at the landward stations). The first mode therefore reflects some mechanism that has

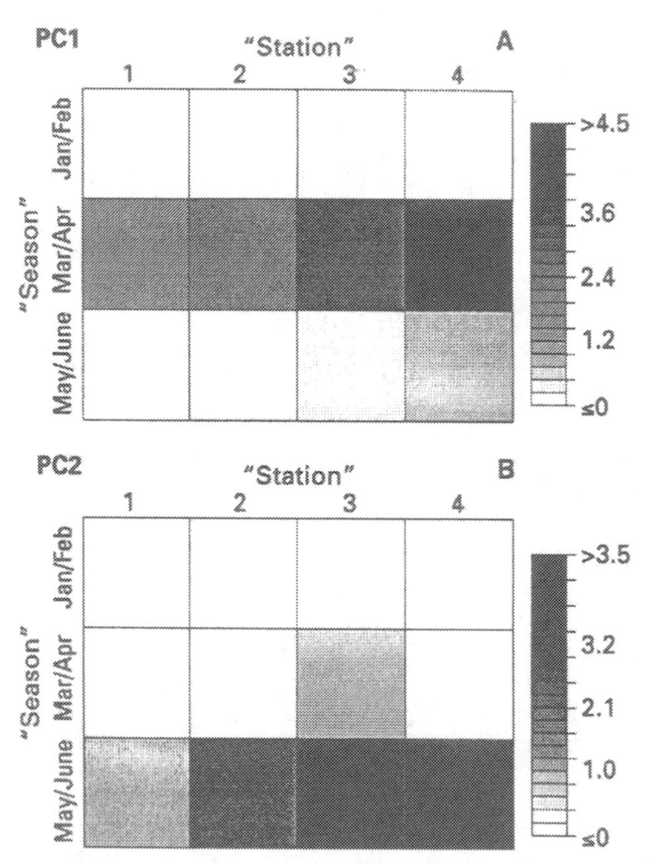

Figure 9-4. Eigenmaps for the first two PCs of chlorophyll variability in SSFB. These can be interpreted as contour plots when shading intensity is proportional to the PC coefficient magnitude for each station-season.

its greatest impact in March–April and toward the landward portion of the estuary. Coefficients of PC2 were largest for May–June, and these also showed a spatial pattern with strongest expression at stations 3 and 4 (Figure 9-4B). This second mode of variability could, therefore, reflect a mechanism associated with prolonging the spring bloom beyond April, especially in the landward estuary. Because principal components are orthogonal, these May–June anomalies (PC2) are uncoupled, to some extent, from the biomass anomalies observed during the bloom peaks in March–April (PC1).

Interannual Variability of the Principal Components

If the dominant patterns of chlorophyll distribution in SSFB can largely be interpreted in the context of two modes of variability, then how do these modes vary over time? Principal component analysis yields a quantity ("amplitude" or "score") that measures the relative strength of each PC for each observation—in this case, an amplitude for each year. Time series of PC amplitudes can thus be used as a simple representation of the year-to-year fluctuation in the strength of each mode of chlorophyll variability in SSFB. Figure 9-5 shows the amplitude time series for PC1 and PC2 after rotation. The first, dominant mode (magnitude of the March–April chlorophyll anomaly) had its largest amplitude in 1986; relatively large PC1 amplitudes were also calculated for 1983, 1988, and 1990 (Figure 9-5A). Each of these years was characterized by unusually high chlorophyll concentrations (i.e., positive anomaly), especially in the landward estuary, during March–April. Conversely, the years with small PC1 amplitudes (negative anomalies), such as 1978, 1979, 1981, 1987, and 1991, were characterized by small March–April blooms (see Figure 9-2). Years such as 1980 and 1982 had PC1 amplitudes near zero, indicating that chlorophyll distributions during March–April of those years were close to the 15-year mean.

The second mode, accounting for 34% of the total variance, had a very different time series (Figure 9-5B). The PC2 amplitude was largest for 1983, and this reflects the exceptionally high chlorophyll concentrations observed during May–June of that year (see Figure 9-2). Hence, the yearly amplitude for PC2 shows the extent to which the spring bloom persisted beyond April; note that 1982 and 1991 were also years of relatively high chlorophyll concentration during May–June. Conversely, years such as 1985 and 1988 had unusually low chlorophyll concentrations (negative PC2 amplitudes) during May–June.

Amplitude time series of principal components can be used to explore hypotheses about the mechanisms underlying the variability (see Jassby et al. 1990, 1992, for examples). If Cloern's (1991a) hypothesized connection between annual phytoplankton dynamics and river flow is true, then we would expect a correlation between biomass anomalies, expressed as PC amplitudes, and some measures of annual river flow. The hydrology of SSFB is complex because this estuary

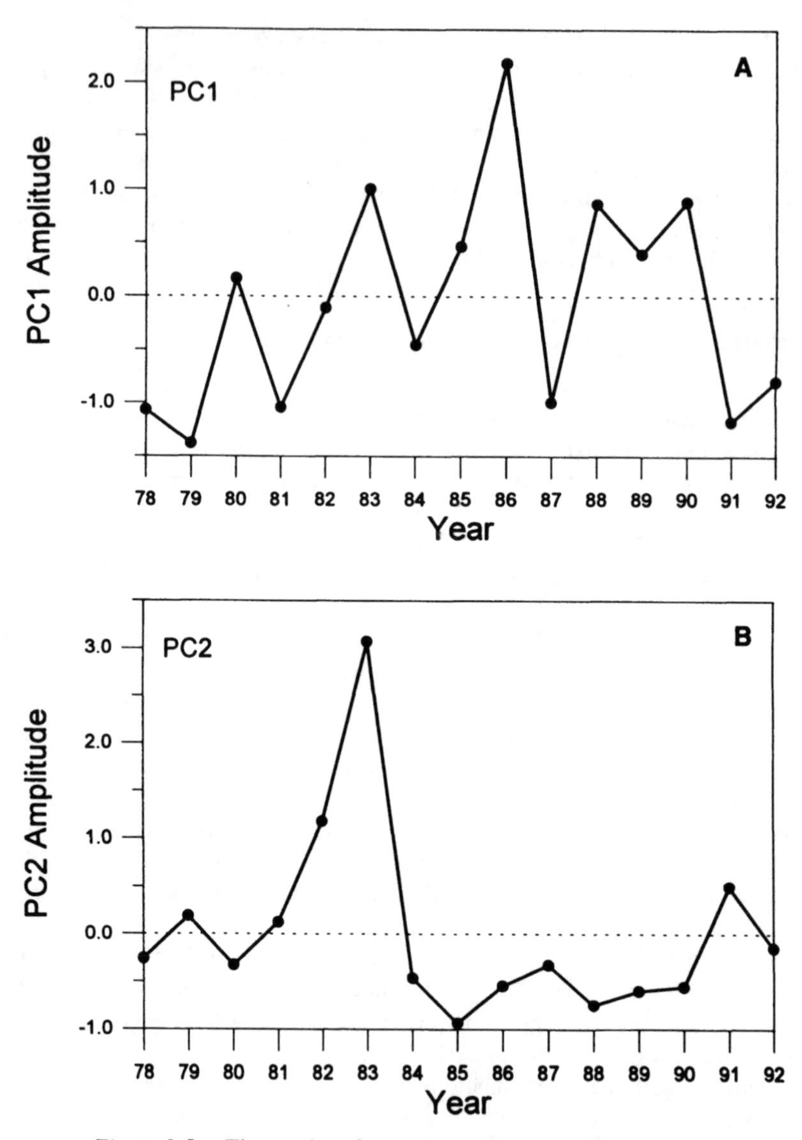

Figure 9-5. Time series of annual amplitudes for PC1 and PC2.

receives runoff from local streams that discharge into the landward estuary, plus runoff originating in the Sacramento-San Joaquin river basins that is delivered to the seaward estuary (see Figure 9-1). Therefore, we used seasonal-mean surface salinity for stations 3 and 4 (landward estuary, influenced by both sources of fresh water) as an index of antecedent river discharge, and then plotted PC amplitudes against this flow index for each year (Figure 9-6). For PC1, which

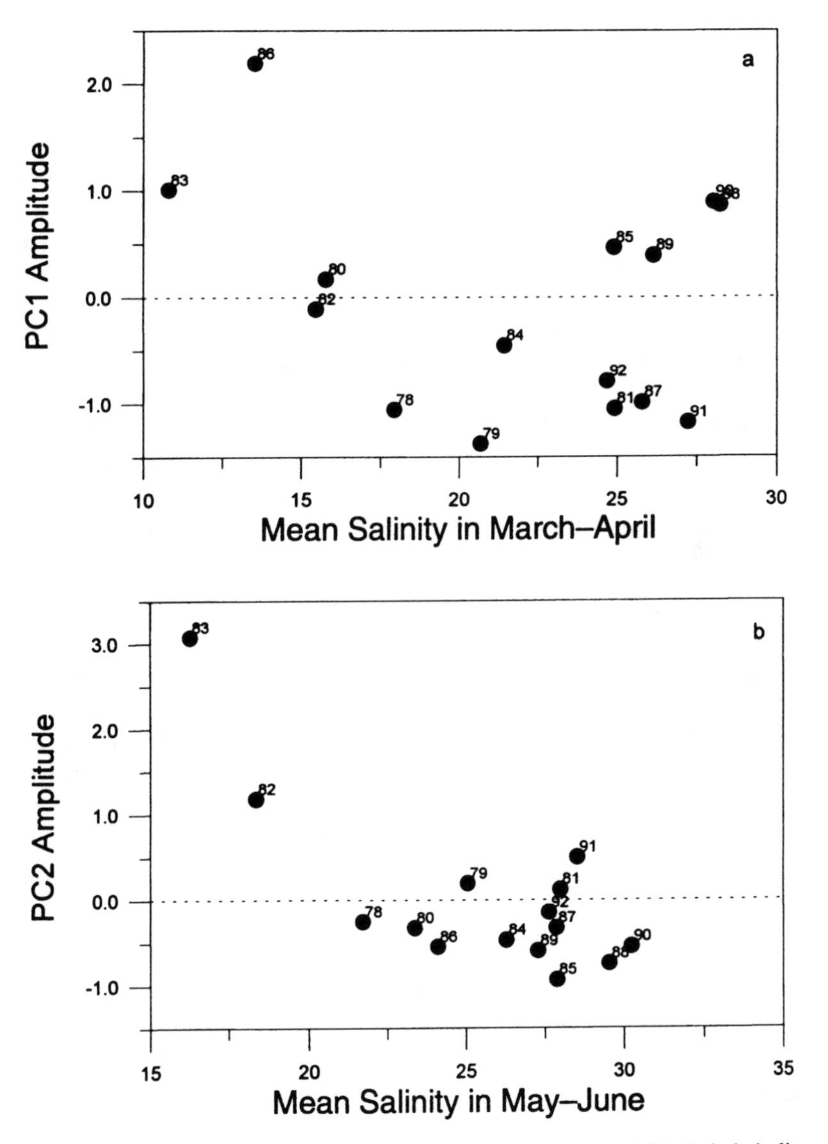

Figure 9-6. Amplitude versus seasonal-mean salinity for PC1 and PC2. Labels indicate the year of each observation.

accounts for 53% of the station-seasons chlorophyll variance, there was no correlation between yearly PC amplitude and the corresponding March–April surface salinity (Figure 9-6A). However, the two years with largest PC1 amplitudes (1986 and 1983) were the two years with the lowest mean surface salinity—i.e., highest antecedent river flow. These years with exceptionally large March–

April blooms and large PC1 amplitudes were also years of exceptionally high winter-spring runoff. Similarly, for the second principal component, there was no clear relation between PC2 amplitude and the corresponding (May–June) mean surface salinity (Figure 9-6B). However, the only two years with PC2 amplitude greater than one (1983, 1982) were, again, the two years having the largest seasonal salinity (flow) anomalies.

These results, based on PCA to quantify annual fluctuations in the dominant modes of chlorophyll variability, are not strictly compatible with the original hypothesis of Cloern (1991a). However, they do suggest an alternative and related hypothesis. The years with largest PC1 and PC2 amplitudes, 1983, 1986, and 1982, were years of extreme climate and hydrology. For example, 1983, a strong ENSO year, was the wettest year of this century in northern California; mean January–June river discharge into the upper San Francisco Bay was 3900 m^3/s, more than three times the long-term (1922–1992) average for these months. Mean surface salinity at the entrance to San Francisco Bay was 13.3 during March of 1983, and this was the lowest recorded monthly mean salinity since measurements began in 1992 (Peterson et al. 1989). The second-largest monthly salinity anomaly occurred there in March 1986, following a series of intense February—March storms, when mean discharge to the upper estuary exceeded 5000 m^3/s. The coherence between these hydrologic anomalies and the chlorophyll anomalies revealed by PCA suggests the following revised hypothesis:

(1) Under a broad range of hydrologic conditions, yearly fluctuations in phytoplankton bloom dynamics of SSFB are driven by mechanisms other than fluctuating river flow.

(2) However, during years of exceptionally high runoff and river flow, spring bloom magnitude (e.g., 1986) and duration (e.g., 1983) can be greatly amplified.

Our new analysis does support a connection between river flow and interannual phytoplankton dynamics, but through relationships that appear to be strongly nonlinear and sensitive only to extreme hydrologic conditions.

Acknowledgments

This work was supported by the U.S. Geological Survey and the U.S. Environmental Protection Agency (grant CE-009604-01) through the Center for Ecological Health Research at the University of California, Davis.

References

Afifi, A. A., and V. Clark. 1990. *Computer-Aided Multivariate Analysis*. Von Nostrand Reinhold, New York.

Chatfield, C., and A. J. Collins. 1980. *Introduction to Multivariate Analysis*. Chapman & Hall, New York.

Cloern, J. E. 1982. Does the benthos control phytoplankton biomass in South San Francisco Bay (USA)? *Marine Ecology Progress Series* **9:**191–202.

———. 1991a. "Annual Variations in River Flow and Primary Production in the South San Francisco Bay Estuary." In M. Elliott and D. Ducrotoy (eds.), *Estuaries and Coasts: Spatial and Temporal Intercomparisons*. Olsen and Olsen, Denmark.

———. 1991b. Tidal stirring and phytoplankton bloom dynamics in an estuary. *Journal of Marine Research* **49:**203–221.

Cloern, J. E., T. M. Powell, and L. M. Huzzey. 1989. Spatial and temporal variability in South San Francisco Bay. II. Temporal changes in salinity, suspended sediments, and phytoplankton biomass and productivity over tidal time scales. *Estuarine, Coastal and Shelf Science* **9:**599–619.

Jassby, A. D., J. E. Cloern, and T. M. Powell. 1993. Organic carbon sources and sinks in San Francisco Bay: Variability induced by river flow. *Marine Ecology Progress Series* **95:**39–54.

Jassby, A. D., C. R. Goldman, and T. M. Powell. 1992. Trend, seasonality, cycle, and irregular fluctuations in primary productivity at Lake Tahoe, California-Nevada, USA. *Hydrobiologia* **246:**195–203.

Jassby, A. D., and T. M. Powell. 1990. Detecting changes in ecological time series. *Ecology* **71:**2044–2052.

Jassby, A. D., T. M. Powell, and C. R. Goldman. 1990. Interannual fluctuations in primary production: Direct physical effects and the trophic cascade at Castle Lake, California. *Limnology and Oceanography* **35:**1021–1038.

Jolliffe, I. T. 1986. *Principal Component Analysis*. Springer-Verlag, New York.

Peterson, D. H., D. R. Cayan, J. F. Festa, F. H. Nichols, R. A. Walters, J. V. Slack, S. W. Hager, and L. E. Schemel. 1989. "Climate Variability in an Estuary: Effects of Riverflow on San Francisco Bay." In D. H. Peterson (ed.), *Aspects of Climate Variability in the Pacific and the Western Americas*. American Geophysical Union, Geophysical Monograph 55, Washington, D.C.

Powell, T. M., J. E. Cloern, and L. M. Huzzey. 1989. Spatial and temporal variability in South San Francisco Bay. I. Horizontal distributions of salinity, suspended sediments and phytoplankton biomass and productivity. *Estuarine, Coastal and Shelf Science* **28:**583–597.

Powell, T. M., J. E. Cloern, and R. A. Walters. 1986. "Phytoplankton Spatial Distribution in South San Francisco Bay—Mesocale and Small-scale Variability." In D. A. Wolfe (ed.), *Estuarine Variability*. Academic Press, New York.

Priesendorfer, R. W. 1988. Principal component analysis in meteorology and oceanography. *Developments in Atmospheric Science* 17. Elsevier, New York.

Wienke, S. M., B. E. Cole, J. E. Cloern, and A. E. Alpine. 1992. Plankton studies in San Francisco Bay. XIII. Chlorophyll distributions and hydrographic properties in San Francisco Bay, 1991. U.S. Geological Survey Open-File Report 92–158.

10

Scales of Variability in a Stable Environment: Phytoplankton in the Central North Pacific

Elizabeth L. Venrick

Since 1968, scientists from Scripps Institution of Oceanography (SIO) have been studying the structure and dynamics of the epipelagic ecosystem of the central North Pacific Ocean. This Central Pacific Study has its roots in the California Cooperative Oceanic Fisheries Investigations (CalCOFI), an interagency study originally established to understand the disappearance of the sardines from the California Current. Since 1949, CalCOFI has studied the physics, chemistry, and biology of this ecosystem through a series of regular cruises. The California Current is the eastern limb of the huge anticyclonic central North Pacific gyre (Figure 10-1). Current waters are derived from four distinct oceanographic environments and are supplemented locally by upwelling. Each water source has its own physical and chemical signature and its own complement of plankton species (Figure 10-2). The flow of the current is characterized by strong meanders, narrow eddies, and regions of rapid flow, squirts, and jets, embedded in a generally southward flow (Figure 10-3). These complex flow patterns mix and stir together disparate species assemblages, which, in turn, are continually responding to the changes in ambient conditions. Thus, ecosystem structure at any place or time is largely determined by the advective history of the water parcel. After nearly 20 years of studies in the California Current, it became obvious to CalCOFI researchers that the unpredictability of this ecosystem makes it an inefficient, even impossible environment in which to address some basic questions about communities—about diversity and stability in epipelagic ecosystems and the relative importance of physical and biological processes in shaping them. To address these types of questions in a more favorable environment, a group of SIO researchers, under the leadership of Professor John A. McGowan, began a study of the epipelagic ecosystem near the axis of the central North Pacific gyre.

The central North Pacific environment was selected for study for several reasons. It is a large, semi-closed circulation system that extends across the

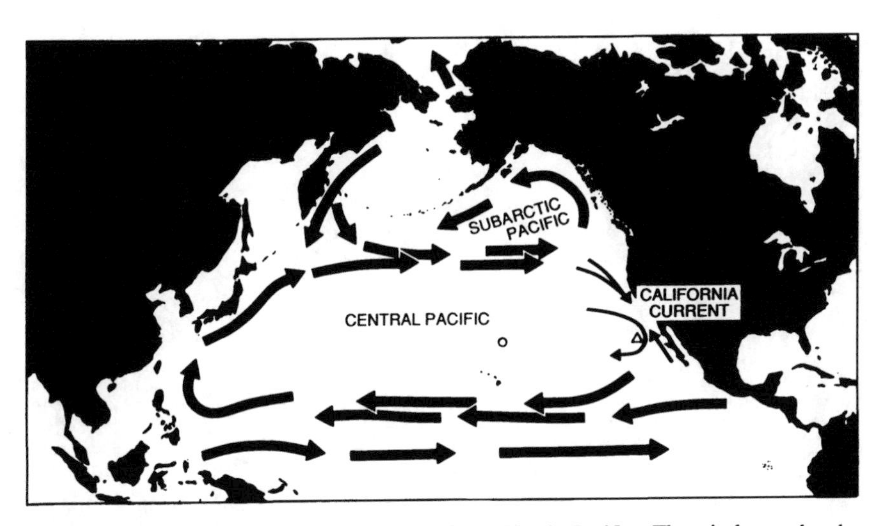

Figure 10-1. Major circulation patterns of the North Pacific. The circle marks the location of the Climax station, near the axis of the central North Pacific gyre. The triangle marks the location of the Edge station, near the eastern edge of the gyre.

Pacific approximately between latitudes 15° N and 40° N. The ecosystem is very old; present boundaries have persisted since the Pliocene or earlier (McGowan and Walker 1985). The environment is internally homogeneous. Zooplankton species found in the central North Pacific tend to occur throughout the gyre, with major species shifts restricted to the periphery (Fager and McGowan 1963; McGowan 1971, 1974). In short, the central North Pacific environment has the characteristics of great age, isolation, and stability that are expected to facilitate the development of a "climax" community. The specific study site (dubbed the "Climax" station) is located at 28° N, 155° W (Figure 10-1), far enough from the Hawaiian Islands to be removed from local effects of shoaling bottom topography and anthropogenic influences.

The initial focus of the Central Pacific Study was the small scale and seasonal variability of the epipelagic environment and component species. The plan was a series of quarterly cruises over a two-and-one-half-year period, with intensive spatial and temporal replication within each cruise. Owing to a variety of circumstances, some favorable, most not, the program has continued through nearly 20 years without ever achieving seasonal coverage within a single year. Many of the samples were collected on ships of opportunity, and within-cruise replication is occasionally limited. Nevertheless, enough data have been accumulated to identify patterns of variability on several spatial and temporal scales.

This chapter reviews the phytoplankton data collected during the Central Pacific Study and examines the major scales of spatial and temporal variability in this "climax" community. Data include both total chlorophyll *a* and phytoplankton species abundances. Chlorophyll is frequently used as an index of phytoplankton

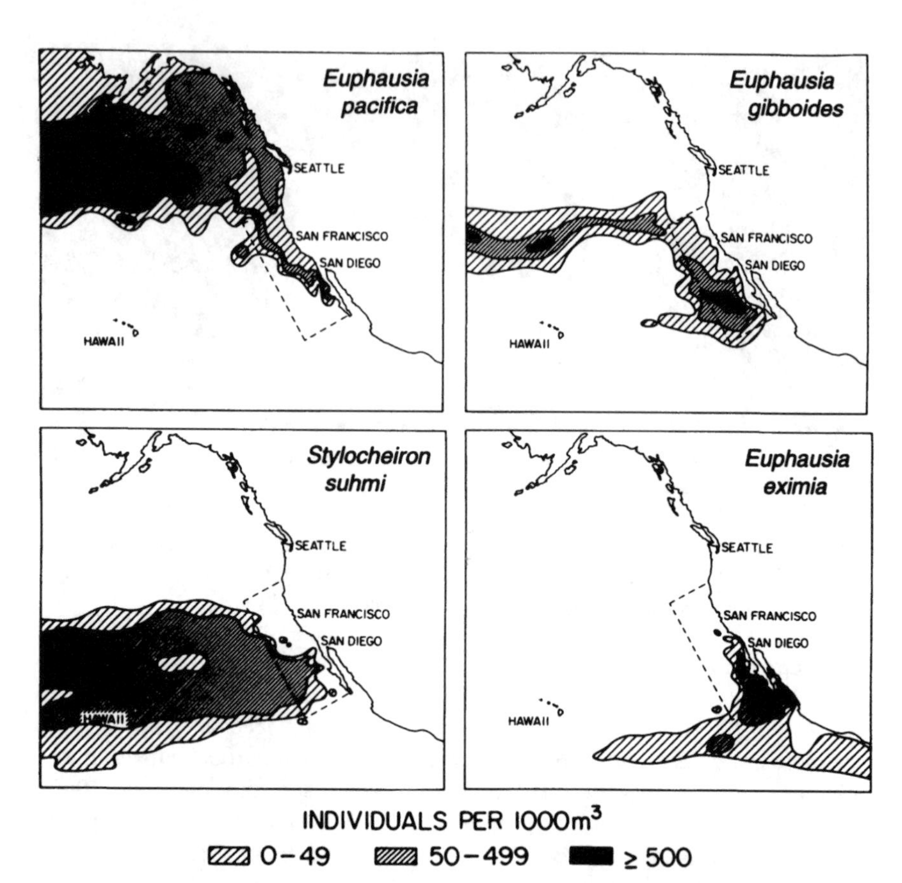

Figure 10-2. Distributions of four species of euphausiids found regularly in the California Current. Clockwise from top left: a subarctic species, a transition species, an equatorial species, and a central species (courtesy of Edward Brinton, SIO). Dashed box indicates CalCOFI area.

biomass. However, as the carbon-to-chlorophyll ratio in phytoplankton is a function of ambient light, such use is valid only when light conditions are stable. During the winter, incident irradiance is reduced, and the deepened winter mixing alters both the average light received by the cells and its temporal variability. Especially during this period, interpretation of chlorophyll in terms of biomass must be done with caution.

Methods

Data were collected between 1968 and 1985 during a total of 21 cruises in a region bounded by 26.5–31° N and 150.5–158° W. The routine procedures have been described in data reports (e.g., Scripps Institution of Oceanography 1974,

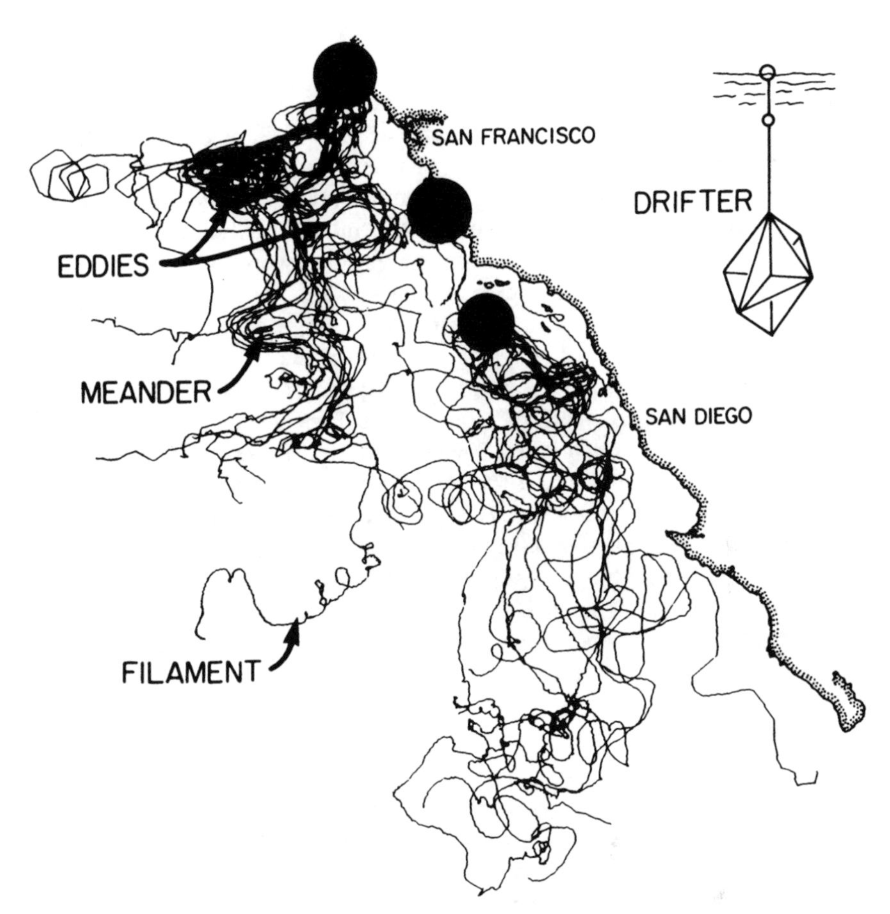

Figure 10-3. Flow patterns in the California Current as indicated by the tracks of satellite-monitored drifters drogued at 15 m depth. Disks indicate drifter release areas (courtesy of Peter Niiler, SIO).

1975, 1978), and the data have been discussed in the many papers referenced in this article.

Chlorophyll

The chlorophyll samples used in these analyses come from a total of 454 hydrocasts, representing from 1 to 41 casts per cruise and up to 18 depths per cast. Chlorophyll samples were filtered onto GF/C glass fiber filters and were either ground or extracted in 90% acetone in the dark and under refrigeration for 24–48 hours (Venrick and Hayward 1984). Comparisons with H/A Millipore filters have shown the mean loss of chlorophyll through GF/C filters to be less than 10% in the central North Pacific (Venrick et al. 1987a).

Floristics

Samples were analyzed for floristics on six summer cruises (June–September) and two winter cruises (January–February) between 1972 and 1985. Floristic data are available from 1–3 stations per cruise and from 8–18 depths per station. Phytoplankton samples were 500-ml water samples preserved with 1% neutralized formalin. From each sample, cells from 265–273 ml were concentrated by settling and enumerated under a phase-contrast inverted microscope. The entire sample was counted at 100X for larger cells, and one-sixth of the sample was counted at 250X for most of the remaining material. A single transect at 400X was counted for the three smallest and most abundant species.

Only individuals that could be identified to genus or species are included in these analyses. There are a total of 310 taxa, which include diatoms, dinoflagellates, coccolithophores, and silicoflagellates. To reduce recognition bias (my increasing ability to differentiate species), not all species are included in any one analysis. The selection of species to be included depends upon the goals of the specific analysis and is discussed in the appropriate publication. No attempt was made to enumerate the smallest cells (picoplankton), which are now thought to be a very important component of oligotrophic ecosystems (e.g., Stockner and Antia 1986), but which cannot be identified to taxon with the methods employed here.

Similarity

As an index of floral similarity, I have used Kendall's nonparametric correlation coefficient (τ). Since the number of species in the combined species list is large, the value of τ is relatively insensitive to the number of species, and the magnitude of τ is a measure of the similarity of the rank order of species abundances in two data sets; it is used here the way the percent similarity index (PSI; Whittaker 1952) is used, in contrast to the usual statistical (probabilistic) use. The advantage of τ over the PSI is that it gives equal weight to each species, regardless of its absolute abundance. I use τ to identify the region of vertical transition between species assemblages, in which case the data being compared are individual samples from adjacent depths. I also use τ to measure change of flora over space or time, in which case the data being compared are species abundances integrated through a depth zone on one station.

Results

The Central North Pacific Environment

The salient features of the central North Pacific environment are illustrated in Figure 10-4. During most of the year, the surface isothermal layer extends to 35–50 m. Levels of nutrients are low or undetectable above a sharp nutricline that begins between 120 m and 150 m. Chlorophyll concentrations are low in the near-surface waters and reach a maximum in a sharp layer just above the

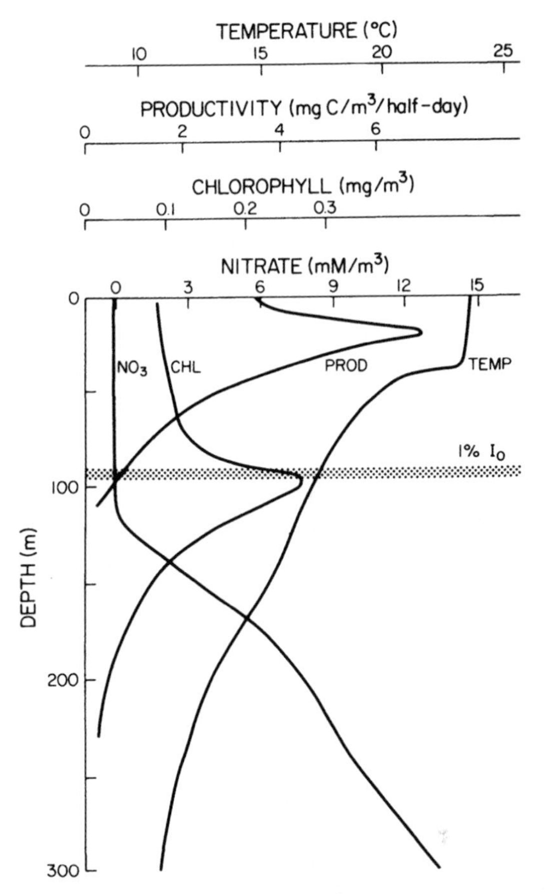

Figure 10-4. Characteristic vertical profiles of physical, chemical, and biological parameters in the central North Pacific. Stippling indicates the depth reached by 1% of the surface irradiance (I_0).

start of the nutricline, at depths reached by less than 1% of the surface light. The increase in chlorophyll in the maximum layer is primarily due to an increase in the concentration of chlorophyll per cell, which is a physiological response to the low light intensities at these depths (Eppley et al. 1973; Beers et al. 1975, 1982). It is not an increase in phytoplankton biomass. Maximum primary production occurs near the bottom of the mixed layer and is separated by 50–100 m from the nutricline. Low but positive productivity has been measured in the chlorophyll maximum layer (Venrick et al. 1973). Although the central North Pacific environment is oligotrophic, with low rates of primary production and low biomasses of phytoplankton and zooplankton, it contains one of the most species-rich marine planktonic communities known (Reid et al. 1978).

These vertical distribution patterns persist with little change throughout the central North Pacific environment and throughout most of the year. For instance, the vertical chlorophyll structure is nearly unchanged for 9000 km along latitude 24° N (Figure 10-5). Near the edge of the central environment, the nutricline and chlorophyll maximum layer shoal and the vertical structure are compressed.

The Vertical Structure of Phytoplankton Species

The dominant spatial pattern of phytoplankton occurs vertically in the water column. Recurrent group analysis defines two associations of species, one occupying the upper euphotic zone, the other the lower (Venrick 1982). There is a narrow region of overlap, usually near 100 m in depth. On every cruise for which floristics are available, this species dichotomy is apparent (Figure 10-6). A few species are among the 20 dominant species in both zones, but even these have consistent abundance maxima in one zone or the other. In no instance is a species more abundant in the shallow zone on one cruise and in the deep zone on another.

The length of the species list and the species dominance structure are similar in both associations (Figure 10-7). Each association includes diatoms, dinoflagellates, coccolithophores, and silicoflagellates in approximately the same proportion. Larger genera have both shallow and deep representatives, but in two- and

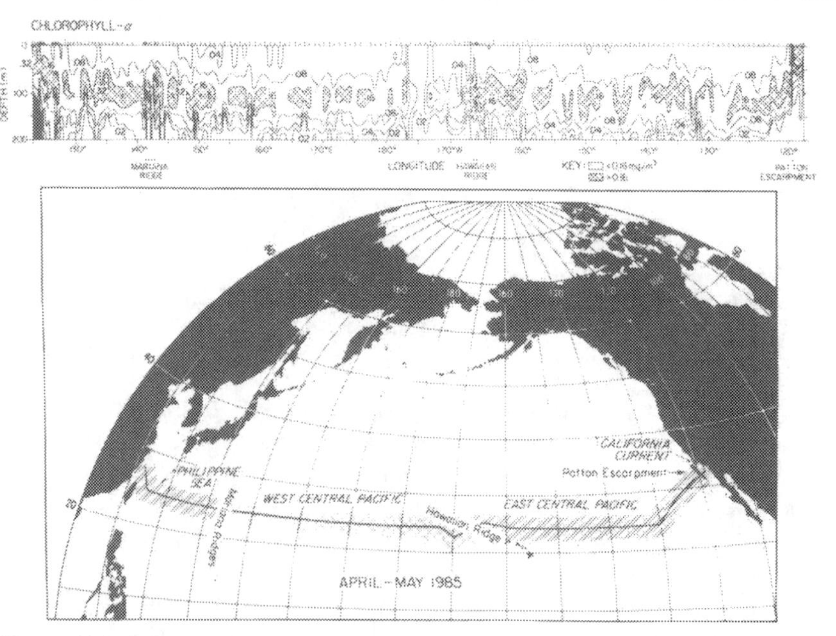

Figure 10-5. Cruise track and vertical distribution of chlorophyll across the central North Pacific. (Adapted from Figs. 1 and 3, in E. L. Venrick, 1991, Mid-ocean ridges and their influence on the large-scale patterns of chlorophyll and production in the North Pacific, *Deep-Sea Research* 38, Supp. 1:S83–S109, with kind permission from Pergamon Press Ltd., Headington Hill Hall, Oxford OX3 OBW, U.K.).

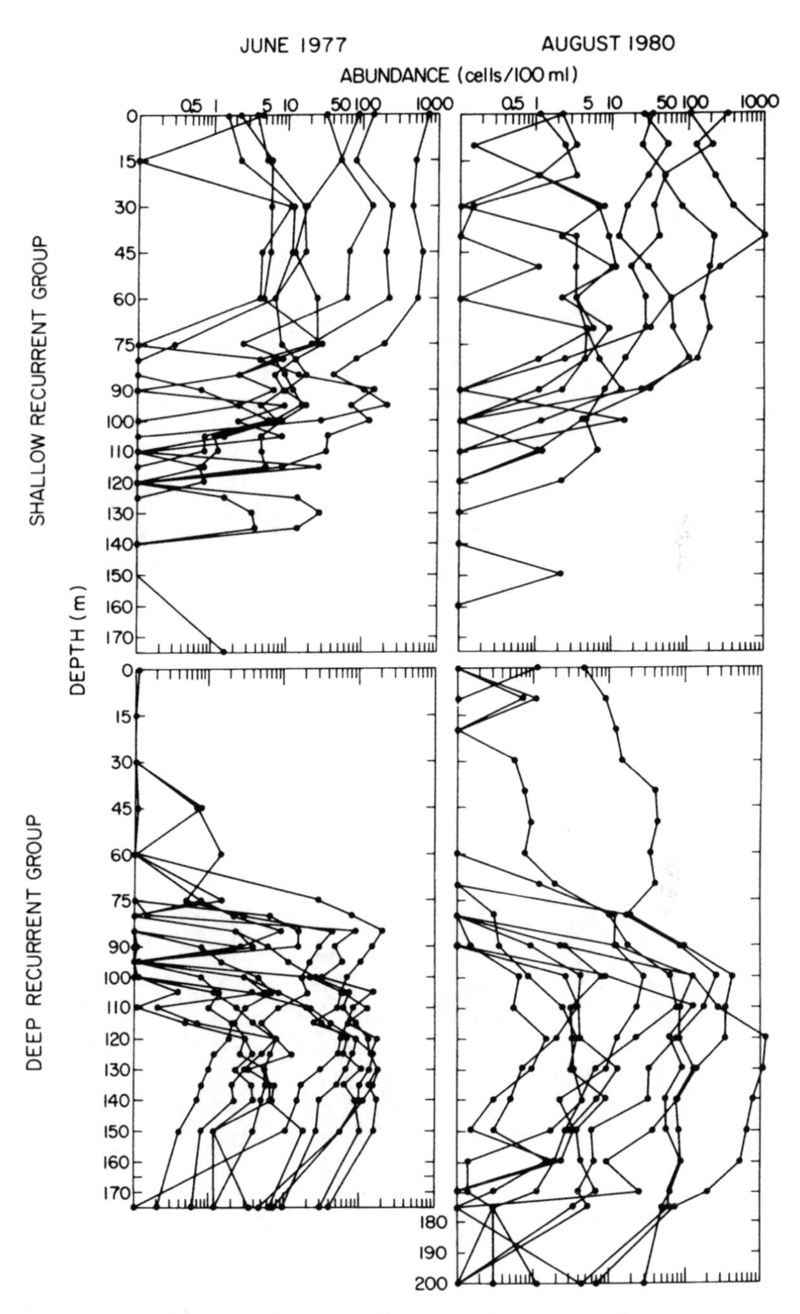

Figure 10-6. Vertical distribution of species classified into the first two recurrent groups in June 1977 and August 1980 (adapted from Fig. 2 in Venrick 1986, by permission).

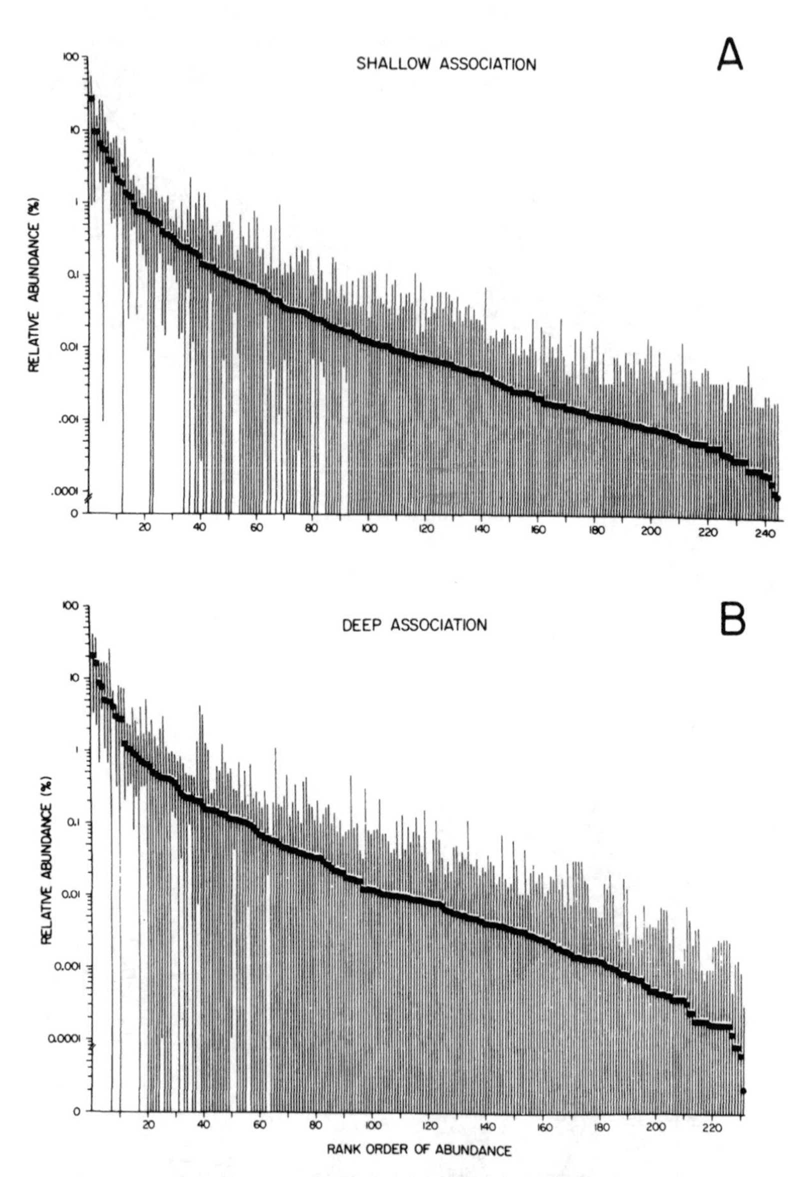

Figure 10-7. Mean species structure of phytoplankton between 1973 and 1985. Vertical bars indicate the range of relative abundances over 10 stations in six summers. (From E. L. Venrick, 1990, Phytoplankton in an oligotrophic ocean: Species structure and interannual variability, *Ecology* 71(4):1547–1563. Copyright © 1990 by the Ecological Society of America. Reprinted by permission.)

three-species genera, congeners tend to be abundant in the same layer. There appears to be no relationship between vertical distribution and cell size or shape.

The location of the transition region between associations can be identified by examining the correlation (Kendall's τ) between the rank order of species abundances in vertically adjacent samples. Within each association, the value of τ is stable. As the interval between samples includes or spans the transition region, a reduction in the value is apparent (Figure 10-8).

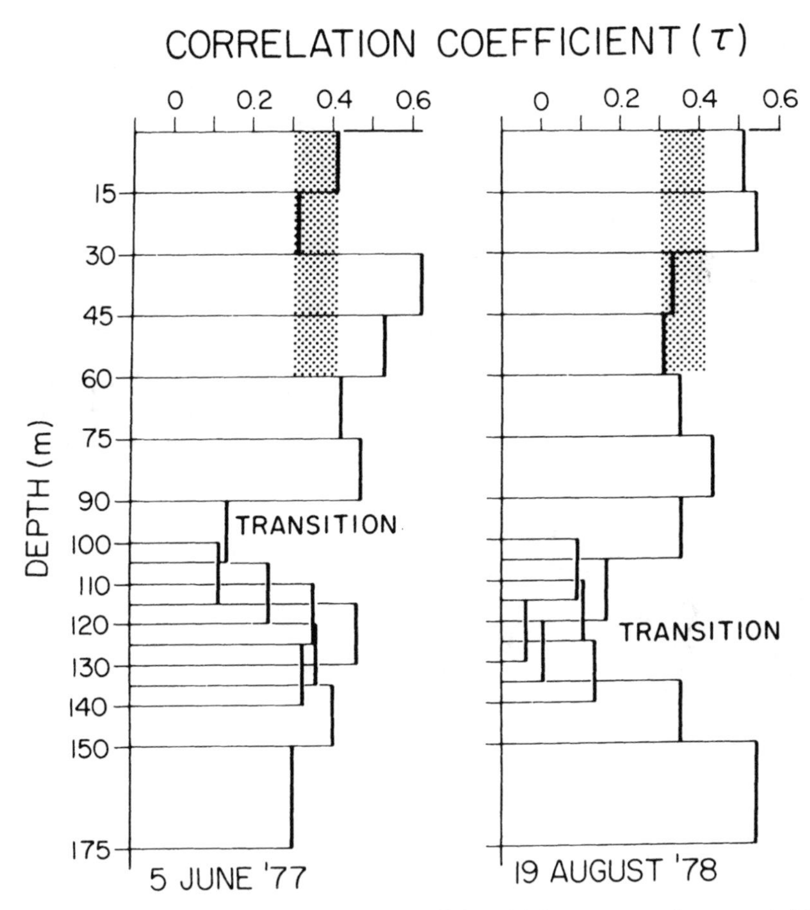

Figure 10-8. Kendall's rank correlation coefficient (τ) between samples separated by depth intervals of 15 m. Shaded area indicates the range of the mean value of τ between pairs of samples from the same depth in the mixed layer from different stations. Thus, the shaded area indicates the range of τ due to horizontal variability. (From E. L. Venrick, 1982, Phytoplankton in an oligotrophic ocean: Observations and questions, *Ecological Monographs* 52(2):129–154. Copyright © 1982 by the Ecological Society of America. Reprinted by permission.)

Within each association, "key species" have been identified that are consistently abundant in that association (Venrick 1988). There are 17 key species in the shallow zone and 20 in the deep. These 37 species account for more than 75% of the total number of cells. The abundances of the shallow key species are concordant along the vertical axis, as are those of the deep key species. Thus, these 37 species are useful indices of the distributions of shallow and deep associations.

The summed abundance of the deep key species closely parallels the increase in chlorophyll concentration at the top of the chlorophyll maximum layer (Figure 10-9). The chlorophyll maximum layer is composed of species characteristic of the deep association, with insignificant numbers of shallow-living species. Thus, even though the maximum layer does not represent a biomass increase, it represents a shift in species composition. The upper edge of the chlorophyll maximum layer is a useful indicator of the transition between shallow and deep associations when floristics are unavailable.

Physiological differences between shallow- and deep-living phytoplankton in oligotrophic oceans have been identified (Dugdale and Goering 1967; Eppley et al. 1973). Shallow phytoplankton are dependent upon recycled nutrients. Estimates of recycling efficiencies in this zone are as great as 80–90%, which appear to be sufficient to maintain relatively high phytoplankton turnover rates. Nevertheless, the source of new nutrients (i.e., nutrients from outside the euphotic zone) necessary to maintain indefinitely the observed

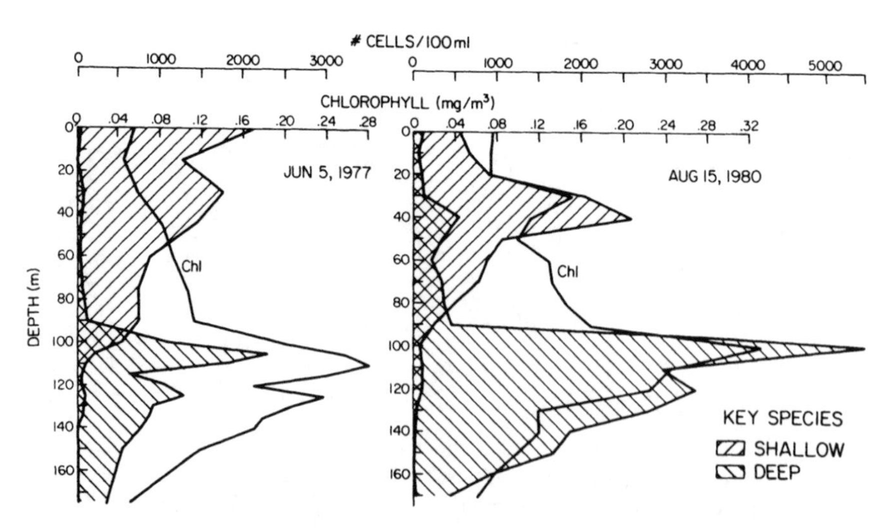

Figure 10-9. Vertical distribution of chlorophyll and key species on two summer stations. (From E. L. Venrick, 1988, The vertical distributions of chlorophyll and phytoplankton species in the North Pacific central environment, *Journal of Plankton Research* 10(5):987–998, by permission of Oxford University Press).

level of primary production in this zone is a hotly debated issue in biological oceanography (Hayward 1991).

In contrast, species in the deep zone receive new nutrients from deeper waters, but their production is light limited. Measured rates of ^{14}C uptake are very low, and gross production must support losses due to respiration, grazing, and sinking. The mechanisms that allow deep-living species to persist are not yet understood, but the data presented in this paper suggest that seasonal mixing may be important.

Because of the species shift between shallow and deep associations, and because of their physiological differences, there is little reason to expect similar patterns of either environmentally or biologically induced variability. Thus, these two associations are considered separately in the remainder of these analyses.

Interannual Changes

Chlorophyll in both zones increases over time (Venrick et al. 1987b). However, this increase is significant only in the deep zone (75–200 m), where the rate of increase is approximately 0.5 mg/m^2/yr, for a net doubling during the 17-year study (Figure 10-10, bottom). We have tentatively hypothesized a relationship between this increase in chlorophyll and an increase in the intensity of winter storms during this same period (Venrick et al. 1987b).

Phytoplankton species data from 10 stations on six summer cruises between 1973 and 1985 have been integrated within each zone. In both zones, there are slow but significant changes in the dominance structure (reduction in τ) during the 12.2-year period (Figure 10-11; Venrick 1990). Linear regression suggests that the rate of change in the deeper layer (0.027 per year) is three times greater than that in the shallow zone (0.009 per year), which is consistent with the more rapid increase in the deep chlorophyll.

These rates of change may be evaluated against three standards. Computer simulation suggests that the variability introduced into the data by subsampling error reduces the maximum expected value of τ from 1.00 to 0.70–0.76 in the shallow assemblage and to 0.52–0.55 in the deep assemblage. The magnitude of this reduction depends upon the dominance structure of the assemblage and thus varies from sample to sample. The values of similarity between samples collected from the same association on the same cruise range between 0.45 and 0.67 (Figure 10-11). This estimates the value of τ in the presence of small-scale field heterogeneity. Clearly a large fraction of this within-cruise variability is due to subsampling error, especially in the deep zone, where small-scale heterogeneity is no greater than the estimated subsampling error. Finally, the value of τ between shallow and deep associations at the same station ranges between 0.00 and 0.31, with a median of 0.12. Thus, the differentiation within a zone that occurs over 12 years is less than that maintained vertically over a spatial interval of 20 m.

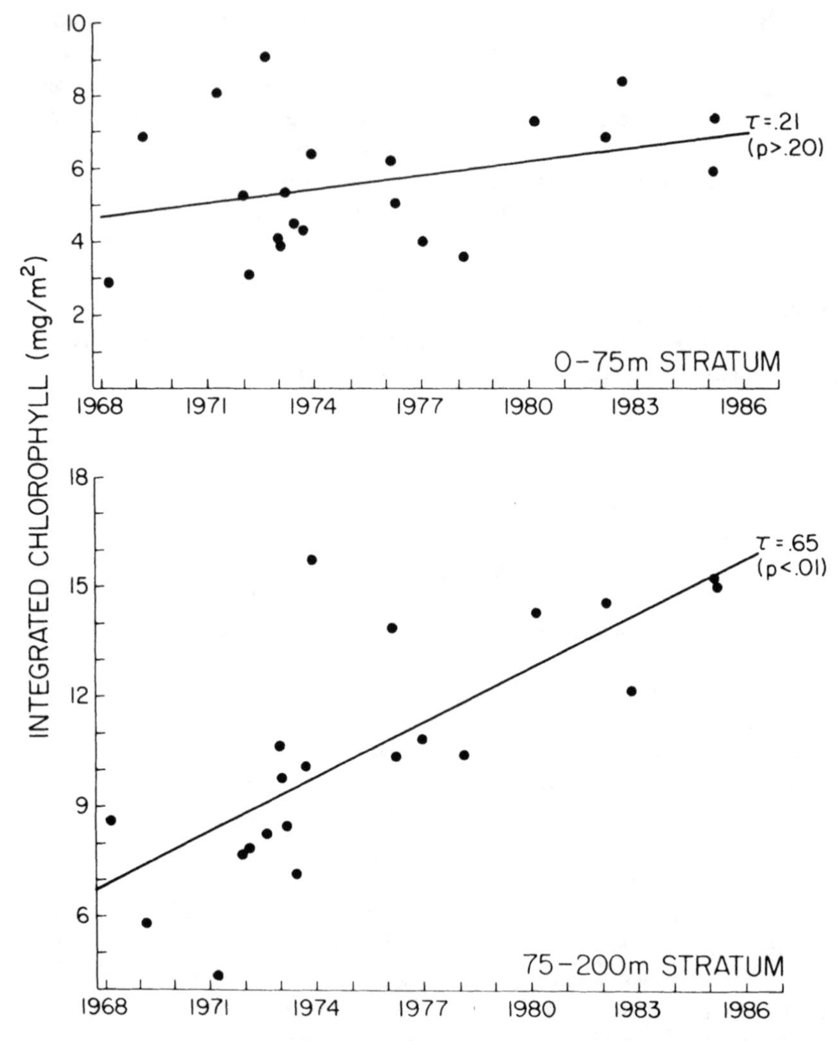

Figure 10-10. The increase of chlorophyll over time. Each dot represents the mean for one cruise, where there are 1 to 41 casts per cruise. The significance of the increase is evaluated by the value of Kendall's nonparametric correlation coefficient (τ), whereas the change is described by linear regression. (From Venrick 1993, by permission.)

When the above analysis is repeated separately for the three major taxa (Figure 10-12), diatoms in the deep zone show the most rapid change in species structure (Venrick 1982). Rates of change of other components, in either zone, are indistinguishable. The apparent differences in the absolute stability of species structure (mean τ) can be attributed to differences in the magnitude of subsampling error.

It must be emphasized that the 12 years spanned by this study represent more

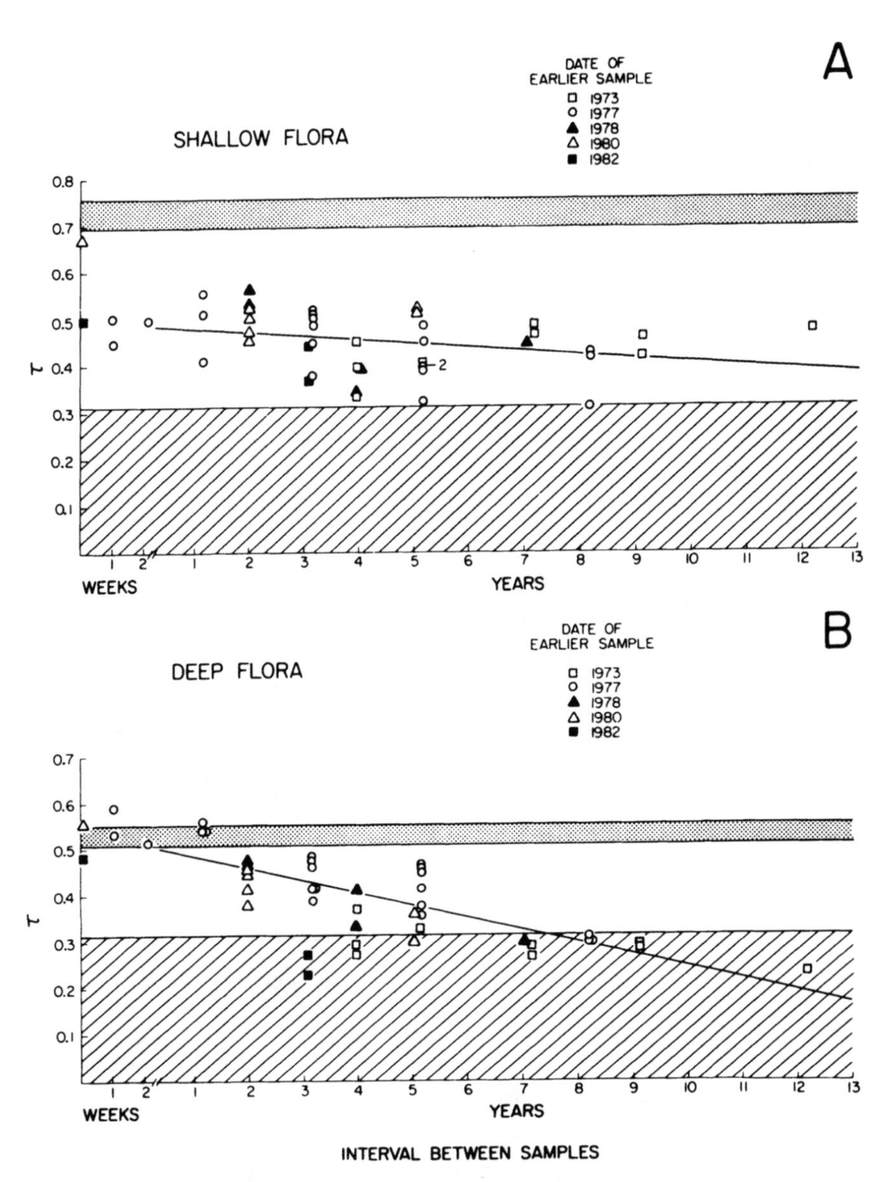

Figure 10-11. The change in species dominance structure over time as measured by the change in Kendall's nonparametric correlation coefficient (τ) between pairs of samples separated by increasing intervals of time. The stippled region indicates the maximum expected value of τ from computer simulation. The hatched region indicates the range of τ between the shallow and deep associations on single cruises. The change in τ over time is described by a linear regression. (From E. L. Venrick, 1990, Phytoplankton in an oligotrophic ocean: Species structure and interannual variability, *Ecology* 71(4):1547–1563. Copyright © 1990 by the Ecological Society of America. Reprinted by permission.)

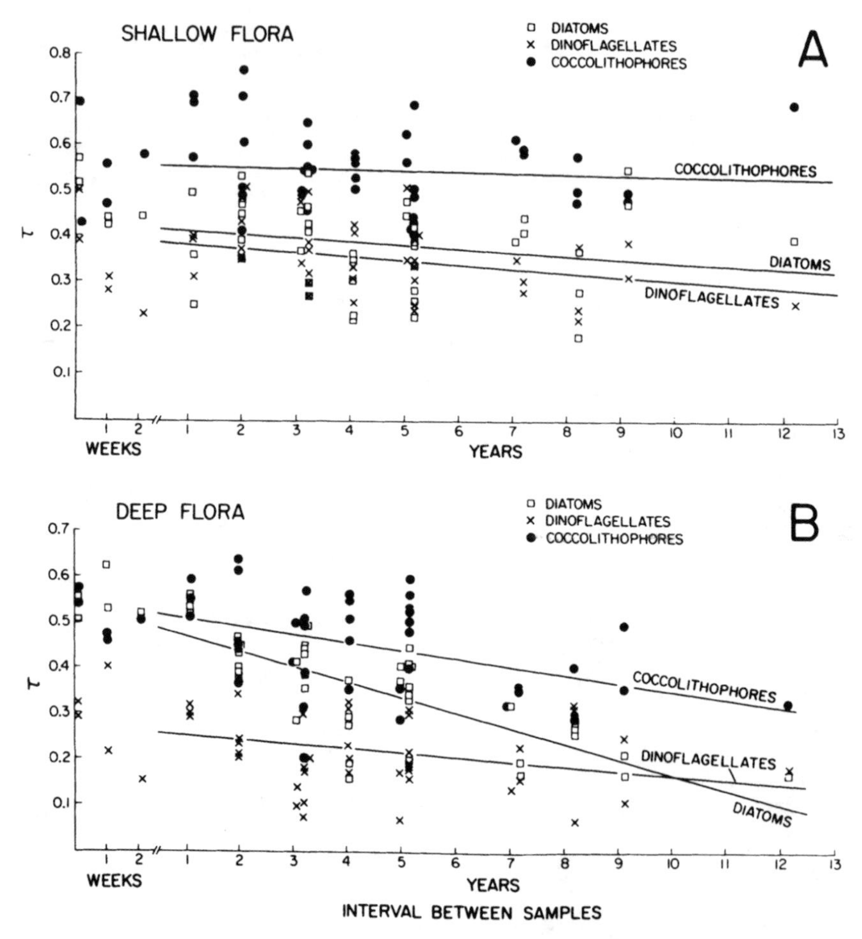

Figure 10-12. As for Figure 10-11, except that the change in species dominance structure over time is evaluated independently for diatoms, dinoflagellates, and coccolithophores. (From E. L. Venrick, 1990, Phytoplankton in an oligotrophic ocean: Species structure and interannual variability, *Ecology* 71(4):1547–1563. Copyright © 1990 by the Ecological Society of America. Reprinted by permission.)

than 1000 phytoplankton generation times. Thus, although changes in species dominance structure are statistically significant, they are very slow by ecological standards.

Large-Scale Spatial Changes

The large size and spatial homogeneity of the central North Pacific environment have been documented for physical and chemical properties and for zooplankton species. There have been numerous studies of central North Pacific phytoplankton species, but differences in methods or in vertical coverage have made it difficult to compare them. Until now, there has been no quantitative evaluation of the stability of phytoplankton species structure over space.

In July 1985, phytoplankton samples were collected 3200 km from the Climax station, near the eastern edge of the central North Pacific environment (the Edge station, 30° 50.2′ N, 121° 20.5′ W; see Figure 10-1). The near-surface waters in this region are cooler and fresher than those of the Climax station, but the magnitudes and vertical profiles of nutrients and chlorophyll are virtually identical (Figure 10-13; Venrick, 1991, 1992). The distribution of τ between vertically adjacent samples indicates that the same two-layered structure is present (Figure 10-14). The τ coefficient is used to compare the species composition of Edge and Climax stations within each zone (Figure 10-15). The similarity of species structure in replicate samples at the Edge station is of the same magnitude as that at the Climax station, indicating small-scale variability of similar magnitudes. In the shallow assemblage, samples separated in space by 3200 km are as similar as samples separated in time by 4–5 years. In the deep assemblage, the spatio-temporal convergence is slower, and 3200 km appears to be equivalent to 7–12

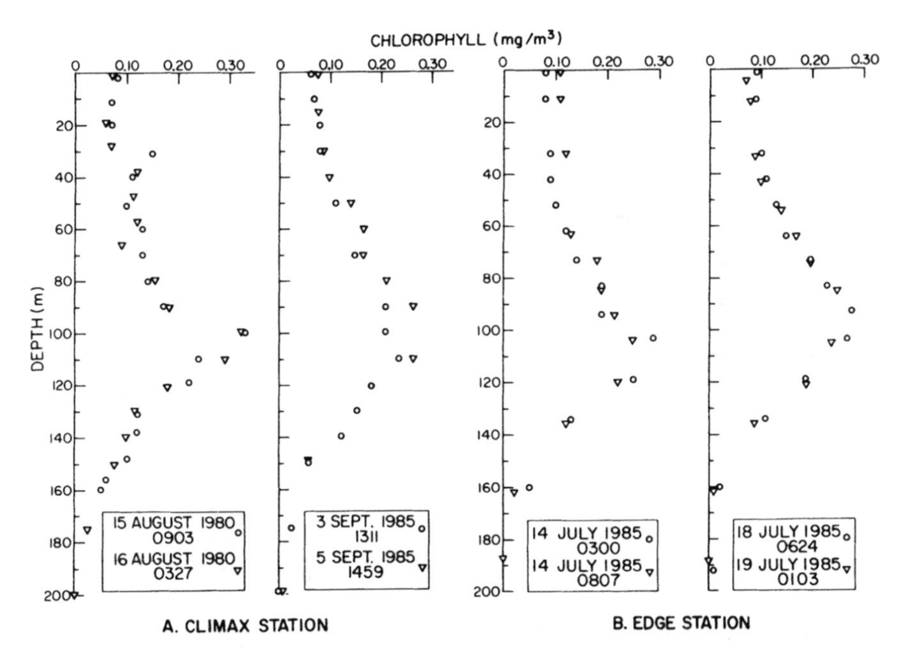

Figure 10-13. Vertical distribution of chlorophyll from the Climax station compared with that at the Edge station. (From E. L. Venrick, 1992, Phytoplankton species structure in the central North Pacific: Is the edge like the center? *Journal of Plankton Research* 14(5):665–680, by permission of Oxford University Press.)

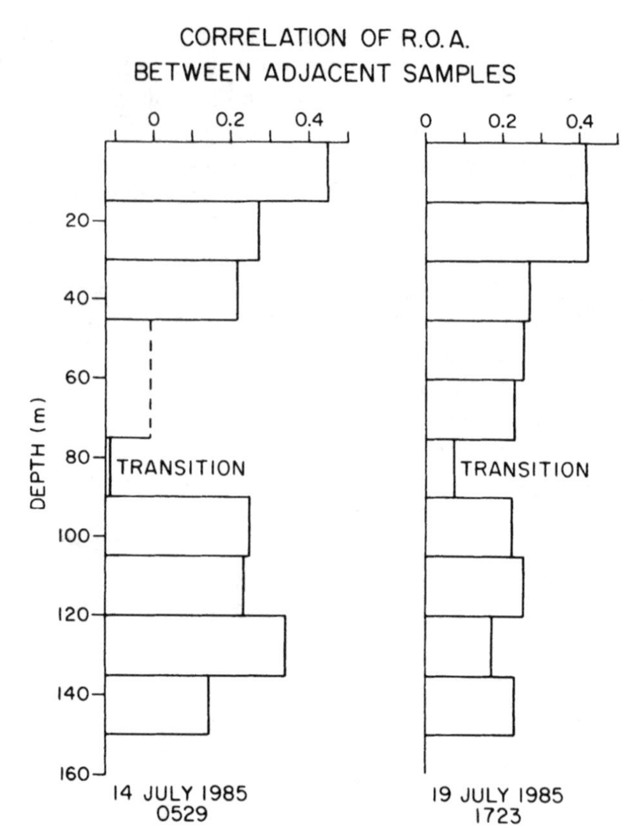

Figure 10-14. Kendall's rank correlation coefficient (τ) between samples from the Edge station separated by depth intervals of 15 m. Note the similarity to Figure 10-8. (From E. L. Venrick, 1992, Phytoplankton species structure in the central Pacific: Is the edge like the center? *Journal of Plankton Research* 14(5):665–680, by permission of Oxford University Press.)

years. Thus, the deep association, which has the less stable species structure in the temporal dimension, also shows the greater large-scale spatial variability. These empirical estimates of spatio-temporal equivalence range between 2.2 and 0.7—a remarkable degree of agreement with the scaling factor of 1 km/day derived from temporal and spatial scales of energy transfer in aquatic habitats (Harris 1986). From the ecological perspective, it is the fact of convergence rather than the rate that is important, as it indicates that the samples are indeed from a single ecosystem with active interchange of species. In the central North Pacific, presumably this interchange is maintained by the basin-wide ocean currents.

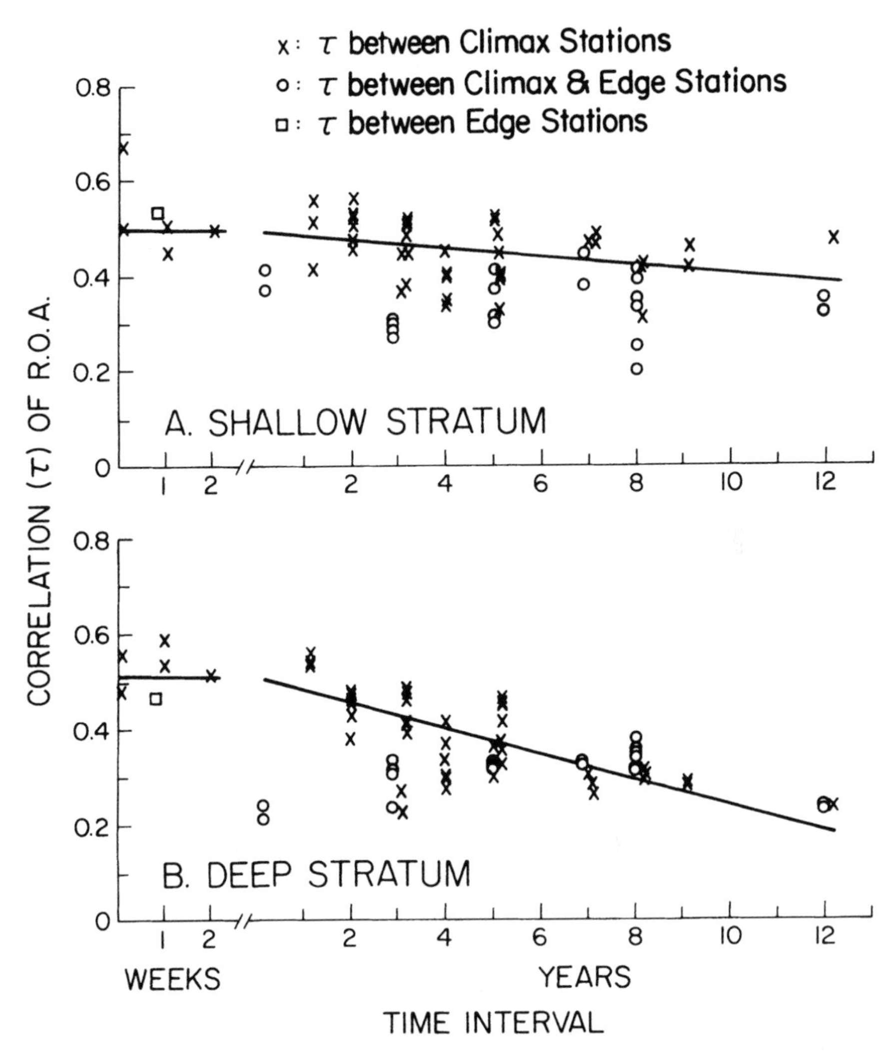

Figure 10-15. A comparison of temporal and spatial changes in species dominance structure within the central North Pacific as measured by changes in Kendall's nonparametric correlation coefficient. Solid line indicates the linear regression of the change in τ over time at the Climax station. (From E. L. Venrick, 1992, Phytoplankton species structure in the central North Pacific: Is the edge like the center? *Journal of Plankton Research* 14(5):665–680, by permission of Oxford University Press.)

The Seasonal Cycle

Identification of interannual changes in chlorophyll and species structure is facilitated by the fact that most of our samples were collected between June and October. Conversely, the sparsity of samples during the rest of the year complicates resolution of seasonality. To date, most studies in the central North Pacific have concluded that seasonality is very low or absent (Hayward et al. 1983).

When the dominant scales of spatial (vertical) and temporal (interannual) variability are removed, some evidence for a seasonal cycle emerges (Venrick 1993). The interannual changes in chlorophyll have been removed independently from the shallow and deep zones by means of linear regression (Figure 10-10). In the shallow zone, the residuals about the regression show a nonsignificant tendency to increase during the summer months (Figure 10-16, top). This trend is enhanced by the addition of data collected between January and March, so that all together the data suggest an abrupt decrease in chlorophyll in the shallow zone in late February through early March.

In the deep zone, there is a significant decrease in chlorophyll during the summer months (Figure 10-16, bottom). The few data collected between January and May suggest a winter-spring increase during late winter. The mean rate of decrease of chlorophyll during the summer, 0.8 mg/m^2/month about a mean concentration of 10.9 mg/m^2, indicates that an average doubling of chlorophyll

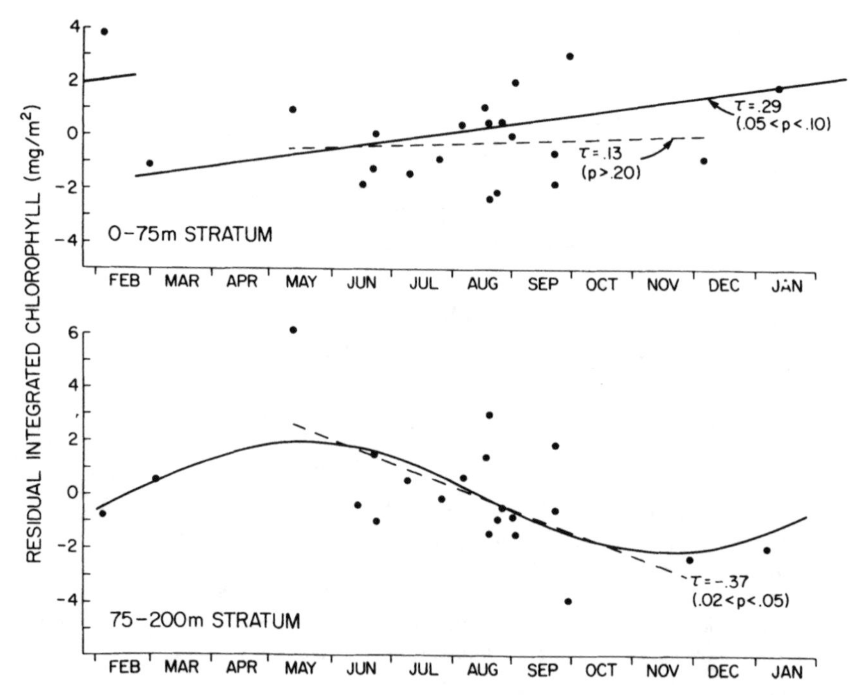

Figure 10-16. Seasonal changes of chlorophyll. Points are the residuals about interannual regression lines (Figure 10-10). Dotted lines represent summer trends; solid lines represent annual cycles. Linear trends are evaluated with Kendall's nonparametric correlation coefficient (τ) and described by linear regression. The seasonal cycle in the deep zone is approximated with a least squares cosine curve. (From Venrick 1993, by permission.)

during the winter would more than offset summer losses and maintain the deep populations from year to year.

Thus, the chlorophyll data suggest that the two phytoplankton associations have different seasonal cycles. The chlorophyll in the shallow zone may have no cycle, or it may increase during the year and decrease abruptly in late winter. The chlorophyll in the deep zone decreases during the summer and fall and increases during the winter. Interpretation of seasonal patterns of chlorophyll as seasonal patterns of biomass is risky; there is likely a change in the rate of chlorophyll per cell in response to diminishing day length and increasing mixed layer depth during winter. Examination of floristic data is needed to determine whether changes in species abundances are consistent with changes of chlorophyll: whether the shallow-living species are more abundant in summer and the deep-living species more abundant in winter.

Floristic data are available from three stations in January 1981 and from two stations in February 1973. These data are discussed fully elsewhere (Venrick 1993). Here, I present only the February data, which were collected during the period of deeper winter mixing. To minimize the influence of interannual changes, the February data are compared with the closest summer data, collected in June 1973.

In February 1973, the study area was occupied by a water mass that was isothermal to 140–150 m. The subsurface chlorophyll maximum had been destroyed, and the distinction between shallow and deep flora had been eliminated by the deep mixing (Figure 10-17, top). Deep flora dominated the water column. By the following June, the typical summer conditions were reestablished. The mixed layer shoaled to 30–40 m, the chlorophyll maximum layer was again developed below 100 m, and the dichotomy of species distributions had been reestablished (Figure 10-17, bottom).

In the summer, fewer species were more abundant in the deep stratum and, over all species, fewer were more abundant in the winter samples (Table 10-1). However, a greater proportion of deep-living species were more abundant in February ($p < 0.025$), as suggested by both the chlorophyll data and the analysis of key species. Conversely, a disproportionate number of shallow-living species were more abundant in June.

These observations are consistent with the seasonal patterns suggested by the chlorophyll data. Growth of shallow-living species may be disrupted by deep winter mixing, which, conversely, may favor growth of deep-living species. When the winter mixed layer is deep enough to reach the light-limited but nutrient-sufficient species in the deep association, these species are mixed upward into higher light levels. Here they undergo a brief period of growth before summer stratification is reestablished and cells settle again to deeper, darker waters. The magnitude of this cycle is low, and the chlorophyll data suggest that a net doubling of biomass may be sufficient to offset the gradual losses incurred during the rest of the year. With doubling times between 0.5 day (Laws et al. 1987)

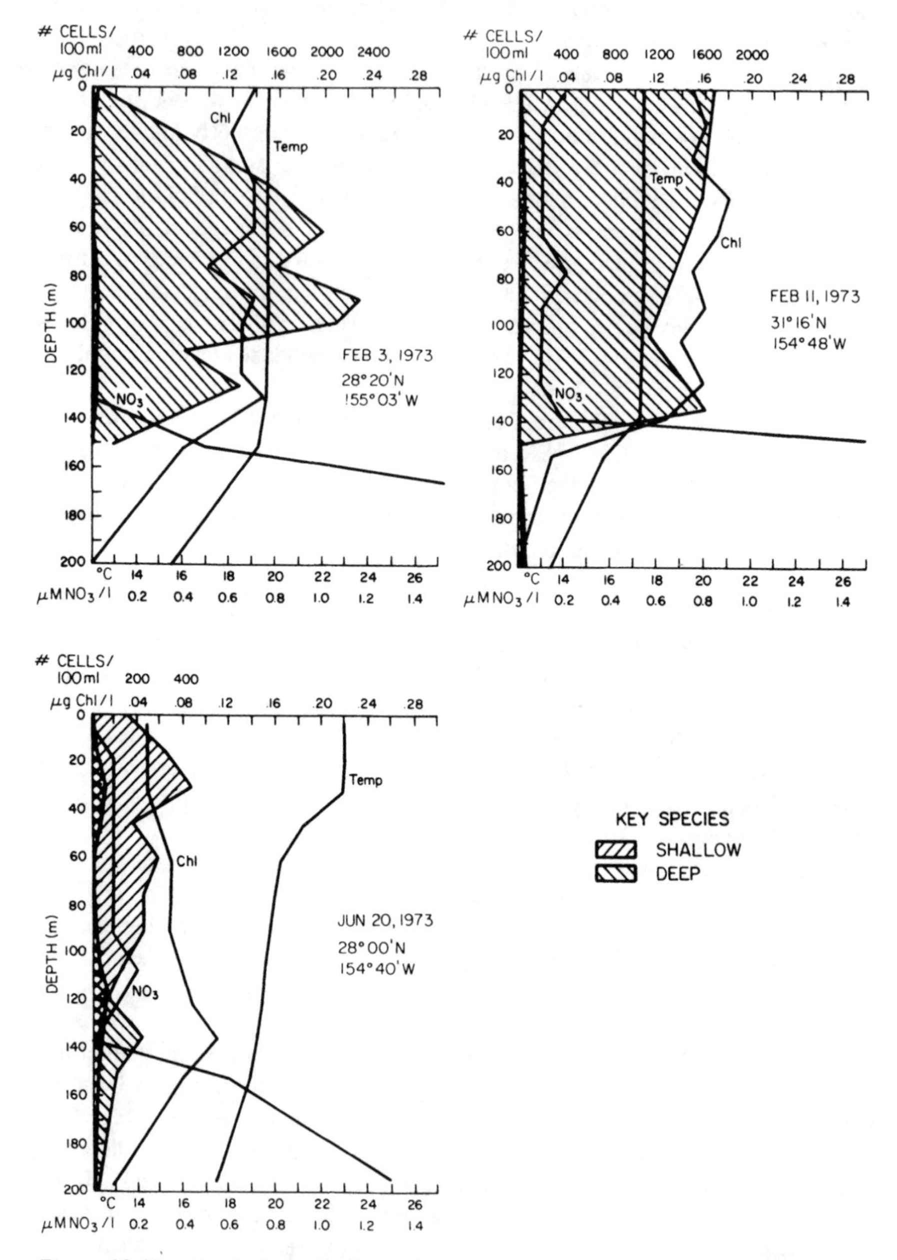

Figure 10-17. Vertical distributions of temperature, nitrate, chlorophyll, and shallow and deep key species in February and June, 1973. (Adapted from Fig. 5 in Venrick 1993, by permission.)

Table 10-1. Contingency table to examine the independence of seasonality and summer depth preference of phytoplankton in 1973

	Greater abundance in summer	Greater abundance in winter	Sums
Shallow species	77 (71.3)	23 (28.7)	100
Deep species	25 (30.7)	18 (12.3)	43
Sums	102	41	143
	$\chi_1^2 = 5.29$ $p < 0.025$		

Note: Shallow (deep) species are those with maximum summer abundance within the shallow (deep) stratum. Values are observed frequencies; those in parentheses are the frequencies expected under the hypothesis of independence.

and 5 days (Eppley et al. 1973), this growing season may need to be only a week or two in length. Thus, the persistence of the deep-living flora may be explained by growth induced by winter conditions, and these may persist too briefly to be sampled routinely.

To evaluate these observations in the context of an annual cycle, it is necessary to know the frequency distribution of the winter mixed layer depths. From the data available through the National Oceanic Data Center, I have extracted all February records of temperature from the study region. More than 60% of the February profiles show mixed layers of 120 m or greater, and the mixed layer exceeds 150 m about 25% of the time (Venrick 1993). In spite of obvious limitations, these data suggest that mixing to 120 m, which is deep enough to bring some of the deep-living cells to the surface, is sufficiently predictable in the central North Pacific that deep-living species may have adapted their life cycles accordingly.

Discussion

The patterns of diversity and stability of species structure of the phytoplankton in the central North Pacific are shared by other components of that ecosystem. Copepods are the most diverse group in the macrozooplankton, with more than 200 species in the upper 600 m (Hayward and McGowan 1979). The diversity structure of copepods is similar to that of phytoplankton (Figure 10-18), and both resemble the diversity structure of tropical rain forests more closely than that of other, less diverse ecosystems (McGowan 1990).

To indicate temporal changes in copepod species structure, McGowan and Walker (1985) have used the percent similarity index (PSI) in much the same way that I have used Kendall's τ. Their profile of changing PSI with increasing time (Figure 10-19) is similar to profiles of τ for phytoplankton species (Figure

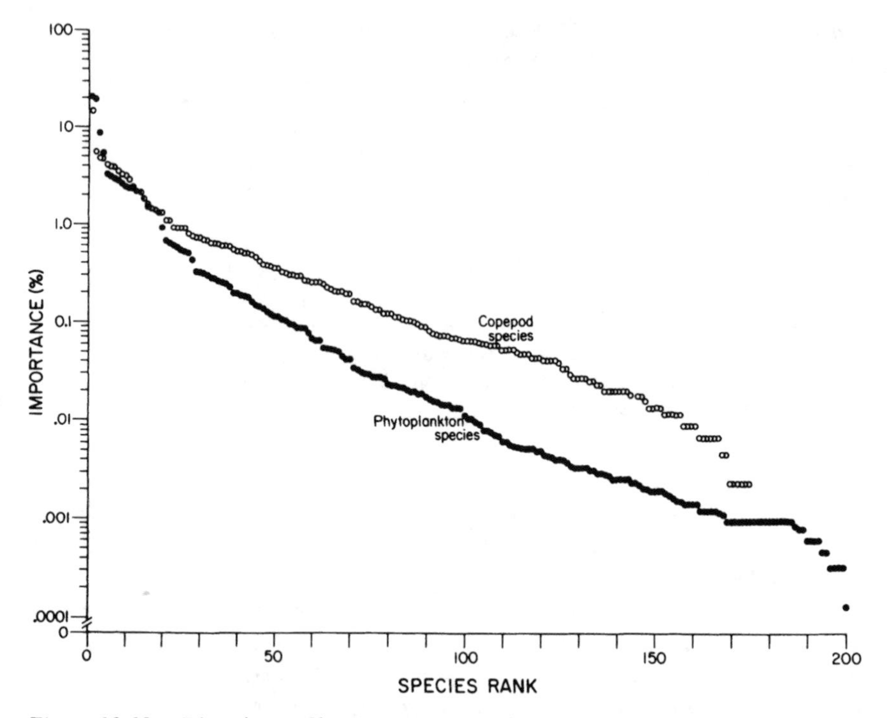

Figure 10-18. Diversity profiles (as percentage of total abundance: importance) of 175 species of copepods compared with 200 species of phytoplankton, both from the Climax area. Copepod data are based upon abundances in net tows integrating 0–300 m from seven cruises, 1964–1973. Phytoplankton data are based upon species abundances integrated between 0 and 140 m from three stations in 1977. The steeper profile of phytoplankton diversity may reflect only the more limited sampling scale. (Adapted from J. A. McGowan and P. W. Walker, in press, "Pelagic Diversity Patterns," in R. E. Rickleffs and D. Schluter, eds., *Species Diversity in Ecological Communities,* by permission of the University of Chicago Press.)

10-11). Small-scale heterogeneity is only slightly greater than subsampling error (measured directly as the variability between replicate aliquots of a sample). Change in species structure over time is slow; PSI decreases from a median of 73% to a median of 60% over a three-and-one-half-year period. Additional analysis extends this series over nine years and shows a continued slow change in species structure (McGowan and Walker 1985; Table 10-1). Thus, two important components of the central North Pacific epipelagic ecosystem have the same high diversity and temporal stability, in spite of the fact that the organisms have ambits of different sizes, generation times of different durations, and, probably, different proximal regulating mechanisms.

In spite of the size, age, and stability of the central North Pacific, variability of species structure on several scales has been identified. The system clearly is

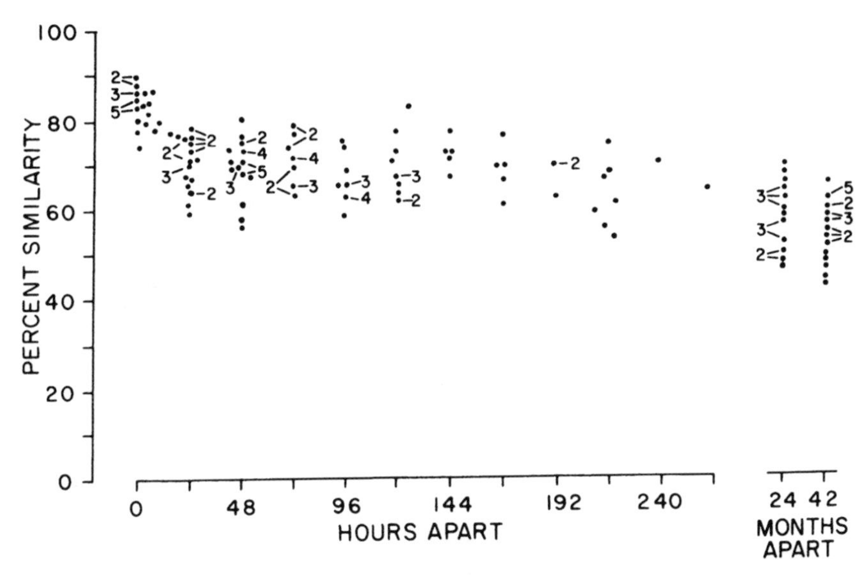

Figure 10-19. Percentage similarity of copepods in the central North Pacific as a function of time between samples. Samples were collected on four cruises with nets that integrated between 0 and 300 m. Separations of 0 hour are comparisons of replicate counts of the same sample. (From J. A. McGowan and P. W. Walker, 1985, Dominance and diversity maintenance in an oceanic ecosystem, *Ecological Monographs* 55(1):103–118. Copyright © 1985 by the Ecological Society of America. Reprinted by permission.)

not at steady state in the pure sense of the term. The stability of biomass and species structure in the central North Pacific has been contrasted with those of the plankton and nekton in the California Current, a continually changing mixture of different water sources and biota, and thus a very different physical regime.

Studies of varved sediments in anoxic basins off southern California have shown that large fluctuations in pelagic fish stocks have occurred for at least 1800 years (Soutar and Isaacs 1969; Baumgartner et al. 1992). Spectral analysis indicates that two of the dominant species share fluctuations with a period of about 57 years, and both have unshared dominant peaks at scales longer than 100 years (Baumgartner et al. 1992). Attempts are under way to correlate these long period fluctuations with large-scale climatological phenomena. It is likely that some of these low-frequency scales are shared by the plankton, but plankton data do not exist to evaluate these scales directly.

Zooplankton data from the California Current extend, more or less continuously, from 1949 to the present. Spatially, zooplankton biomass may vary over three or more orders of magnitude over intervals less than 40 nautical miles (Bernal and McGowan 1981; Smith 1971). The time series of zooplankton biomass has been the subject of a series of analysis since 1979. The variance spectrum of biomass over time is dominated by fluctuations with periods longer than 20 months, and these events are coherent throughout the California Current system (Bernal 1979). Periods of high biomass are more closely related to periods of strong southward flow than to periods of upwelling (Figure 10-20; Bernal and McGowan 1981). Conversely, periods of low biomass are related to the decrease in the southward flow, which is a manifestation of the northward propagation

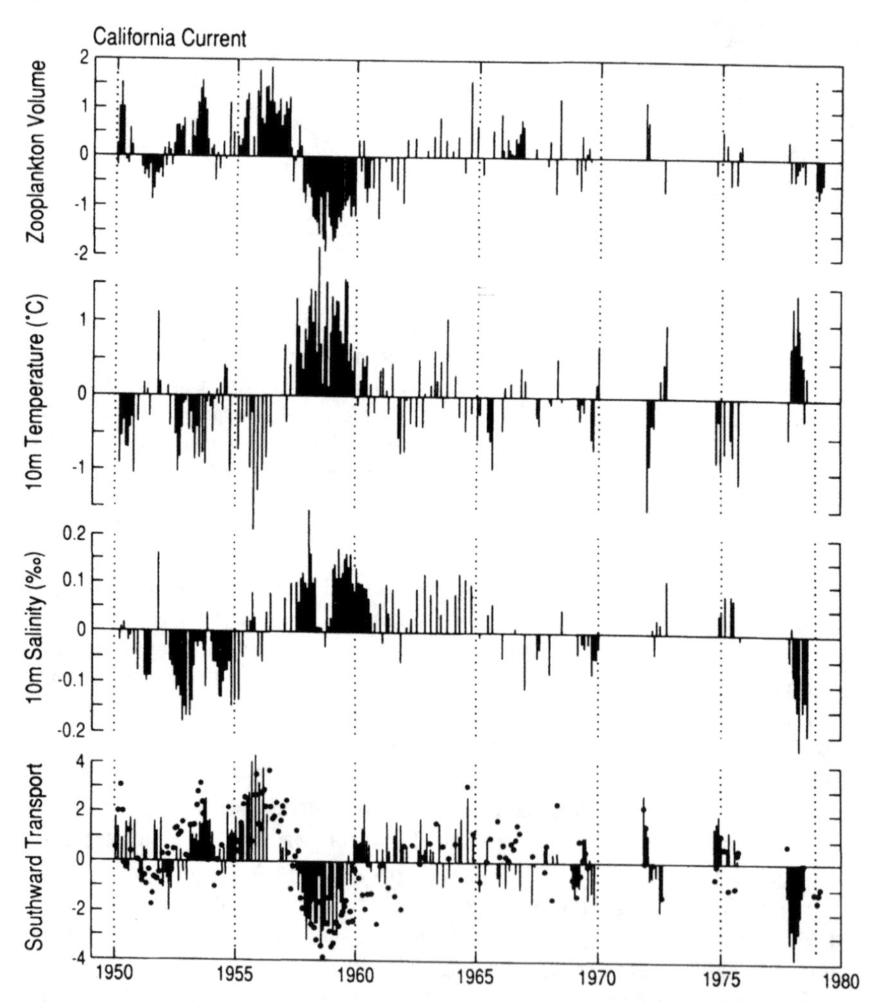

Figure 10-20. Time series of nonseasonal, standardized anomalies for spaced-averaged zooplankton volume and three environmental properties in the California Current. From top to bottom: zooplankton volume in the upper 140 m, 10 m temperature, 10 m salinity, principal EOF of 0/500 steric height (transport). The dots on the bottom panel repeat the zooplankton data (Adapted from Fig. 6 in Chelton et al. 1982, reprinted by permission.)

of the equatorial El Niño phenomenon (Bernal 1979; Chelton et al. 1982). Both advection and zooplankton biomass also show a semiannual signal. The linkages between current flow and biomass appear to be dominated by advection of biomass in the northern portion of the California Current and by local production stimulated by advection of nutrients in the southern portion (Roesler and Chelton 1987).

Studies of individual zooplankton taxa have shown large changes in species distributions associated with El Niño in the California Current (Brinton 1991; several papers in CalCOFI Reports 7, 1960). As a consequence, at any location there are strong interannual shifts in species dominance structure. In a study of 104 species of copepods, the median PSI in samples taken 10–15 months apart is only 20% (Figure 10–21). Comparison of Figures 10-19 and 10-21 shows that the stability of copepod species structure in the California Current is much less than in the central Pacific on even the smallest scales, consistent with the more complex, less predictable physical environment in the California Current.

The most extensive data on phytoplankton biomass (chlorophyll) in the California Current are near-surface concentrations estimated from satellite (CZCS) imagery between 1979 and 1986 (Pelaez and McGowan 1986; Strub et al. 1990; Thomas and Strub 1990). These data indicate that phytoplankton biomass, like

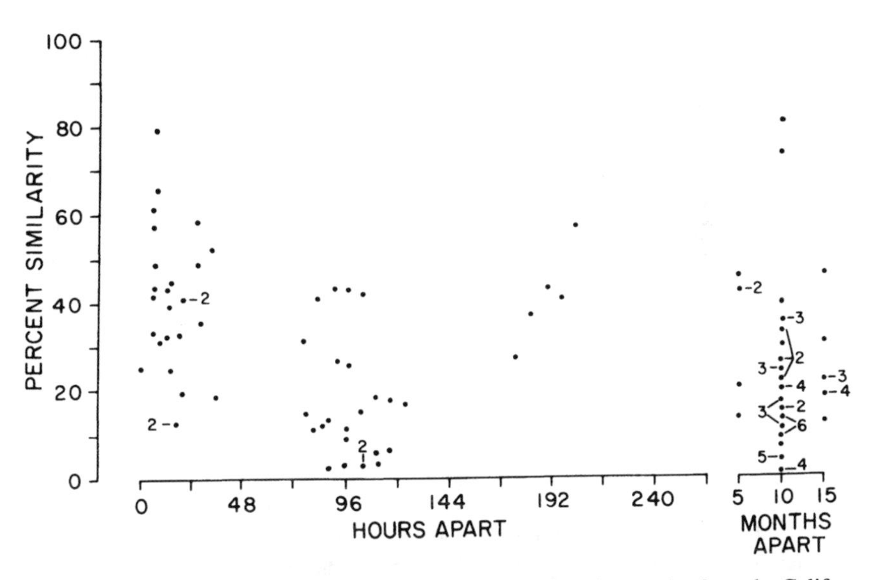

Figure 10-21. Percentage similarity of 104 copepod species in samples from the California Current as a function of time between samples. The smallest intervals represent closely spaced tows. (From J. A. McGowan and P. W. Walker, 1985, Dominance and diversity maintenance in an oceanic ecosystem, *Ecological Monographs* 55(1):103–118. Copyright © 1985 by the Ecological Society of America. Reprinted by permission.)

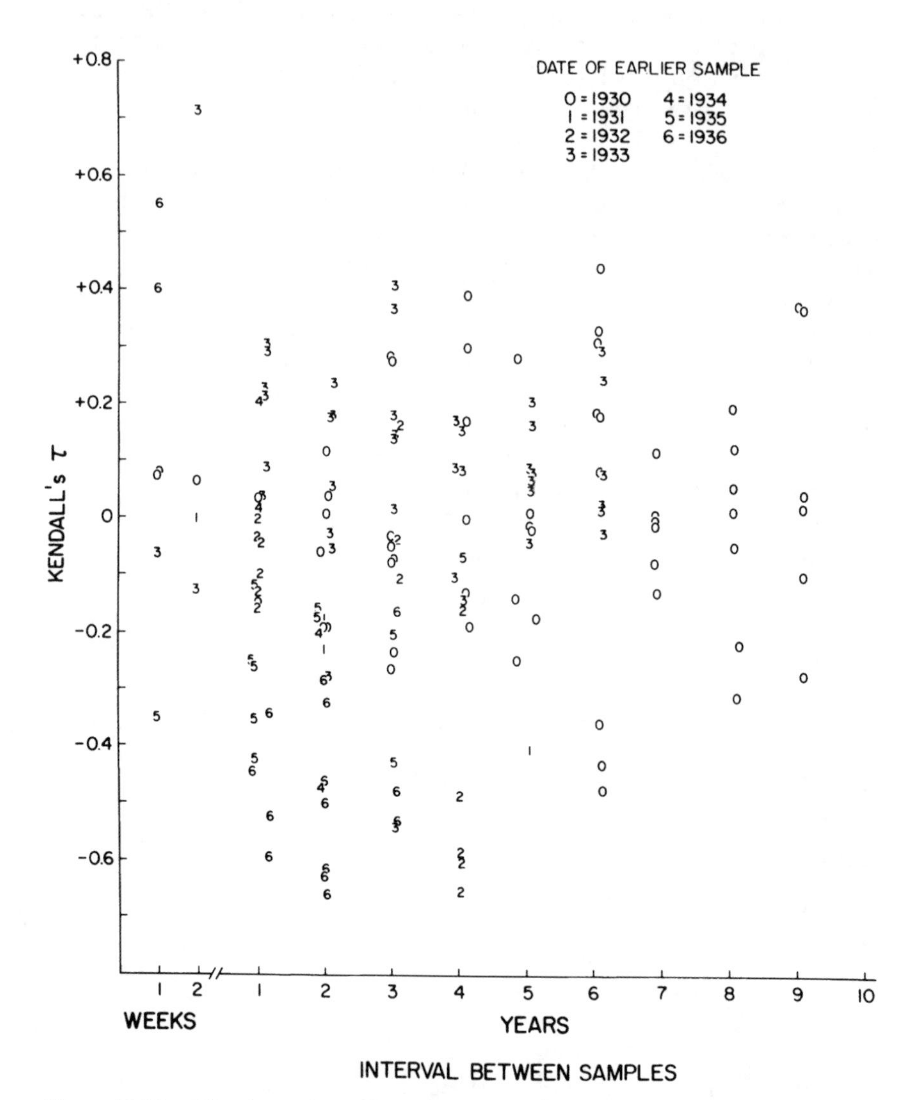

Figure 10-22. The change in species dominance structure over time as measured by the change in Kendall's nonparametric correlation coefficient (τ) between pairs of samples separated by increasing intervals of time. Data are for diatoms from the end of the SIO pier between 1930 and 1939. (From E. L. Venrick, 1990, Phytoplankton in an oligotrophic ocean: Species structure and interannual variability, *Ecology* 71(4): 1547–1563. Copyright © 1990 by the Ecological Society of America. Reprinted by permission.)

zooplankton biomass, is spatially heterogeneous and shows large interannual variations.

Recent data on phytoplankton species composition in the California Current are sparse. However, there are phytoplankton data collected between 1920 and 1939 off the end of the SIO pier (Allen 1936, 1939). With some modification, these may be compared with diatoms from the shallow zone of the central North Pacific (Figure 10-12; Venrick 1990). The Allen data are the weekly mean abundances of diatoms. To make the temporal coverage of the Allen data more comparable to that from the central North Pacific, I have selected a random two-week interval between June and September for each year between 1930 and 1939. To reduce the inflation of the variance of τ due to small sample size, I have eliminated all pairs of samples that included fewer than 10 species.

The species structure of diatoms off the Scripps pier has no stability on any time scale (Figure 10-22). Negative correlations occur over time periods as short as one week; and over the entire analysis, negative correlations are significantly more frequent than positive ones (binomial $p = 0.002$). Allen's data illustrate his conclusions of 50 years ago (Allen 1939): ". . . our records [from the California Current] show no two years alike in the twenty, no two months alike, and no two weeks alike."

The relative stability of the species structures within these two environments is consistent with the findings from other environments: the dominance structure and the stability of that structure seem to be related to environmental stability. However, the concepts of diversity and stability are relative ones, and a final judgment must rest on a comparison over many different systems. Unfortunately, such comparisons become more difficult as they become more useful—that is, as they span increasingly dissimilar organisms from increasingly dissimilar environments. The next challenge for plankton ecologists is to relate their observations directly to observations from other types of ecosystems, such as terrestrial and benthic ecosystems, which have stronger observational bases and more extensive theoretical development. However, because of the difficulties of adjusting the relative scale of observations to the relative scale of the component species (e.g., copepods and warblers, or phytoplankton and trees), this challenge is unlikely to be easily met.

Acknowledgments

This work was supported by the Marine Life Research Group of Scripps Institution of Oceanography and partially by the Office of Naval Research.

References

Allen, W. E. 1936. Occurrence of marine plankton diatoms in a ten-year series of daily catches in southern California. *American Journal of Botany* **23**:60–63.

———. 1939. Summary of results of twenty years of researches on marine phytoplankton. *Proceedings of the Sixth Pacific Science Congress* **3:**577–588.

Baumgartner. T. R., A. Soutar, and V. Ferreira-Bartrina. 1992. Reconstruction of the history of Pacific sardine and northern anchovy populations over the past two millennia from sediments of the Santa Barbara Basin, California. *CalCOFI Reports* **33:**24–40.

Beers, J. R., F. M. H. Reid, and G. L. Stewart. 1975. Microplankton of the North Pacific Central Gyre. Population structure and abundance, June 1973. *Internationale Revue der gesamten Hydrobiologie* **60:**607–638.

———. 1982. Seasonal abundance of the microplankton population in the North Pacific central gyre. *Deep-Sea Research* **29:**227–245.

Bernal, P. A. 1979. Large-scale biological events in the California Current. *CalCOFI Reports* **20:**89–101.

Bernal, P. A., and J. A. McGowan. 1981. "Advection and Upwelling in the California Current." In F. A. Richards (ed.), *Coastal Upwelling*. American Geophysical Union, Washington, D.C., pp. 381–399.

Brinton, E. R. 1991. Euphausiid distributions in the California Current during the warm winter-spring of 1977–78, in the context of a 1949–1966 time series. *CalCOFI Reports* **22:**135–154.

CalCOFI Reports. 1960. Volume 7.

Chelton, D. B., P. A. Bernal, and J. A. McGowan. 1982. Large-scale interannual physical and biological interaction in the California Current. *Journal of Marine Research* **40:**1095–1125.

Dugdale, R. C., and J. J. Goering. 1967. Uptake of new and regenerated forms of nitrogen in primary productivity. *Limnology and Oceanography* **12:**196–206.

Eppley, R. W., E. H. Renger, E. L. Venrick, and M. M. Mullin. 1973. A study of plankton dynamics and nutrient cycling in the central gyre of the North Pacific Ocean. *Limnology and Oceanography* **18:**534–551.

Fager, E. W., and J. A. McGowan. 1963. Zooplankton species groups in the North Pacific. *Science* **140:**453–460.

Harris, G. P. 1986. *Phytoplankton Ecology, Structure, Function and Fluctuations*. Chapman & Hall, New York.

Hayward, T. L. 1991. Primary production in the North Pacific central gyre: A controversy with important implications. *Trends in Ecology and Evolution* **6:**281–284.

Hayward, T. L., and J. A. McGowan. 1979. Patterns and structure in an oceanic zooplankton community. *American Zoologist* **19:**1045–1055.

Hayward, T. L., E. L. Venrick, and J. A. McGowan. 1983. Environmental heterogeneity and plankton community structure in the central North Pacific. *Journal of Marine Research* **41:**711–729.

Laws, E. A., G. R. Di Tullio, and D. G. Redalje. 1987. High phytoplankton growth and production rates in the North Pacific subtropical gyre. *Limnology and Oceanography* **32:**905–918.

McGowan, J. A. 1971. "Oceanic Biogeography of the Pacific." In B. M. Funnel and

W. R. Riedel (eds.), *The Micropalaeontology of Oceans*. Cambridge University Press, London, pp. 3–73.

———. 1974. "The Nature of Oceanic Ecosystems." In C. B. Miller (ed.), *The Biology of the Oceanic Pacific*. Oregon State University Press, Corvallis, OR, pp. 9–28.

———. 1989. Pelagic ecology and Pacific climate. *Geophysical Monographs* **55:**141–150.

———. 1990. "Species Dominance-Diversity Patterns in Oceanic Communities." In G. M. Woodwell (ed.), *The Earth In Transition*. Cambridge University Press, London, pp. 395–421.

McGowan, J. A., and P. W. Walker. 1979. Structure of the copepod community of the North Pacific Central Gyre. *Ecological Monographs* **49:**195–226.

———. 1985. Dominance and diversity maintenance in an oceanic ecosystem. *Ecological Monographs* **55(1):**103–118.

———. In press. "Pelagic Diversity Patterns." In R. E. Rickleffs and D. Schluter (eds.), *Species Diversity in Ecological Communities: Historical and Geographical Perspectives*. University of Chicago Press, Chicago.

Pelaez, J., and J. A. McGowan. 1986. Phytoplankton pigment patterns in the California Current as determined by satellite. *Limnology and Oceanography* **31:**951–968.

Reid, J. L., E. R. Brinton, A. Fleminger, E. L. Venrick, and J. A. McGowan. 1978. "Ocean Circulation and Marine Life." In H. Charnock and G. Deacon (eds.), *Advances in Oceanography*. Plenum, New York, pp. 65–130.

Roesler, C. S., and D. B. Chelton. 1987. Zooplankton variability in the California Current, 1951–1982. *CalCOFI Reports* **28:**59–96.

Scripps Institution of Oceanography. 1974. Physical, chemical and biological data, CLIMAX I Expedition, 19 September–28 September 1968. SIO Ref. 74–20. Scripps Institution of Oceanography, University of California at San Diego, La Jolla, CA.

———. 1975. Physical, chemical and biological data, CLIMAX II Expedition, 27 August–13 October 1969. SIO Ref. 75–6. Scripps Institution of Oceanography, University of California at San Diego, La Jolla, CA.

———. 1978. Physical, chemical and biological data, INDOPAC Expedition, 23 March 1976–31 July 1977. SIO Ref. 78–21. Scripps Institution of Oceanography, University of California at San Diego, La Jolla, CA.

Smith, P. E. 1971. Distributional atlas of zooplankton volume in the California Current region, 1951 through 1966. *CalCOFI Atlas* 13.

Soutar, A., and J. D. Isaacs. 1969. History of fish populations inferred from fish scales in anaerobic sediments off California. *CalCOFI Reports* **13:**63–70.

———. 1974. Abundance of pelagic fish during the 19th and 20th centuries as recorded in anaerobic sediment off the Californias. *Fishery Bulletin* **72:**257–273.

Stockner, J. G., and N. J. Antia. 1986. Algal picoplankton from marine and freshwater ecosystems: A multidisciplinary perspective. *Canadian Journal of Fisheries and Aquatic Sciences* **43:**2472–2503.

Strub, P. T., C. James, A. C. Thomas, and M. R. Abbott. 1990. Seasonal and nonseasonal variability of satellite-derived surface pigment concentration in the California Current. *Journal of Geophysical Research* **95:**11501–11530.

Thomas, A. C., and P. T. Strub. 1990. Seasonal and interannual variability of pigment concentrations across a California Current frontal zone. *Journal of Geophysical Research* **95:**13023–13042.

Venrick, E. L. 1982. Phytoplankton in an oligotrophic ocean: Observations and questions. *Ecological Monographs* **52(2):**129–154.

———. 1986. "Patchiness and the Paradox of the Plankton." In A. C. Pierrot-Bults, S. van der Spoel, B. J. Zahuranec, and R. K. Johnson (eds.), *Pelagic Biogeography*. UNESCO, pp. 261–265.

———. 1988. The vertical distributions of chlorophyll and phytoplankton species in the North Pacific central environment. *Journal of Plankton Research* **10(5):**987–998.

———. 1990. Phytoplankton in an oligotrophic ocean: Species structure and interannual variability. *Ecology* **71(4):**1547–1563.

———. 1991. Mid-ocean ridges and their influence on the large-scale patterns of chlorophyll and production in the North Pacific. *Deep-Sea Research* 38, Suppl. **1:**S83-S109.

———. 1992. Phytoplankton species structure in the central North Pacific: Is the edge like the center? *Journal of Plankton Research* **14(5):**665–680.

———. 1993. Phytoplankton seasonality in the central North Pacific: The endless summer reconsidered. *Limnology and Oceanography* **38(6):**1135–1149.

Venrick, E. L., J. A. McGowan, and A. W. Mantyla. 1973. Deep maxima of photosynthetic chlorophyll in the Pacific Ocean. *Fishery Bulletin* **71:**41–52.

Venrick, E. L., and T. L. Hayward. 1984. Determining chlorophyll on the 1984 CalCOFI surveys. *CalCOFI Reports* **25:**74–79.

Venrick, E. L., S. L. Cummings, and C. A. Kemper. 1987a. Picoplankton and the resulting bias in chlorophyll retained by traditional glass-fiber filters. *Deep-Sea Research* **34:**1951–1956.

Venrick, E. L., J. A. McGowan, D. R. Cayan, and T. L. Hayward. 1987b. Climate and chlorophyll *a*: Long-term trends in the central North Pacific Ocean. *Science* **238:**70–72.

Whittaker, R. H. 1952. A study of summer foliage insect communities in the Great Smoky Mountains. *Ecological Monographs* **22:**1–44.

11

Ocean Time Series Research Near Bermuda: The Hydrostation S Time Series and the Bermuda Atlantic Time-Series Study (BATS) Program

Anthony F. Michaels

Long-term time series observations are a powerful tool for investigating biogeochemical processes in the ocean. Oceanic ecosystems are variable on a wide range of time and space scales (Dickey 1991). This variability arises from a combination of physical and biological processes and has important consequences for the measurement and interpretation of the upper ocean carbon cycle. The seasonal cycle of ocean mixing and phytoplankton primary production are the most obvious temporal patterns. Interannual variations in the seasonal cycle provide natural experiments on the relationship between the physical forcings and the biological response. Maintaining a high-quality oceanic time series study with consistent data quality over many decades is a scientific and logistical challenge. The ocean near Bermuda is unique in terms of the diversity and duration of the ocean and atmospheric time series that have been successfully maintained over decadal time scales.

In 1988, the Bermuda Atlantic Time-Series Study (BATS) was initiated as part of the U.S. Joint Global Ocean Flux Study (U.S. JGOFS) program. The Bermuda station is one of two time series efforts in the U.S. JGOFS, the other being the Hawaii Ocean Time-series Station (HOTS). Both of these time series stations are funded by the National Science Foundation, with additional support from other agencies. The purpose of BATS and HOTS is coincident with the larger goals of JGOFS, namely ". . . to determine and understand . . . the processes controlling the time-varying flux of carbon and associated biogenic elements in the ocean. . . ." (Scientific Committee on Ocean Research 1987). The focus of the BATS and HOTS efforts is on understanding the "time-varying" components of the ocean carbon cycle. Within this context, there are a number of more specific goals, including: (1) to understand the seasonal and interannual variations in ocean physics, chemistry, and biology; (2) to understand the relationships between the external physical forcings and the biological rate processes; and (3) to understand the processes that determine sea surface pCO_2, including thermodynamics, particle export, and gas exchange. The overall program is a mixture of traditional time series monitoring of ocean processes and the application of specific process-oriented studies of ocean biogeochemistry within the time series framework.

Bermuda was chosen for a U.S. JGOFS time series in part to build upon the diversity of existing, long-term time series in this area. Perhaps the most prominent is the biweekly hydrographic sampling at Hydrostation S started by Henry Stommel and co-workers in 1954 (Schroeder and Stommel 1969). This is one of the longest-running ocean time series operations in the world. From 1957 to 1963, this hydrographic study was supplemented by a program of biological and chemical measurements to determine the seasonal cycle of ocean production (Menzel and Ryther 1960, 1961). This five-year program provided the first detailed study of oceanic biogeochemistry in the Sargasso Sea. The data are still widely cited and reused to address hypotheses on the magnitude and controls of oceanic production (e.g., Platt and Harrison 1985; Fasham et al. 1990). In addition to Hydrostation S, there is a 15-year time series of deep-ocean sediment trap collections (Deuser 1986) and a 12-year time series of atmospheric measurements through the WATOX and AEROCE programs (Galloway et al. in press). This rich time series history and the diversity of existing measurement programs provide a valuable framework for the near-surface biogeochemistry investigations in the more recent BATS program.

This chapter discusses the Hydrostation S and BATS projects, including summaries of the measurement program, the dataset, the observed temporal patterns, and some very practical issues in the operation of a long-term ocean time series study. Most of the chapter focuses on the BATS time series, including a discussion of some of the data for the first three years. Papers on the first two years' results have been published elsewhere (Lohrenz et al. 1992; Michaels et al. in press). Those discussions are extended to a third year in this chapter, and some facets of the dataset are discussed in greater detail. The BATS data illuminate some of the mechanisms that lead to the annual spring phytoplankton bloom in the northern Sargasso Sea (Menzel and Ryther 1960, 1961). In addition, the oxygen and nutrient data can be used to provide independent estimates of the rates of new production associated with the 1989 spring phytoplankton bloom. High-frequency variability in the time series records, likely due to mesoscale eddies, indicates some fundamental limitations of a biweekly measurement program for resolving ocean processes. I also present some direct observations of mesoscale hydrographic features that suggest that both horizontal and vertical processes are involved in the delivery of nutrients to the surface. Adequate resolution of these features will require new and expanded sampling strategies and technologies (moorings and autonomous vehicles) to be resolved adequately. Finally, the Hydrostation S time series can be used to evaluate decadal-scale variability in ocean mixing and to infer variability in biological production on these longer time scales. The BATS and Hydrostation S data provide important guidelines for the development of future time series programs, particularly in the context of an extensive program like the proposed Global Ocean Observing System (GOOS).

The Measurement Program

Hydrographic sampling at Hydrostation S (Figure 11-1) began in 1954 with the routine measurement of temperature, salinity, and oxygen at 24 depths (0–2600 m). Traditionally, temperature was measured using reversing thermometers, salinity by conductivity (recently with a Guildline, AutoSal), and oxygen by Winkler titration. In 1988, temperature and salinity measurements were modified by the addition of a Conductivity-Temperature-Depth (CTD) system to provide near-continuous profiles of temperature and salinity.

The Bermuda Atlantic Time-Series Study (BATS) commenced monthly sampling in October 1988 near the site of the Ocean Flux Program (OFP) station located 70 km southeast of Bermuda (Figure 11-1). The OFP station is the current

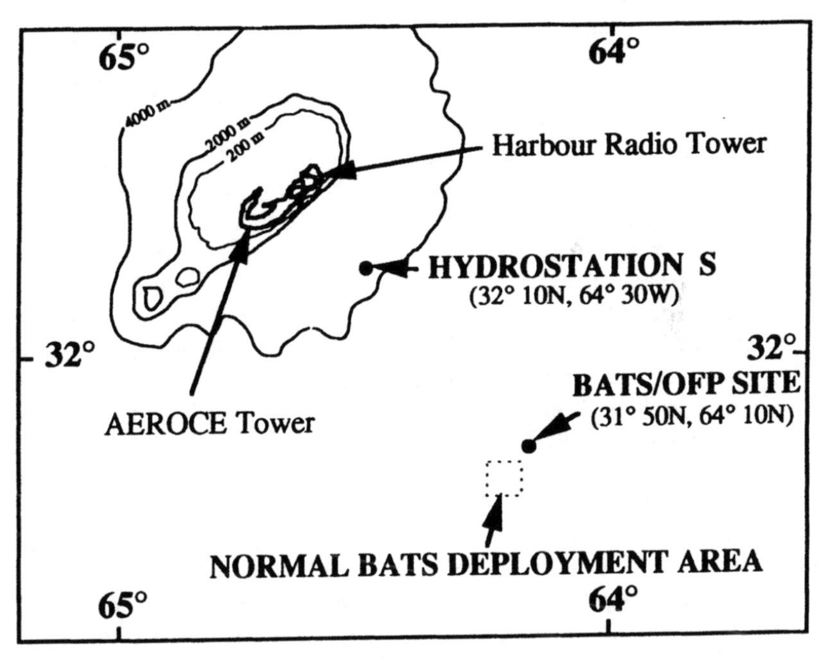

Figure 11-1. Map of Bermuda and the surrounding waters illustrating the location of the Bermuda Atlantic Time-Series Study (BATS) site, the site of the Hydrostation S sampling (1954 to the present), the location of the Ocean Flux Program site (Deuser 1986), and the locations of three of the atmospheric sampling stations on the island of Bermuda.

site of Dr. Werner Deuser's (WHOI) long-term deep-sea sediment trap mooring (Deuser 1986). On each BATS cruise, a free-drifting sediment trap array is deployed approximately 7 km southeast of the OFP station and all of the hydrocasts are started near this array. The 400-m array typically drifts between 10 and 50 km (occasionally 150 km) during the three-day deployment. On each cruise, hydrocasts and water bottle collections are made at 36–48 depths during a series of 4–7 casts from the surface to 4200 m. Details of the sampling scheme, analytical methods, and quality control procedures are available in the BATS Methods Manual and Data Reports (Knap et al. 1991, 1992, 1993a,b) and in published papers (Michaels et al. 1993, in press; Lohrenz et al. 1992). The list of BATS measurements is presented in Table 11-1, and more detail of each are presented in the text.

A Seabird CTD with additional sensors is used to measure continuous profiles of temperature, salinity, dissolved oxygen, downwelling irradiance, beam attenuation, and *in situ* fluorescence. This instrument package is mounted on a 12-position General Oceanics Model 1015 rosette, which is typically equipped with twelve, 12-liter Niskin bottles. The salinity and dissolved oxygen concentrations as calculated from the CTD sensors are calibrated with the discrete salinity and oxygen measurements collected from the Niskin bottles on the rosette (Michaels et al. in press; Knap 1993b).

Water samples for the discrete chemical and biological analyses are collected in Teflon-coated Niskin bottles on a General Oceanics rosette. Gas samples can be easily compromised by contact with the atmosphere. Therefore, the samples for dissolved oxygen are sampled and processed before any other measurements. They are drawn into individually numbered, 115-ml BOD bottles and analyzed using the Winkler titration chemistry. The samples are titrated with an automated titration and endpoint detection system. Automated systems improve analytical precision and reduce the influence of variability between analysts. The duplication of oxygen samples at every depth allows a routine check on the precision of the oxygen measurement (see next section). Oxygen saturation is calculated from the Weiss algorithms (Weiss 1970). Samples for dissolved inorganic carbon (DIC) and alkalinity are drawn after the oxygen samples or are sampled first on a different cast. The DIC samples are analyzed with an automated coulometric system, and alkalinity is determined by titration. Salinity samples are drawn after the gas samples and are analyzed within one week of collection on a Guideline AutoSal 8400A salinometer. The samples are standardized to IAPSO Standard Seawater regularly during each sample run. Samples for nitrate+nitrite and phosphate are filtered with Whatman GF/F glass fiber filters and frozen. Samples for silicate are filtered through 0.4-μm Nucleopore filters and refrigerated at 4° C. The samples are analyzed within two weeks using traditional colorimetric techniques (Strickland and Parsons 1972) or a Technicon Nutrient Autoanalyzer (Knap et al. 1993b).

Chlorophyll samples are filtered onto Whatman GF/F glass fiber filters, frozen

Table 11-1. List of core measurements made in the BATS program

Continuous electronic measurements: Parameter	Depth Range (m)	Technique/Instrument
Temperature	0–4200	Thermister on SeaBird SBE-9 CTD
Salinity	0–4200	Conductivity sensor on SeaBird SBE-9 CTD
Depth	0–4200	Digiquartz pressure sensor on SeaBird SBE-9 CTD
Dissolved oxygen	0–4200	SeaBird polarographic oxygen electrode
Beam attenuation	0–4200	SeaTech, 25 cm transmissometer
Fluorescence	0–500	SeaTech fluorometer
PAR	0–500	Biospherical scalar irradiance sensor, 400–700 nm

Discrete measurements from Niskin bottles on CTD: Parameter	Depth Range (m)	Technique/Instrument
Salinity	0–4200	Conductvity on Guideline AutoSal 8400A
Oxygen	0–4200	Winkler titration, automated endpoint detection
Total CO_2	0–250	Automated coulometric analysis
Alkalinity	0–250	High precision titration
Nitrate	0–4200	CFA colorimetric with Technicon AA
Nitrite	0–4200	CFA colorimetric with Technicon AA
Phosphate	0–4200	CFA colorimetric with Technicon AA
Silicate	0–4200	CFA colorimetric with Technicon AA
Dissolved organic carbon	0–4200	High-temperature, catalytic oxidation
Particulate organic carbon	0–4200	High-temperature combustion, CHN analyzer
Particulate organic nitrogen	0–4200	High-temperature combustion, CHN analyzer
Particulate silica	0–4200	Chemical digestion, colorimetric analysis
Fluorometric chlorophyll *a*	0–250	Acetone extraction, Turner fluorometer
Phytoplankton pigments	0–250	HPLC, resolves 19 pigments
Bacteria	0–3000	DAPI stained, fluorescence microscopy

Rate measurements: Parameter	Depth Range (m)	Technique/Instrument
Primary production	0–140	Trace-metal clean, *in situ* incubation, ^{14}C uptake
Bacterial activity	0–140	Thymidine incorporation
Particle fluxes	150,200,300	Free-drifting cylindrical trap (MultiPITs)
Mass flux		Gravimetric analysis
Total carbon flux		Manual swimmer removal, CHN analysis
Organic carbon flux		Manual swimmer removal, acidification, CHN
Organic nitrogen flux		Manual swimmer removal, CHN analysis

in liquid nitrogen, and analyzed by both fluorometric and HPLC techniques. Particulate organic carbon and nitrogen samples are filtered onto precombusted GF/F glass fiber filters, dried and acidified under vacuum, analyzed on a Control Equipment Corp. 240-XA Elemental Analyzer, and standardized using acetanilide as a reference standard. Particulate matter is defined operationally by the pore size of the filters. Whatman GF/F glass fiber filters have a nominal pore size of 0.7 μm. These typically remove nearly all of the phytoplankton, but nearly half of the bacteria and probably most of the ocean viruses pass through these filters. Inclusion of the bacteria in the particulate carbon values would increase the values by 20–40% (Gundersen and Michaels, unpublished data).

Primary production is measured by the uptake of ^{14}C in dawn-to-dusk *in situ* incubations. All the sampling and sample processing employs the new trace-metal clean procedures (Fitzwater et al. 1982). These procedures have resulted in a two- to fourfold increase in the specific production rates (mg C/mg chl/hr) measured in this area compared with the data from the 1950s and 1960s (Michaels et al. in press). Total production rates are only slightly higher (Lohrenz et al. 1992), perhaps because the more vigorous winter mixing in the 1950s and 1960s led to larger spring blooms than have occurred during the BATS era. Incubations usually are conducted *in situ* at eight standard depths (20-m vertical spacing) in the upper 140 m. Details of the primary production methods are found in the BATS methods manual and in Lohrenz et al. (1992). Bacterial productivity is estimated at the same depths using a modified version (Knap 1993b) of the tritiated-thymidine technique.

The configuration of the sediment trap array is described in Lohrenz et al. (1992) and follows the basic MultiPIT design (Knauer et al. 1979). The sediment traps are constructed of 3-inch I.D., 27-inch-long polycarbonate tubes with 0.5-inch I.D. baffles on the top. Multiple tubes (up to 12) are mounted on PVC crosses. These crosses are attached to the array line at depths of 150, 200, 300, and 400 m. The flotation system includes both subsurface and surface floats with a 5-m section of bungee cord in between, to reduce (but not eliminate; see Gust et al. in press) the transfer of surface wave motions to the subsurface line. Zooplankton swimmers (animals that actively enter the trap) are removed by examining the filtered samples at 25–50X magnification and manually removing the zooplankton with a very fine forceps. The removed zooplankton are saved in preservative so that they can be compared with zooplankton counts from replicate traps that have not been filtered. This allows some estimate of the effectiveness of swimmer removal (Michaels et al. 1990). Total mass flux is determined gravimetrically, and the carbon and nitrogen content of the samples are determined with a CHN analyzer.

Data Quality and Continuity

A critical concern with any time series study is that the data quality (accuracy and precision) remain constant (or are known) for the duration of the study. This ensures that changes in the dataset over time are actually due to changes in the natural system being measured rather than arising from changes in the measurement technique or data quality. This problem is fundamental to the interpretation of any time series record and in practice can be very difficult. Optimally, we would hope that every measurement could be traced directly back to a primary certified standard, that measurements of analytical precision would be collected for every sample, and that all of the documentation of the data quality would be included in the overall dataset. For most oceanographic measurements, however, there are no primary certified standards, and the methods themselves are often under constant modification as demonstrated problems are solved. We use a variety of practices to help ensure that the BATS data quality is consistent, that our accuracy and precision remain the same, and that we can document the accuracy and precision of the measurements through time. Some of these practices are described below. This is not an all-inclusive list of quality assurance techniques, but represents a practical subset for a routine oceanographic time series.

People are the most significant issue in the creation and maintenance of a long-term time series. The maintenance of high data quality over years to decades is possible only when the people who make the measurements are dedicated, motivated, highly skilled, and truly interested in the science. For very long time series, the dedication of the measurement team is the most difficult challenge. By its very nature, a time series is a repetitive process that can become extremely boring for the people who actually do the work. At BATS we are fortunate in having a very dedicated group of skilled technicians who are also interested in the scientific goals of the program. These interests are encouraged by having many of the technicians simultaneously enrolled in graduate programs. The technicians participate in scientific meetings where they present papers and posters, and they are encouraged to write scientific papers on aspects of the core program or individual research topics that they investigate in the context of the time series study. Without this group of people and their high level of personal interest in the success of the program, mechanistic solutions to ensure data quality, such as those described below, would be unsuccessful.

Replication of 10–100% of the Samples in Each Cast

The precision of each measurement is determined directly for each cruise by making replicate measurements on a routine basis. There are a number of possible sources of variability, some related to the quality of the measurement and others related to the representativeness of the measurement. We routinely consider four sources of variability for an individual sample or profile: (1) analytical, (2) sampling, (3) between-bottle, and (4) natural spatial and temporal heterogeneity on a time/space scale shorter/smaller than the nominal sampling interval between cruises. The first two sources of variability, analytical and sampling, concern the quality of an individual measurement. The second two are related to the representativeness of the measurements.

We routinely take replicate samples from the same Niskin sampling bottle for 10–20% of the sampled depths. For the measurement of dissolved oxygen, where sampling and measurement have proved more variable, we routinely draw duplicate samples from every sampling bottle. This within-bottle replication provides a direct measurement of the sum of the variability due to analysis and sampling, a value that we consider the precision of the measurement on that cruise. In addition, we always make replicate analyses of the standards for each measurement. This gives an estimate of the analytical precision for direct comparison with the overall precision. These data are reported directly in the data reports, so that the information on precision for each cruise is always present.

Natural variability on the scale of Niskin bottle sampling and on larger scales is not well resolved. Replicate Niskin bottles at the same depth are regularly sampled at one to two depths each cruise, usually below 2000 m. Complete replicate casts are made infrequently, primarily because of funding constraints. Replicate casts on each cruise would provide data on small-scale (kilometer) and short-term (hours/days) variability and could lead to information on the representativeness of a single cast for the individual cruise. Space scales of 20–100 km are resolved in explicit experiments four times each year to assess mesoscale hydrographic features (see below).

Reference to Absolute Standards

Wherever possible, we refer the sample analysis back to an absolute reference material or standard. Unfortunately, there are no reference standards for most oceanographic measurements. In the few cases where reference materials exist, we document the link to the reference materials in the data reports. Thermistors and pressure sensors on the CTD system are sent to an appropriate calibration facility, where the calibration is checked against a known reference. Salinity samples are standardized against a consensus reference material, IAPSO water. For some wet chemical measurements (e.g., nutrients), there are reference materials that can be used to check for accuracy; even in these cases, however, there is no consensus in the field as to the appropriateness of these reference materials.

Standard and Sample Carry-Over Between Time Points

In the absence of absolute standards, we have developed a set of techniques to guard against measurement inaccuracies arising from random or systematic variations in the accuracy of the standards between cruises or analysis periods. These internal standards are created in batches, then analyzed over a period of months to years. Many ocean properties are fairly consistent in the deep sea, which acts as a further check on measurement accuracy. For these internal standards, we collect a large number of replicate samples from the same depth or make a large amount of a standard material. This material is stored and a small portion of the reference materials are run with each subsequent set of samples. This method works only where there is a good preservation technique. To guard against problems with preservation, we sometimes have two separate batches of reference materials of different sources and source times running concurrently. Systematic offsets in the internal standards, the normal standards, and deepwater values are clues that the inaccuracy of the standard is leading to variability in the measurements.

Internal Consistency Between Measurements

Datasets are also examined closely for internal consistency. As mentioned above, deepwater properties are less variable, and systematic offsets in deepwater properties are frequently a good indicator of accuracy and precision problems. Many ocean properties covary, particularly those of the biologically labile compounds. Elemental ratios of nutrients and oxygen have fairly well-documented average behaviors (the Redfield ratios), and large deviations from these ratios may be an indication of a quality problem with one measurement. As with any single dataset, this analysis must be treated with caution, however, as there can be substantial real deviations from the Redfield ratios.

Overlap of New Sampling Methods and People Before Changeover

Methods change and improve with time. In the BATS time series, we strive to make each measurement in the best possible way. This means that as techniques improve, we must change our methods to compensate. Conversion to a new method is an opportunity for long-term biases to be introduced to the dataset if the new method produces a different value than the old one. This offset occurs quite readily when the new method involves an increase in sample resolution or a change in the detection limits. We routinely schedule an overlap period of up to one year during which the measurement is made with both techniques on the same samples. Data quality can likewise be altered systematically when there is turnover among the people who make the measurements. We try to have some overlap in analysts and to have replicate samples analyzed by both outgoing and incoming analysts to measure any offsets during the change in personnel. We also try to have several people conducting each measurement, so that the sudden departure of a single person does not compromise measurement quality before a new person is hired. These overlap procedures allow us to determine and document the magnitude of any systematic differences between the two techniques and between two people.

Routine Participation in Methods Intercomparisons with Other Labs

Intercomparisons and intercalibration exercises are another means of documenting that the time series techniques yield results that are similar to those measured

Table 11-2. Time series research programs on or near Bermuda

Principal Investigator	Topic	Funding
Ancillary research at BATS and Hydrostation S		
Ammerman (TAMU)	Bacteria dynamics and phosphate cycling	NSF
Bidigare (UHawaii), Andersen (Bigelow), Keller (Bigelow)	Pigments and phytoplankton	NSF
Boysen (UCSC)	Coccoliths as flux tracers	Post-doc
Brzezinski (UCSB) and Nelson (OSU)	Silica cycling and flux	NSF
Buesseler (WHOI)	Thorium and colloid dynamics	NSF
Buesseler (WHOI), Michaels (BBSR), and Knap (BBSR)	Calibration of traps with thorium	NSF
Chisholm (MIT) and Olson (WHOI)	Picoplankton dynamics by flow cytometry	NSF
Dacey (WHOI), Wakeham (SkIO) and Michaels (BBSR)	DMS dynamics	NSF
Ducklow (HPEL) and Carlson (HPEL)	Bacteria dynamics	NSF
Ferrari (Smithsonian)	Copepod sex ratios	Smithsonian
Giovannoni (OSU)	Bacteria molecular biology	NSF
Gundersen (BBSR)	Bacteria nitrogen cycling	Grad student
Gust (UHamburg)	Sediment trap hydrodynamics	NSF
Jickells (UEA)	Iodine, iodate dynamics	NERC
Johnson (BBSR)	Biological-physical modeling	Grad student
Keeling (SIO)	Surface CO_2 time series	NSF
Knap (BBSR) and Bates (BBSR)	Profile CO_2 and alkalinity	NOAA
Lipschultz (BBSR)	Nitrogen cycling	NSF
Madin (WHOI)	Zooplankton dynamics	NSF
Michaels (BBSR), Caron (WHOI), and Swanberg (UBergen)	Symbiotic protozoa and carbon fluxes	NSF
Moffett (WHOI)	Trace metal photo-oxidation, chelation	ONR
Moran (WHOI)	Colloids and trace metals	Post-doc
Sholkovitz (WHOI)	Rare earth element dynamics	NSF
Sherriff-Dow (BBSR)	Bacterial metabolism of DOC	Grad student
Siegel (UCSB), Smith (UCSB), and Michaels (BBSR)	Bio-optical profiling	NSF
Siegel (UCSB) and Michaels (BBSR)	SeaWIFS Cal/Val and algorithm development	NASA
Siegel (UCSB), Michaels (BBSR), Brzezinski (UCSB), and Hansell (BBSR)	Inherent optical properties	NASA

(Continued)

Table 11-2. (Continued)

Principal Investigator	Topic	Funding
Thierstein (ETH)	Coccolithophore distributions	ETH
Wakeham (SkIO)	^{13}C of organic compounds	NSF
Waser (WHOI)	Natural ^{32}P and ^{33}P cycling	Grad student
Zafiriou (WHOI)	Deep ocean nitrite dynamics	NSF
Independent research programs		
Deuser (WHOI)	Deep-ocean Sediment Trapping	NSF
Madin (WHOI)	Salp Ecology and Dynamics	NSF
Sayles (WHOI)	Deep-ocean Benthic Geochemistry	NSF
AEROCE (Many P.I.s)	Atmospheric Chemistry Time-series	NSF
Knap and Simmons	Bermuda Air-quality Time-series	Bermuda

by other labs in the field at the same time. We conduct informal intercomparisons with the other JGOFS time series study in Hawaii, and we also try to participate in as many formal intercomparison exercises as possible.

Ancillary Programs

One of the greatest strengths of the BATS program is the relatively large number of other scientists who are funded to conduct research in this area (Table 11-2). Many are funded for ancillary projects on the BATS cruises; others have their own independent programs. Some of these independent programs (most notably Werner Deuser's moored sediment trap program and the atmospheric sampling programs) predate the BATS program. The ancillary users at BATS are scientists who make measurements on the BATS cruises; in many cases, the samples are actually collected by the BATS technicians. These samples are colocated in time and space with the core BATS measurements and allow for a great deal of comparison between the datasets. The ancillary measurements and scientists also investigate processes that may help in the interpretation of the core measurement program.

Seasonal Patterns of Physics, Chemistry, and Biology at BATS and Hydrostation S

The BATS data for the first three years are presented as time versus depth contour plots for some of the measured parameters in the upper 400 m of the water column (Figures 11-2 and 11-3). Upper ocean seawater values were variable and showed an obvious seasonal pattern of winter mixing and a resultant spring bloom. The depth of the mixed layer showed a spring maximum in all three years and was followed by a rapid stabilization to a shallow summer mixed layer. Despite this similarity, there were significant differences between years in the intensity of winter mixing, with the intensity significantly greater in 1989 and

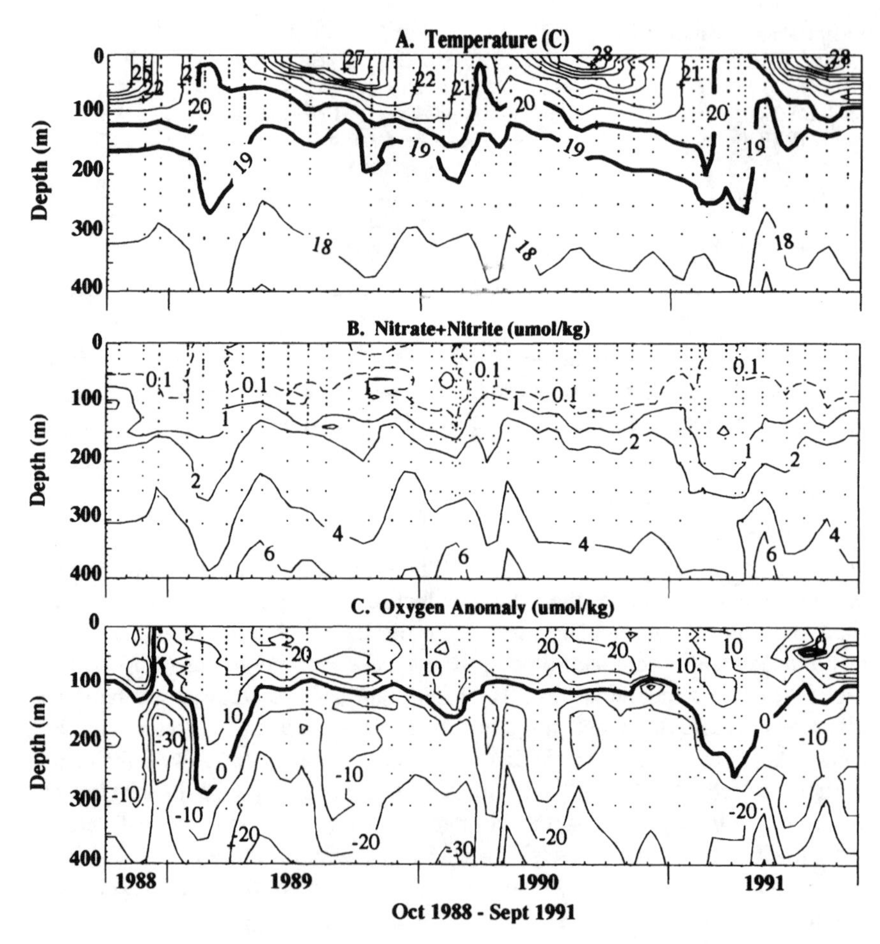

Figure 11-2. Contour plots of temperature, oxygen anomaly, and nitrate at the BATS site during the first three years of the program (October 1988–September 1991): (A) potential temperature (° C); (B) nitrate+nitrite (μmol/kg); (C) oxygen anomaly (μmol/kg). Oxygen anomaly is defined as the difference between the measured oxygen concentration and the calculated saturation concentration of oxygen at the *in situ* temperature and salinity.

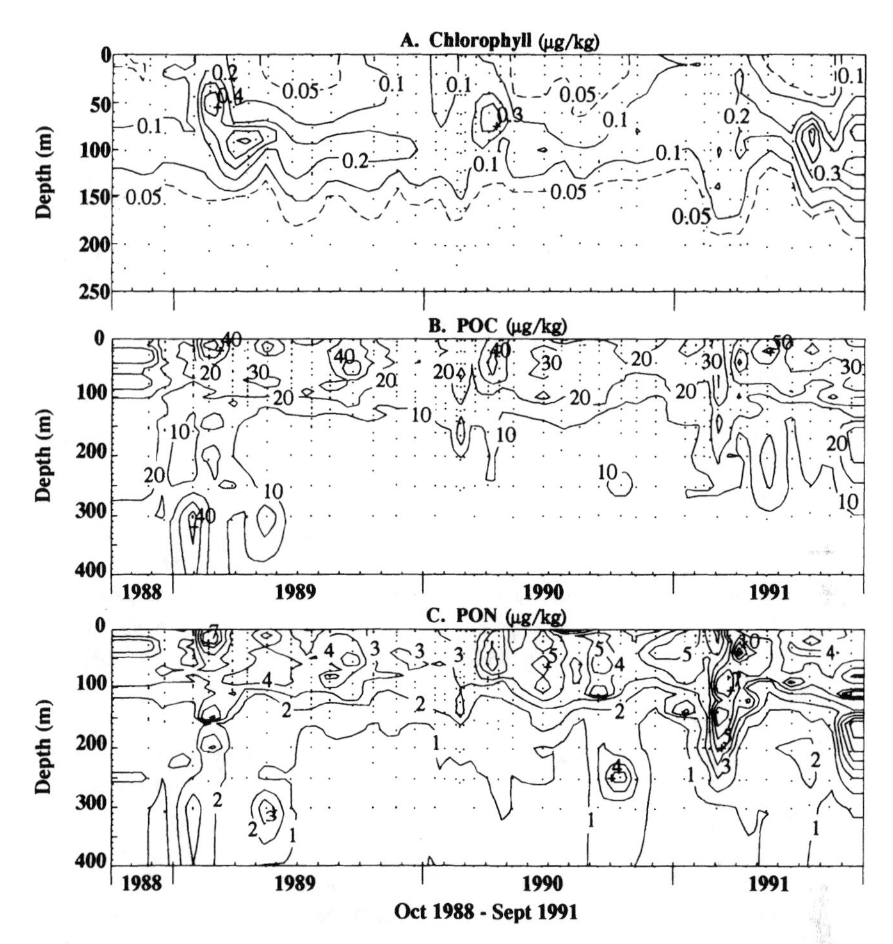

Figure 11-3. Contour plots of chlorophyll, particulate organic carbon, and nitrogen at the BATS site during the first three years of the program (October 1988–September 1991): (A) chlorophyll *a* (μg/kg); (B) particulate organic carbon (μg/kg); (C) particulate organic nitrogen (μg/kg).

1991 than in 1990. These interannual differences are reflected in the biological parameters.

During the February 1989 spring bloom, oxygen concentrations were supersaturated from the surface to 250 m depth, whereas in December 1988 the water was undersaturated with oxygen at all depths (Figure 11-2C). Following the 1989 bloom, oxygen anomalies continued to increase in the stratified surface waters, and they decreased below saturation in the lower euphotic and upper mesopelagic zones. Nutrient concentrations in the upper euphotic zone were uniformly low during most of the year (Figure 11-2B). During the 1989 bloom sampling, the concentration of nitrate+nitrite in the mixed layer reached values of 0.5 to 1.0 μmol/kg. Phosphate and silicate concentrations did not show a seasonal increase during this period. During the winter and spring of 1990, the oxygen concentration in the mixed layer was always above saturation. Despite the 170-m mixed layer, which should have entrained water with measurable nutrients, nitrate+nitrite was near the detection limit of 0.05 μmol/kg during the spring 1990 cruises. There was some depletion of nitrate between 100 and 150 m during the February 1990 cruise in the period of deepest mixing. The bloom in 1991 was of similar intensity to that in 1989 but lasted for a much longer period. Oxygen concentrations were slightly above saturation early in the bloom and increased as it progressed. As with 1989, the supersaturation extended to the base of the 200-m mixed layer. There was measurable nitrate in the euphotic zone for more than three months.

In addition to the spring mixing events, there were small surface maxima in nitrate+nitrite concentrations in September, October, and November of 1989 (Figure 11-2B). These elevated surface nutrients may be due to rainfall inputs from summer storms and hurricanes (Michaels et al. 1993), although they occurred after the summer minimum in salinity. Coincident with these surface nutrients, there were anomalously high nutrients near 60 m in both October and November of 1989.

Each spring bloom was characterized by the presence of elevated stocks of chlorophyll *a*, particulate organic carbon (POC), and particulate organic nitrogen (PON) (Figure 11-3). Despite the relatively recent mixing, the concentration of chlorophyll *a* in February 1989 was a factor of 10 higher than the upper euphotic zone chlorophyll concentration during the rest of the year (Figure 11-3A). By the next sampling period (March 26, 1989), the subsurface chlorophyll maximum had descended to 100 m. This maximum became less pronounced through the summer. Particulate organic carbon and nitrogen were also elevated by a factor of 10 at the surface during the 1989 bloom. As with chlorophyll, the surface POC and PON returned to typical values by March. However, unlike the chlorophyll, POC and PON did not show a marked subsurface maximum.

The spring bloom in 1990 also was characterized by elevated concentrations of chlorophyll *a* and particulate organic matter between February and April 1990. The peak 1990 chlorophyll concentration was 40% of the peak 1989 value of 0.5 μg/kg. Peak POC and PON concentrations were 20–35% lower than the peak values in 1989.

The bloom in 1991 was very different in its pattern of particulate materials. Although the chlorophyll concentration did increase to values of greater than 0.5 μg/kg, this increase occurred in the June–September period, well after the depletion of nutrients. Particulate organic carbon and nitrogen peak values were more synchronous with the nutrient depletion and, again, were concentrated near the surface.

The rates of phytoplankton primary production as measured by the uptake of ^{14}C and particulate carbon fluxes measured with the sediment traps also showed a seasonal pattern synchronous with the winter mixing (Figure 11-4). However, the seasonal variations in the production rates were at odds with the relative strength of each bloom as interpreted from the intensity of the mixing. The period of high production was short in 1989 (a single cruise), while it was more prolonged during 1990, the winter with the shallowest mixed layer. Production in spring 1991 was the highest of the three spring periods, but again, elevated production lasted past the period of nutrient inputs. In addition, there were periods of high production during the late summer, traditionally considered a time of low production. The June–July production peaks in 1989 preceded the anomalous surface nutrients observed that fall. The high summer production in 1990 lasted much of the fall period and was not accompanied by observable nutrient inputs.

Sediment trap fluxes at 150 m showed peak fluxes that are usually synoptic with the periods of high production (Figure 11-4). Fluxes were high in 1990, the year with the smallest nutrient inputs, as inferred from one-dimensional profiles; and they were relatively low in 1991, when there was a prolonged period of nutrient inputs and elevated production. The annual average carbon and nitrogen fluxes estimated by the sediment trap are 0.8 mol C/m^2/yr and 0.1 mol N/m^2/yr. Sediment trap fluxes are usually less than 10% of the rates of primary production, suggesting that most of the primary production is recycled in the upper water column.

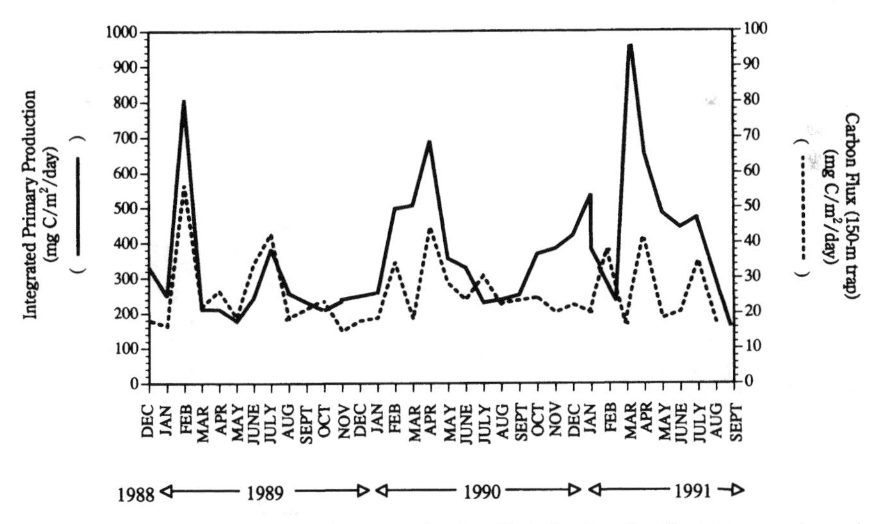

Figure 11-4. The integrated primary production (0–140 m) and sediment trap estimated carbon flux at 150 m at the BATS site. Data for the first two years are also found in Lohrenz et al. (1992).

Measurement and Inference of Biogeochemical Rate Processes

Menzel and Ryther (1960, 1961) first documented the seasonal pattern of mixing, biomass, and production in the Sargasso Sea at Hydrostation S, 45 km northwest of the BATS site. Although the BATS dataset differs in the intensity of deep mixing and the magnitude of primary production, the overall patterns and their explanation are the same. Surface cooling and wind mixing in the winter and spring cause the formation of a mixed layer that erodes into the nutrient-rich layers below the euphotic zone. This deep mixing introduces nutrients into the euphotic zone. After the water column stabilizes in early spring, these nutrients stimulate a bloom of phytoplankton and a period of increased primary production that lasts from one to three months. In both datasets, we can see substantial interannual variability in the timing and intensity of the bloom, related to variations in the intensity and timing of deep mixing in the winter. In the BATS dataset, we can also see that particle exports are often synoptic with the productivity peaks and that nutrient levels are very rapidly drawn back below detection limits. The rapid disappearance of nutrients and covariance of carbon fluxes suggest a rapid transfer of inorganic material into biomass, and perhaps an equally rapid transport of organic material out of the euphotic zone.

The marked seasonality and decadal variability in ocean mixing of the ocean near Bermuda are firmly established from the 37-year Hydrostation S record. There is a significant decadal scale variability in the record, with a period of lower ventilation in the mid-1970s bracketed by periods of strong ventilation in the early 1960s and the late 1970s (Jenkins and Goldman 1985). A more recent analysis shows that the mid-1980s were also a period of low ventilation with winter mixed layers of less than 80 m (Figure 11-5). The mixed layers of approximately 150 m in the years immediately before the BATS program were shallow compared to the historical values.

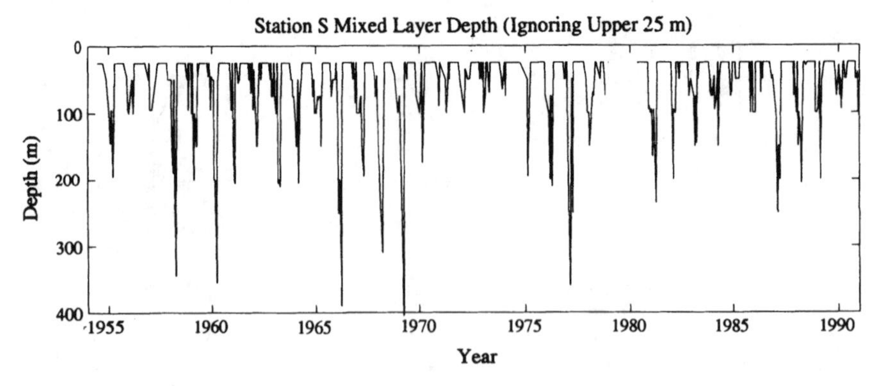

Figure 11-5. Depth of the mixed layer at Hydrostation S from 1955 to 1990. (Figure also appears in Michaels et al. in press.)

For most of the year, nutrients are below detection limits and the supply of nutrients is thought to limit the production of the upper ocean. Nutrients for plant growth can come from two types of sources: internal recycling, or from outside the euphotic zone. New production is defined as the phytoplankton primary production supported by the uptake of exogenous nutrients (Dugdale and Goering 1967), primarily nitrates from below the euphotic zone. Over time scales where steady state assumptions are appropriate, the amount of new production will equal the export of organic nitrogen from the euphotic zone. These export processes include the sinking of organic particles (Eppley and Peterson 1979), the downward mixing of dissolved organic nitrogen (Jackson 1988; Toggweiler 1989), or the vertical migration of zooplankton (Longhurst and Harrison 1988). The relatively consistent stoichiometric relationships among oxygen, carbon, and nutrients in oceanic particulate matter (the Redfield ratios) can be used to compare the cycling of the different elements. Annual average new production rates have been inferred from the seasonal cycles of oxygen and nitrate and by incorporating data on the ventilation time scale of the main thermocline from tritium dating (Jenkins 1982, 1988; Jenkins and Goldman 1985; Spitzer and Jenkins 1989). These rates consistently range from 0.4–0.7 mol $N/m^2/yr$. Tracer methods for inferring annual average new production rates integrate over relatively large spatial scales (hundreds of kilometers) and long time scales (years to decades) and are a robust description of the average new production.

The particle fluxes and new production rates estimated with the sediment traps (0.1 $mol/N/m^2/yr$) are approximately 20% of the 0.4–0.7 mol $N/m^2/yr$ estimated by Jenkins and co-workers. This discrepancy between short-term measurements (the traps) and the long-term tracer studies must be resolved to interpret fully the carbon and nutrient cycles in this area. First, particle fluxes may not be measured accurately by sediment traps. There are many known problems with trapping technology, and we cannot rule out that the technique itself is flawed (Buesseler et al. submitted; Knauer and Asper 1989). Second, there may be other processes that transfer organic carbon to depth. The vertical migration of zooplankton (Longhurst and Harrison 1988) and the downward mixing of elevated DOC and DON concentrations near the surface (Jackson 1988; Toggweiler 1989) have both been suggested, though neither has been shown to account for 80% of the export in any region. Third, particle export processes may be occurring on time and space scales that are not sampled adequately by the short trap deployments (three days out of each month) used in the BATS program. Finally, the new production rates estimated by Jenkins and co-workers are from data sets collected in the 1960s and in the early 1980s (Jenkins 1982, 1988; Jenkins and Goldman 1985; Spitzer and Jenkins 1989). It is possible (though unlikely given the mixed layer depth chronology) that new production rates in the early 1990s are dramatically different than in earlier periods.

At the BATS station, new production rates have also been estimated from the changes in oxygen and nitrate concentrations during the February 1989 bloom

(Michaels et al. in press). During winter mixing, the nitrate concentrations in the euphotic zone increase, and all of this euphotic zone nitrate disappears by late spring. Consequently, the stock of nitrate in the upper 100 m at its highest concentrations is a minimum estimate of the new production rate for that year. A similar rationale can be used to look at the stocks of oxygen during the bloom. These rates of new production for the 1989 bloom were 19–44% of the annual average new production of 0.5 mol N/m^2/yr estimated by Jenkins and co-workers (Michaels et al. in press). The estimates of new production in this single event also equal or exceed the annual particle export as estimated from short-term sediment trap deployments (Lohrenz et al. 1992) at the same station. With both nitrate and oxygen, these estimates are underestimates of the new production associated with this event, as they ignore continued mixing of nutrients into the system as the surface waters are depleted. Similar short-term production events are also noticeable in the BIOWATT mooring data from a location west of Bermuda (Dickey 1991). Clearly, if they are actually a product of temporal variation, spring bloom events like those observed at BATS could account for a significant proportion of the annual new production.

We can also use this technique and the Hydrostation S dataset to estimate the minimum new production during each spring bloom over the past four decades. Figure 11-5 shows the chronology of mixed layer depths since 1955. We can estimate the nitrate content of the mixed layer during the period of deep mixing from the temperature of the mixed layer (e.g., Siegel et al. 1990). Between 17° and 19.5° C, there is a strong correlation between the nitrate concentration and the temperature of the water in the main thermocline. As the surface cools and the mixed layer deepens, it begins to entrain nutrients from these deeper, nutrient-rich layers. Thus, we can use the temperature of the mixed layer at its deepest extent to infer the nitrate concentration. We must assume that the nitrate-temperature relationship has remained unchanged over the past four decades. We also assume that all of the nitrate mixed into the upper 100 m is then used by phytoplankton (i.e., the depth of the nutricline has been constant at 100 m). Using these assumptions, we can calculate minimum new production rates for every year that the mixed layer temperature drops below 19.5° C. This result is presented in Figure 11-6. Again, there is substantial interannual variability in the minimum new production, with rates up to 0.45 mol/N/m^2/yr during the 1960s and periods of lower new production in the early 1970s and mid-1980s. This analysis supports new production estimates on the order of 0.5 mol N/m^2/yr and further suggests that the sediment trap values of 0.1 mol N/m^2/yr must underestimate the total new production.

Short-Term Variability in Phytoplankton Community Composition

The HPLC pigment data indicated major differences in phytoplankton community structure between biweekly cruises in 1990 (Figure 11-7). The interpretation of

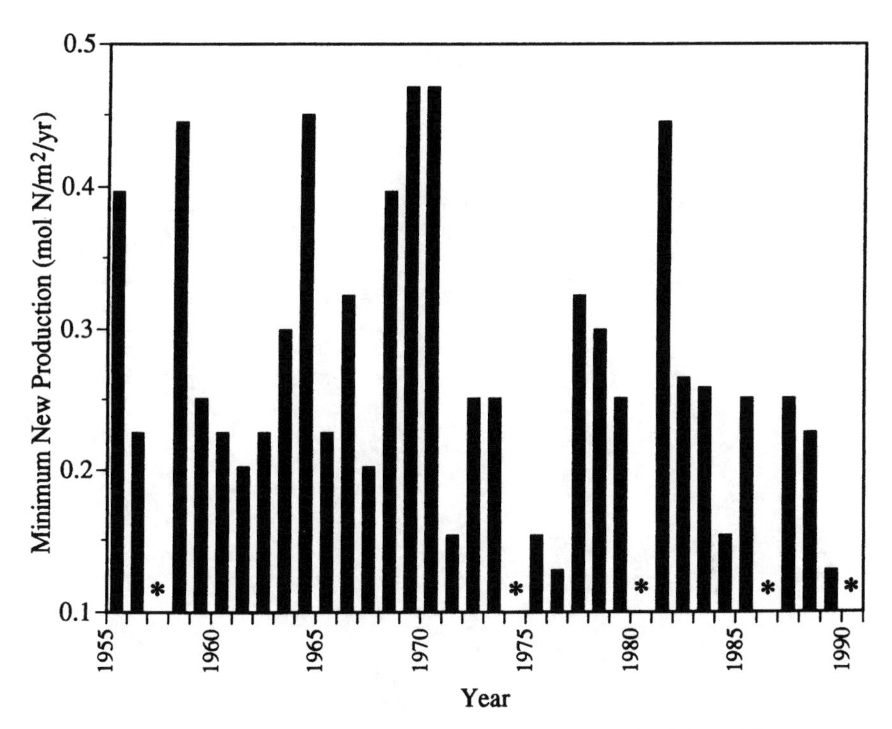

Figure 11-6. Minimum new production associated with the spring bloom. New production is calculated from the temperature of the deepest measured spring mixed layer and a regression of temperature and nitrate concentration (see text). Asterisks show years when mixed layer temperatures exceeded 19.5° C, outside the range over which the temperature-nitrate regression holds. Nitrate was probably undetectable throughout the spring in these years (see 1990 data in Figure 11-2B).

these pigment profiles is presented in detail elsewhere (Michaels et al. in press). The important features for the purposes of this chapter are the large variations in the vertical structure of different pigments between the biweekly cruises. Each of these pigments is characteristic of one or more groups of phytoplankton. For example, in early May, the chlorophyll *b* peak at 100 m was associated with elevated zeaxanthin concentrations at the same depth, suggesting the presence of prochlorophyte-like phytoplankton. In early April, there is no corresponding zeaxanthin peak at 100 m, indicating that the chlorophyll *b* was associated with eukaryotic chlorophytes and/or prasinophytes, and not prokaryotic prochlorophyte-like phytoplankton. Some pigments (fucoxanthin, chlorophyll c_1+c_2, chlorophyll c_3, diadinoxanthin, and diatoxanthin) were present throughout the bloom period. Sometimes they co-occurred with 19′-hexanoyloxyfucoxanthin and 19′-butanoyloxyfucoxanthin, suggesting that prymnesiophytes and chrysophytes (respectively) were present. At other times, elevated levels of fucoxanthin,

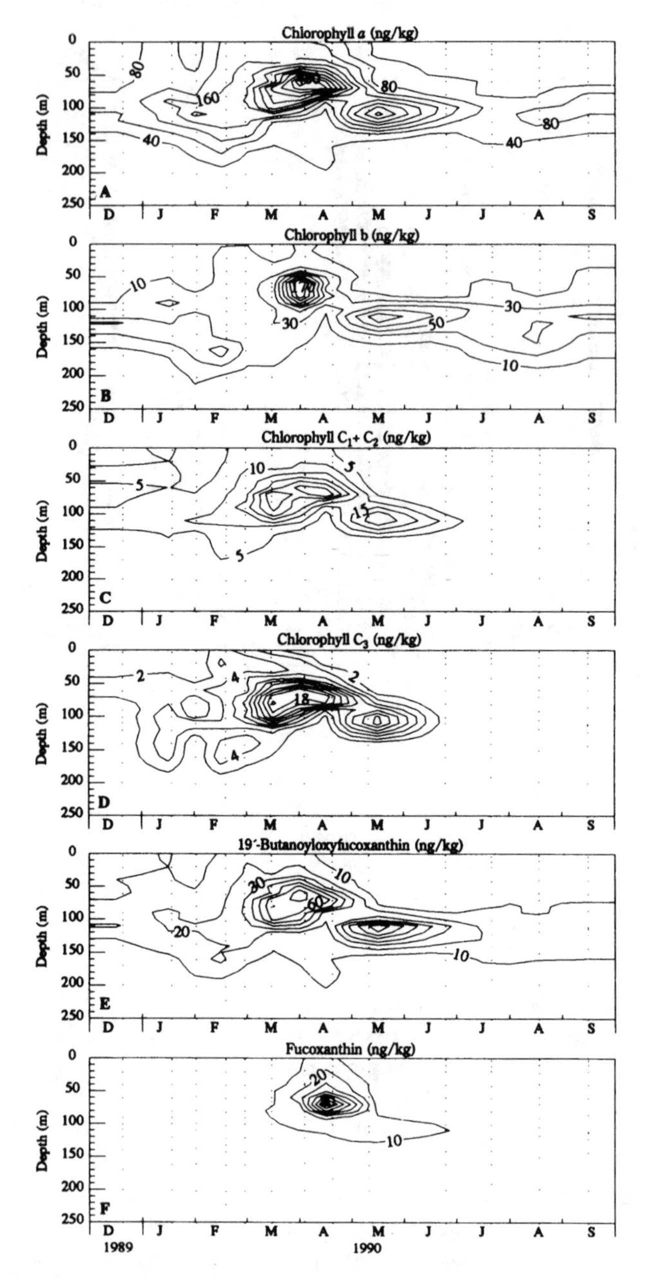

Figure 11-7. Contour plot of HPLC-determined pigment concentrations (ng/km) at the BATS site during December 1989–June 1990): (A) chlorophyll *a;* (B) chlorophyll *b;* (C) chlorophyll $c_{1+}c_2$; (D) chlorophyll c_3; (E) 19′-butanoyloxyfucoxanthin; (F) fucoxanthin; (G) 19′-hexanoyloxyfucoxanthin; (H) zeaxanthin; (I) peridinin; (J) prasinoxanthin; (K) diadinoxanthin; (L) diatoxanthin. (Figure modified from Michaels et al. in press).

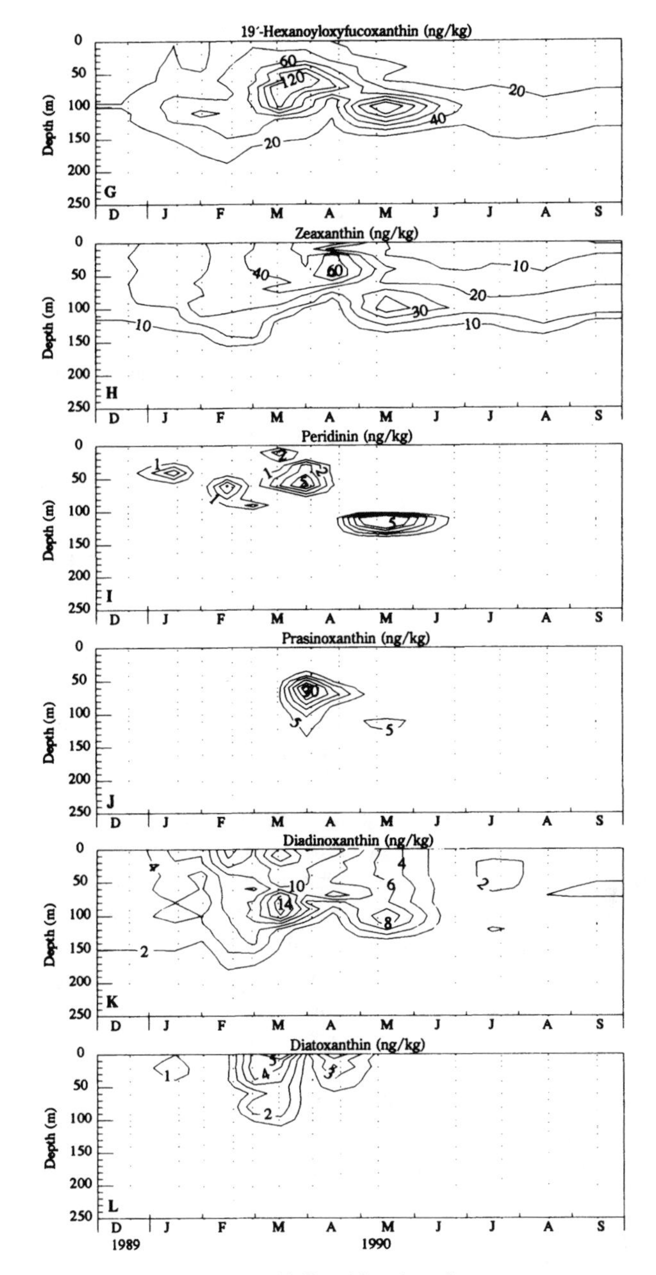

Figure 11-7. (Continued).

diadinoxanthin, and diatoxanthin concentrations were observed without high levels of 19′-hexanoyloxyfucoxanthin and 19′-butanoyloxyfucoxanthin, a possible indication of diatom-dominated blooms.

Two interpretations of these high-frequency variations in pigment structure are possible. The variations may be due to the infrequent sampling of a very dynamic phytoplankton population, or they may arise from the haphazard sampling of a spatially heterogeneous ocean. It seems fundamental to the goals of the time series program to be able to resolve this difference, as it goes to the heart of the types of processes that will control upper ocean biogeochemistry. The oceans are patchy on spatial scales of tens to hundreds of kilometers. The variability in pigment structure could arise from the sequential sampling of different water masses in a heterogeneous environment. Conversely, phytoplankton populations have specific growth rates of up to a few doublings each day and could undergo dramatic transformations over a period of two weeks. In reality, both processes must contribute to the time series variability, and it will be a significant challenge to sort out the relative importance of the two.

Spatial Heterogeneity

Horizontal patchiness in ocean physics and biology has other important implications for the measurement and interpretation of a one-dimensional (vertical) biogeochemistry time series. In November 1991, we occupied a 4×5 grid of 20 stations 20 minutes apart in both longitude and latitude (Figure 11-8) and made profiles of hydrography and nutrients at each station. There was considerable spatial heterogeneity in these profiles, even on the same density surface. Figure 11-8A shows the nitrate concentration on the 26.16 sigma-theta density surface, and Figure 11-8B shows the depth of that density surface. There are changes in the nitrate concentration of more than 1 μmol/kg over horizontal distances of 15–30 km. At some of these locations, this density surface changes depth by as much as 40 m in the vertical scale over the same horizontal scale. There will be transport of nutrients along these gradients on the isopycnal surface, and, in some cases, the depth changes of the surface will cause a vertical nutrient flux by isopycnal mixing. The horizontal gradients on these density surfaces are a thousandfold smaller than vertical gradients in the same area (on the order of 0.1 μmol/kg/km compared with approximately 0.1 μmol/kg/m). However, horizontal eddy diffusivities are often 10^4 to 10^5 times greater than vertical diffusivities, and the net vertical nutrient flux along a sloping isopycnal could exceed the diapycnal flux. Thus, horizontal advection/diffusion processes may be an important source of nutrients to the euphotic zone. These horizontal processes cannot be measured or estimated with the traditional one-dimensional (vertical) time series.

In summary, horizontal patchiness confounds time series measurements in two

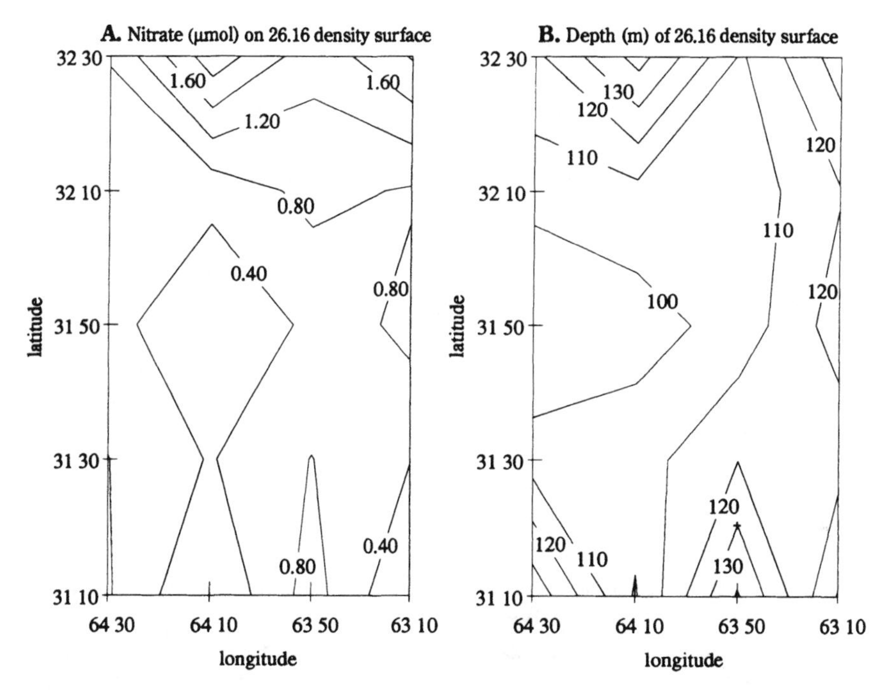

Figure 11-8. (A) Distribution of nitrate + nitrite on the 26.16 sigma-theta density surface over a 100-km × 140-km area. (B) Depth of the 26.16 sigma-theta surface over the same area.

ways. First, for a sequential set of measurements at one location, we cannot discern the difference between temporal evolution of the water and the sequential sampling of different water masses. This confuses the interpretation of the time series records as a temporal sequence. Second, a one-dimensional time series (two-dimensional dataset, depth and time) will not resolve any processes that are two- or three-dimensional. The goal of BATS and JGOFS is to understand the carbon cycle in the ocean (Scientific Committee on Ocean Research 1987). To the extent that processes like isopycnal mixing of nutrients may contribute to the control of ocean carbon cycling, a one-dimensional time series will fail to resolve these processes fully.

The Design of an Ocean Time Series Sampling Program

The BATS and Hydrostation S sampling programs yield a consistent view of the large-scale seasonal patterns in the upper ocean carbon cycle, and they probably yield good estimates of interannual and decadal variability in some of these processes. Between the BATS and Hydrostation S sampling programs, there are approximately 40 hydrographic cruises each year to the area southeast of Bermuda. Yet despite this intense sampling effort, it is apparent that some important biogeochemical features are not adequately resolved. Some processes likely occur on shorter time scales than can be resolved with weekly to biweekly sampling. The spatial heterogeneity of the environment causes some aliasing of the temporal signal. In addition, there are processes that are fundamentally horizontal, such as the isopycnal nutrient inputs to the euphotic zone, and these cannot be resolved with a one-dimensional time series program.

Some of the new and emerging ocean sampling technologies may significantly improve our resolution of ocean variability and resolve some of the time and space scales that alias a traditional time series program. Moorings can provide very high-resolution temporal datasets to show the contribution of uncommon, short-duration events. Moorings provide a true time series in that they sample more frequently than the inherent time scale of most biogeochemical processes. Moorings tend to have poor depth resolution, and a single mooring will still be unable to distinguish between a rapid evolution of ocean properties and the transition from one water mass to another. Satellites provide extensive spatial and temporal coverage of near-surface properties with some weather limitations. Robotic submarines called Autonomous Underwater Vehicles (AUVs) may allow for repeated, high-frequency, three-dimensional surveys of a region. AUVs have the additional advantage of being able to return water samples routinely to the laboratory for analysis. This expands the range of measurements that can be made with an AUV compared to truly remote technologies (i.e., moorings). For many measurements, however, there are still no appropriate remote sensors, so human beings and ships are required.

Prior to the establishment of a time series sampling program, estimates must be made of the temporal and spatial scales that are important for the scientific questions of the study and the specific region of study involved. Logistical constraints must also be considered. From this information, a sampling strategy can be determined that both addresses the scientific questions and resolves the relevant temporal and spatial scales. For some environments and scientific questions, a truly one-dimensional approach may be appropriate, and some combination of moorings and traditional hydrographic sampling could be employed. In heterogeneous areas, satellites and three-dimensional mapping using AUVs or ships may also be required to interpret and understand the one-dimensional datasets. The required frequency of sampling as determined from the temporal scale of the processes might determine the choice of traditional shipboard sampling, remote platform sampling or AUV sampling. In all of these decisions, logistics (especially distance to the station and availability of a research vessel) and costs must be taken into account so that the sampling strategy yields the most relevant information for the dollars spent.

The existing U.S. JGOFS time series stations in Bermuda and Hawaii are valuable examples of the power and potential of time series observations for addressing globally important questions. Understanding the patterns and controlling the processes of temporal changes in ocean biogeochemistry are necessary components of any attempt to understand the role of the oceans in global processes. These manned stations are also natural testbeds for the development of remote technologies for time series research. Efforts to test and evaluate mooring and AUV technologies at the U.S. JGOFS time series stations are part of the future plans for these sites, and some such efforts have already begun. These simultaneous validations of remote and traditional technologies are an important first step in developing the sampling strategies that will be the heart of the future Global Ocean Observing System.

Acknowledgments

I thank the members of the BATS team, Anthony Knap (Principal Investigator), Rachel Sherriff-Dow, Rodney Johnson, Kjell Gundersen, Jens Sorensen, Ann Close, Frances Howse, Margaret Best, Melodie Hammer, Nick Bates, Alice Doyle, Tye Waterhouse, and the crews of the RV *Weatherbird I*, RV *Henlopen*, and RV *Weatherbird II* for collecting and analyzing the samples and for their valuable assistance and support. Bill Jenkins, Jim Swift, Steve Emerson, Tom Hayward, and George Heimerdinger served on a U.S. JGOFS Time-Series Oversight Committee in the early stages of the BATS program. This group provided many valuable suggestions to improve data quality, and we are very grateful. We acknowledge the support of the NSF through grants OCE-8613904 and OCE-8801089 to Anthony Knap, and grants OCE-9017173 and OCE-9016990 to Anthony Michaels. This is BBSR Contribution number 1354. The processed data from the Bermuda Atlantic Time-Series Study (BATS) are available as data reports through the U.S. JGOFS Planning Office (Woods Hole Oceanographic Institution, Woods Hole, MA 02543, U.S.A.). Upon request, a diskette containing the CTD profiles and all of the discrete data can be acquired from George Heimerdinger at the National Ocean Data Center (NODC, Woods Hole Oceanographic Institution), or from the author of this chapter.

References

Buesseler, K. O., A. F. Michaels, A. H. Knap. Submitted. A 3-D time-dependent approach to calibrating sediment trap fluxes. *Global Biogeochemical Cycles*.

Deuser, W. G. 1986. Seasonal and interannual variations in deep-water particle fluxes in the Sargasso Sea and their relation to surface hydrography. *Deep-Sea Research* **33**:225–246.

Dickey, T. D. 1991. The emergence of concurrent high-resolution physical and bio-optical measurements in the upper ocean and their applications. *Review of Geophysics* **29**:383–413.

Dugdale, R. C., and J. J. Goering. 1967. Uptake of new and regenerated forms of nitrogen in primary productivity. *Limnology and Oceanography* **12:**196–206.

Eppley, R. W., and B. W. Peterson. 1979. Particulate organic matter flux and planktonic new production in the deep ocean. *Nature* **282:**677–680.

Fasham, M. J. R., H. W. Ducklow, and S. M. McKelvie. 1990. A nitrogen-based model of plankton dynamics in the oceanic mixed layer. *Journal of Marine Research* **48:**591–639.

Fitzwater, S. E., G. A. Knauer, and J. H. Martin. 1982. Metal contamination and its effect on primary production measurements. *Limnology and Oceanography* **27:**544–551.

Galloway, J. N., D. L. Savioe, W. C. Keene, and J. M. Prospero. In press. The temporal and spatial variability of scavenging ratios for nss-sulfate, nitrate, methanesulfonate and sodium in the atmosphere over the North Atlantic Ocean. *Atmospheric Environment.*

Gust, G., A. F. Michaels, R. Johnson, W. G. Deuser, W. Bowles. In press. Mooring line motions and sediment trap hydromechanics: *In situ* intercomparison of three common deployment designs. *Deep-Sea Research.*

Jackson, G. A. 1988. Implications of high dissolved organic matter concentrations for oceanic properties and processes. *Oceanography* **1:**28–33.

Jenkins, W. J. 1982. Oxygen utilization rates in North Atlantic subtropical gyre and primary production in oligotrophic systems. *Nature* **300:**246–248.

———. 1988. Nitrate flux into the euphotic zone near Bermuda. *Nature* **331:**521–523.

Jenkins, W. J., and J. C. Goldman. 1985. Seasonal oxygen cycling and primary production in the Sargasso Sea. *Journal of Marine Research* **43:**465–491.

Knap, A. H., A. F. Michaels, R. L. Dow, R. J. Johnson, K. Gundersen, G. A. Knauer, S. E. Lohrenz, V. A. Asper, M. Tuel, H. Ducklow, H. Quinby, and P. Brewer. 1991. *U.S. Joint Global Ocean Flux Study. Bermuda Atlantic Time-Series Study. Data Report for BATS 1–BATS 12, October, 1988–September, 1989.* U.S. JGOFS Planning Office, Woods Hole, MA.

Knap, A. H., A. F. Michaels, R. L. Dow, R. J. Johnson, K. Gundersen, J. C. Sorensen, A. Close, G. A. Knauer, S. E. Lohrenz, V. A. Asper, M. Tuel, H. Ducklow, H. Quinby, and P. Brewer. 1992. *U.S. Joint Global Ocean Flux Study. Bermuda Atlantic Time-Series Study. Data Report for BATS 13–BATS 24, October, 1989–September, 1990.* U.S. JGOFS Planning Office, Woods Hole, MA.

Knap, A. H., A. F. Michaels, R. L. Dow, R. J. Johnson, K. Gundersen, J. C. Soresen, A. Close, A. Doyle, G. A. Knauer, S. E. Lohrenz, V. A. Asper, M. Tuel, H. Ducklow, H. Quinby, and P. Brewer. 1993a. *U.S. Joint Global Ocean Flux Study. Bermuda Atlantic Time-Series Study. Data Report for BATS 25–BATS 36, October, 1990–September, 1991.* U.S. JGOFS Planning Office, Woods Hole, MA.

Knap, A. H., A. F. Michaels, R. L. Dow, R. J. Johnson, K. Gundersen, J. C. Sorensen, A. Close, F. Howse, M. Hammer, N. Bates, A. Doyle, and T. Waterhouse. 1993b. *BATS Methods Manual.* U.S. JGOFS Planning Office, Woods Hole, MA.

Knauer, G. A., and V. Asper. 1989. *Sediment trap technology and sampling, U.S. GOFS Planning Rep. #10.* U.S. JGOFS Planning Office, Woods Hole, MA.

Knauer, G. A., J. H. Martin, and K. W,. Bruland. 1979. Fluxes of particulate carbon, nitrogen, and phosphorus in the upper water column of the northeast Pacific. *Deep-Sea Research* **26:**97–108.

Lohrenz, S. E., G. A. Knauer, V. L. Asper, M. Tuel, A. F. Michaels, and A. H. Knap. 1992. Seasonal and interannual variability in primary production and particle flux in the northwestern Sargasso Sea: U.S. JGOFS Bermuda Atlantic Time-Series. *Deep-Sea Research* **39:**1373–1391.

Longhurst, A. R., and W. G. Harrison. 1988. Vertical nitrogen flux from the oceanic photic zone by diel migrant zooplankton and nekton. *Deep-Sea Research* **35:**881–889.

Menzel, D. W., and J. H. Ryther. 1960. The annual cycle of primary production in the Sargasso Sea off Bermuda. *Deep Sea Research* **6:**351–366.

———. 1961. Annual variations in primary production of the Sargasso Sea off Bermuda. *Deep-Sea Research* **7:**282–288.

Michaels, A. F., A. H. Knap, R. L. Dow, K. Gundersen, R. J. Johnson, J. Sorensen, A. Close, G. A. Knauer, S. E. Lohrenz, V. A. Asper, M. Tuel, and R. R. Bidigare. In press. Seasonal patterns of ocean biogeochemistry at the U.S. JGOFS Bermuda Atlantic Time-series Study site. *Deep-Sea Research.*

Michaels, A. F., D. A. Siegel, R. Johnson, A. H. Knap, and J. N. Galloway. 1993. Episodic inputs of atmospheric nitrogen to the Sargasso Sea: Contributions to new production and phytoplankton blooms. *Global Biogeochemical Cycles* **7(2):**339–351.

Michaels, A. F., M. W. Silver, M. M. Gowing, and G. A. Knauer. 1990. Cryptic zooplankton "swimmers" in upper ocean sediment traps. *Deep-Sea Research* **37:**1285–1296.

Platt, T., and W. G. Harrison. 1985. Biogenic fluxes of carbon and oxygen in the ocean. *Nature* **318:**55–58.

Schroeder, E., and H. Stommel. 1969. How representative is the series of Palinuris stations of monthly mean conditions off Bermuda? *Progress in Oceanography* **5:**31–40.

Scientific Committee on Ocean Research. 1987. *The Joint Global Ocean Flax Study: Background, Goals, Organization and Next Steps,* Report of the International Scientific Planning and Coordination Meeting for Global Ocean Flux Studies, Paris, 2/17–19/87. SCOR, Dalhousie University, Halifax, Nova Scotia, Canada.

Siegel, D. A., R. Itturiaga, R. R. Bidigare, R. C. Smith, H. Pak, T. D. Dickey, J. Marra, and K. S. Baker. 1990. Meridional variations of the spring-time phytoplankton community in the Sargasso Sea. *Journal of Marine Research* **48:**379–412.

Spitzer, W. S., and W. J. Jenkins. 1989. Rates of vertical mixing, gas exchange and new production: Estimates from seasonal gas cycles in the upper ocean near Bermuda. *Journal of Marine Research* **47:**169–196.

Strickland, J. D. H., and T. R. Parsons. 1972. *A Practical Handbook of Seawater Analysis.* Fisheries Research Board of Canada, Ottawa.

Toggweiler, J. R. 1989. "Is the Downward Dissolved Organic Matter (DOM) Important in Carbon Transport?" In W. H. Berger, V. S. Smetacek, and G. Wefer (eds.), *Productivity of the Oceans: Past and Present*. Wiley and Sons Ltd., New York.

Weiss, R. F. 1970. The solubility of nitrogen, oxygen and argon in water and seawater. *Deep-Sea Research* **17**:721–735.

12

Planning Long-Term Vegetation Studies at Landscape Scales

Thomas J. Stohlgren

Long-term ecological research is receiving more attention now than ever before. Two recent books, *Long-term Studies in Ecology: Approaches and Alternatives,* edited by Gene Likens (1989), and *Long-term Ecological Research: An International Perspective,* edited by Paul Risser (1991), prompt the question, "Why are these books so thin?" Except for data from paleoecological, retrospective studies (see below), there are exceptionally few long-term data sets in terrestrial ecology (Strayer et al. 1986; Tilman 1989; this volume). In a sample of 749 papers published in *Ecology,* Tilman (1989) found that only 1.7% of the studies lasted at least five field seasons. Only one chapter in each of the review books dealt specifically with expanding both the temporal *and the spatial* scales of ecological research (Berkowitz et al. 1989; Magnuson et al. 1991). Judging by the growing number of landscape-scale long-term studies, however, such as the Long-Term Ecological Research (LTER) Program (Callahan 1991), the U.S. Environmental Protection Agency's Environmental Monitoring and Assessment Program (EMAP; Palmer et al. 1991), the U.S. Army's Land Condition-Trend Analysis (LCTA) Program (Diersing et al. 1992), and various agencies' global change research programs (CEES 1993), there is a growing interest to expand ecological research both temporally and spatially. In fact, many "intellectual frontiers in ecology" identified in the Sustainable Biosphere Initiative (Lubchenco et al. 1991) deal with expanding the spatial scale of investigations to solve complex environmental problems (i.e., global change, loss of biological diversity, and sustainable ecological systems).

Strayer et al. (1986) listed the contributions of long-term studies as: (1) providing society with critical data on many practical issues, (2) contributing to the development of ecological theory, (3) contributing to general ecological knowledge, and (4) contributing to education. Levin (1992 and Chapter 14 of this volume) points out the discontinuity between the scale of most ecological studies (i.e., plots tens of meters on a side) and some ecological models, such as general circulation models, that operate on scales of hundreds of kilometers on a side. Except for some ecosystem studies and selected resource inventory and monitoring programs, there is little linkage of plot-level (usually < 1 hectare) data to landscale scales (up to hundreds of kilometers on a side). Landscape-scale, long-term studies provide a means to: (1) quantify patterns of resources and processes at multiple spatial scales, (2) detect spatial and temporal trends in resources, and (3) generalize findings over broader areas (and over longer time scales) with known precision and accuracy.

Planning long-term vegetation studies at landscape scales is difficult because the scale of perturbations (e.g., volcanism, fire, insect outbreaks, global change) can exceed typical sizes of study units, and because responses to perturbations may vary locally. Conventional sampling strategies that randomly select a few study plots in relatively homogeneous study units may not be adequate to describe long-term vegetation dynamics at landscape scales (see Hinds 1984 for a review). First, the physical environment and living organisms are rarely uniformly or randomly distributed. Instead, resources are aggregated in patches, or they form gradients or other kinds of spatial structures (Legendre and Fortin 1989). Second, it is difficult to predict where on the landscape responses to exogenous perturbations will be distributed (e.g., global change effects on biological diversity, Peters and Lovejoy 1992). Third, the intensity of responses to endogenous perturbations may vary with local or regional conditions, and it varies temporally (Pickett and White 1985). Fourth, subtle changes in vegetation dynamics at landscape scales cannot be detected without a long-term record (Strayer et al. 1986; this volume). Last, ecologists have only a cursory understanding of coupled spatial and temporal variability (Haury et al. 1978; Powell 1989 and Chapter 8 of this volume; Steele 1989; Levin 1992 and Chapter 14 of this volume). Thus, innovative sampling strategies and new analytical tools may be needed to extrapolate information from plots to landscapes and, later, from landscapes to regions.

New analytical tools also may be needed to extrapolate information from plots to landscapes and from the past and present to the future. The few reference texts with fitting titles, such as *Spatial Time Series* (Bennett 1979) and *Spatial and Temporal Analysis in Ecology* (Cormack and Ord 1979), rely, unfortunately, on only a few real-life examples, usually with spatial heterogeneity ignored. The lack of appropriate long-term data sets may be thwarting the use of existing analytical tools and impeding the development of new ones.

As discussed later in more detail, landscape-scale, long term vegetation studies require considerable planning for spatio-temporal analyses that depend on: (1) carefully conceiving and properly designing studies, (2) selecting study sites that reflect resource heterogeneity, aggregations, and gradients, and (3) developing a better understanding of the spatial and temporal autocorrelation of terrestrial resources. For now, let us consider how ecologists have dealt with these concerns when assessing long-term vegetation change.

Ways to Assess Long-Term Vegetation Change

Three families of studies adapted from Likens (1989) may (with many assumptions and caveats) lend themselves to time series analyses that address long-term vegetation changes: (1) retrospective studies, (2) chronosequence studies that substitute space for time, and (3) long-term vegetation monitoring.

Retrospective studies try to infer past vegetation characteristics based on pollen stratigraphy, macrofossils, or tree ring chronologies. Retrospective studies serve as a baseline for modern observations. They can observe very slow processes and measure responses to rare events (Davis 1989), but these studies are usually too coarse grained compared to long-term ecological studies, calibration is difficult, and past direction and magnitude of change may not be an indicator of changes in the future (Davis 1989). Green (1982, 1983) used pollen stratigraphy and time series analysis to study succession following fire in the forests of southwest Nova Scotia. In this study, spatial heterogeneity was ignored by treating the landscape (pollen catchment basin) as one point. The exact size of a pollen catchment was reportedly unknown (i.e., pollen catchments often exceed the size of a study lake or watershed), and the pollen record may have been diluted or greatly influenced by upwind areas.

Graumlich et al. (1989) studied past rates of forest net primary productivity (based on a one-time analysis of tree age structure and radial growth rates). In this analysis, several assumptions are made on past production losses by tree mortality (undetected because of decomposition) and on the use of species-specific regression equations to estimate components of total biomass. The forest stands also differed with respect to species composition, elevation, and time since establishment. Although the stands could not be viewed as true replicates, they could be used to imply a response to regionwide changes in climate. Graumlich et al. (1989) found that annual net productivity had increased 60% since 1880. Continued monitoring of the stands would further support the dendrochronology studies.

Although retrospective studies may provide general information on past temporal trends, landscape boundaries for such trends may be nebulous (Green 1983, COHMAP Members 1988). Thus, it is difficult or impossible to extrapolate from site-specific time series analyses to broader areas. More important, most paleoecological studies have shown that species respond individualistically (Davis 1983; Brubaker 1986, 1988; Jacobson et al. 1987), so the validation of current and future trends in species composition change still requires long-term vegetation studies.

Chronosequence studies (substituting space for time) attempt to infer a temporal trend by creating or studying different-aged sites. These methods allow relatively convenient and inexpensive sampling. They may show general trends in plant succession, they may generate hypotheses about patterns and mechanisms, and they may be complementary to long-term studies (Pickett 1989). For example, Christensen and Peet (1984) compared five successional age-classes of upland forests in North Carolina and provided important insights to secondary succession that would otherwise have required over 80 years of study. Duffy and Meier (1992) studied the diversity of herbaceous understory plants in a chronosequence of forest stands in the southern Appalachians to show that plant diversity was far greater in old-growth compared to ~ 90-year-old clearcut stands. Chronosequence studies, however, rely heavily on enormous, untested assumptions, including: that the sites had the same initial conditions, that boundary conditions (i.e., drought cycles, insect outbreaks, vegetation dynamics) were/are comparable, and that priority effects (i.e., past ecosystem processes on the sites) were similar (Pickett 1989). Chronosequence studies are receiving increasing scrutiny by the scientific community (Pickett 1989; Elliott and Loftis 1993).

Establishing long-term vegetation plots for forest monitoring is expensive and, by definition, time consuming. Yet measurements can be made uniformly over time with testable precision and accuracy. Such measurements (e.g., tree growth rates and changes in basal area, changes in species composition, etc.) provide valuable data for spatio-temporal analyses. The power of long-term, landscape-scale studies increases when combined with short-term experiments that can determine cause-effect relationships, retrospective studies that help interpret the rate and magnitude of past change, and chronosequence studies that can describe general temporal trends. Long-term studies are essential to detect, assess, and validate predicted vegetation changes at landscape scales, but they are not without constraints and limitations.

Constraints and Limitations of Long-Term Landscape-Scale Vegetation Studies

Large-scale field investigations are expensive. They attempt to sample many variables over large spatial scales for many years. Time is always a major constraint. Typical graduate programs, fellowships, and grants last only two to five years. Is it any wonder that Tilman (1989) found that only 1.7% of ecological studies lasted at least five field seasons, given that funding agencies are more apt to fund short-term, experimental studies (Pickett 1989)?

Long-term project management, logistical support for continued consistent measurements, and maintenance of a data management system require a substantial organizational structure (Strayer et al. 1986; Franklin et al. 1990). Field crews need adequate training, supplies must be ordered before the field season, and a permanent research staff is necessary to maintain continuity of field measurements over time.

The monitoring of long-term trends in terrestrial ecosystems is still in its infancy, and it is hampered most by inadequate sampling procedures (Hinds

1984). To date there has been little standardization of techniques for landscape-scale long-term vegetation studies (see below). Methods used are often site- and study-specific. Critical analysis of sampling techniques and designs are just now surfacing (National Science Foundation 1977; Hinds 1984; Legendre and Fortin 1989; Fortin et al. 1989).

Exiting data on natural landscapes commonly are poor (Stohlgren and Quinn 1992). The development of practical field methods capable of detecting subtle, long-term changes is difficult in the absence of baseline information of vegetation cover type, soils, geology, microclimate, disturbance history, and so forth. Long-term vegetation studies can have huge start-up costs associated with acquiring baseline data and testing sampling procedures.

There are also theoretical limitations in planning long-term studies. Vegetation (stand) models, for example, often are based on poor (best available) information (Shugart 1989). Vegetation models with a "spatial component" (geographic information system models, boundary models, gap models, etc.) are just now being developed (Scott et al. 1993). Sampling design theory for landscape ecology is new (Fortin et al. 1989). Emphasis has been placed on analysis of homogeneous study sites and not on heterogeneous landscapes.

Lastly, there are scientific limitations. Hinds (1984) states that "field sampling variability is both large and poorly understood in relation to the rest of any long-term effects detection effort." Poorly replicated permanent plots may be put in atypical locations. Detecting subtle changes is difficult. Simple regression models (e.g., McDonald and Brown 1992) may not be adequate to predict complex effects of global climate change on flora and fauna. Rarely can cause-effect relationships be proven directly by long-term monitoring (Tilman 1989); monitoring must be accompanied by other studies and experiments.

Selected Ongoing Long-Term Vegetation Studies

The most famous long-term vegetation study is the > 100 years' agricultural research at the Rothamsted Experimental Station in England (Johnston 1975; Johnston et al. 1986). The proven value of this data series is exemplified by a recent analysis of long-term acid deposition at the site (Goulding et al. 1987)—an impact that could not be anticipated when data collection began. Curiously, the best example of a long temporal *and* large spatial series of vegetation is the continuous phytoplankton surveys of the North Atlantic (Colebrook 1982). Time series analysis on this > 50-year dataset correlated phytoplankton (and zooplankton) distributions to large-scale climatic variations (McGowan 1990). I am unaware of analogous examples of long-term (50+ years), landscape-scale studies of terrestrial vegetation.

Most long-term terrestrial vegetation studies are conducted at the scale of one to two hectares (ha) (Armstrong-Colaccino 1990; Levin 1992 and Chapter 14 of this volume). Many long-term study examples below are designed to address directly local phenomena (e.g., succession, forest practices, tree mortality). Keep in mind that new research priorities, including global change, biological diversity, and sustainable ecological systems, will require landscape-scale inventory and monitoring of biological resources and processes (Lubchenco et al. 1991). This strongly implies the need to expand both the temporal and the spatial scales of new research programs.

Forest "Reference Stands"

One commonly accepted technique (made fashionable by Dr. Jerry Franklin of the University of Washington) for monitoring changes in forest structure and demography over time is to establish large (1 ha or greater) "reference stands" in representative forest types (Hawk et al. 1978). In the future, time series analysis on forest dynamics will be possible because each tree is tagged and mapped, mortality is assessed annually, and tree diameters are remeasured at 5-year intervals (Riegel et al. 1988). A typical layout (Figure 12-1A) shows the systematic design of a 1-ha reference stand in Sequoia National Park (Parsons et al. 1992). Permanent forest plots like these have been used successfully to describe forest structure (Riegel et al. 1988) and to monitor bole growth, tree establishment, mortality, and pathogen effects (Franklin and DeBell 1988; Parsons et al. 1992) and forest litterfall and nutrient dynamics (Stohlgren 1988a,b).

Ecological processes, such as disturbance, and theoretical models also can be evaluated with a series of permanent plots. Franklin et al. (1985), for example, established long-term vegetation plots in the blast zone of Mount St. Helens to follow annual patterns of succession. The study, now in its thirteenth year, is finding that multiple pathways of succession are acting on the landscape. Had the initial study design focused on a single theoretical model (e.g., the relay floristic model of Egler), the researchers might have selected some parameters as important "indicators" of change at the beginning of the study that may not have been suited for interpreting results in the long term.

Site selection for reference stands is often subjective. Strayer et al. (1986), noting funding agency and research institution as constraints, suggested that some long-term studies have been conducted on uniquely attractive sites and may not be representative of the landscape. When I analyzed the spatial patterns of giant sequoias (*Sequoiadendron giganteum*) in previously logged (Stohlgren 1992a) and unlogged sequoia groves (Stohlgren 1992b), I found that core areas of the groves had different structural characteristics than peripheral areas of the groves. Without an extensive analysis of size distributions and spatial patterns, it would be difficult to quantify how typical (in terms of sequoia size distributions) any single 1- to 3-ha area of the grove was compared to other areas within the 55- to 80-ha groves. I agree somewhat with Strayer et al. (1986) that reference

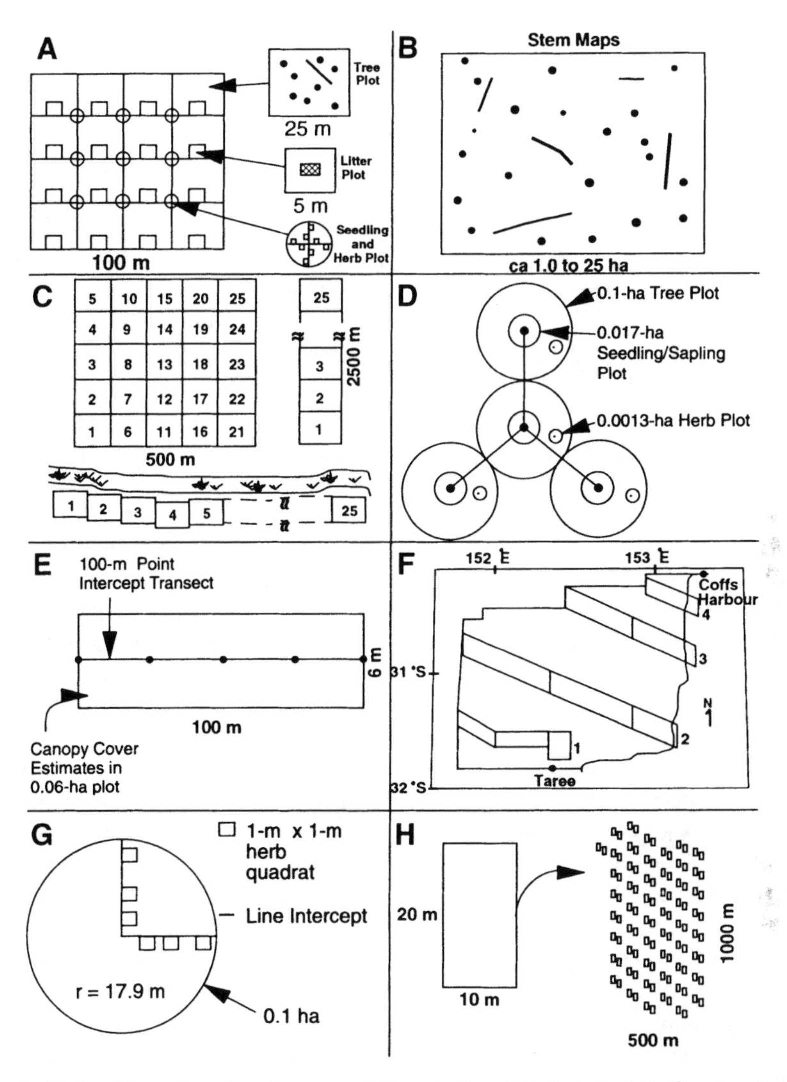

Figure 12-1. Examples of various vegetation studies. (A) A typical "reference stand" layout in a sequoia-mixed conifer forest in Sequoia National Park (adapted from Parsons et al. 1992). (B) Typical tree and log "stem map" (adapted from Dallmeier 1992). (C) Alternative permanent plot placement for biodiversity monitoring (adapted from Dallmeier et al. 1992). (D) Plot layout for EPA's EMAP program and USDA Forest Service's Forest Health Monitoring program (adapted from Palmer et al. 1991 and Burkman and Hertel 1992). (E) Vegetation plot layout for the U.S. Army's LCTA Program (adapted from Diersing et al. 1992). (F) The "gradsect sampling" approach (adapted from Austin and Heyligers 1991). (G) Plot design for natural resource survey in Sequoia and Kings Canyon National Parks (adapted from Graber et al., in press). (H) Systematic-clustering placement of forest plots (adapted from Fortin et al. 1989).

stands may contribute to knowledge of a specific site without greatly adding to general knowledge of landscape ecology.

Measuring Biological Diversity at Permanent Plots

Similar to the reference stand approach, the Smithsonian Institution/Man and the Biosphere Biological Diversity Program (SI/MAB) is establishing permanent vegetation plots in representative tropical forests (and in other Biosphere Reserve areas worldwide) to inventory and monitor changes in biological diversity (Dallmeier 1992; stem maps in Figure 12-1B). A 25-ha area is divided into 25 plots of 1 ha each. The plots may be arranged contiguously in a 500-m × 500-m square, or as a 100-m × 2500-m transect as shown in Figure 12-1C. As with reference stands, each tree > 10 cm dbh (diameter at breast height) is mapped, tagged, and identified. The plots are recensused annually.

Besides the important baseline information on forest structure at each site, one 1-ha plot was established in Puerto Rico one year before Hurricane Hugo passed over the site in September 1989. The pre- and post-disturbance data are proving invaluable in understanding the role of hurricanes in controlling the dynamics of tabonuco forest (Dallmeier et al. 1992). Most permanent plots with stem maps, however, are < 2 ha (Armstrong-Colaccino 1990), but there are exceptions, including a 50-ha permanent plot on Barro Colorado Island, Panama (Hubbell and Foster 1987). Most apparent is the extreme lack of consistency in plot sizes, shapes, numbers, and pattern of placement on the landscape (Armstrong-Colaccino 1990). Although this is not surprising given the different primary objectives of most of the studies, greater consistency in data collection would allow for greater among-site comparisons and far greater overall unity of individual studies.

The major underlying assumption of both the reference stand and the biological diversity permanent plot approaches is that the plots are typical or representative of the forest type or broader area. Krebs (1989) refers to this sampling design as "judgmental sampling" because the investigator selects study sites on the basis of experience. He points out that this *may* give the correct results, but that statisticians reject this type of sampling because it cannot be evaluated by theorems of probability theory. Because plots are expensive to establish and monitor, establishment of replicate plots is usually restricted (Kareiva and Anderson 1988). Within the reference stands, important parameters may be overlooked initially. The take-home message from Berkowitz et al. (1989) bears repeating—that results from long-term study plots may be difficult to extrapolate because of: (1) fundamental differences between the study system and surrounding areas, (2) unknowable boundary conditions or intrusion of unique events, (3) unknown uncertainty or bias in results due to poor replication, (4) the nature of underlying process/phenomenon (i.e., different perturbations may act on the system in different ways through time), and (5) inappropriate methods or poor data.

Landscape-Scale Vegetation Inventory and Monitoring Programs

There are several inventory and monitoring methodologies in use today on public lands. In the United States, national or regional surveys include the U.S. Environmental Protection Agency's Environmental Monitoring and Assessment Program (EMAP) (Palmer et al. 1991; Messer et al. 1991), the USDA Forest Service's Forest Health Monitoring Program (Burkman and Hertel 1992; Conkling and Byers 1992), and the U.S. Army's Land Condition-Trend Analysis (LCTA) Program (Diersing et al. 1992).

The EPA's EMAP program is a nationwide systematic survey of natural resources. Within a honeycomb pattern of 40 km^2 hexagons that span the entire United States, several groupings of vegetation plots are established. The groupings consist of nested plots: four fixed-area circular plots including 0.1-ha plots for trees, 0.017-ha plots for seedlings and saplings, and 0.0013-ha plots for herbaceous vegetation (Figure 12-1D). The Forest Service's Forest Health Monitoring Program piggy-backed on the EMAP survey design for long-term monitoring by selecting specific EMAP forest plots as permanent forest health monitoring plots (Burkman and Hertel 1992).

The U.S. Army's LCTA Program is designed to standardize inventory and monitoring procedures for plant and wildlife on 4.9 million ha of public land. The program's permanent plots (100-m × 6-m transects) are allocated in a stratified random pattern based on soils, color reflectance categories, and vegetative cover using a geographic information system (Diersing et al. 1992; Figure 12-1E).

Several other surveys of vegetation under way also have monitoring potential. The North American Sugar Maple Decline Project (Miller et al. 1991) is conducting an inventory and monitors trends in *Acer saccharum* demography and health. For this study, 0.1-ha areas containing clumps of sugar maples are subjectively selected for study. The USDA Forest Service's Forest Response Program uses a stratified random design with 0.04-ha plots (Zedaker and Nicholas 1990).

Recently, Austin and Heyligers (1991) introduced a new approach to large-scale vegetation surveys with a design called "gradsect sampling" (Figure 12-1F). Replicate transects are deliberately placed on the steepest environmental gradients in the sample area. In one example, four gradsects were established, encompassing 32% of the 20,000 km^2 study area on the North Coast of New South Wales, Australia. Each gradsect was divided into segments for geographical replication, then further stratified into seven to nine similar classes of altitude, rock type, and rainfall. Similar areas were then sampled proportionately with 50-m × 20-m plots. For the three dominant tree species, canopy trees and saplings were ranked according to their abundance. This gradient approach might be effective for long-term monitoring if it were scaled down spatially (i.e., applied in a smaller area with permanently established and monitored plots) and scaled up in intensity (i.e., increasing the accuracy and precision of measurements in each plot).

Other vegetation surveys under way with potential for long-term monitoring include that of David Graber's research team, which has established more than 500 0.1-ha circular vegetation plots uniformly at 1-km grid points throughout Sequoia and Kings Canyon National Parks (Graber et al., in press; Figure 12-1G). Similar to the gradsect technique, this survey is geographically broad but may miss rare habitats. Nevertheless, the level of replication in this systematic survey, and the detailed information collected on each plot, make this an exemplary baseline database for assessing long-term changes in plant species distribution and diversity.

GAP analysis (Scott et al. 1993) uses existing data (e.g., coarse-scale vegetation maps, wildlife observations, species range maps, etc.) as a means to identify the gaps in representation and protection of biodiversity at state and regional spatial scales. The authors acknowledge that GAP analysis is not a substitute for a biological inventory. I mention GAP analysis only to illustrate that it is not a protocol for designing landscape-scale vegetation studies, though it may provide a means to stratify areas for thorough biological inventory.

The lack of standardization of techniques in these examples is an inherent consequence of different goals, objectives, and scales of study. Still, ecologists planning long-term studies at landscape scales face similar decisions, constraints, and limitations. Addressing design considerations in a systematic way is a crucial first step toward implementing extensive field studies.

GOSSIP: A Framework for the Design of Long-Term, Landscape-Scale Vegetation Studies

Regardless of the initial intent of the long-term study or the size of the study area, initial decisions on Goals, Objectives, Scale, Sampling design, Intensity of sampling, and Pattern of sampling (GOSSIP) affect the long-term value of datasets for time series and other analyses (Figure 12-2). Following such a framework may not lead to increased standardization of field techniques, but it may help in planning long-term, landscape-scale studies and in evaluating the strengths and weaknesses of alternate study designs and field techniques.

Two new research programs in Rocky Mountain National Park, Colorado, provide examples of how the GOSSIP framework is being employed. These long-term vegetation studies are being designed to: (1) assess the potential effects of global change on forest species distribution and productivity, and (2) assess changes in biodiversity (using vascular plant diversity as an indicator) at landscape scales.

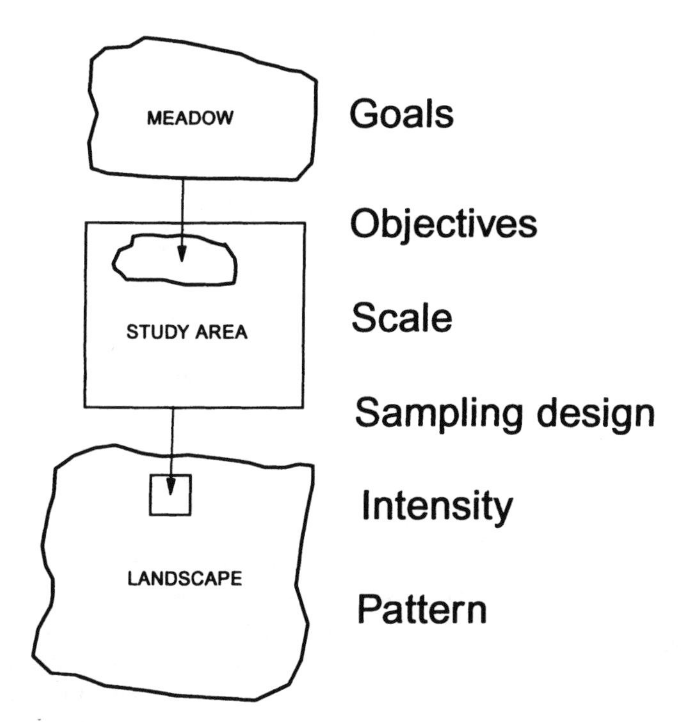

Figure 12-2. GOSSIP: a framework for decision making when planning long-term vegetation studies.

Goals

First and foremost, study goals must be articulated clearly. An age-old question in ecology is what to measure (Krebs 1989). Resolving what the theoretical questions are, early in a study, will help define study objectives and incorporate the practical and theoretical constraints. Goals may be lofty—for example, to monitor changes in biological diversity, to detect the effects of acid deposition or global change, or to evaluate the processes influencing the structure and species composition of forests. Unlike archetypal short-term, single-hypothesis field experiments, long-term studies need not be (and perhaps should not be) designed to address a well-defined hypothesis (Taylor 1989). Like most field experiments, long-term studies face conceptual problems—such as environmental variability, multiple stable equilibria, indirect or feedback effects, and transient dynamics (Tilman 1989)—that cannot be addressed easily at the beginning of a study. Careful planning and years of data may be needed to develop an accurate conceptual model of the system under study.

The Franklin et al. (1985) example of multiple successional pathways on Mount St. Helens reminds us that it may be foolish to contrive goals to address a single ecological theory. Long-term landscape-scale vegetation researchers may wish to adopt a pioneering spirit. In setting goals for long-term studies, ecologists should take stock in the resources and theories on hand, select a clear direction and trail, acknowledge the unknowns, and keep the goal in mind throughout their travels.

One goal of our global change research program in Rocky Mountain National Park is to develop an understanding of the abiotic and biotic controls on forest distribution and productivity as a basis for assessing potential vegetation change for a range of projected climate scenarios (Stohlgren et al. 1993a). The primary goals of our biodiversity project are to understand the spatial components of vascular plant diversity and to evaluate changes in patterns of species richness over time.

Objectives

The objectives of the study detail what is feasible in the field to approach the stated goals. Jones (1986) describes the process as "scoping" and "problem definition," where general problems are reduced to specific ones, where specific issues are identified, and where priorities are set for specific inventory and monitoring data needs. It is particularly important, when setting objectives for long-term, landscape-scale studies, that existing data (and auxiliary data collected by others) be evaluated fully for spatial and temporal completeness, accuracy, and precision. Note that this is usually no small task, but it does help to identify the types and levels of data that are needed. Narrower objectives will usually reduce the scope of a monitoring project by eliminating extraneous data (MacDonald et al. 1991). As mentioned above, priorities for data collection are based on the objectives and the practical and theoretical constraints and limitations of landscape-scale, long-term studies. At this point, it is helpful to visualize data products (i.e., tables and figures), and it is important to acknowledge the potential limits to which study results can be extrapolated spatially and temporally (see Berkowitz et al. 1989). MacDonald et al. (1991) assert that careful formulation of objectives is essential to preclude unrealistic expectations.

Specific objectives also help in the selection of appropriate spatial scale, sampling design, and field methodologies. For example, the vegetation portion of the global change research program in the Colorado Rockies, initiated in March 1992, is an interdisciplinary study that addresses vegetation dynamics, soil characteristics, microclimate, effects of disturbance, and nutrient and water use efficiency at all major forest ecotones. The objectives of the program are to: (1) investigate the potential for vegetation change at the forest-tundra ecotone, (2) assess abiotic and biotic controls on forest distribution and productivity at upper and lower treeline and intermediate ecotones, (3) evaluate the effects of climate variation on disturbance regimes and the temporal dynamics of montane forests in the Colorado Front Range, and (4) provide a synthesis of these three studies to assess the biospheric feedback of vegetation to regional/mesocale climate and hydrologic models. These objectives build on existing, lower-resolution data sets (Peet 1981). Although the objectives are broad, they are kept realistic by a series of specific hypotheses under each of the major objectives listed above (Stohlgren et al. 1993a). Budget and organizational constraints are minimized by secure long-term funding by the National Park Service (NPS). Continuity of measurements is ensured with permanent research staff. There still is some uncertainty associated with the analytic tools needed to integrate climatic, hydrological, and vegetation data collected at different scales and levels of resolution (Stohlgren et al. 1993a), but most of the typical constraints and limitations have been adequately addressed.

Initiated in May 1992, the biodiversity project has as its objectives: (1) to develop and test alternate field methods for collecting plant diversity information, and (2) to establish long-term plots to assess changes in plant diversity in Rocky Mountain National Park. This seemingly monumental task is not without its constraints and limitations. Species-level plant distribution data at landscape scales are expensive to collect. Most species are rare, and complete surveys of large areas are generally cost prohibitive (Stohlgren and Quinn 1992). Sampling techniques for biodiversity are not well developed (see the section on sampling design below). Still, patterns of plant diversity can be elucidated only by systematic surveys. Constraints and limitations on achieving the project objectives can be minimized by developing cost-efficient sampling techniques and by setting reasonable limits on "completeness" (e.g., 80% to 95% complete surveys). For large nature reserves, botanical surveys might be considered taxonomically complete if the survey is geographically complete (not missing huge areas), ecologically complete (not missing important or rare community types), and temporally complete (covering seasonal and annual variation in species occurrence, Stohlgren and Quinn 1992).

Field work for the global change study and the biodiversity study is conducted primarily in Rocky Mountain National Park. Additional studies nearby (Peet 1981; Allen et al. 1991) will aid in extrapolating results to broader areas. Predictive models of vegetation change due to climate change may be completed in a few years, whereas validation of predictions may take decades. Determining the temporal limits of the biodiversity projects will depend on species discovery/accumulation rates and the rate of observed changes (Heltsche and Forrester 1983; Miller and Wiegert 1989).

Investigators must evaluate periodically whether specific objectives are addressed directly by the field data. This feedback loop ensures that the overall study goals are also met or approached. By clearly articulating the goals and objectives of the study, the ecologist can begin to address issues of scale (Figure 12-2).

Scale

Levin (1992 and Chapter 14 of this volume) points out that large-scale ecological studies require the interfacing of phenomena that occur at very different scales of space, time, and organization. The scale selected for study sets the limits as to which ecological phenomena can be studied. Many long-term vegetation studies (e.g., the reference stand and contiguous biodiversity plot approaches described above) take many measurements at relatively few sites. Thus, time series analyses would focus on ecological trends perhaps accurately assessed but restricted spatially. Because unique events and peculiar site characteristics and land-use histories are fused into the analyses, temporal trends may not be extrapolated confidently to other areas.

Landscape-scale field surveys, in contrast, may collect moderately few measurements at many sites. These measurements may adequately assess spatial variation but may not detect temporal trends, because of the few measurements at each site. This illustrates the major tradeoff between small-scale and landscape-scale long-term vegetation studies: an inevitable loss of information (or accuracy) as the scale of the study area increases, or a diminishing ability to extrapolate as it decreases.

Whereas time series studies ask how long is long enough, long-term landscape-scale studies additionally ask how large an area should be considered. For landscape-scale questions of temporal trends, the ecologist must consider the temporal *and* spatial environmental variability of plot-level, stand-level, and landscape-level elements. A nested sample plot design could be used as a compromise in some situations. Here, many measurements taken at a few small sites are nested in larger areas where fewer measurements are taken. In this way, results of intensive studies on temporal trends can be extrapolated to larger areas with known estimates of accuracy and precision (Haslett and Raftery 1989). Also, the ecologists gain increased understanding of the representativeness of the intensive study sites.

Our global change research program is using nested plot techniques (Stohlgren et al. 1993b). Because one objective is to increase our understanding of forest species distribution, we focused our attention on ecotones. Ecotones often represent the physiological (or competitive) limit-of-distribution of species, and so they may be sensitive indicators of the potential effects of climate change (Holland et al. 1991; Cornelius and Reynolds 1991). At the plot and stand levels, our ecotone transect (Figure 12-3) is designed to collect more-detailed information in the 20-m × 20-m plots at the transect ends and in the middle (i.e., the ecotone), and less-detailed information along the 200-m to 400-m transect. Because transect locations are selected randomly and because we have replicate transects along major environmental gradients (e.g., slopes, aspects, elevation, etc.), we are able to extrapolate from plot-level findings to the landscape level. Similar to gradsect approach (Austin and Heyligers 1991), the transects cover heterogeneous

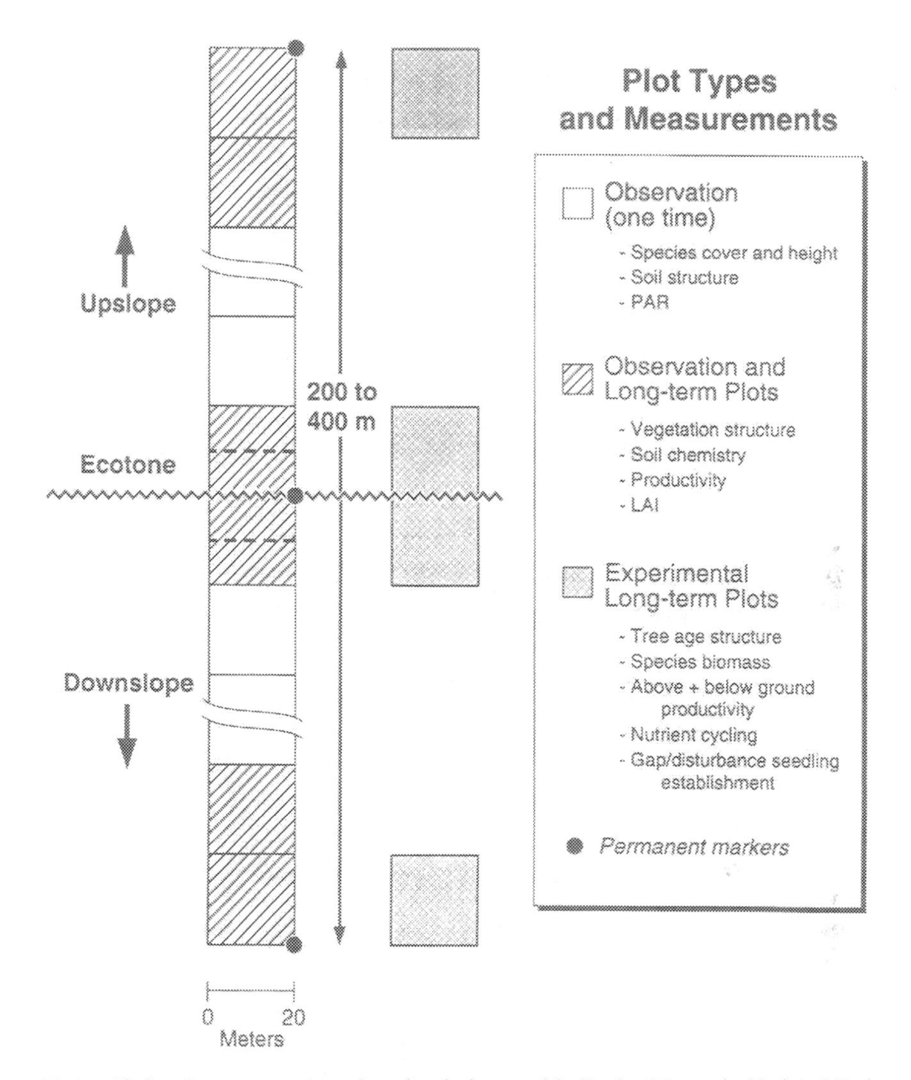

Figure 12-3. Long-term vegetation plot design used in Rocky Mountain National Park to detect the potential effect of global climate change on forest species distribution and productivity. PAR is photosynthetically active radiation, LAI is leaf area index.

areas to understand better the biotic and abiotic factors controlling that heterogeneity. Borrowing from the reference stand approach, more detailed information on trees (including stem maps) will be collected in selected plots.

At the stand to landscape levels (Figure 12-4), the study sites of several investigators will be co-located to allow extrapolation of results of process/mechanistic studies to larger areas (Stohlgren et al. 1993b). Our information will augment a previous analysis of environmental gradient analysis research in the

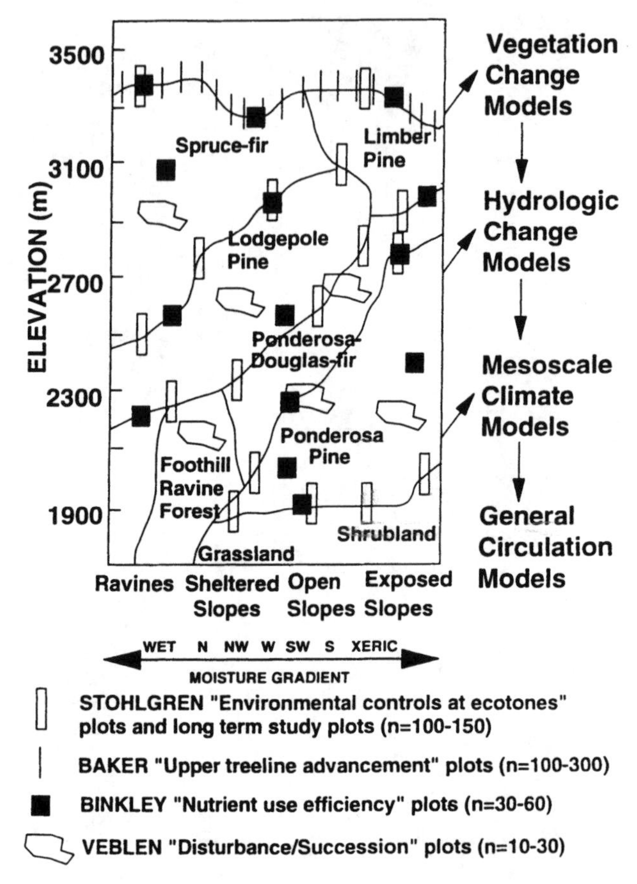

Figure 12-4. Planned intensity and pattern of long-term vegetation plots in the Rocky Mountain National Park global change research program. Gradient analysis information from Peet (1981).

Front Range of the Colorado Rockies that was conducted over large areas (> 1000 km^2) and identified elevation, topographic position, and moisture gradients to be major determinants of plant community distributions (Peet 1981; Allen et al. 1991). We are investigating resource use efficiency and species-environment relationships at the plot level to provide a mechanistic understanding of landscape-level vegetation patterns.

Sampling at multiple spatial scales is also an important component of plant diversity studies (Whittaker et al. 1979; Shmida 1984; see Figure 12-5). Because it is virtually impossible to inventory an entire landscape for vascular plant diversity (much less for biodiversity), sampling species richness at multiple spatial scales allows for mathematical estimates of total diversity. Palmer (1990, 1991) compared several methods for estimating species richness, including num-

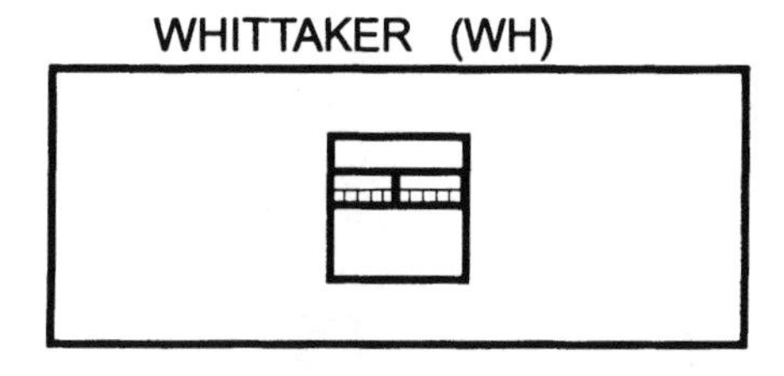

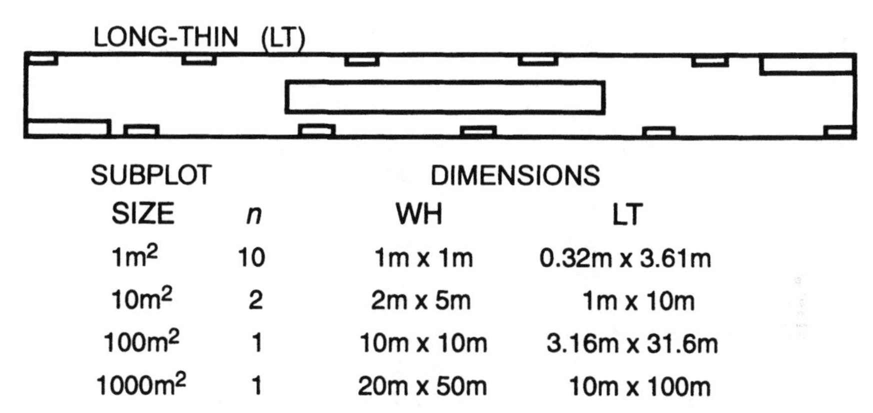

SUBPLOT SIZE	*n*	DIMENSIONS WH	DIMENSIONS LT
1m^2	10	1m x 1m	0.32m x 3.61m
10m^2	2	2m x 5m	1m x 10m
100m^2	1	10m x 10m	3.16m x 31.6m
1000m^2	1	20m x 50m	10m x 100m

Figure 12-5. Comparison of two plot designs used to assess plant diversity. The Whittaker plot design is adapted from Shmida (1984). The long-thin plot design is the author's.

ber of observed species, extrapolation from species-area curves, integration of the lognormal distribution, and nonparametric estimators. He cautioned that species-area curves may have different forms at different scales. Thus, monitoring changes in biodiversity from extrapolated values may not be accurate. Trend analysis from monitoring of species richness from a series of carefully designed, strategically placed long-term study plots may be the only viable option for detecting trends in biodiversity.

Sampling Design

Strayer et al. (1986) suggest that design features of long-term studies include (1) the importance of a simple, accommodating design, (2) rigorously defined objectives, (3) experimentation, and (4) cooperation with real-world management programs. They also stress the need to develop new methods (i.e., the experimental nature of long-term studies), and emphasize that appropriate attention be given to sample archiving, data management, and statistical considerations.

Decisions must be made about plot shape and size, the parameters selected for study, and the frequency and precision and accuracy of measurements, which all affect the value of the long-term data set (Zedaker and Nicholas 1990). The Whittaker biodiversity plot collects species richness data at multiple spatial scales

(Figure 12-5), but it may have three distinct design flaws. First, if the habitat is not strictly homogeneous, species richness is influenced by plot shape. A circular plot (with a minimal perimeter-to-surface area ratio) will have fewer species, in general, than a long-thin rectangle covering a more heterogeneous area (Figure 12-6, top). Second, plot size influences species richness (Figure 12-6, bottom). Note that the Whittaker multiscale plot design shifts from 1-m × 1-m squares to 2-m × 5-m rectangles to a 10-m × 10-m square, then back to a 20-m × 50-m rectangle, which confounds the influences of plot shape and size (Figure 12-5). Third is the problem of spatial autocorrelation. Not only are the ten 1-m × 1-m plots contiguous in one small area of the 20-m × 50-m plot (i.e., high spatial autocorrelation), the successively larger plots are superimposed on the smaller plots (i.e., the plots are not independent in terms of species richness). Thus, a species-rich area in one of the 1-m × 1-m plots affects the species richness reported in the larger-sized plots.

The alternate biodiversity plot design (the long-thin plot, Figure 12-5) being tested in Rocky Mountain National Park covers a more heterogeneous habitat, keeps the proportions of the various-sized plots constant, and reduces the problems of spatial autocorrelation and superimposed plots. We completed a vascular plant survey of the Beaver Meadows study area (Figure 12-7) to conduct preliminary tests of the two biodiversity plot techniques in the field. We randomly located the southwest corner (of north-south oriented plots) and superimposed three Whittaker (WH) plots and three long-thin (LT) plots in the meadow, ecotone, and forest areas of the study site. In all habitat types, and for all plot sizes, the long-thin plot design performed better (returned higher species richness values) than the Whittaker design (Figure 12-8). That is, the species richness sampling with the long-thin plot design more accurately reflected the total species richness recorded in the completed vascular plant survey of the areas. While additional field tests are needed in a variety of habitat types, and perhaps mathematical simulations as well, the long-thin plot technique looks promising.

In contrast to biodiversity research, investigating global change effects on forest species distributions requires contiguous plots (Figure 12-3). Here the purpose of long-term vegetation plots is to document current species-environment relationships and detect changes in forest species distribution and productivity along environmental gradients. Only by straddling ecotones do these permanent study transects provide a means to develop and validate predictive models of change in vegetation distribution in the Colorado Rockies. Changes in forest productivity in the plots along the transect extremes, combined with the plots from other investigators (Figure 12-4), should document the potential rate of change. The long-term vegetation transects, in a sense, are plot-to-landscape-level time capsules of forest change.

Intensity

Another important consideration is the intensity of sampling (number and size of plots on the landscape), which influences the amount of variation accounted

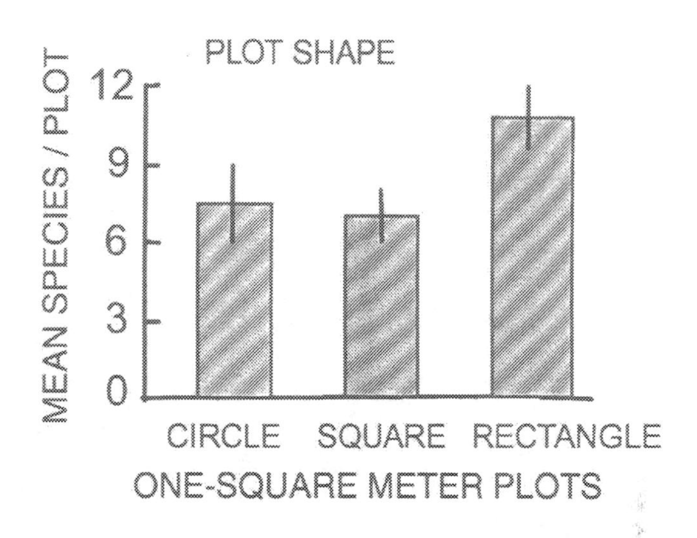

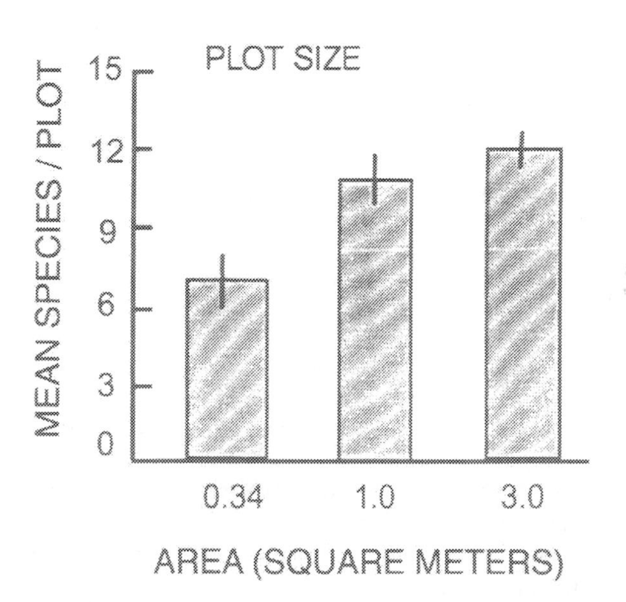

Figure 12-6. Effects of quadrat shape (top) and size (bottom) in estimating vascular plant species richness in Beaver Meadow, Rocky Mountain National Park, Colorado. For each case, $n = 30$ randomly selected plots in the same 1-ha area. Vertical bar is the standard error.

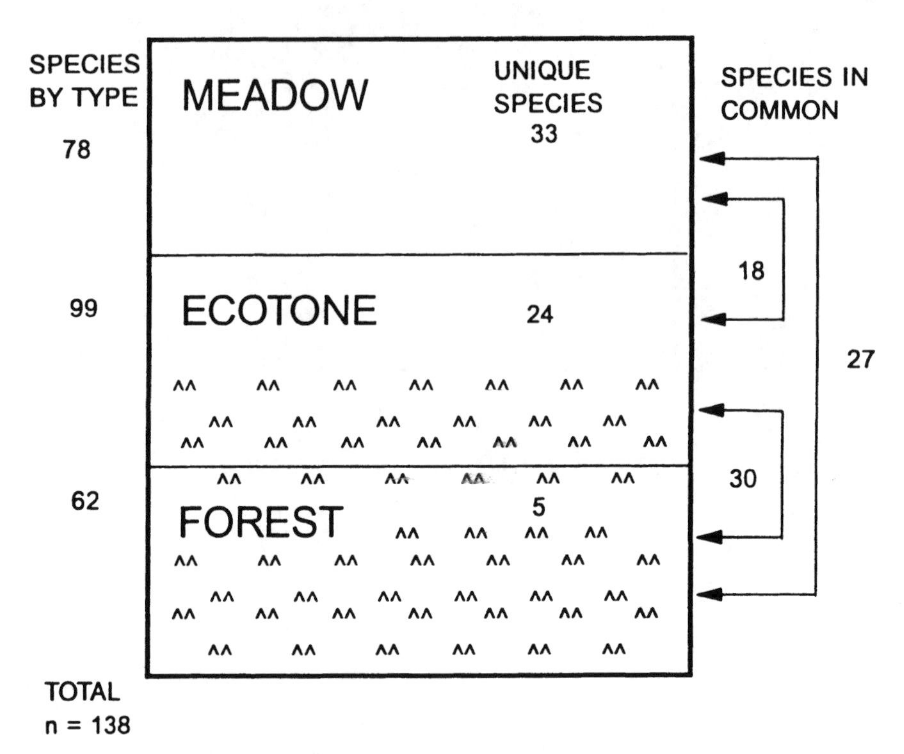

Figure 12-7. Example: Vascular plant species richness in a 300-m × 450-m area of Beaver Meadow, Rocky Mountain National Park, Colorado. The area can be viewed as half pine forest and half montane meadow, or divided into thirds to include a separate ecotone (or ecocline) area.

for by the study and the ability to detect trends accurately over time. Hinds (1984) stated that "the major statistical difficulty is specifying appropriate replication standards in a world that is full of unique places." While the debate continues about replication and pseudoreplication (discussed below), in most cases there are rarely enough replicates in ecological studies. Kareiva and Anderson (1988) showed that: (1) as plot size increased in ecological studies, the number of replicates decreased, and (2) once plot size reached about 3 m^2, the number of replicates was consistently less than five.

Sample size determination in long-term landscape-scale studies is tricky business. Typically, appropriate sample sizes are determined after evaluating between-plot variance from initial field tests (Krebs 1989, pp. 173–199). The appropriate sample size is then determined by correspondence between the variance and a predetermined level of accuracy. Besides systematic error (bias in

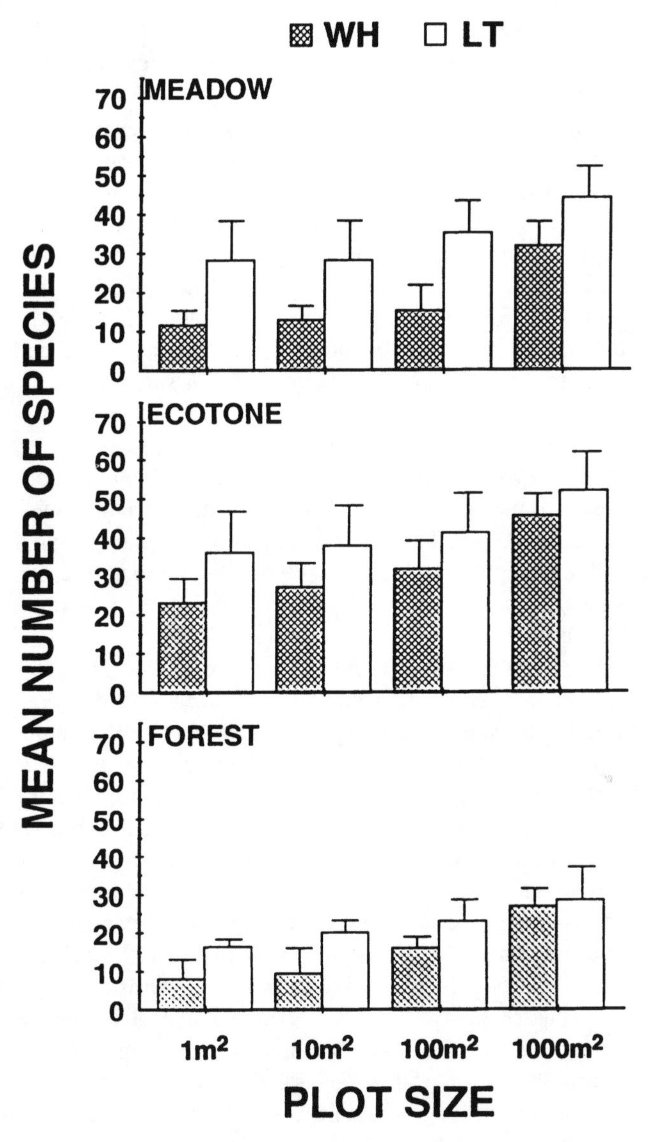

Figure 12-8. Mean number of species recorded in different vegetation communities in Beaver Meadow, Rocky Mountain National Park using the Whittaker (WH) plot design (Shmida 1984) and the long-thin (LT) plot design. Mean of three replicate trials; vertical bar is the standard error.

measuring device) and sampling error (bias in the methodology employed), long-term landscape-scale measurements may have considerable spatial variability and temporal variability: the "signal-to-noise ratio" may be very small. Disturbances of some kind or another frequently occur on most landscapes; thus, studies designed to quantify disturbance effects must anticipate the need for large sample sizes. Interacting natural processes on an already heterogeneous landscape are further complicated by a range of species-specific and site-specific responses.

Spatial variability is often addressed by selecting smaller, more homogeneous areas to study (e.g., the "reference stand" approach). But for global change and biodiversity research, the heterogeneous areas may be more important (i.e., most likely to detect change). Temporal variability can be addressed only by evaluating several years of data. Sample size determination in long-term, landscape-scale vegetation studies may be an iterative process of evaluating spatial and temporal variability simultaneously in the first several years of study and then adjusting sample sizes accordingly.

Replication and pseudoreplication are bewildering concepts for many landscape ecologists (Hurlbert 1984; Hawkins 1986). Pseudoreplication occurs when samples are not dispersed in space (or time) in a manner appropriate to the hypothesis being tested (Hurlbert 1984). For example, repeated sampling in one (or a few) subjectively placed neighboring plots may not be representative of the forest stand, vegetation community type, or landscape implied by the hypothesis. How does this notion influence the potential for replicate measurements and replicate sites in long-term, landscape-scale studies? Hargrove and Pickering (1992) argue that pseudoreplication cannot (and need not) be avoided in landscape or regional ecology considering the multivariate, spatially and temporally complex facet of a landscape. They suggest greater reliance on combinations of "natural experiments" (i.e., chronosequence studies), "quasi-experiments" (i.e., before/after impact studies), and meta-analysis (a controversial statistical technique that produces a unified result from diverse, contradictory studies).

In contrast to Hargrove and Pickering (1992), I believe replication is possible (and a suitable requirement) of long-term, landscape-scale studies. For the global change research program in Rocky Mountain National Park to achieve its study objectives, there must be adequate replication of long-term transects along all major ecotones (Figure 12-4). We are not far enough along in the research program to quantify sample size, but we anticipate needing 10 to 15 200+-m transects per ecotone (100 to 150 transects; 1000 to 1500 20-m × 20-m plots overall), and additional plots (see Binkley, Veblen, and Baker plots in Figure 12-4).

Sample size requirements to evaluate long-term changes in biodiversity are unknown for many reasons. First, methodologies for biodiversity research are still being developed and tested (e.g., the Whittaker versus long-thin plot tests). Second, it is unrealistic to attempt to survey and monitor all species: most species are rare, and the cost of monitoring most or all of them would be prohibitive

(Magurran 1988). Third, statistical techniques to determine completeness are in the developmental stage (Heltshe and Forrester 1983; Miller and Wiegert 1989; Palmer 1990, 1991). Last, but not least, sample size, which is determined by sample variance (Krebs 1989), is unequivocally dependent on the pattern of plot placement, the last element of the GOSSIP design framework.

Pattern

A premier challenge in ecological research is to develop field techniques to detect and quantify patterns in space and time and to explicate underlying mechanisms (Lubchenco et al. 1991). This section describes briefly how the spatial patterning of sampling influences results and provides selected examples of how ecologists are dealing with this seemingly incapacitating problem.

In a homogeneous environment, the pattern of sampling (spatial arrangement of study plots on the landscape) is largely irrelevant. For example, estimating biomass in a field of corn is accomplished simply with a random or systematic sampling design. In natural systems and landscapes, however, plot placement is *everything*. Natural landscapes are complex mosaics of vegetation types and structures, soils, geology, and animals. Superimposed on these resources are temporally and spatially variable responses to disturbance, herbivory, competition, and pathogens. The resulting complex landscape is complicated further by spatially and temporally variable land-use history and external threats (e.g., air pollution, invasion by non-native plants and animals). Determining the optimal pattern of sampling required to describe and understand such a landscape is far from a trivial problem. The first step is to quantify and understand the spatial dependence of information (Cressie 1991; Cullinan and Thomas 1992).

Oceanographers (Colebrook 1982; McGowan 1990; Magnuson et al. 1991), limnologists (Nero and Magnuson 1992), and meteorologists (Haslett and Raftery 1989) may be more advanced than terrestrial vegetation ecologists at synthesizing information from widely scattered sampling sites. Recently, however, research has focused on the problems of spatial autocorrelation and optimizing the spatial arrangements of vegetation plots on forested landscapes. For example, Legendre and Fortin (1989) and Fortin et al. (1989) tested several different patterns of plot placement to evaluate which one was more capable of detecting the spatial structure (i.e., high-density areas) of a sugar maple (*Acer saccharum*) forest. By collecting data from a larger number of plots, they subsampled the data using various spatial patterns and interpolated data from the subsamples (using an analytical approach called kriging). They then compared the spatial patterns of each subsample pattern to the population pattern. They showed that a systematic-cluster design (uniform plots each with a paired offset plot; Figure 12-1(H) was superior to a strict systematic or random sampling pattern using the same number of plots.

I used the systematic-cluster design to describe ponderosa pine (*Pinus pon-*

derosa) stands in Lava Beds National Monument in northeastern California (Stohlgren 1993). My two objectives were to quantify the spatial patterns of high basal area pine stands and to estimate the number of bald eagle winter roost trees (old-growth trees) in the study area. Although the sampling design detected the general distribution of stands of large pines, it failed to determine accurately the number of rare, heavily clustered bald eagle winter roost trees in the study site (Stohlgren 1993). It seems that more research in experimental design is needed to optimize the pattern of sampling required to detect rare but important landscape features.

Pattern of sampling also influences species enumeration for evaluations of biodiversity (Sokal and Thomson 1987). After establishing more than 500 0.1-ha plots at uniform 1-km grid intersections in Sequoia and Kings Canyon National Parks (0.015% of the land base), Graber et al. (in press) report encountering only two-thirds of the vascular plant species known in the parks. In this case, the uniform placement of plots likely contributed to missing the rare but important landscape features in the parks (i.e., species-rich areas such as riparian zones, small meadows, and serpentine outcrops).

The second major component of pattern is temporal: how frequently should data be collected at each site? The pattern of sampling of long-term study plots determines the ability to evaluate accurately spatial and temporal changes in vegetation at landscape scales. The major analytical challenges are determining: (1) the appropriate dimensionless metrics of temporal and spatial variability (Magnuson et al. 1991), (2) the magnitude of variation due to "location," "year," and "other" (Magnuson et al. 1991), and (3) the number of replicate samples (measurements, plots, transects, etc.) required to study spatial and temporal trends.

Magnuson et al. (1991) point out the difficulty of comparing diverse ecosystems without "dimensionless parameters." These are parameters that can be compared across a wide array of locations, or for long time periods (e.g., precipitation and temperature, nutrient concentrations, and species richness). The resulting data can be analyzed with a Model II ANOVA (Sokal and Rohlf 1981) to determine the magnitude of variation due to "location," "year," and "other" (Magnuson et al. 1991). For example, in a study from the Northern Temperate Lakes Long-Term Ecological Research (LTER) site, Kratz et al. (1991) found a relationship between the among-year variability exhibited by a lake and that lake's topographic position in the landscape. It will be difficult in many terrestrial vegetation studies, however, to factor out measurement errors from the interaction between the main effects of "year" and "location." Without adequate replication of study sites, the interaction and error terms cannot be separated.

For the global change research program in Rocky Mountain National Park, landscape-scale hypotheses are based on the park's vegetation cover map and the vegetation gradient models of Peet (1981) and Allen et al. (1991). The placement and pattern of plots (transects) are selected randomly along vegetation boundaries by the computer (using a randomizing geographic information system, or GIS, function). The transects are located in the field with a combination of USGS maps, aerial photographs, and a global positioning system. Three to five replicate long-term transects are selected randomly for each of the four cardinal directions (i.e., aspects) along all major ecotones (Figure 12-4). Less intensive measurements from additional randomly located plots and transects (i.e., by Binkley, Veblen, and Baker, Figure 12-4) within and between forest community types will help quantify the representativeness of the long-term plots. Because we view sampling design, intensity of plots, and pattern of plot placement to be an iterative process, we will likely make several adjustments in the first few years of the research program.

There is still much to learn about the interaction of sampling design, intensity of plots, and pattern of plot placement in long-term, landscape-scale studies. Ecologists need creative study designs relying on nested sampling techniques with an unbiased selection of study plot locations. There are no off-the-shelf protocols dealing with the selection and patterning of study sites on heterogeneous landscapes (Drost and Stohlgren 1993). To inventory and monitor large nature reserves for biodiversity, for example, an optimal sampling strategy might include a systematic-cluster pattern (Figure 12-1H) of long-thin plots (Figure 12-5) with the intensity of sampling determined by the species accumulation rates, combined with a stratified random sampling of rare habitat types and species-rich ecotones (Figure 12-7). Mathematical models of species discovery/accumulation rates (Heltshe and Forrester 1983; Miller and Wiegert 1989; Palmer 1990, 1991) could be used to assess inventory completeness (Stohlgren and Quinn 1992).

Quantifying Trends in Space and Time: Research Needs

There are many theoretical and analytical barriers to synthesizing information on temporal trends from widely scattered sampling sites. Powell (1989 and Chapter 8 of this volume) states that further progress on coupling spatial and temporal data on physical and biological systems is difficult to conceive without a substantial infusion of theoretical guidance. The kind of theoretical attention paid to time series analysis (as noted throughout this volume) and spatial data analysis (Cressie 1991) must likewise be paid to coupled spatial and temporal dynamics.

Many analytical tools are in the development and testing phase. For example, upscaling information from plots (or points) to the landscape level involves spatial interpolation. Yet there is disagreement as to whether kriging (e.g., Legendre and Fortin 1989) or a Bayesian alternative (Le and Zidek 1992) is the better geostatistical tool for the interpolation of environmental data. Where repeated measures are taken on a few trees or plots, treatment of the data requires a full consideration of the statistical issues involved (see Moser et al. 1990).

Haslett and Raftery (1989) provide an excellent example of space-time modeling of environmental data. They estimate the long-term average wind power output at a site in Ireland for which few data are available. Their synthesis includes deseasonalization, kriging, autoregressive moving average modeling, and fractional differencing in ways that can be applied directly in terrestrial ecology (e.g., to estimate spatio-temporal patterns in primary production).

In many cases, new research is needed in the design of long-term, landscape-scale studies. Extensive field surveys and long-term monitoring must be accurate, precise (i.e., repeatable) and cost effective (Margules and Austin 1991). For the systematic inventory of biotic resources at landscape scales, ecologists will need more-detailed remotely sensed data (Bian and Walsh 1993) and more field examples of the influence of sampling pattern (plot placement) on inventory and monitoring results (Fortin et al. 1989). Research is needed on developing optimal sampling strategies for monitoring changes in biodiversity. Landscape-scale inventories will require combinations of systematically placed plots and searching techniques to produce inventories that are geographically, ecologically, and taxonomically near-complete (Stohlgren and Quinn 1992). New generations of temporal-spatial models must be developed to produce mechanistic models of spatially and temporally variable resources and complex ecosystem processes that work at large spatial scales (e.g., Haslett and Raftery 1989; Twery et al. 1991; Davey and Stockwell 1991). Again, spatially dispersed permanent plots and long-term monitoring will be needed to validate the new models. Research is also needed on feedbacks between biotic and abiotic portions of ecosystems and landscapes (Lubchenco et al. 1991). Finally, enlarging ecological perspectives to include alternate land management strategies (Matson and Carpenter 1990) and human values (National Research Council 1990) will be the key to attracting funding and to the overall success of long-term landscape-scale vegetation studies.

Conclusion

Long-term, landscape-scale vegetation studies are relatively new to ecology. Conventional sampling strategies of randomly selecting a few study plots in a relatively homogeneous study unit may not be adequate when assessing large-scale processes and subtle long-term vegetation change. We should be cognizant of the assumptions and limitations of existing field methods. Techniques and methods are still in the experimental mode. The technology (e.g., geographic information systems, global positioning devices) is advancing more rapidly than field methodologies. Regardless of the initial intent of the long-term study or the size of the study area, initial decisions on study Goals, Objectives, Scale, Sampling design, Intensity of sampling, and Pattern of sampling (GOSSIP) affect the long-term value of datasets for time series and other analyses. Because of a backlog in research needs, taking an experimental approach and using the GOSSIP framework for decision making will help plan landscape-scale, long-term vegetation studies. Long-term landscape-scale studies must be combined with large-scale experiments and ecological modeling. Still, the only way to validate interpretations of short-term experiments and ecological models is by careful long-term monitoring of many, widely dispersed permanent plots.

Acknowledgments

The U.S. National Park Service has provided me the opportunities to plan long-term, landscape-scale vegetation studies, develop and test methods, and assist in service-wide inventory and monitoring efforts and global change research. Maurya Falkner, Amy Johnson, Greg Newman, and Connor Stohlgren provided field assistance for the biodiversity project. Dan Binkley, Tom Veblen, and Bill Baker helped design the research program on the effects of global change on vegetation in the Colorado Rockies. Maurya Falkner assisted gleefully with the graphics. John Steele provided many editorial and substantive comments on an earlier version of the manuscript. To all I am grateful.

References

Allen, R. B., R. K. Peet, and W. L. Baker. 1991. Gradient analysis of latitudinal variation in southern Rocky Mountain forests. *Journal of Biogeography* **18:**123–139.

Armstrong-Colaccino, A. (ed.). 1990. *Ecological Data Exchange (EDEX) Catalog*. Third edition. Yale School of Forestry and Environmental Studies, New Haven, CT.

Austin, M. P., and P. C. Heyligers. 1991. "New Approach to Vegetation Survey Design: Gradsect Sampling." In C. R. Margules and M. P. Austin (eds.), *Nature Conservation: Cost Effective Biological Surveys and Data Analysis*. CSIRO, Australia, pp. 31–36.

Bennett, R. J. 1979. *Spatial Time Series: Analysis-Forecasting-Control*. Pion Limited, London.

Berkowitz, A. R., K. Kolosa, R. H. Peters, and S. T. A. Pickett. 1989. "How Far in Space and Time Can the Results from a Single Long-Term Study Be Extrapolated?" In G. E. Likens (ed.), *Long-Term Studies in Ecology: Approaches and Alternatives*. Springer-Verlag, New York, pp. 192–198.

Bian, L., and S. J. Walsh. 1993. Scale dependencies of vegetation and topography in a mountainous environment of Montana. *Professional Geography* **45:**1–11.

Brubaker, L. B. 1986. Tree population responses to climate change. *Vegetatio* **67:**119–130.

———. 1988. "Vegetation History and Anticipating Future Vegetation Change." In J. Agee and D. Johnson (eds.), *Ecosystem Management for Parks and Wilderness*. University of Washington Press, Seattle, WA, pp. 41–61.

Burkman, W. G., and G. D. Hertel. 1992. Forest health monitoring. *Journal of Forestry*, September.

Callahan, J. T. 1991. "Long-Term Ecological Research in the United States: A Federal Perspective." In P. G. Risser (ed.), *Long-Term Ecological Research: An International Perspective*. SCOPE 47. John Wiley & Sons, New York, pp. 9–22.

CESS (Committee on Earth and Environmental Sciences). 1993. *Our Changing Planet: The FY 1993 U.S. Global Change Research Program*. CEES, National Science Foundation, Washington, DC.

Christensen, N. L., and R. K. Peet. 1984. Convergence during secondary forest succession. *Journal of Ecology* **72:**25–36.

COHMAP Members. 1988. Climatic changes of the last 18,000 years: Observations and model simulation. *Science* **241:**1043–1052.

Colebrook, J. M. 1982. Continuous plankton records: Seasonal variations in the distribution and abundance of plankton in the North Atlantic Ocean and the North Sea (*Calanus finmarchicus*). *Journal of Plankton Research* **4:**435–462.

Conkling, B. L., and G. E. Byers (eds.). 1992. Forest Health Monitoring Field Methods Guide. Internal Report. U.S. Environmental Protection Agency, Las Vegas, NV.

Cormack, R. M., and J. K. Ord (eds.). 1979. *Spatial and Temporal Analysis in Ecology*. International Co-operative Publishing House, Fairland, MD.

Cornelius, J. M., and J. F. Reynolds. 1991. On determining the statistical significance of discontinuities within ordered ecological data. *Ecology* **72:**2057–2070.

Cressie, N. 1991. *Statistics for Spatial Data*. John Wiley & Sons, New York.

Cullinan, V. I., and J. M. Thomas. 1992. A comparison of quantitative methods for examining landscape pattern and scale. *Landscape Ecology* **7:**221–227.

Dallmeier, F. (ed.). 1992. *Long-term Monitoring of Biological Diversity in Tropical Forest Areas: Methods For Establishment and Inventory of Permanent Plots*. MAB Digest 11. United Nations Educational, Scientific, and Cultural Organization (UNESCO), Paris, France.

Dallmeier, F., C. M. Taylor, J. C. Mayne, M. Kabel, and R. Rice. 1992. "Effects of Hurricane Hugo on the Bisley Biodiversity Plot, Luquillo Biosphere Reserve, Puerto Rico." In F. Dallmeier. (ed.), *Long-Term Monitoring of Biological Diversity in Tropical Forest Areas: Methods for Establishment and Inventory of Permanent Plots*. MAB Digest 11. United Nations Educational, Scientific, and Cultural Organization (UNESCO), Paris, France, pp. 47–72.

Davey, S. M., and D. R. B. Stockwell. 1991. Incorporating wildlife habitat into an AI environment: Concepts, theory, and practicalities. *AI Applications* **5:**59–104.

Davis, M. B. 1983. Quaternary history of deciduous forest of eastern North America and Europe. *Annals of the Missouri Botanical Garden* **70:**550–563.

———. 1989. "Retrospective Studies." In G. E. Likens (ed.), *Long-term Studies in Ecology: Approaches and Alternatives*. Springer-Verlag, New York, pp. 71–89.

Diersing, V. E., R. B. Shaw, and D. J. Tazik. 1992. US Army Land Condition-Trend Analysis (LCTA) program. *Environmental Management* **16:**405–414.

Drost, C., and T. J. Stohlgren. 1993. Natural resource inventory and monitoring bibliography. Cooperative Parks Studies Unit Technical Report NPS/WRUC/NRTR-93/04, University of California, Davis, CA.

Duffy, D. C., and A. J. Meier. 1992. Do Appalachian herbaceous understories ever recover from clearcutting? *Conservation Biology* **6:**196–201.

Elliott, K. J., and D. L. Loftis, 1993. Vegetation diversity after logging in the Southern Appalachians. *Conservation Biology* **7:**220–221.

Fortin, M. J., P. Drapeau, and P. Legendre. 1989. Spatial autocorrelation and sampling design in plant ecology. *Vegetatio* **83:**209–222.

Franklin, J. F., C. S. Bledsoe, and J. T. Callahan. 1990. Contributions of the Long-Term Ecological Research Program. *BioScience* **40:**509–523.

Franklin, J. F., and D. S. DeBell. 1988. Thirty-six years of tree population change in an old-growth Pseudotsuga-Tsuga forest. *Canadian Journal of Forest Research* **18:**633–639.

Franklin, J. F., J. A. MacMahon, F. J. Swanson, and J. R. Sedell. 1985. Ecosystem responses to the eruption of Mount St. Helens. *National Geographic Research* **1:**198–216.

Goulding, K. W. T., A. E. Johnston, and P. R. Poulton. 1987. "The Effect of Acid Deposition, Especially Nitrogen, on Grassland and Woodland at Rothamsted Experiment Station, England, Measured over More Than 100 Years." In *Proceedings of the International Symposium on the Effects of Air Pollution on Terrestrial and Aquatic Ecosystems*. Grenoble, France.

Graber, D. M., S. A. Haultain, and J. E. Fessenden. In press. "Conducting a Biological Survey: A Case Study from Sequoia and Kings Canyon National Parks." In *Proceedings of the Fourth Biennial Conference on Science in California's National Parks (1993).* Cooperative Parks Studies Unit, University of California, Davis, CA.

Graumlich, L. J., L. B. Brubaker, and C. C. Grier. 1989. Long-term trends in forest net primary productivity: Cascade Mountains, Washington. *Ecology* **70:**405–410.

Green, D. G. 1982. Fire and stability in the postglacial forests of southwest Nova Scotia. *Journal of Biogeography* **9:**29–40.

———. 1983. The ecological interpretation of fine resolution pollen records. *New Phytologist* **94:**459–477.

Hargrove, W. W., and J. Pickering. 1992. Pseudoreplication: A *sine qua non* for regional ecology. *Landscape Ecology* **6:**251–258.

Haslett, J., and A. E. Raftery. 1989. Space-time modelling with long-memory dependence: Assessing Ireland's wind power resource. *Applied Statistics* **38:**1–21.

Haury, L. R., J. A. McGowen, and P. H. Wiebe. 1978. "Patterns and Processes in the Time-Space Scales of Plankton Distributions." In J. H. Steele (ed.), *Spatial Pattern in Plankton Communities*. Plenum, New York, pp. 277–327.

Hawk, G. M., J. F. Franklin, W. A. McKee, and R. B. Brown. 1978. H. J. Andrews Experimental Forest reference stand system: Establishment and use history. USDA Forest Service Bulletin 12, US International Biosphere Program.

Hawkins, C. P. 1986. Pseudo-understanding of pseudoreplication: A cautionary note. *Bulletin of the Ecological Society of America* **67:**185.

Heltshe, J. F., and N. E. Forrester. 1983. Estimating species richness using the jackknife procedure. *Biometrics* **39:**1–12.

Hinds, W. T. 1984. Towards monitoring of long-term trends in terrestrial ecosystems. *Environmental Conservation* **11:**11–18.

Holland, M. M., P. G. Risser, and R. J. Naiman (eds.). 1991. *Ecotones*. Chapman & Hall, New York.

Hubbell, S. P., and R. B. Foster. 1987. "The Spatial Context of Regeneration in a Neotropical Forest." In A. J. Gray, M. J. Crawley, and P. J. Edwards (eds.), *Colonization, Succession and Stability*. Blackwell Scientific Publications, Oxford, U.K.

Hurlbert, S. H. 1984. Pseudoreplication and the design of ecological field experiments. *Ecological Monographs* **54:**187–211.

Jacobson, G. L., T. Webb III, and E. C. Grimm. 1987. "Patterns and Rates of Vegetation Change During the Glaciation of Eastern North America." In W. F. Ruddiman and E. H. Wright (eds.), *North America and Adjacent Oceans During the Last Deglaciation*. Geological Society of America, Boulder, CO, pp. 277–288.

Johnston, A. E. 1975. Experiments on Stackyard Field, Woburn, 1876–1974. I. History of the field and details of the cropping and manuring in the Continuous Wheat and Barley experiments. Rothamsted Experimental Station Report for 1974. Part **1:**29–44.

Johnston, A. E., K. W. T. Goulding, and P. R. Poulton. 1986. Soil acidification during more than 100 years under permanent grassland and woodland at Rothamsted. *Soil Use and Management* **2:**3–10.

Jones, K. B. 1986. "The Inventory and Monitoring Process." In A. Y. Cooperrider, R. J. Boyde, and H. R. Stuart (eds.), *Inventory and Monitoring of Wildlife Habitat*. USDI Bureau of Land Management. Service Center, Denver, CO, pp. 1–10.

Kareiva, P. M., and M. Anderson. 1988. "Spatial Aspects of Species Interactions: The Wedding of Models and Experiments." In A. Hastings (ed.), *Community Ecology*. Lecture Notes in Biomathematics 77. Springer-Verlag, Berlin, pp. 35–50.

Kratz, T. K., B. J. Benson, E. R. Blood, G. L. Cunningham, and R. A. Dahlgren. 1991. The influence of landscape position on temporal variability of four North American Ecosystems. *American Naturalist* **138:**355–378.

Krebs, C. J. 1989. *Ecological Methodology*. Harper & Row, New York.

Le, N. D., and J. V. Zidek. 1992. Interpolation with uncertain spatial covariances: A Bayesian alternative to Kriging. *Journal of Multivariate Analysis* **43:**351–374.

Legendre, P., and M. J. Fortin. 1989. Spatial pattern and ecological analysis. *Vegetatio* **80:**107–138.

Levin, S. A. 1992. The problem of pattern and scale in ecology. *Ecology* **73:**1943–1967.

Likens, G. E. (ed.). 1989. *Long-term Studies in Ecology: Approaches and Alternatives*. Springer-Verlag, New York.

Lubchenco, J., A. M. Olson, L. B. Brubaker, S. R. Carpenter, M. M. Holland, S. P. Hubbell, S. A. Levin, J. A. MacMahon, P. A. Matson, J. M. Melillo, H. A. Mooney,

C. H. Peterson, H. R. Pulliam, L. A. Real, P. J. Regal, and P. G. Risser. 1991. The sustainable biosphere initiative: An ecological research agenda. *Ecology* **72:**371–412.

MacDonald, L. H., A. W. Smart, and R. C. Wissmar. 1991. *Monitoring Guidelines to Evaluate Effects of Forestry Activities on Streams in the Pacific Northwest and Alaska*. USEPA Report EPA/910/9-91-001, Seattle, WA.

Magnuson, J. J., T. K. Kratz, T. M. Frost, C. J. Bowser, B. J. Benson, and R. Nero. 1991. "Expanding the Temporal and Spatial Scales of Ecological Research and Comparison of Divergent Ecosystems: Roles for LTER in the United States." In P. G. Riser (ed.), *Long-Term Ecological Research: An International Perspective*. SCOPE 47. John Wiley & Sons, New York, pp. 45–70.

Magurran, A. E. 1988. *Ecological Diversity and Its Measurement*. Princeton University Press, Princeton, NJ.

Margules, C. R., and M. P. Austin. (eds.). 1991. *Nature Conservation: Cost Effective Biological Surveys and Data Analysis*. Commonwealth Scientific and Industrial Research Organization (CSIRO), Australia.

Matson, P. A., and S. R. Carpenter (eds.). 1990. Statistical analysis of ecological response to large-scale perturbations. *Ecology* **71:**2037–2068 (special issue).

McDonald, K. A., and J. H. Brown. 1992. Using montane mammals to model extinctions due to global change. *Conservation Biology* **6:**409–415.

McGowan, J. A. 1990. Climate and change in oceanic systems: The value of time series data. *Trends in Ecology and Evolution* **5:**293–299.

Messer, J. J., R. A. Linthurst, and W. S. Overton. 1991. An EPA program for monitoring ecological status and trends. *Environmental Monitoring and Assessment* **17:**67–78.

Miller, I., D. Lachance. W. G. Burkman, and D. C. Allen. 1991. *North American Sugar Maple Decline Project: Organization and Field Methods*. USDA Forest Service Northeastern Field Experiment Station, General Technical Report NE-154.

Miller, R. I., and R. G. Wiegert. 1989. Documenting completeness, species-area relations, and the species-abundance distribution of a regional flora. *Ecology* **70:**16–22.

Moser, E. B., A. M. Saxton, and S. R. Pezeshki. 1990. Repeated measures analysis of variance: Application to tree research. *Canadian Journal of Forest Research* **20:**524–535.

National Research Council. 1990. *Forest Research: A Mandate for Change*. National Academy Press, Washington, D.C.

National Science Foundation (NSF). 1977. *Long-Term Measurements: Report of a Conference*. National Science Foundation Directorate for Biological, Behavioral, and Social Sciences, Washington, D.C.

Nero, R. W., and J. J. Magnuson. 1992. Effects of changing spatial scale on acoustic observations of patchiness in the Gulf Stream. *Landscape Ecology* **6:**279–291.

Palmer, M. W. 1990. The estimation of species richness by extrapolation. *Ecology* **71:**1195–1198.

——— 1991. Estimating species richness: The second-order jackknife reconsidered. *Ecology* **72:**1512–1513.

Palmer, C. J., K. H. Ritters, J. Strickland, D. C. Cassell, G. E. Byers, M. L. Papp, and C. I. Liff. 1991. *Monitoring and Research Strategy for Forests—Environmental Monitoring and Assessment Program (EMAP)*. EPA/600/4-91/012. US Environmental Protection Agency, Washington, D.C.

Parsons, D. J., A. C. Workinger, and A. E. Esperanza. 1992. *Composition, Structure, and Physical and Pathological Characteristics of Nine Forest Stands in Sequoia National Park*. USDI National Park Service, CPSU, Technical Report No. NPS/WRUC/NRTR 92/50. University of California, Davis, CA.

Peet, R. K. 1981. Forest vegetation of the Colorado front range. *Vegetatio* **45**:3–75.

Peters, R. L., and T. E. Lovejoy (eds.). 1992. *Global Warming and Biological Diversity*. Yale University Press, New Haven, CT.

Pickett, S. T. A. 1989. "Space-for-Time Substitution as an Alternative to Long-term Studies." In G. E. Likens (ed.), *Long-term Studies in Ecology: Approaches and Alternatives*. Springer-Verlag, New York, pp. 110–135.

Pickett, S. T. A., and P. S. White (eds.). 1985. *The Ecology of Natural Disturbance and Patch Dynamics*. Academic Press, New York.

Powell, T. M. 1989. "Physical and Biological Scales of Variability in Lakes, Estuaries, and the Coastal Ocean." In J. Roughgarden, R. M. May, and S. A. Levin (eds.), *Perspectives in Ecological Theory*. Princeton University Press, Princeton, NJ, pp. 157–176.

Rigel, G. M., S. E. Green, M. E. Harmon, and J. F. Franklin. 1988. *Characteristics of Mixed Conifer Forest Reference Stands at Sequoia National Park, California*. USDI National Park Service, CPSU Technical Report No. 32. University of California, Davis, CA.

Risser, P. G. (ed.). 1991. *Long-Term Ecological Research: An International Perspective*. SCOPE 47, John Wiley & Sons, New York.

Scott, J. M., F. Davis, B. Csuti, R. Noss, B. Butterfield, C. Groves, H. Anderson, S. Caicco, F. D'Erchia, T. C. Dewards, Jr., J. Ulliman, and R. G. Wright. 1993. GAP analysis: A geographic approach to protection of biological diversity. *Wildlife Monographs* **123**:1–41.

Shmida, A. 1984. Whittaker's plant diversity sampling method. *Israel Journal of Botany* **33**:41–46.

Shugart, H. H. 1989. "The Role of Ecological Models in Long-Term Ecological Studies." In G. E. Likens (ed.), *Long-term Studies in Ecology: Approaches and Alternatives*. Springer-Verlag, New York, pp. 90–109.

Sokal, R. R., and F. J. Rohlf. 1981. *Biometry: The Principles and Practice of Statistics on Biological Research*. Second edition. W. H. Freeman, San Francisco.

Sokal, R. R., and J. D. Thomson. 1987. "Applications of Spatial Autocorrelation in Ecology." In P. and L. Legendre (eds.), *Developments in Numerical Taxonomy*. NATO ASI Series, Volume 14, Springer-Verlag, Berlin, pp. 431–466.

Steele, J. H. 1989. "Discussion: Scale and Coupling in Ecological Systems." In J. Roughgarden, R. M. May, and S. A. Levin (eds.), *Perspectives in Ecological Theory*. Princeton University Press, Princeton, NJ, pp. 177–180.

Stohlgren, T. J. 1988a. Litter dynamics in two Sierran mixed conifer forests. I. Litterfall and decomposition rates. *Canadian Journal of Forest Research* **18:**1127–1135.

———. 1988b. Litter dynamics in two Sierran mixed conifer forests. II. Nutrient release in decomposing leaf litter. *Canadian Journal of Forest Research* **18:**1136–1144.

———. 1992a. Resilience of a heavily logged grove of giant sequoia (*Sequoiadendron giganteum*) in Sequoia and Kings Canyon National Parks. *Forest Ecology and Management* **54:**115–140.

———. 1992b. Spatial pattern analysis of giant sequoia (*Sequoiadendron giganteum*) in Sequoia and Kings Canyon National Parks, California. *Canadian Journal of Forest Research* **23:**120–132.

———. 1993. Bald eagle winter roost characteristics in Lava Beds National Monument, California. *Northwest Science* **67:**44–54.

Stohlgren, T. J., J. Baron, and T. G. F. Kittel. 1993a. "Understanding Coupled Climatic, Hydrological, and Ecosystem Responses to Global Climate Change in the Colorado Rockies Biogeographical Area." In *Proceedings of the Seventh Conference on Research and Resource Management in Parks and on Public Lands*. The George Wright Society, Washington, D.C. (In review.)

Stohlgren, T. J., D. Binkley, T. T. Veblen, and W. L. Baker. 1993b. The potential effect of global change on vegetation in the Colorado Rockies: A long-term study approach. (In prep.).

Stohlgren, T. J., and J. F. Quinn. 1992. An assessment of biotic inventories in western U.S. national parks. *Natural Areas Journal* **12:**145–154.

Strayer, D., J. S. Glitzenstein, C. G. Jones, J. Kolasa, G. E. Lichens, M. J. McDonnell, G. G. Parker, and S. T. A. Pickett. 1986. *Long-term Ecological Studies: An Illustrated Account of their Design, Operation, and Importance to Ecology*. Institute of Ecosystem Studies Occasional Publication Number 2. Millbrook, NY.

Taylor, L. R. 1989. "Objective and Experiment in Long-term Research." In G. E. Likens (ed.), *Long-term Studies in Ecology: Approaches and Alternatives*. Springer-Verlag, New York, pp. 20–71.

Tilman, D. 1989. "Ecological Experiments: Strengths and Conceptual Problems." In G. E. Likens (ed.), *Long-term Studies in Ecology: Approaches and Alternatives*. Springer-Verlag, New York, pp. 136–157.

Twery, M. J., G. A. Elmes, and C. B. Yuill. 1991. Scientific exploration with an intelligent GIS: Predicting species composition from topography. *AI Applications* **5:**45–53.

Whittaker, R. H., W. A. Niering, and M. O. Crisp. 1979. Structure, pattern, and diversity of a mallee community in New South Wales. *Vegetatio* **39:**65–76.

Zedaker, S. M., and N. S. Nicholas. 1990. *Quality Assurance Methods Manual for Forest Site Classification and Field Measurements*. US EPA Document EPA/600/3-90/082, Corvallis, OR.

13

Time Series Compared Across the Land-Sea Gradient

Susan C. Warner, Karin E. Limburg, Arturo H. Ariño, Michael Dodd, Jonathan Dushoff, Konstantinos I. Stergiou, and Jacqueline Potts

The environmental sciences of ecology, limnology, and oceanography are relatively young. Until recently, emphasis has been placed on short-term inquiry into structure and function of physical, chemical, and biotic entities and processes. Relatively few studies have been sustained longer than 10 years, but the exceptions have revealed a seemingly ever-increasing range of system behaviors, associated with cascades of time scales.

Steele (1985, 1991a,b) has suggested that marine and terrestrial systems exhibit biogeochemical responses to environmental change at quite different time scales. Noting that atmosphere and oceans, while obeying the same rules of fluid dynamics, operate at vastly different time scales, he postulated that fundamental biotic differences are constrained by the different physics on land (subject to atmospheric influences) versus in the sea. Oceanographers have indicated that biotic processes in the ocean may be coupled to physical forces (Denman and Powell 1984), whereas terrestrial systems are more controlled by biotic interactions (Steele 1991b). In the open ocean, the organisms can move about freely to avoid unfavorable conditions; but terrestrial organisms, being less mobile, have evolved in order to tolerate unfavorable conditions.

The effect that trophic structure plays in an ecosystem's responses has been examined by Steele (1991a) and Pimm (1991). A major difference between marine and terrestrial systems is the location of the largest organisms in the trophic structure (Steele 1991a). In terrestrial systems, the largest organisms are at the bottom of the food chain, whereas in the marine systems, the largest organisms are at the top of the trophic structure. (Although kelp may be more comparable to terrestrial trees than phytoplankton, kelp make up only a small portion of the marine primary producers.)

Jacqueline Potts, of the Rothamsted Experimental Station, provided the Rothamsted data series.

It is often difficult to make meaningful comparisons across major ecosystem boundaries, but the current need to understand global change processes requires that these problems be overcome. To a large extent, the processes that receive the most study are those that respond most quickly. This enables researchers to get "publishable" results, but it may also mean that slower processes are overlooked, although they may be just as important. Another difficulty in making comparisons across ecosystems is that each ecosystem has its own suite of researchers, who interact primarily with others working in the same area. Moreover, certain processes, even those common to all ecosystems, may be studied intensively in some systems and not at all in others.

As previously noted, the scales of terrestrial and aquatic food chains appear to run in reverse; in terrestrial systems the most important primary producers are both the longest-lived and the largest organisms, whereas in aquatic systems the primary production is dominated by the smallest organisms, which also have the shortest life spans. Several authors have noted the importance of the scale of spatial and temporal processes in the planning of ecological investigations (Haury et al. 1978; Wiens 1989; Magnuson 1990 and Chapter 19 of this volume; Dunning et al. 1992). Conclusions drawn from studies that are too short to capture the likely response time of the process may well be invalid. Additionally, the "down-stream" relationship of freshwater and estuarine systems to terrestrial ones is an important aspect, implying that, at some temporal scales, these systems will be coupled and may share some features. In contrast, pelagic system dynamics may be divorced or only slightly affected by terrestrial-estuarine dynamics. The importance of long-term episodic events (such as El Niño) versus gradual change is recognized as having a major influence on system structure and function in most ecosystems. Other problems that arise in making these comparisons are often crucial. These include the spatial scale and the time scale at which the sampling was done. Sampling that is too infrequent may result in missing most of the system's response [referred to as *aliasing* by Platt and Denman (1975)]. This has been a problem in sampling plankton, for instance; sampling at yearly or even monthly intervals misses most of the variation in plankton communities, whose species' life spans are typically less than 14 days. Another problem is knowing what variables to compare between ecosystems to ensure that the comparisons are valid.

One rather discouraging fact, which has been noted as well by Weatherhead (1986), is the typical brevity of most ecological studies. Study length appears to be driven by political funding cycles and/or graduate student tenure (Likens 1989), rather than by some ecologically meaningful scale (e.g., the slowest turnover or generation time in a system or organism of interest). Even when a short study is appropriate, the frequency of sampling may be inappropriate for the process or organism being studied (Smith 1978). When studies exceed the average window of 2–5 years, they often are pursued by an unusually dedicated person and do not survive beyond his or her research career. Indeed, with the notable exception of the Rothamsted experiment in England (Rothamsted 1991), the longest datasets (in terms of years) are collected for economic studies (e.g., fisheries, timber harvests, pelts) and may reflect economic processes (changes in consumer preferences, technological innovations) as well as natural variability. These problems should raise a cautionary flag about comparison of datasets; it is clear that almost any dataset will reflect a number of biases that could influence its interpretation.

This chapter reports on comparisons of time series datasets across the terrestrial-to-marine gradient, motivated by the hypotheses proposed by Steele. Out of the potentially dozens of ways to draw comparisons, we have chosen to concentrate on a few descriptive measures. The simplest approach is to look at the variability within the datasets. Is variability expressed in fundamentally different ways on land versus in water? Periodicities and evidence of density dependence were also selected for comparison.

Finally, comparisons are made with respect to time and spatial scale. A particularly striking difference is the radically different time scale for primary production: days to weeks for aquatic phytoplankton, versus years to centuries for trees. This may be called a "real-time" comparison. Other ways to compare time scales may be equally important—for instance, the generation time or life span of an organism, or the turnover time of a process or abiotic pool. We also consider the spatial scale over which the datasets were collected, and over which the variables exist. In order to examine these aspects of time series in different systems, we compared a variety of time series from different environments.

Selection of Datasets and Analyses

The heterogeneity of available datasets might well be a subject of study in itself. Their origins are extremely diverse, and long-term datasets are scattered throughout nearly all the natural science journals. There are also some very interesting datasets that have been patiently collected over the years by devoted scientists who never considered them complete or worth publishing. Finally, some datasets appear in internal or otherwise obscure technical reports of government or international bodies and are not intended for the general scientific public (and may even require prior clearance for access). In general, however, most biological datasets share one characteristic: the longer the series (in years), the more likely it is to have been recorded for commercial or industrial purposes.

The content of datasets is also extremely variable. The data may consist of direct recordings, such as counts of individuals or temperature records, or indirect estimations, such as population size deduced from feeding scars or atmospheric carbon dioxide suggested by shell deposits. They may even be model projections based on actual data, such as fish stock. The time span usually is not long in the case of direct recordings, but it may be hundreds or thousands of years for geophysical data from ice cores. The time unit may range from hours or days for plankton data to years for vertebrate population data; correspondingly, the number of data points in the series may be very few (i.e., 30 yearly points for population data) or a great many (thousands of points for a few months of hourly plankton counts).

The datasets considered here are listed in Table 13-1. Most of the data were obtained in one of two ways: either directly from published papers or reports or from people involved in their collection; or from a database created by Witteman et al. (1990) and later expanded and managed by Ariño and Pimm (submitted). In all, we had access to some 220 datasets having a minimum of 20 data points, though most were longer. The distribution of the data was very uneven. On the biological side, we had an abundance of data on population estimations of game birds, commercial mammals, and insect pests, but much less on fish and very little on freshwater organisms. On the abiotic side, many sets are available from automatic stations or devices, such as continuous data recorders or weather stations, which tend to be much less numerous in undeveloped regions of the world. Consequently, the data are biased toward organisms of interest to humans and/or toward systems amenable to measurement by automatic recording devices located in economically developed areas.

The initial criterion for selection was to include at least two long series from any of the cells in our comparison design, which was intended to represent four dimensions:

(1) terrestrial, freshwater, and marine environments;
(2) microorganisms, plants, invertebrates, ectothermic vertebrates, birds, and mammals;
(3) physical/chemical data from equivalent environments;
(4) tropical, temperate, and arctic regions.

We searched for suitable series and selected the ones that seemed most appropriate. We discarded datasets on the basis of one or more of the following:

- unreliability of the data;
- data produced by models not fully accepted;
- major changes in methods within the set that might show spurious effects;
- data acquired by digitizing published plots in which the accuracy of the digitizing process was not within an error margin of 1% or better.

The final eliminations left us with the longest (in terms of data points) of the data series available that had no missing values, or the longest stretch of a dataset without missing values. (Of all the datasets used, only one series had any missing data points.) A total of 63 time series recorded at 38 locations were analyzed

Table 13-1. Data series used in comparisons

Location	Variables	No. of Data Points	Freq.*	Environment	References
Sun	Sunspot no.	283	ba	celestial	Shumway 1988
British Isles	Camberwell Beauty abundance (*Nymphalis antiopa*)	101	a	terrestrial	Williams 1958
England	Scarlet Tiger moth abundance (*Panaxia dominula*)	40	a	terrestrial	Jones 1989
Southern Scotland	Red grouse abundance (*Lagopus lagopus scoticus*)	105	a	terrestrial	Mackenzie 1952
Southern Scotland	Black grouse abundance (*Tetrao tetrix*)	89	a	terrestrial	Mackenzie 1952
Yorkshire, England	Hare abundance (*Lepus europaeus*)	90	a	terrestrial	Middleton 1934
Canada	Lynx pelts (*Lynx canadensis*)	114	a	terrestrial	Elton and Nicholson 1942
Southern Sweden	Pine cone production (*Pinus silvestris*)	44	a	terrestrial	Hagner 1965 (in Harper 1977)
Central Sweden	Pine cone production (*Pinus silvestris*)	44	a	terrestrial	Hagner 1965 (in Harper 1977)
Northern Sweden	Pine cone production (*Pinus silvestris*)	44	a	terrestrial	Hagner 1965 (in Harper 1977)
Southern Sweden	Spruce cone production (*Picea abies*)	61	a	terrestrial	Svärdsson 1957 (in Remmert 1980)
Central Sweden	Spruce cone production (*Picea abies*)	61	a	terrestrial	Svärdsson 1957 (in Remmert 1980)
Northern Sweden	Spruce cone production (*Picea abies*)	61	a	terrestrial	Svärdsson 1957 (in Remmert 1980)
Colorado, U.S.A.	Porcupine feeding scars† (*Erethizon epixanthum*)	121	a	terrestrial	Spencer 1964

Rothamsted, England	Ave. summer precip.	139	a	terrestrial	Rothamsted 1991
	Biomass	136			
	Max. summer temp.	114			
	Min. summer temp.	114			
England and Wales	Heron population (*Ardea cinerea*)	50	a	terrestrial	Stafford 1971; Reynolds 1979
Thessaloniki, Greece	Air temp., 1924–1973	600	m	terrestrial	Flocas and Arseni-Papadimitriou 1974
Bergen, Norway	Air temp.	108	a	terrestrial	Stergiou 1991
Woodwalton, England	Dragonfly populations (*Ischnura elegans* and *Sympetrum striolatum*) (2 series)	27	a	fresh water	Moore 1991
Lovojarvi Lake, Finland	Diatom profiles (*Asterionella formosa* and *Synedra acus*) (2 series)	69	a	fresh water	Simola 1984
Lake Constance, Switzerland	Hrs. of sun	23	a	fresh water	Eckman 1988
	April temp.	23			
	zooplankton (small & large)	23			
	D.O. conc.	23			
	SRP, wind	23			
	Coregonus YCS	21			
Mývatn Lake, Iceland	Total catch (2 species)	61	a	fresh water	Gudmundsson 1979; Jonasson 1979 (both in Remmert 1980)
Lake Windemere (South basin)	Pike population (*Esox lucius*) (2–18 yrs.)	37	a	fresh water	Kipling 1983
Hubbard Brook, NH, U.S.A.	Nitrate conc.	119	m	fresh water	Likens et al. 1977
Hudson River, NY, U.S.A.	Water temp.	780	m	fresh water	obtained from NOAA
Crystal River, FL, U.S.A.	Insolation, P:R ratio, temp., GPP, salinity	90	bw	estuarine	Montague et al. 1981
Corpus Christi, TX, U.S.A.	Benthic biomass	109	m	estuarine	Flint 1985
	benthic abundance	109			
	biomass:abundance	109			

continued

Table 13-1. (Continued).

Location	Variables	No. of Data Points	Freq.*	Environment	References
Chesapeake Bay, MD, U.S.A.	Temp.	150	m	estuarine	obtained from Holland as discussed in Holland et al. 1987
	bottom salinity	96			
	D.O. conc.	96			
	benthic biomass	96			
	benthic abundance				
Chesapeake Bay, MD, U.S.A.	Striped bass CPUE (*Morone saxatilis*)	43	a	estuarine	Summers et al. 1985
Maryland coast, U.S.A.	Striped bass population (*Morone saxatilis*)	42	a	marine	Van Winkle et al. 1979
Apalachicola Bay, FL, U.S.A.	Temp.	147	m	estuarine	Livingston 1987
	epibenthic CPUE	148			
	fish CPUE	148			
	bottom salinity	148			
Miramichi, North Atlantic	Salmon population (*Salmo salar*)	64	a	marine	Allee et al. 1949
North Sea	*Calanus finmarchicus* density	552	m	marine	obtained from R. Williams
Georges Bank, NW Atlantic Ocean	Cod population (*Gadus morhua*)	34	a	marine	Loucks and Sutcliffe 1978
North Sea	Hake population (*Merluccius merluccius*)	30	a	marine	Holden 1991
North Sea	Sea surface temp.	408	m	marine	obtained from R. Williams
Massachusetts fishery, U.S.A.	Summer flounder catch (*Paralichthys dentatus*)	48	a	marine	Mid-Atlantic Fishery Management Council 1993
New Jersey fishery, U.S.A.	Summer flounder catch (*Paralichthys dentatus*)	48	a	marine	Mid-Atlantic Fishery Management Council 1993

*Freq.: a=annual, ba=biannual, m=monthly, bw=biweekly.

†Data corrected for tree mortality.

‡Evenly spaced series obtained by polynomial interpolation from unevenly spaced series.

and compared. Unfortunately, not all cells in the original comparison design are represented; we do not have any series of microorganisms or any from tropical areas.

Each of the data series was plotted against time to look for obvious characteristics. The series from the Crystal River had been sampled unevenly, so evenly spaced series were obtained by polynomial interpolation utilizing Newton divided difference coefficients (IBM 1992). Where necessary, logarithmic transformation was used to stabilize the variance. General statistical parameters such as mean, standard deviation, and coefficient of variation were obtained and recorded for each series. Each monthly or biweekly series was then examined for evidence of seasonality, both visually and with a rank test for significant seasonality (Kendall and Ord 1990); if present, seasonality was removed by subtracting the monthly or biweekly mean from each data point. The deseasonalized data series was then plotted again. Trends were detected by linear or quadratic regression, and significant trends were removed. Stationarity was checked with the use of a Dickey-Fuller test (Kendall and Ord 1990). Autocorrelation function plots were used, along with spectral analysis and a phase length test (Kendall and Ord 1990), to look for significant cycling.

Comparisons

Scale

The scale of ecological time series can be characterized by the number of years in the series and the area over which the information was collected. Scale can also be viewed in other ways, such as generation time or the frequency of data collection.

Many authors have tried to combine time and space scales on logarithmic scatter plots; the time axis usually stretches from hours to millennia, and the space scale shows kilometers through to global scale. Steele (1991b and Chapter 1 of this volume) and McDowell et al. (1990 and Chapter 15 of this volume) produced time-space diagrams for climate and ocean systems with events ranging from thunderstorms to ice ages. Cloern and Nichols (1985), Champ et al. (1987), McDowell et al. (1990 and Chapter 15 of this volume), and Steele (1991b and Chapter 1 of this volume) produced time-space diagrams for major biological phenomena; and Franklin et al. (1990) and Magnuson et al. (1991) produced diagrams for physical reset events of ecological systems such as fire or El Niño. Holling (1992), in an extensive treatment of scale and dynamics of ecosystems, produced a number of time-space diagrams, including those for different landscape elements and for decisions made by wading birds. Magnuson (1990 and Chapter 19 of this volume) and Swanson and Sparks (1990) related the time and space scales for the major physical and biological events to ecological research time scales.

It is clear that most ecological research is on a much shorter time scale than the major natural physical and biological phenomena affecting the planet. Our biological datasets were divided into aquatic and terrestrial, then plotted on time-space diagrams for those series for which we could obtain estimates of the scale of data collection plus the life span and size of the species involved (Figure 13-1). The series used for these plots and the values of life span, size, and scale used for each are listed in Table 13-2. The plots (Figure 13-1) include only those series for which we had all of these estimates. In each case the length of the series in years is plotted against the scale (or area) in hectares over which the data were collected, each on a logarithmic scale. The coefficients of determination for these regressions (log transformed) were only 0.13 for the aquatic series and 0.02 for the terrestrial series, with slopes of −0.02 and 0.01, respectively.

The same graphs also show the life span of the organism plotted against its typical size, similar to a plot of generation time versus size by Bonner (1965). The coefficient of determination for the regression of life span (log transformed) on size (log transformed) was 0.95 for the aquatic series and 0.91 for the terrestrial series, with slopes of 0.90 and 0.85, respectively. Although the spatial scale over which the data were collected covers five orders of magnitude, our datasets cover only a small part of the range of time scale in the generalized diagrams. It would have been possible to increase our time scale considerably by using paleoecological records. However, the time resolution of these records is usually on the order of 300–500 years. Even fine-scale paleoecological measurements such as those by Gear and Huntley (1991) are still very coarse in comparison to the rest of our data series; thus, they were not used. There are particularly few datasets that span the time scales between 50 and 300 years adequately, although this time scale may be very relevant to many ecosystem functions.

A good spread of spatial scale is represented in the aquatic and terrestrial systems in Figure 13-1. In the terrestrial system (Figure 13-1B), the larger scales are represented by hunting or forestry records covering large areas or by a migratory butterfly, which is recorded in the British Isles and in some years flies long distances across Europe. The smallest spatial scale in Figure 13-1B is that of the population records for a moth. (We have data series with smaller spatial scales, but they do not pertain to single species, so life span estimates would not be accurate.) Though it was not possible to make annual records of natural vegetation on a large scale before the advent of satellite technology, it would be possible to obtain annual estimates of plant productivity over large areas by using agricultural crop records, in the form of market prices that indicate production and are available for some species for several hundreds of years.

Aquatic systems (Figure 13-1A) tend to be more homogeneous than the terrestrial systems, so the plant and animal records are usually representative of a larger area. The longest time series (in years) and largest area covered are usually for commercial fisheries, although there are also two 69-year records of algae (diatoms) from an arctic lake soil core. Zooplankton and phytoplankton complete

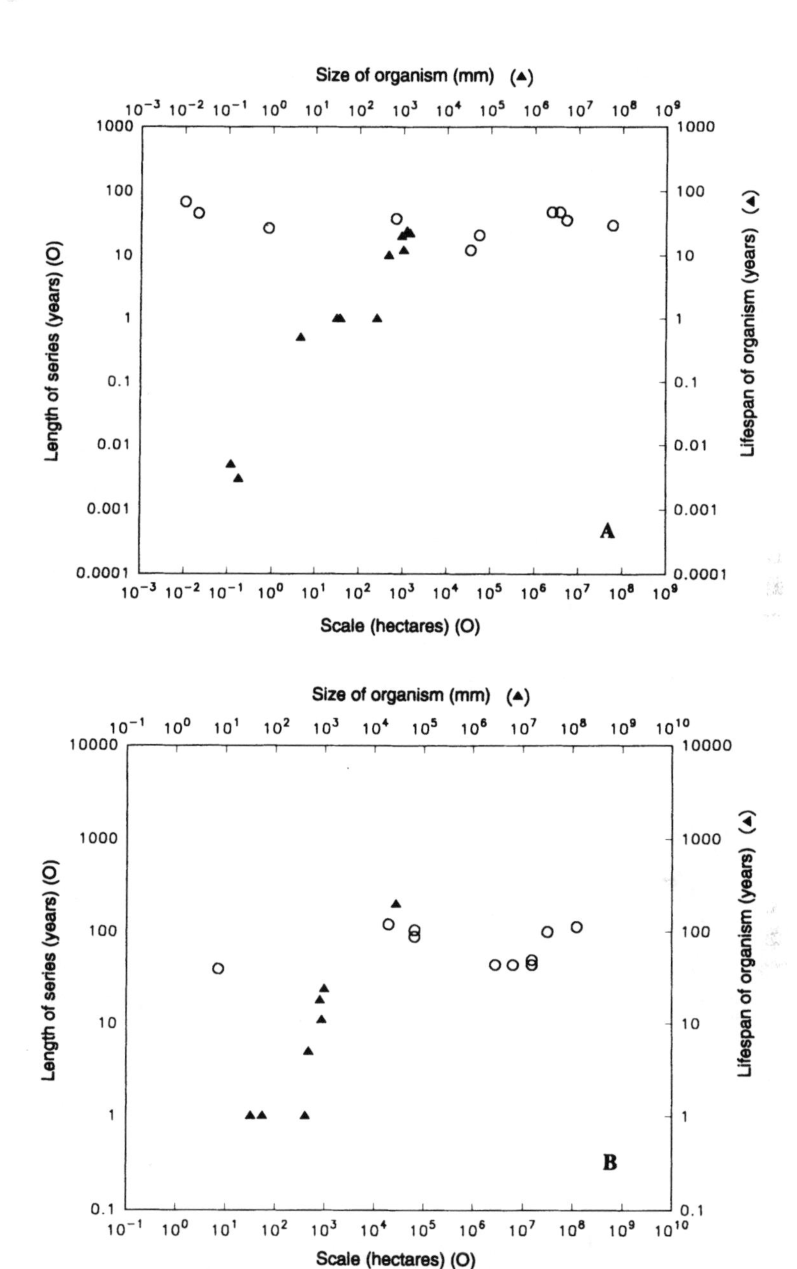

Figure 13-1. Plot of time versus spatial aspects on a logarithmic scale. (A) aquatic series (marine, estuarine, and freshwater); (B) terrestrial series. Points designated with a circle (○) are the length of the series (in years) versus the scale over which the data are collected. Points designated with a triangle (▲) are the life span versus the size of the organism.

Table 13-2. Natural history information for the biotic series used to construct Figure 13-1. (This information was also used in the principal component analysis.)

Series	Length of series (yrs)	Life span of organism (yrs)	Reference	Size of organism (mm)	Reference	Area over which the data was collected (ha)	Reference
Camberwell Beauty abundance	101	1	Chinery 1989	32	Higgens and Riley 1970	3.1×10^7	Williams 1958; and Bartholomew et al. 1985
Scarlet Tiger moth abundance	40	1	Skinner 1985	55	Skinner 1985	7	Jones 1989
Red grouse abundance	105	1	Johnsgard 1983	400	Johnsgard 1983	6.5×10^4	Mackenzie 1952
Black grouse abundance	89	5	Johnsgard 1983	470	Johnsgard 1983	6.5×10^4	Mackenzie 1952
Hare abundance	90	7	Corbet and Harris 1991	540	Southern 1964	—	—
Lynx pelts	114	11	Spector 1956	864	Caras 1967	1.2×10^8	Elton and Nicholson 1942
Pine cone production (Southern Sweden)	44	200	Carlisle and Brown 1968	2.7×10^4	Spector 1956	2.8×10^7	Sporrong and Wennstrom 1990
Pine cone production (Central Sweden)	44	200	Carlisle and Brown 1968	2.7×10^4	Spector 1956	6.3×10^6	Sporrong and Wennstrom 1990
Pine cone production (Northern Sweden)	44	200	Carlisle and Brown 1968	2.7×10^4	Spector 1956	1.5×10^6	Sporrong and Wennstrom 1990
Spruce cone production (Southern Sweden)	61	400	Spector 1956	3.7×10^4	Spector 1956	—	—
Spruce cone production (Central Sweden)	61	400	Spector 1956	3.7×10^4	Spector 1956	—	—
Spruce cone production (Northern Sweden)	61	400	Spector 1956	3.7×10^4	Spector 1956	—	—
Porcupine feeding scars	121	18	Roze 1989	80	Grzimek 1975	1.9×10^4	Spencer 1964 and USGS maps
Rothamsted-biomass	136	—	—	50	this chapter	0.0001	Rothamsted 1991
Gray heron abundance	50	24	Spector 1956	980	del Hoyo et al. 1992	1.5×10^7	Bartholomew et al. 1985
Dragonfly (*Sympetrum striolatum*) abundance	27	1	Moore 1991	37	McGeeney 1986	0.8	Moore 1991
Dragonfly (*Ischnura elegans*) abundance	27	1	Moore 1991	31	McGeeney 1986	0.8	Moore 1991

Diatom profiles (*Asterionella formosa*)	69	0.005	Lund 1949	0.12	Dodd 1987	0.0001	Simola 1984
Diatom profiles (*Synedra acus*)	69	0.003	Eppley 1977	0.18	Dodd 1987	0.0001	Simola 1984
Lake Constance - *Coregonus* YCS	21	10	Wheeler 1969	460	Wheeler 1969	4.8×10^4	Numann 1972
Mývatn Lake	61	—	—	—	—	3.7×10^3	Jónasson 1979
Pike population	37	24	Spector 1956	1200	Altman and Ditmer 1962	672	Talling 1988
Striped bass (Chesapeake Bay)	43	24	Spector 1956	1270	Wheeler 1975	—	—
Striped bass (MD)	42	24	Spector 1956	1270	Wheeler 1975	—	—
Apalachicola Bay (Fish CPUE)*	12.3	1	Livingston 1987; Hoese and Moore 1977	250	Livingston 1987; Hoese and Moore 1977	3.4×10^4	Livingston 1987 and USGS maps
Salmon population	64	13	Spector 1956	760	Thompson 1980	—	—
Calanus finmarchicus abundance	46	0.5	Marshall and Orr 1972	4.5	Marshall and Orr 1972	0.02	Glover 1967
Cod population	34	22	Altman and Ditmer 1962	1420	Backus and Bourne 1987	5.3×10^6	Backus and Bourne 1987
Hake population	30	12	Altman and Ditmer 1962	1000	Wheeler 1975	5.8×10^7	Holden 1978; Bartholomew et al. 1985
Summer flounder (MA)	48	20	Mid-Atlantic Fishery Management Council 1993	910	Mid-Atlantic Fishery Management Council 1993	3.7×10^6	Mid-Atlantic Fishery Management Council 1993; Backus and Bourne 1987
Summer flounder (NJ)	48	20	Mid-Atlantic Fishery Management Council 1993	910	Mid-Atlantic Fishery Management Council 1993	2.4×10^6	Mid-Atlantic Fishery Management Council 1993; Backus and Bourne 1987

*Natural history data for the most common species, *Leiostomus xanthurus*, was used.

many generations even in the relatively short time series available; the animals and plants in the center of Figure 13-1A complete 1 to 40 generations in their time series, whereas the trees in terrestrial systems (Figure 13-1B) have not even been recorded for one generation. The major difference between terrestrial and aquatic systems is the generation time and size of the primary producers. In the oceans, microscopic phytoplankton are the major primary producers, but in the terrestrial systems production comes mainly from trees, shrubs, and perennial herbs; many of these plants are long-lived either as individuals or as clones.

If the generation times on land and in the sea are different, then the frequency of measurement should also be different. Indeed, many of the marine datasets for organisms with a rapid turnover were recorded monthly rather than annually. The seasonal cycle becomes increasingly important with distance from the equator, in which case annual records may not be adequate, especially for species with rapid population dynamics (e.g., insects, which may require monthly or even daily monitoring). A good example of daily monitoring is the Rothamsted insect survey, which has used suction traps throughout Britain to monitor aerial populations of aphids for the last 30 years. The frequency scale for recording physical variables—for example, daily water or air temperature—may be the same for the aquatic and terrestrial systems. The daily figures may be amalgamated to produce monthly or annual means. However, in the life history of many organisms there are "critical periods"; so it may be possible to explain more of the year-to-year variability in population density or yield by using physical data from a portion of the year than by using annual or monthly averages. For example, the variability in Rothamsted hay yield is better explained by weekly rainfall and temperature in certain months (during rapid spring growth) than by annual means (Cashen 1947). Even with the monthly or weekly data, however, the amount of explainable year-to-year variability in hay yield is limited. An alternative approach is to use larger spatial scales; for hay yield, this is possible by using agricultural records collected from farmers. Smith (1960) found that by averaging hay yield over four geographical regions of England (each of the order of 20,000 km^2), he was able to achieve a 95% correlation between actual and predicted yield.

The problem of studying processes at appropriate scales has been frequently noted [see Levin (1992 and Chapter 14 of this volume) for a discussion]. Although the problem is recognized, actually altering the methods and the spatial and time scales in which studies are carried out is a difficult challenge, but one that must be met if we are to make meaningful comparisons in the future and obtain valid answers to the questions we pose.

Density Dependence

The importance of density in regulating population growth is a ubiquitous question in ecological theory. We used our dataset to compare evidence of density dependence in populations from different environments and of different trophic levels. Although a wide variety of techniques has been used to investigate the nature of density dependence in populations, here we restricted ourselves to two simple techniques, because of the large number of datasets of different origins. We used the first method of Bulmer (1975) and a test for nonlinearity based on a lag of one year.

We tested 22 of our datasets for density dependence. All 22 consisted of population data for a single species that reflected abundance in some way, and all were from series longer than 12 years. All but one (the *Calanus*) series were annual data.

Bulmer's Method

This method tests whether there is a tendency of a population to fluctuate around some mean value. When the population size is high, growth in population size is inhibited; but when the population size is small, there is no inhibition of increase in population size. Populations that are not density dependent undergo a random walk through time. This method is essentially the same as testing for a first-order autoregressive process (AR1) (A. Solow, pers. comm.). The results of our density dependence tests are presented in Table 13-3. With Bulmer's method, statistically significant density dependence is indicated by a value of $R_{(cal)}$ that is less than the value of R at $P = 0.05$ ($R_{(0.05)}$). Significant cycling was removed before testing for density dependence with this method. This included 3-year cycling and seasonality in the *Calanus* series, as well as a cycling of 10 years in the lynx series.

It should be noted that this method looks for any tendency to return to the mean value. It might be expected that density dependence would be extremely common under this definition. In fact, some of the abiotic series showed significant "density dependence" with this test. Wolda and Dennis (1993) used a similar test for density dependence and found rainfall data to be "density dependent." They recommended that both positive and negative results should be interpreted with caution. Positive results may be obtained with sloppy data as well as with series for which the idea of density dependence is absurd, such as the rainfall data. Negative results may simply mean that the series are not long enough for density dependence to be detected (Wolda and Dennis 1993).

Despite the broadness of this definition of density dependence, Stiling (1988) found in a review of insect studies that only about half of the species studied exhibited density dependence. He suggested that many insect species might exhibit essentially density-independent behavior within their normal range and exhibit density dependence only as the population approaches extreme values.

Hassell et al. (1989), however, reached a different conclusion using the same data that Stiling used. They found that as the length of the data series increased, so did the probability that density dependence would increase, which led them to suggest that for many of the species studied, density dependence was not necessarily absent but the data series may have been too short for statistical significance.

Table 13-3. Results of the tests for density dependence.

Series	Bulmer's method		Quadratic regression	
	$R_{(0.05)}$	$R_{(cal)}$	Sign (+/−)	*P*-value
Calanus finmarchicus	20.3800	0.7385*	−	0.0001*
Heron (*Ardea cinerea*)	2.0068	2.2123	+	0.5347
Porcupine (*Erethizon epixanthum*)	4.6054	1.0271*	−	0.4525
Atlantic salmon (*Salmo salar*)	2.5192	1.1643*	−	0.7705
Striped bass (*Morone saxatilis*) - Maryland	1.7140	1.5771*	−	0.0418*
Diatoms (*Asterionella formosa*)	2.7022	0.5716*	−	0.7301
Diatoms (*Synedra acus*)	2.7022	0.8329*	−	0.3847
Dragonfly (*Ischnura elegans*)	1.1650	0.5016*	+	0.5776
Dragonfly (*Sympetrum striolatum*)	1.1650	1.0521*	+	0.8338
Lake Constance - *Coregonus*	0.9454	0.5735*	+	0.7779
Pike (*Esox lucius*)	1.5676	3.0179	+	0.2628
Cod (*Gadus morhua*)	1.4944	2.0549	+	0.8795
Hake (*Merluccius merluccius*)	1.2748	1.3439	+	0.6442
Striped bass (*Morone saxatilis*) - Chesapeake Bay	1.7506	7.6757	−	0.0752
Summer flounder (*Paralichthys dentatus*) -Massachusetts	1.9336	3.4007	−	0.4645
Summer flounder (*Paralichthys dentatus*) - New Jersey	1.9336	2.4665	−	0.0984
Camberwell Beauty butterflies (*Nymphalis antiopa*)	3.8734	0.7970*	−	0.7623
Red grouse (*Lagopus lagopus scoticus*)	4.0198	2.3975*	−	0.0004*
Black grouse (*Tetrao tetrix*)	3.4342	0.9259*	+	0.8625
Hare (*Lepus europaeus*)	3.4708	1.5263*	−	0.0368*
Lynx (*Lynx canadensis*)	4.3492	2.3981*	−	0.0494*
Scarlet Tiger moth (*Panaxia dominula*)	1.6408	1.4653*	−	0.3727

*Significant density dependence at $P \leq 0.05$.

Density dependence was detected in the *Calanus* data, as well as in 14 of the 21 series of annual data. This contradicts the contention of Solow and Steele (1990) that 30 generations and moderately strong density dependence are needed to reject density independence with this method. Six of the seven series that were not density dependent were fish population series. Most of these series are biomass data from commercial fisheries, not actual population counts. Two other fish populations (Atlantic salmon and striped bass from Maryland) did show density dependence with this method. Bulmer (1975) offers an adjustment to the method to be used for data that are estimates rather than actual count data, but den Boer and Reddingius (1989) contend that it would be much better to correct the errors in the estimates than to use Bulmer's alternative method. This was not done, as there seemed to be no reliable way to correct the data.

Although we have some reservations as to both the detection of density dependence by this method and the limited number of series in each category, we note that seven of the eight series from terrestrial systems showed density dependence while only three of the eight from marine environments did. All but one of the freshwater series showed density dependence by this method. It should be noted that all of the series that failed to show density dependence by Bulmer's method have a ratio of series length to life span of 2.5 or less, suggesting that with additional years of data these series may have indicated significant density dependence. Two series (striped bass from Maryland and the Lake Constance *Coregonus*) with a series length-to-life span ratio of less than 2.5 did indicate density dependence by Bulmer's method. More series, especially series of greater length, are needed to test adequately the indication of a possible ecosystem effect.

All of the series that failed to indicate density dependence are for secondary consumers (i.e., trophic level 3). Organisms at this trophic level are generally larger organisms with longer life spans. If longer series (i.e., covering more life spans) were obtained, then these series might also indicate density dependence.

Nonlinearity

Although the linear model described above provides population regulation, a linear relationship between this year's population and next year's does not allow for an "optimal" population size, above which the expected value for next year declines with increases in this year's population. A linear model also does not allow for complex dynamics under a deterministic dynamic, such as periodic orbits, or for deterministic chaos. Both of these features appear frequently in discussions of density dependence, and these features, more than simple population regulation, are called to mind for many ecologists by the term *density dependence,* largely thanks to the pioneering work by May (e.g., May 1974).

To test for nonlinearity in the relationship between this year's population size and next year's, we used a quadratic regression model. This method resulted in fewer of the series showing density dependence than the Bulmer's method. Of the 22 series tested, the following series had a significantly negative quadratic term (P=0.05 or less), indicating density dependence: *Calanus,* striped bass from Maryland, red grouse from southern Scotland, European hares from Yorkshire, U.K., and the lynx from Canada (Table 13-3). All of these series were found to be density dependent with Bulmer's method also. The remaining 17 series did not indicate density dependence. Using this nonlinear method of detecting density dependence resulted in three of the eight series' from terrestrial systems showing density dependence as compared with only two of the eight from marine systems. None of the freshwater series indicated the presence of density dependence by this method. The limited number of series and the confounding factor of life span prevent us from drawing further conclusions.

Variation

Ecologists are concerned with population variation for a number of reasons. Dramatic changes in populations can result in outbreaks (which in turn may produce major effects within ecosystems) or in sudden decreases to the point of danger of extinction. Population variation can also provide insight into how species interact with each other and with the environment. Although there is no single definition of "variability," it may be possible to classify the term into some categories, each with more or less specific features but all contributing to the whole picture. For instance, we may look at behavior of the extreme values within the series themselves, and the periodicities and trends along the series. We used two measures to compare variability within our datasets: the coefficients of variation, and the Hurst exponent.

The coefficient of variation (CV) is the standard deviation × 100 divided by the mean, and it provides a unitless measure of how broadly a quantity varies. Coefficients of variation are shown in Figure 13-2. The CV was high for the Camberwell Beauty butterflies, and that series shows up as an outlier on each one of the plots (trophic level 2, invertebrate, terrestrial series, and biotic series). No significant differences were found in the CV based on variable type or environment. A high CV means great changes in terms of percentage of the average, or a very ragged series with alternating very high and very low values.

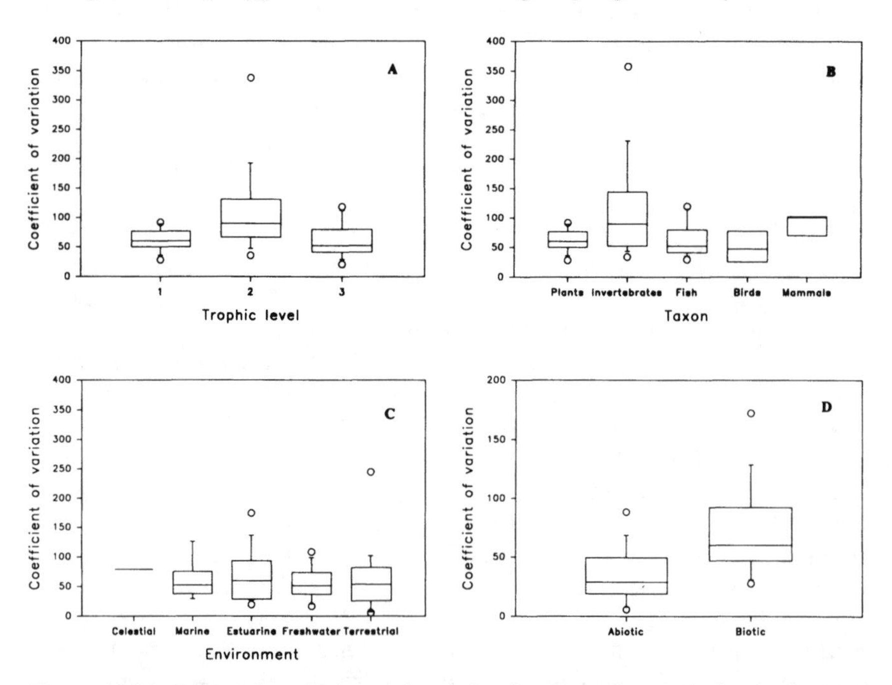

Figure 13-2. Range of coefficients of variation for (A) each trophic level; (B) each taxon; (C) each environment; and (D) abiotic and biotic series. The box represents the middle 50% of the series. The median is indicated by the horizontal line through the box. The ends of the box are the twenty-fifth and seventh-fifth percentiles. The error bars indicate the tenth and ninetieth percentiles of data. Extreme values are shown as circles.

However, it says nothing of cycling or trendy (and therefore predictable) behavior, or about the probability of hitting some extreme values.

Of course, the extreme values may be among the most interesting values of a biological time series, as they inform us about uncommon events or critical situations (extinctions, pest outbreaks, and the like). How they behave can probably be analyzed by calculating the Hurst exponent,[1] which is related to the fractal dimension of the series. In general, series having a high Hurst exponent will have a greater probability of "breaking a record," i.e., reaching a value not previously recorded (either too high or too low) for a given time. This happens because such series have what has been termed "reddened" noise, in which the variance increases with the time even though the differences between successive parameter estimations seem to be random.

The Hurst exponent for white noise is 0.5. A value in this range suggests a population that quickly explores the limits of its total range. The Hurst exponent for Brownian motion is 1. A value near 1 suggests a population that continually wanders and continually reaches new extremes.

The Hurst exponents for the series studied are shown in Figure 13-3. Note that the exponents for animals are substantially above 0.5 (Figure 13-3A, trophic levels 2 and 3), indicating a tendency for these populations to continue wandering over time. The Hurst exponent was found to be significantly negatively correlated with life span ($P = 0.005$) and the size of the organism ($P = 0.003$). The Hurst exponent was not significantly affected by environment or taxon (Figure 13-3B and C). No difference was found between abiotic and biotic series (Figure 13-3D). There was a nonsignificant ($P=0.14$) negative correlation with series length, which is of theoretical interest because a negative correlation with series length suggests the possibility that variation may stabilize over the very long term.

Sugihara and May (1990) used a measure similar to the Hurst exponent to indicate a population's vulnerability to extinction. (This is calculated differently than the Hurst exponent discussed here, but the idea is similar.) In a comparison of two bird series, one species was found to be more prone to local extinctions.

---

[1]The Hurst exponent is calculated as the slope of the regression between the logarithm of the rescaled range of the series and the logarithm of time, after a short transient of some 15 time units. Rescaled range was defined by Hurst (1951) and Mandelbrot and Wallis (1969) as the relative range divided by the standard deviation, both for the *d*th time point; the relative range is calculated algorithmically as

$$Q_d = \frac{R_d}{S_d},$$

$$\text{where } R_d = max_0^d\left(X_t^* - \frac{i}{d}(X_d^*)\right) - min_0^d\left(X_t^* - \frac{i}{d}(X_d^*)\right)$$

$$S_d^2 = \frac{X_{d^{\cdot}}^2}{d} - \frac{X_d^{*2}}{d^2}$$

$$X_t^* = \sum_{i=1}^{t} X_i.$$

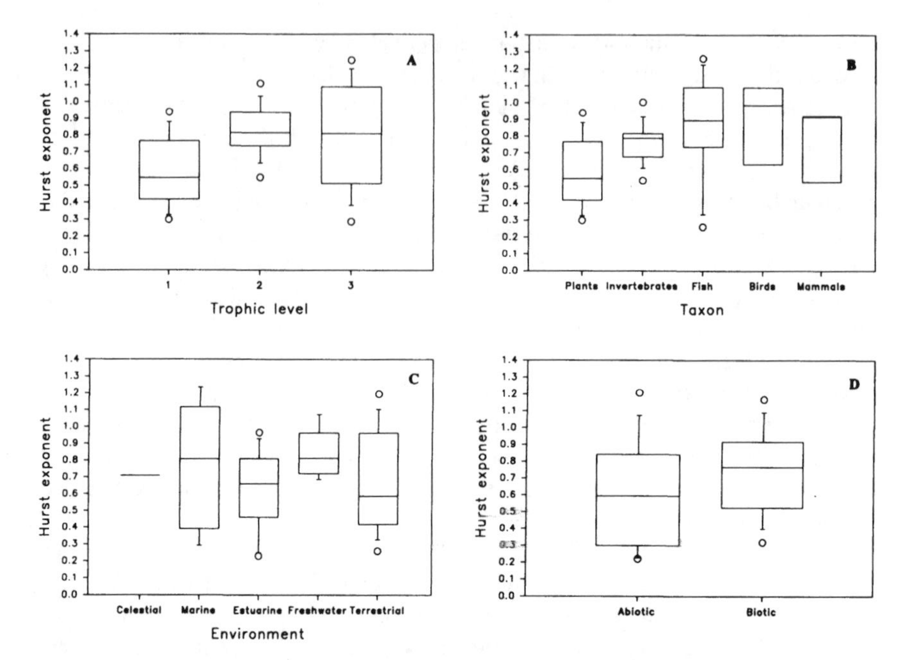

Figure 13-3. Range of Hurst exponents for (A) each trophic level; (B) each taxon; (C) each environment; and (D) abiotic and biotic series. The box represents the middle 50% of the series. The median is indicated by the horizontal line through the box. The ends of the box are the twenty-fifth and seventy-fifth percentiles. The error bars indicate the tenth and ninetieth percentiles of data. Extreme values are shown as circles.

Sugihara and May (1990) suggested that this also meant that it was not regulated by density-dependent processes. We did not see such a relationship with the series that we used.

Periodicities

Many of the series sampled at monthly or biweekly intervals had significant seasonality. Only four of these series (Chesapeake Bay benthic biomass, Corpus Christi benthic biomass and the biomass-to-abundance ratio, and Crystal River P:R ratio) were not significantly seasonal. One may interpret the abundance series (Chesapeake Bay and Corpus Christi) as seasonal; but one should not interpret the corresponding biomass as indicating that when fewer organisms are present, those that remain make up for the loss and increase in size. The other series, such as temperature and salinity in the systems, would be expected to show strong seasonality, which they do. The estuarine series are at the mouths of rivers where the flow of fresh water from the river varies seasonally. There is also an effect on salinity due to the tidal cycle, but this is at a periodicity shorter than that at which the sampling was done, and so we do not detect it. The effect of this problem with sampling is discussed by Platt and Denman (1975) and may include incorrect periodicities. This problem must be addressed when the sampling program is planned, either by taking more frequent samples (Platt and Denman recommend at least four samples within the shortest period fluctuation) or by using an averaging process during the sampling.

Very few of the series that we looked at showed significant periodicities (as indicated on the autocorrelation function plots) other than seasonality, though many series may simply be too short to have captured enough cycles. Those series that showed significant cycling were the sunspot series (22 periods or 11 years), the *Calanus* (3-year period in addition to seasonality), and the lynx series (9.6 years). In some of the other series there was an indication of possible cycling, but these indications were not significant at the 0.05 level. These included the black grouse (5 years), herons (16.7 years), and the sunspot series (50 years). Significance was also checked with a phase-length test (Kendall and Ord 1990). Pimm (1991) found a probable 16-year cycle in the heron series, but the series is only long enough to have three periods, and the drastic reductions in population numbers are known to be due to harsh winters. The cycle may well be significant if the pattern continues as more data are collected. Although the spectral analyses indicated other possible cycles in some of the series, they were not significant. Additional years of data would clarify whether this is due to the shortness of the series.

A Fisher-kappa test was done in order to look at the series and determine whether anything other than white noise was present. This test should detect a sinusoidal component that is buried in white noise (SAS 1984; Fuller 1976). The first ordinate is not used in this test. The test indicated that there was something other than white noise present for many of the series. If the significance level is categorized so that differences <0.01, between 0.01 and 0.05, between 0.05 and 0.10, and >0.10 are used and examined by environment, there is a significant ($P<0.05$) difference between environments by a chi-square test. It is harder to reject the hypothesis of only white noise for terrestrial and freshwater systems than it is for estuarine or marine systems; that is, marine and estuarine systems are more likely to have something other than white noise. This coincides with the idea of Steele (1985) that marine systems have reddened spectra at shorter periods. As more and longer series are collected for terrestrial systems, it may become easier to reject the white noise hypothesis for these systems as well.

Overall Contributing Factors

How are all the parameters listed above related to each other? More specifically, is there some general relationship or set of relationships between the measurable parameters and the *a priori* classification of the series? To gain a general idea, we analyzed the data in a multivariate design that was expected to indicate, for each of the series, the main relations between parameters—such as CV, Hurst exponent, length, standard deviation, latitude, life span (for biotic series), size of organism (for biotic series), and scale—and parameterized categorical variables, such as type of trend. Principal component analysis (PCA) was done separately for the abiotic and the biotic series. This allowed us to use the life span and size of organism data for the biotic series without causing the analysis to drop all the abiotic series, for which such information is not applicable. PCA will drop any series that has a missing value. This is a problem, because in some cases we do not have a good estimate of the scale over which the data are (or were) collected. (It should be noted that PCA treats categorized variables as numeric.)

The first two principal components are plotted against each other in Figures 13-4 and 13-5 for the abiotic series and the biotic series, respectively. The contribution of each principal component for the abiotic series is shown in Table 13-4. The first principal component accounts for almost 50% of the standardized variation for these series. The first principal component has several variables with similar loadings (Table 13-5). It is positively affected most by the environment (i.e., marine, estuarine, terrestrial, etc.), with scale also having a considerable effect. On the negative side, latitude and length have the largest contribution.

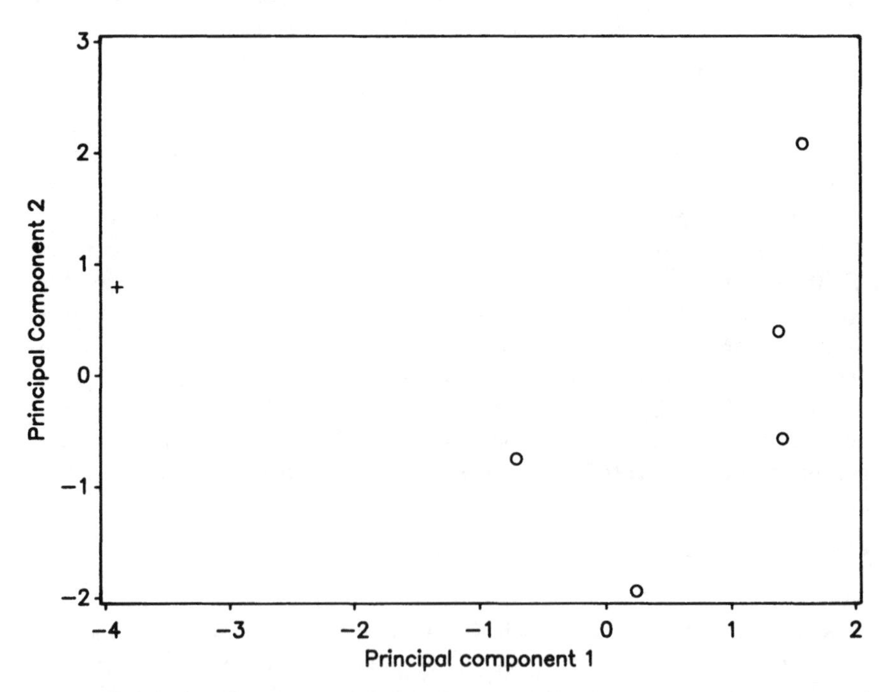

Figure 13-4. Plot of the second principal component against the first principal component for the abiotic series. (+ Marine, ○ Estuarine)

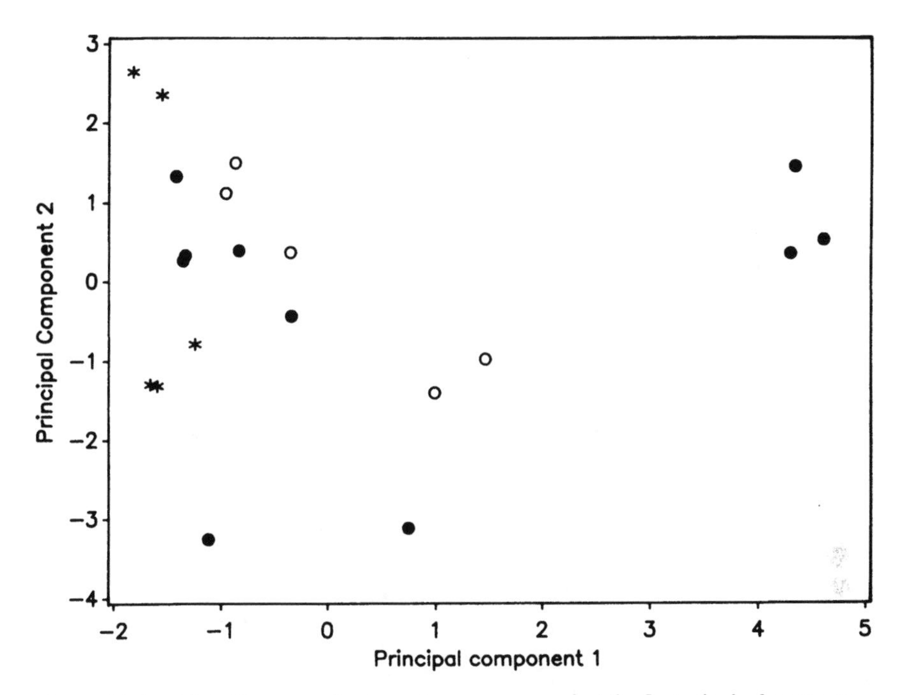

Figure 13-5. Plot of the second principal component against the first principal component for the biotic series. Estuarine series were dropped. (* Marine, ○ Freshwater, • Terrestrial)

The second principal component, which consists primarily of the measures of variances that we used, accounts for an additional 22% of the variance. Additional principal components and their loadings are shown in Tables 13-4 and 13-5. The first four principal components account for 95% of the variance. It can be seen from Figure 13-4 that the estuarine series all fall on the right side of the plot, with the one marine series on the left. Other abiotic series were dropped because of missing information. There are a variety of methods to determine how many principal components should be considered (see Jackson 1993 for a review). We have shown those that contribute up to 95% of the variance. Using the Kaiser-Guttman criteria ($\lambda > 1.0$), only the first three would be considered, whereas only the first two would be included using the broken-stick model. Certainly the first two principal components are more interpretable and more important than the later ones.

The biotic principal component analysis is more complex because it contains the same input variables plus the life history variables, such as life span, size of organism, generation time, taxon, and trophic level. The plot of the first two components is given in Figure 13-5. The estuarine series were dropped because of missing values, such as life span, because these series are not of a single species. The terrestrial series are scattered throughout the plot, but the aquatic

Table 13-4. Amount of influence of principal components of the PCA of the abiotic series.

Principal component	Eigenvalue	Proportion of variation accounted for	Cumulative variation accounted for
Principal component 1*†	4.449	0.494	0.494
Principal component 2*†	1.955	0.217	0.712
Principal component 3*	1.322	0.147	0.859
Principal component 4	0.808	0.090	0.948

*Significant with the Kaiser-Guttman criteria.

†Significant with the broken-stick model.

Table 13-5. Loadings of the first four principal components for the PCA for the abiotic series.

Variable	Principal component			
	First	Second	Third	Fourth
Length	−0.412	0.283	0.051	−0.114
CV	0.225	0.575	−0.111	0.330
Hurst	0.212	0.370	0.528	0.418
Latitude	−0.437	0.167	−0.220	0.195
Trend	−0.285	−0.112	0.619	0.216
Scale	0.364	−0.107	0.384	−0.329
SD	0.318	0.476	−0.215	−0.237
Environment	0.430	−0.199	0.007	−0.018
Significance level of Fisher-kappa test	0.209	−0.366	−0.287	0.673

series are somewhat clustered. Eight principal components were needed to account for 95% of the variance (Table 13-6). The first principal component, which accounts for 30% of the variance, is composed primarily of life history variables, such as size of the organism, life span, and generation time on the positive side, and trophic level and taxon category on the negative side. The second principal component, consisting primarily of the variance measures used (Table 13-7), accounts for an additional 17% of the variance. The environment has a fairly large loading on the first principal component, and then again on the fourth principal component. The biotic series appears to be primarily affected by the various life history parameters and secondarily by the other factors, such as the environment. Although eight principal components were needed to account for 95% of the variance, Kaiser-Guttman criteria would include the first six and the broken-stick model would include only the first three. It is difficult to say how more than the first three or four should be interpreted.

Table 13-6. Amount of influence of principal components of the PCA of the biotic series.

Principal component	Eigenvalue	Proportion of the variance accounted for	Cumulative amount of the variance accounted for
Principal component 1*†	4.428	0.295	0.295
Principal component 2*†	2.538	0.169	0.464
Principal component 3*†	2.068	0.138	0.602
Principal component 4*	1.472	0.098	0.700
Principal component 5*	1.237	0.082	0.783
Principal component 6*	1.039	0.069	0.852
Principal component 7	0.783	0.052	0.904
Principal component 8	0.761	0.051	0.955

*Significant with the Kaiser-Guttman criteria.

†Significant with the broken-stick model.

Table 13-7. Loadings of the first eight principal components for the principal component analysis for the biotic series.

	Principal component					
Variable	First	Second	Third	Fourth	Fifth	Sixth
Length	−0.081	−0.156	−0.327	0.195	−0.449	0.488
CV	−0.007	−0.394	−0.313	0.107	0.024	0.073
Hurst	−0.210	0.372	−0.211	0.110	−0.160	−0.167
Latitude	0.272	−0.090	−0.011	0.438	0.156	0.291
Trend	−0.172	0.430	−0.136	0.359	−0.128	−0.020
Scale	−0.056	−0.227	0.396	0.432	0.137	0.341
SD	−0.164	−0.322	0.462	−0.200	−0.123	−0.008
Environment	0.220	−0.089	−0.034	0.450	0.332	−0.501
Sig. of Fisher-kappa test	0.234	0.166	−0.122	−0.250	0.541	0.284
Trophic level	−0.384	0.061	0.267	−0.098	0.155	0.143
Taxon	−0.342	−0.020	0.239	0.329	−0.016	−0.249
Size of organism	0.422	−0.140	0.194	0.019	−0.169	0.038
Life span	0.409	−0.148	0.225	0.002	−0.191	0.017
Generation time	0.211	−0.330	0.351	0.056	−0.280	0.068
Sig. of neg. quad. term	−0.239	−0.379	0.079	0.061	0.355	0.326

Discussion

We have examined a variety of datasets in order to compare long-term time series across different types of ecosystems. There are a number of limitations in making these comparisons. The time series we have chosen to analyze for this chapter are only a small selection of those available. They illustrate a number of the typical pitfalls or limitations of this kind of data. The main point is the shortness of many of the series. This is particularly important for organisms with long life spans or with generation times or processes that have a long cycle. Unfortunately, this problem is difficult to overcome, inasmuch as observers and techniques change with time.

The marine catch records try to account for changing technology by correcting for fishing effort. Other uncorrected factors may affect any time series analysis, however, including differences in fish migration, government controls on over-fishing in certain areas, and pollution. The grouse time series, for example, appear to be from natural ecosystems, but in reality the heather moors are burnt and managed to achieve high populations of game birds, and shooting may be restricted if the populations fall too low. An alternative approach, to avoid problems of hunting or catching, would be to use only time series from designed experiments. Long-term trials are usually less than 25 years, so they can indicate only short-term periodicities. One striking exception is the Park Grass at Rothamsted (136 years), which reflects the enthusiasm of its originators and the low turnover rate among Rothamsted directors. Even this time series, however, has been affected by changes in methodology, the most notable of which occurred in the 1960s when the technique of dry-matter determination changed. This had little effect on total yield but significantly increased the amount of year-to-year variability explainable by weather variables.

An interesting case is the data series on the Camberwell Beauty, a migratory butterfly, which is recorded by amateur entomologists all over the British Isles (though in most years few butterflies are seen). Occasionally large numbers of the butterfly move up across Europe and reach as far as Britain. It is not clear whether this is due to a population explosion or just particularly good conditions for migration. During the 101 years of the Camberwell Beauty time series, there have also been massive changes to the agricultural systems and loss of natural habitat over which it flies. It is hard to understand how this species, counted where it is (British Isles), could be density dependent there, since it does not feed or breed there. As pointed out by Wolda and Dennis (1993), just because the test indicates statistically significant density dependence, care must be used in its interpretation. The use of time series data to look for density dependence should be examined further.

Another problem with many of the series is that they are either collected on the wrong frequency, such as the zooplankton series from Lake Constance, or for too short a time. If a series is collected at an inappropriate sampling frequency, errors may be made in its interpretation, and periodicities will not be apparent.

Keeping these limitations in mind, we note that principal component analysis indicates that the abiotic series are most affected by factors such as environmental category and latitude, whereas the biotic series seem to be most affected by the life history parameters of the organisms themselves.

A more complete survey of long time series may yield somewhat different results, but in any case it is clear that we need to take greater care to collect data at more meaningful scales (both time and space) for the questions that we are asking. This will involve taking into account species life spans (or generation times) or the relevant time frame for abiotic processes. Individuals with longer life spans (years) will be counted each year in an annual census, but those with shorter life spans (<1 year) will not. One would expect less variation for organisms with long life spans, and this may have been reflected in the negative correlation of life span and the Hurst exponent (although this effect was not seen in relation to the coefficient of variation). Appropriate sampling design requires knowledge of these parameters that in some cases has not been attained, as well as greater cooperation among individuals, especially to carry out experiments over long time scales. Despite improvement in this regard with the establishment of the Long-Term Ecological Research (LTER) sites, the Committee on Environmental Research of the National Research Council recently reported that "the research establishment is currently poorly structured to deal with complex, interdisciplinary research on the large spatial scales and long-term temporal scales" (Grossblatt 1993). Now that the problem has been recognized, we hope that efforts to address it will continue, in the form of more studies on long time scales.

Acknowledgments

The authors kindly thank Bob Williams for providing the North Sea temperature and *Calanus finmarchicus* data from the continuous plankton recorders. We also thank Fred Holland and A. Ransasinghe for providing data from Chesapeake Bay. Stuart Pimm kindly allowed us to use series from his database. We thank John Steele, Tom Powell, Andy Solow, Mitch McClaran, Jason Stockwell, and Tormod Burkey for their continuous support and helpful advice on this chapter. Finally, we thank all of the students and lecturers of this course for their constant encouragement.

References

Allee, W. C., A. E. Emerson, O. Park, T. Park, and K. P. Schmidt. 1949. *Principles of Animal Ecology*. W. B. Saunders, Philadelphia.

Altman, P. L., and D. S. Ditmer. 1962. *Growth Including Reproduction and Morphological Development*. Federation of American Societies for Experimental Biology, Washington, D.C.

Ariño, A. H., and S. L. Pimm. Submitted. On the nature of population extremes. *Proceedings of the National Academy of Sciences*.

Backus, R. H., and D. W. Bourne (eds.). 1987. *Georges Bank*. MIT Press, Cambridge, MA.

Bartholomew, J. C., J. H. Christie, A. Ewington, P. J. M. Geelan, H. A. G. Lewis, P.

Middleton, and B. Winkleman (eds.), 1985. *The Times Atlas of the World*. Times Books Limited, London.

Bonner, J. T. 1965. *Size and Cycle, an Essay on the Structure of Biology*. Princeton University Press, Princeton, NJ.

Bulmer, M. G. 1975. The statistical analysis of density dependence. *Biometrics* **31:**901–911.

Caras, R. A. 1967. *North American Mammals: Fur-Bearing Animals of the United States and Canada*. Meredith Press, New York.

Carlisle, A., and A. H. F. Brown. 1968. Biological flora of the British Isles. *Pinus sylvestris* L. *Journal of Ecology* **56:**269–307.

Cashen, R. O. 1947. The influence of rainfall on the yield and botanical composition of permanent grass at Rothamsted. *Journal of Agricultural Science* **37:**1–10.

Champ, M. A., D. A. Wolfe, D. A. Flemer, and A. J. Mearns. 1987. Long-term biological data sets: Their role in research, monitoring, and management of estuarine and coastal marine systems. *Estuaries* **10:**181–193.

Chinery, M. 1989. *Butterflies and Day-Flying Moths of Britain and Europe*. University of Texas Press, Austin, TX.

Cloern, J. E., and F. H. Nichols. 1985. Time scales and mechanisms of estuarine variability, a synthesis from studies of San Francisco Bay. *Hydrobiologia* **129:**229–237.

Corbet, G. B., and S. Harris (eds.). 1991. *The Handbook of British Mammals*. Third edition. Blackwell Scientific Publications. Oxford, U.K.

del Hoyo, J., A. Elliot, and J. Sargatel. 1992. *Handbook of the Birds of the World. Volume 1. Ostrich to Ducks*. Lynx Editions, Barcelona, Spain.

den Boer, P. J., and J. Reddingius. 1989. On the stabilization of animal numbers. Problems of testing. 1. Power estimates and estimation errors. *Oecologia* **78:**1–8.

Denman, K. L., and T. M. Powell. 1984. Effects of physical processes on planktonic ecosystems in the coastal ocean. *Oceanography and Marine Biology—An Annual Review* **22:**125–168.

Deuel, D., D. McDaniel, and S. Taub. 1989. *Atlantic Coastal Striped Bass: Road to Recovery*. U.S. Department of the Interior, Fish and Wildlife Service, and U.S. Department of Commerce, National Oceanic and Atmospheric Administration, Washington, D.C.

Dodd, J. J. 1987. *The Illustrated Flora of Illinois—Diatoms*. Southern Illinois University Press, Carbondale, IL.

Dunning, J. B., B. J. Danielson, and H. R. Pulliam. 1992. Ecological processes that affect populations in complex landscapes. *Oikos* **65(1):**169–175.

Eckmann, R., U. Gaedke, and H. J. Wetzlar. 1988. Effects of climatic and density dependent factors on year-class strength of *Coregonus lavaretus* in Lake Constance. *Canadian Journal of Fisheries and Aquatic Science* **45:**1088–1093.

Elton, C., and M. Nicholson. 1942. The ten-year cycle in numbers of the lynx in Canada. *Journal of Animal Ecology* **11:**215–244.

Eppley, R. W. 1977. "The Growth and Culture of Diatoms." In D. Werner (ed.), *Biologi-*

cal Monographs—Vol. 13. The Biology of Diatoms. University of California Press, Berkeley.

Flint, R. W. 1985. Long-term estuarine variability and associated biological response. *Estuaries* **8:**158–169.

Flocas, A. A., and A. Arseni-Papadimitriou. 1974. On the annual variation of air temperature in Thessaloniki. *Science Annals, Faculty of Physics and Mathematics, University of Thessaloniki* **14:**129–149.

Franklin, J. F., C. S. Bledsoe, and J. T. Callahan. 1990. Contributions of the long-term ecological research program. *BioScience* **40:**509–523.

Fuller, W. A. 1976. *Introduction to Statistical Time Series*. John Wiley & Sons, New York.

Gear, A. J., and B. Huntley. 1991. Rapid changes in the range limits of Scots Pine 4000 years ago. *Science* **251:**544–546.

Glover, R. S. 1967. The continuous plankton recorder survey of the North Atlantic. *Symposium of the Zoological Society of London* **1967(19):**189–210.

Grossblatt, N. (ed.). 1993. *Report of the Committee on Environmental Research—Executive Summary*. Commission on Life Sciences, National Research Council. National Academy Press, Washington, D.C.

Grzimek, B. (ed.). 1975. *Grzimek's Animal Life Encyclopedia. Volume 11 (Mammals II)*. Van Nostrand Reinhold, New York.

Gudmundsson, F. 1979. The past status and exploitation of the Myvatn waterfowl populations. *Oikos* **32:**232–249.

Hagner, S. 1965. Cone crop fluctuations in Scots Pine and Norway spruce. *Studia Forestalia Suecica* **33:**3–21.

Harper, J. L. 1977. *Population Biology of Plants*. Academic Press, New York.

Hassell, M. P., J. Latto, and R. M. May. 1989. Seeing the wood for the trees: Detecting density dependence from existing life-tables studies. *Journal of Animal Ecology* **58:**883–892.

Haury, L. R., J. A. McGowan, and P. H. Wiebe. 1978. "Patterns and Processes in the Time-Space Scales of Plankton Distributions." In J. H. Steele (ed), *Spatial Patterns in Plankton Communities*. Plenum Press, New York.

Higgins, L. G., and N. D. Riley. 1970. *A Field Guide to the Butterflies of Britain and Europe*. Houghton Mifflin, Boston.

Hoese, H. D., and R. H. Moore. 1977. *Fishes of the Gulf of Mexico—Texas, Louisiana, and Adjacent Waters*. Texas A & M University Press, College Station, TX.

Holden, M. J. 1978. Long-term changes in landings of fish from the North Sea. *Rapports et Procès-Verbaux des Réunions* **172:**11–26.

———. 1991. North Sea cod and haddock stocks in collapse or the end of the "gadoid outburst"? ICES CM **1991/G:**41.

Holland, A. F., A. T. Shaughnessey, and M. H. Hiegel. 1987. Long-term variation in mesohaline Chesapeake Bay macrobenthos: Spatial and temporal patterns. *Estuaries* **10:**227–245.

Holling, C. S. 1992. Cross-scale morphology, geometry and dynamics of ecosystems. *Ecological Monographs* **62(4)**:447–502.

Hurst, H. E. 1951. Long-term storage capacity of reservoirs. *Transactions of the American Society of Civil Engineers* **116:**770.

IBM. 1992. *IBM Engineering and Scientific Subroutine Library Version 2 Guide and Reference*. IBM Corporation, Kingston, NY.

Jackson, D. A. 1993. Stopping rules in principal components analysis: A comparison of heuristical and statistical approaches. *Ecology* **74(8)**:2204–2214.

Johnsgard, P. A. 1983. *The Grouse of the World*. University of Nebraska Press, Lincoln, NB.

Jónasson, P. M. (ed.). 1979. Ecology of eutrophic, subarctic Lake Myvatn and River Laxa. *Oikos* **32:**1–308.

Jones, D. A. 1989. 50 Years of studying the scarlet tiger moth. *Trends in Ecology and Evolution* **4(10)**:298–301.

Kendall, M. G., and J. K. Ord. 1990. *Time Series*. Third edition. Oxford University Press, London and New York.

Kipling, C. 1983. Changes in the population of Pike (*Esox lucius*) in Windermere from 1944 to 1981. *Journal of Animal Ecology* **52:**989–999.

Levin, S. A. 1992. The problem of pattern and scale in ecology. *Ecology* **73(6):**1943–1967.

Likens, G. E. 1989. "Preface." In G. E. Likens. (ed.), *Long-term Studies in Ecology: Approaches and Alternatives*. Springer-Verlag, New York.

Likens, G. E., F. H. Bormann, R. S. Pierce, J. S. Eaton, and N. M. Johnson. 1977. *Biogeochemistry of a Forested Ecosystem*. Springer-Verlag, New York.

Livingston, R. J. 1987. Field sampling in estuaries: The relationships of scale to variability. *Estuaries* **10:**194–207.

Loucks, R. H., and W. H. Sutcliffe, Jr. 1978. A simple fish-population model including environmental influence, for two western Atlantic shelf stocks. *Journal of the Fisheries Research Board of Canada* **35:**279–285.

Lund, J. W. G. 1949. Studies on *Asterionella* I. The origin and nature of the cells producing seasonal maxima. *Journal of Ecology* **37:**389–419.

Mackenzie, J. M. D. 1952. Fluctuations in the numbers of British tetranoids. *Journal of Animal Ecology* **21:**128–153.

Magnuson, J. J. 1990. Long term ecological research and the invisible present. *BioScience* **40:**495–502.

Magnuson, J. J., T. K. Kratz, T. M. Frost, C. J. Bowser, B. J. Benson, and R. Nero. 1991. "Expanding the Temporal and Spatial Scales of Ecological Research and Comparison of Divergent Ecosystems: Roles for LTER in the United States." In *Long-Term Ecological Research: An International Perspective*. John Wiley & Sons, Chichester, U.K.

Mandelbrot, B., and J. R. Wallis. 1969. Robustness of the rescaled range R/S in the measurement of noncyclic long run statistical dependence. *Water Resources Research* **5(1)**:967–988.

Mann, D. G. 1988. "Why Didn't Lund See Sex in *Asterionella?* A Discussion of the Diatom Life Cycle in Nature." In F. E. Round (ed.), *Algae and the Aquatic Environment*. Biopress, Bristol, U.K.

Marshall, S. M., and A. P. Orr. 1972. *The Biology of a Marine Copepod,* Calanus finmarchicus (Gunnerus). Springer-Verlag, New York.

May, R. M. 1974. Biological populations with non-overlapping generations: Stable points, stable cycles and chaos. *Science* **186**:645–647.

McDowell, P., T. Webb III, and P. J. Bartlein. 1990. "Long-term Environmental Change." In B. L. Turner II, W. C. Clark, R. W. Kates, J. F. Richards, J. T. Mathews, and W. B. Meyer (eds.), *The Earth as Transformed by Human Action*. Cambridge University Press, London.

McGeeney, A. 1986. *A Complete Guide to British Dragonflies*. Jonathan Cape Limited, London.

Mid-Atlantic Fishery Management Council. 1993. Amendment 2 to the fishery management plan for the Summer flounder fishery. Mid-Atlantic Fishery Management Council, Dover, DE.

Middleton, A. D. 1934. Periodic fluctuations in British game populations. *Journal of Animal Ecology* **3**:231–249.

Montague, C. L., J. W. Caldwell, R. L. Knight, K. A. Benkert, A. M. Watson, W. F. Coggins, J. J. Kosik, and K. E. Limburg. 1981. *Record of Metabolism of Estuarine Ecosystems at Crystal River, Florida*. Final annual report to the Florida Power Corp., Department of Environmental Engineering Sciences, University of Florida, Gainesville, FL.

Moore, N. W. 1991. The development of dragonfly communities and the consequences of territorial behavior: A 27 year study on small ponds at Woodwalton Fen, Cambridgeshire, United Kingdom. *Odonatologica* **20(2)**:203–231.

Nümann, W. 1972. The Bodensee: Effects of exploitation and eutrophication on the Salmonid community. *Journal of the Fisheries Research Board of Canada* **29**:833–847.

O'Conner, R. M. 1986. "Reproduction and Age Distribution of Female Lynx in Alaska, 1961–1971—Preliminary Results." In S. D. Miller and D. D. Everett (eds.), *Cats of the World: Biology, Conservation, and Management*. National Wildlife Federation, Washington, D.C.

Pimm, S. L. 1991. *The Balance of Nature?* University of Chicago Press, Chicago.

Platt, T., and K. L. Denman. 1975. Spectral analysis in ecology. *Annual Review of Ecology and Systematics* **6**:189–210.

Remmert, H. 1980. *Arctic Animal Ecology*. Springer-Verlag, Berlin.

Reynolds, C. M. 1979. The heronries census: 1972–1977 population changes and a review. *Bird Study* **26**:7–12.

Robbins, C. R., and G. C. Ray. 1986. *A Field Guide to the Atlantic Coast Fishes of North America*. Houghton Mifflin, Boston, MA.

Rothamsted 1991. *Rothamsted Experimental Station Report*. Harpenden, U.K.

Roze, U. 1989. *The North American Porcupine*. Smithsonian Institution Press, Washington, D.C.

SAS Institute Inc. 1984. *SAS/ETS User's Guide,* Version 5 edition. Cary, NC.

Shumway, R. H. 1988. *Applied Statistical Time Series Analysis*. Prentice-Hall, Englewood Cliffs, NJ.

Simola, H. 1984. Population dynamics of plankton diatoms in a 69-year sequence of annually laminated sediment. *Oikos* **43:**30–40.

Skinner, B. 1985. *Moths of the British Isles*. Viking Press, U.K.

Smith, L. P. 1960. The relation between weather and meadow hay yields in England. *Journal of the British Grassland Society* **15:**203–208.

Smith, P. E. 1978. Biological effects of ocean variability: Time and space scales of biological response. *Rapports et Procès-Verbaux des Réunions* **173:**117–127.

Solow, A. R., and J. H. Steele. 1990. On sample size, statistical power, and the detection of density dependence. *Journal of Animal Ecology* **59:**1973–1076.

Southern, H. N. 1964. *The Handbook of British Mammals*. Blackwell Scientific Publications, Oxford, U.K.

Spector, W. S. (ed.). 1956. *Handbook of Biological Data*. W. B. Saunders, Philadelphia.

Spencer, D. A. 1964. Porcupine population fluctuations in past centuries revealed by dendrochronology. *Journal of Applied Ecology* **1:**127–149.

Sporrong, U., and H. F. Wennström (eds.). 1990. *The National Atlas of Sweden. Volume 1. Maps and Mapping*. SNA Publishing, Italy.

Stafford, J. 1971. The heron population of England and Wales, 1928–1970. *Bird Study* **18:**218–221.

Steele, J. H. 1985. A comparison of marine and terrestrial ecological systems. *Nature* **313:**355–358.

———. 1991a. Marine functional diversity. *BioScience* **41:**470–474.

———. 1991b. Can ecological theory cross the land-sea boundary? *Journal of Theoretical Biology* **153:**425–436.

Stergiou K. I. 1991. Possible implications of climatic variability on the presence of Capelin (*Mallotus villosus*) off the Norwegian Coast. *Climatic Change* **19:**369–391.

Stiling, P. 1988. Density-dependent processes and key factors in insect populations. *Journal of Animal Ecology* **57:**581–594.

Sugihara, G., and R. M. May. 1990. Applications of fractals in ecology. *Trends in Ecology and Evolution* **5(3):**79–86.

Summers, J. K., T. T. Polgar, J. A. Tarr, K. A. Rose, D. G. Heimbuch, J. McCurley, R. A. Cummins, G. F. Johnson, K. T. Yetman, and G. T. DiNardo. 1985. Reconstruction of long-term time series for commercial fisheries abundance and estuarine pollution loadings. *Estuaries* **8(2A):**114–124.

Svärdsson, G. 1957. The "invasion" type of bird migration. *British Birds*. **50:**314–343. (See also Remmert 1980.)

Swanson, F. J., and R. E. Sparks. 1990. Long-term ecological research and the invisible place. *BioScience* **40:**503–508.

Talling, J. F., and S. I. Heaney. 1988. "Long-term Changes in some English (Cumbrian) Lakes Subjected to Increased Nutrient Inputs." In F. E. Round (ed.), *Algae and the Aquatic Environment*. Biopress, Bristol, U.K.

Thompson, P. 1980. *The Game Fishes of New England and Southeastern Canada*. Down East, Camden, ME.

VanWinkle, W., B. L. Kirk, and B. W. Rust. 1979. Periodicities in Atlantic coast Striped bass (*Morone saxatilis*) commercial fisheries data. *Journal of the Fisheries Research Board of Canada* **36:**54–62.

Varley, M. E. 1967. *British Freshwater Fishes*. Fishing News (Books) Limited, London.

Weatherhead, P. J. 1986. How unusual are unusual events? *American Naturalist* **128:**150–154.

Wheeler, A. 1969. *The Fishes of the British Isles and North-West Europe*. Michigan State University Press, East Lansing, MI.

———. 1975. *Fishes of the World, an Illustrated Dictionary*. Macmillan, New York.

Wiens, J. A. 1989. Spatial scaling in ecology. *Functional Ecology* **3:**385–397.

Williams, C. B. 1958. *Insect Migration*. N. N. Collins, London.

Williams, R. 1993. Plymouth Marine Laboratory, Plymouth, U.K. Personal communication with the authors (July).

Witteman, G. J., A. Redfearn, and S. L. Pimm. 1990. The extent of complex population changes in nature. *Evolutionary Ecology* **4:**173–183.

Woiwood, I., and R. Harrington. 1993. Flying in the face of change—the Rothamsted Insect Survey. *Rothamsted 50th Anniversary Conference Proceedings*.

Wolda, H., and B. Dennis. 1993. Density dependence tests, are they? *Oecologia* **95:**581–591.

PART III

Processes and Principles

14

The Problem of Pattern and Scale in Ecology

Simon A. Levin

This book is the second of two volumes in a series on terrestrial and marine comparisons, focusing on the temporal complement of the earlier spatial analysis of patchiness and pattern (Levin et al. 1993). The issue of the relationships among pattern, scale, and patchiness has been framed forcefully in John Steele's writings of two decades (e.g., Steele 1978). There is no pattern without an observational frame. In the words of Nietzsche, "There are no facts . . . only interpretations."

The description of pattern varies with the scale of observation, and spatial, temporal, and organizational scales are ultimately coupled. This fact is of both ecological and evolutionary importance. The implication is that on any particular scale of observation, we must average over phenomena occurring on shorter scales in order to reproduce the pattern and regularity that we recognize, just as over longer time scales, species evolve mechanisms such as dispersal, dormancy, and iteroparity for averaging over variability and reducing uncertainty.

Some evolutionary mechanisms result in simple, unweighted arithmetic averages of finer-scale phenomena, but most involve weighted nonlinear averages. The determination of how fine-scale information is transferred to broad scales is the essence of understanding the determinants of pattern and has been the focus of research throughout the sciences. It is the topic of the following paper, reprinted from *Ecology*, in which I treat ecological and evolutionary aspects of scale and pattern.

This chapter is adapted from "The Problem of Pattern and Scale in Ecology," the Robert H. MacArthur Award Lecture presented by Simon A. Levin in August 1989 at Toronto, Ontario, Canada, and published in *Ecology* (1992) 73(6):1943–1967. Reprinted by permission of the Ecological Society of America.

Introduction

Choosing the topic of this lecture was a difficult experience. The occasion was a unique opportunity to advance a personal view of the most fascinating questions in ecology, and several themes seemed appealing. First and foremost, as the lecture honors the contributions of Robert MacArthur, it seemed fitting to discuss the role of theoretical ecology: its historical roots, the influence of MacArthur in transforming it, and how it has changed in the years since his death. For one committed to demonstrating the importance of theory as an essential partner to empiricism, this challenge seemed almost a mandate.

However, another topic that has occupied much of my attention for the past decade seemed equally compelling: the interface between population biology and ecosystems science. The traditions in these two subdisciplines are so distant that few studies seem able to blend them. Conservation biology and ecotoxicology manage to span the middle ground; but the chasm between evolutionary biology and ecosystems science is a wide one, and there is little overlap between the two in journals or scientific meetings. Yet neither discipline can afford to ignore the other: evolutionary changes take place within the context of ecosystems, and an evolutionary perspective is critical for understanding organisms' behavioral and physiological responses to environmental change. Furthermore, cross-system patterns that make the study of ecosystems more than simply the accumulation of unrelated anecdotes can only be explained within a framework that examines the evolutionary forces that act upon individual populations (e.g., MacArthur 1968; Orians and Paine 1983; Roughgarden 1989). The importance of bringing these two schools of thought together should be universally attractive.

Other topics also presented themselves as candidate themes: the interface between basic and applied science, the dynamics of structured populations, life history responses to variable environments, diffuse coevolution, the development of ecological pattern, metapopulations, and the problem of scale. Indeed, as I looked back over my career, which had included flirtations with each of these problems, I was struck by what a patchwork it seemed. What was the thread, if any, that had guided my wanderings? In retrospect, it became clear that a fascination with scale had underlain all of these efforts; it is, I will argue, the fundamental conceptual problem in ecology, if not in all of science.

Theoretical ecology, and theoretical science more generally, relate processes that occur on different scales of space, time, and organizational complexity. Understanding patterns in terms of the processes that produce them is the essence of science and the key to the development of principles for management. Without an understanding of mechanisms, one must evaluate each new stress on each new system *de novo,* without any scientific basis for extrapolation; with such understanding, one has the foundation for understanding and management. A popular fascination of theorists in all disciplines, because of the potential for mechanistic understanding, has been with systems in which the dynamics at one level of organization can be understood as the collective behavior of aggregates of similar units. Statistical mechanics, interacting particle systems, synergetics, neural networks, hierarchy theory, and other subjects all have concerned themselves with this problem, and I shall direct considerable attention to it in this chapter.

Addressing the problem of scale also has fundamental applied importance. Global and regional changes in biological diversity, in the distribution of greenhouse gases and pollutants, and in climate all have origins in and consequences for fine-scale phenomena. The general circulation models that provide the basis for climate prediction operate on spatial and temporal scales (Figure 14-1) many orders of magnitude greater than the scales at which most ecological studies are carried out (Hansen et al. 1987; Schneider 1989); satellite imagery and other means of remote sensing provide spatial information somewhere in between the two, overlapping both. General circulation models and remote sensing techniques also must lump functional ecological classes, sometimes into very crude assemblages (e.g., the "big leaf" to represent regional vegetation), suppressing considerable ecological detail. To develop the predictive models that are needed for management, or simply to allow ourselves to respond to change, we must learn

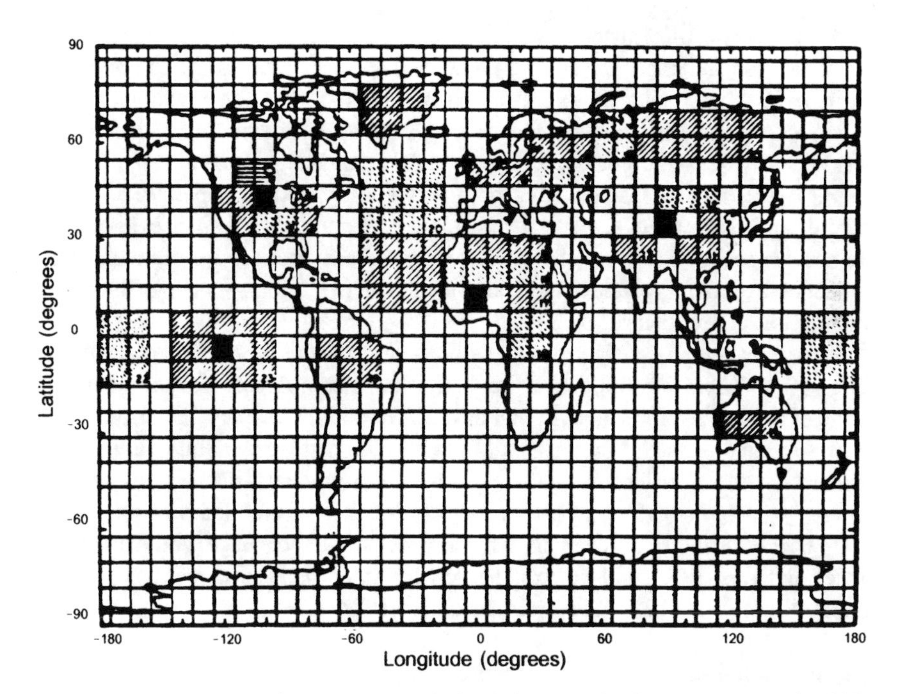

Figure 14-1. Typical 10° grid for general circulation models (from Hansen et al. 1987, by permission). Current models allow much higher resolution, but still much coarser than typical scales of ecological investigation.

how to interface the disparate scales of interest of scientists studying these problems at different levels.

To scale from the leaf to the ecosystem to the landscape and beyond (Jarvis and McNaughton 1986; Ehleringer and Field 1993), we must understand how information is transferred from fine scales to broad scales, and vice versa. We must learn how to aggregate and simplify, retaining essential information without getting bogged down in unnecessary detail. The essence of modeling is, in fact, to facilitate the acquisition of this understanding, by abstracting and incorporating just enough detail to produce observed patterns. A good model does not attempt to reproduce every detail of the biological system; the system itself suffices for that purpose as the most detailed model of itself. Rather, the objective of a model should be to ask how much detail can be ignored without producing results that contradict specific sets of observations, on particular scales of interest. In such an analysis, natural scales and frequencies may emerge, and in these rests the essential nature of system dynamics (Holling 1992a).

The reference to "particular scales of interest" emphasizes a fundamental point: there is no single "correct" scale on which to describe populations or ecosystems (Greig-Smith 1964; Steele 1978, 1989; Allen and Starr 1982; Meentenmeyer and Box 1987; Wiens 1989). Indeed, as I shall discuss in the next section, the forces governing life history evolution, shaped by competitive pressures and coevolutionary interactions, are such that each species observes the environment on its own unique suite of scales of space and time (see, for example, Wiens 1976). Moreover, the attention in the evolutionary literature to the distinction between diffuse and tight coevolution (e.g., Ehrlich and Raven 1964; Feeny 1976, 1983) makes clear that this point extends to biotic complexity as well. Where the linkage between species is tight, coevolutionary responses typically are species specific. On the other hand, where species interact weakly with large collections of other species, the biotic scale of evolutionary change is much broader and more diffuse.

Even for a given species, some evolutionary responses will be to a narrow range of environmental influences, and others will be diffusely linked to a broad range of influences. Indeed, as I shall discuss later in this chapter, the distribution of any species is patchy on a range of scales, and different evolutionary forces will act on those different scales. Specific coevolutionary interactions can be intense on certain scales and not on others, because of the match or mismatch of species distributions; even intraspecific density-dependence will vary with scale, and this effect will be exaggerated for interspecific interactions (Wiens 1986; Wiens et al. 1986; Sherry and Holmes 1988).

The Evolution of Life History Phenomena

When we observe the environment, we necessarily do so on only a limited range of scales; therefore, our perception of events provides us with only a low-dimensional slice through a high-dimensional cake. In some cases, the scales of observation may be chosen deliberately to elucidate key features of the natural system; more often (Figure 14-2), the scales are imposed on us by our perceptual capabilities, or by technological or logistical constraints (Steele 1978). In particular, the observed variability of the system will be conditional on the scale of description (Stommel 1963; Haury et al. 1978; Figure 14-3).

All organisms face the same dilemma: for particular life history stages, the realized environmental variability will be a consequence of the scales of experience. Various life history adaptations, such as dispersal, dormancy, and iteroparous reproduction, have the effect of modifying the scales of observation, and hence the realized variability. For example (Schaffer 1974), the evolution of reproductive schedules and energy allocation will depend on the relative degrees

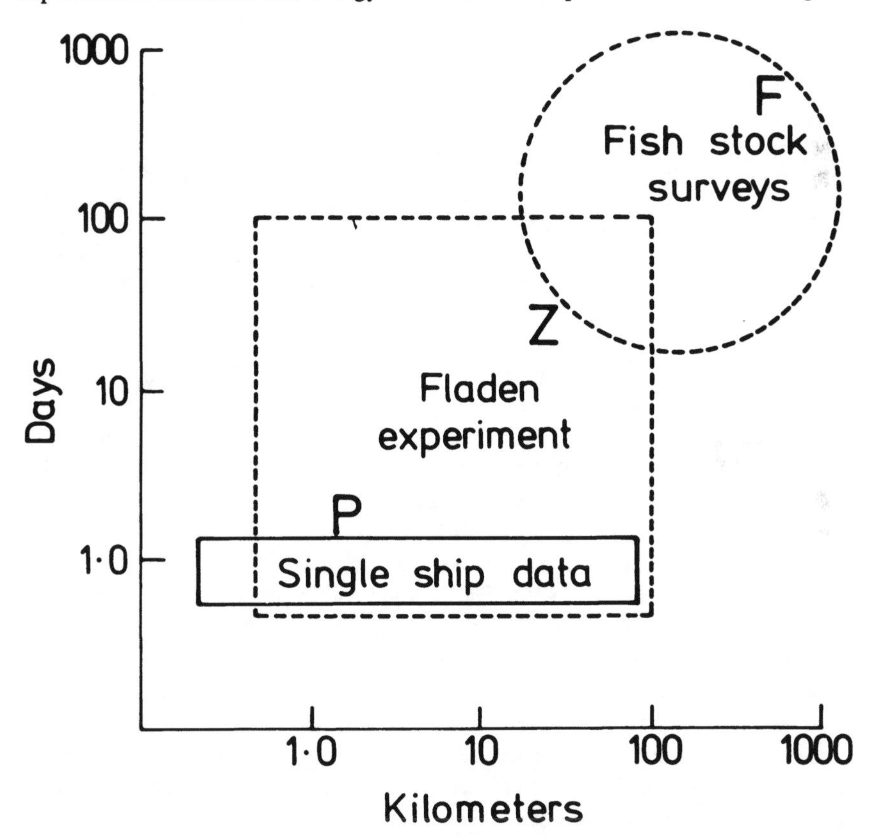

Figure 14-2. The limitation of sampling programs to provide information on particular scales of space and time. F = fish; Z = zooplankton; P = phytoplankton. (From J. H. Steele, 1978, "Some Comments on Plankton Patches, in J. H. Steele (ed.), *Spatial Pattern in Plankton Communities*. NATO Conference Series IV, Volume 3, Plenum Press, New York and London. Reprinted by permission of Plenum Press.)

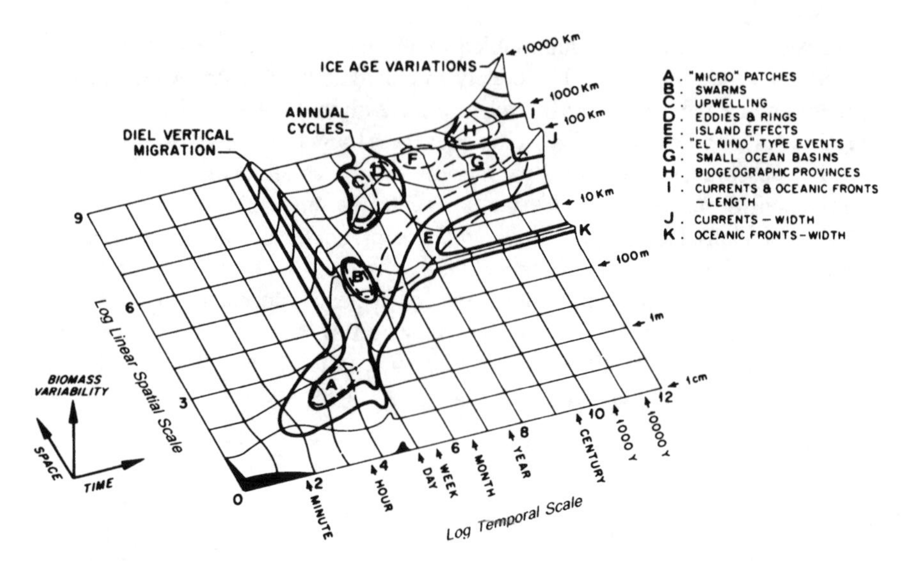

Figure 14-3. Stommel diagram of spatial and temporal scales of zooplankton biomass variability. (From L. R. Haury, J. A. McGowan, and P. H. Wiebe, 1978, "Patterns and Processes in the Time-Space Scales of Plankton Distributions," in J. H. Steele (ed.), *Spatial Pattern in Plankton Communities*. NATO Conference Series IV, Volume 3, Plenum Press, New York and London. Reprinted by permission of Plenum Press.)

of uncertainty experienced by juveniles versus adults, which experience the environment on different scales.

In the case of dispersal or dormancy, the dispersion of a genotype in space or time has the advantage of "spreading the risk" (Reddingius 1971; den Boer 1971), i.e., converting a geometric mean of variation into an arithmetic mean (e.g., Strathmann 1974) and thereby reducing the variability faced by the subpopulation of individuals that share that genotype. (Of course, the potential is substantial here for parent-offspring conflict.) Environmental variability drives the evolution of such life history traits; these, in turn, modify the scale of experience, and hence the observed environmental variability. Because variability is not an absolute, but has meaning only relative to a particular scale of observation, the interaction of dispersal, dormancy, and similar traits can best be thought of in terms of coevolutionary processes between an organism and its environment. The consequence of differential responses of species to variability is a partitioning of resources, and enhanced coexistence.

The linkage between dispersal and dormancy, as ways to deal with environmental variability, can be explored theoretically. In Levin et al. (1984) and Cohen and Levin (1987), a theoretical model is developed for determining evolutionarily stable dispersal and dormancy strategies for annual plants, in relation to environmental variability; similar models for dispersal were discussed earlier by Hamilton and May (1977). Not surprisingly (Figure 14-4), increasing environmental variability selects for higher rates of dispersal; analogous results were obtained for dormancy (e.g., Ellner 1985a,b; Cohen and Levin 1987). Furthermore, in the presence of dormancy, selection for dispersal is reduced (Figure 14-5), and vice versa. When one considers both strategies simultaneously (Figure 14-6), tradeoffs become evident; dispersal and dormancy are alternative ways to reduce the experienced variability, and selection for one reduces the selective pressure on the other because it changes the scale at which the environment is observed.

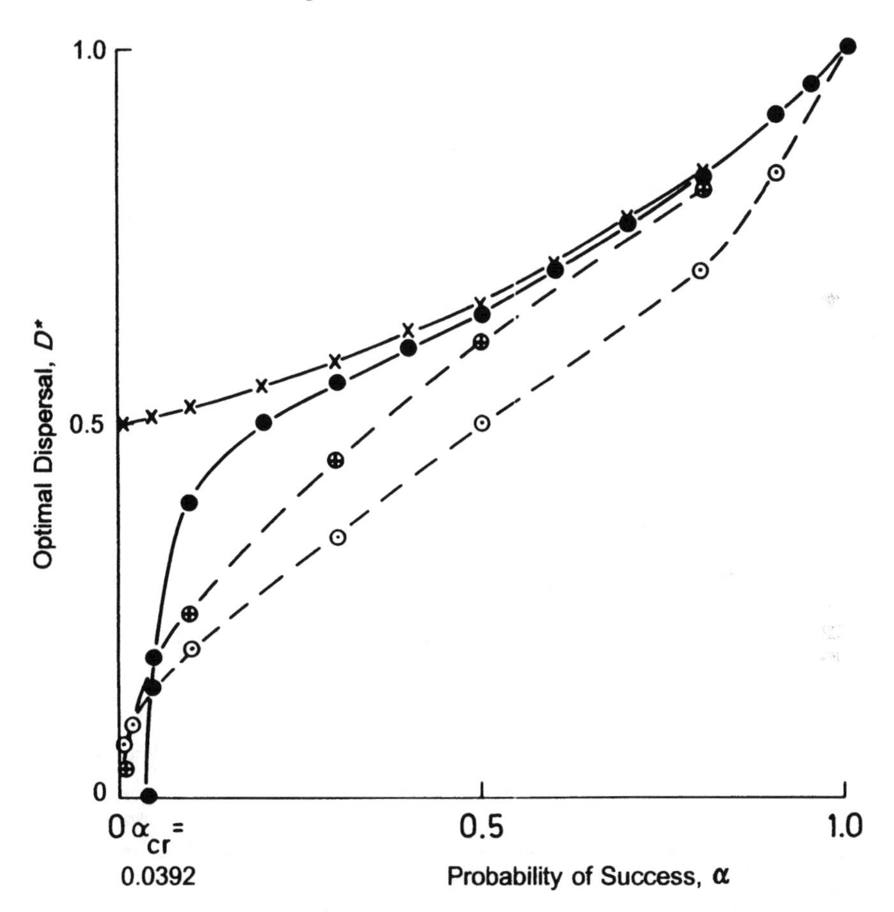

Figure 14-4. Evolutionarily stable dispersal fractions for annual plants, for differing patterns of temporal environmental variation. α = probability of success in finding new site; α_{cr} = critical value of α. (Reprinted from L. H. Weber, S. Z. El-Sayed, and I. Hampton, 1986, "The variance spectra of photoplankton, krill and water temperature in the Antarctic Ocean south of Africa, *Deep-Sea Research* 33:1327–1343. Copyright 1986 by Pergamon Press, Inc., with kind permission from Elsevier Science Ltd., The Boulevard, Langford Lane, Kidlington OX5 1GB, U.K.)

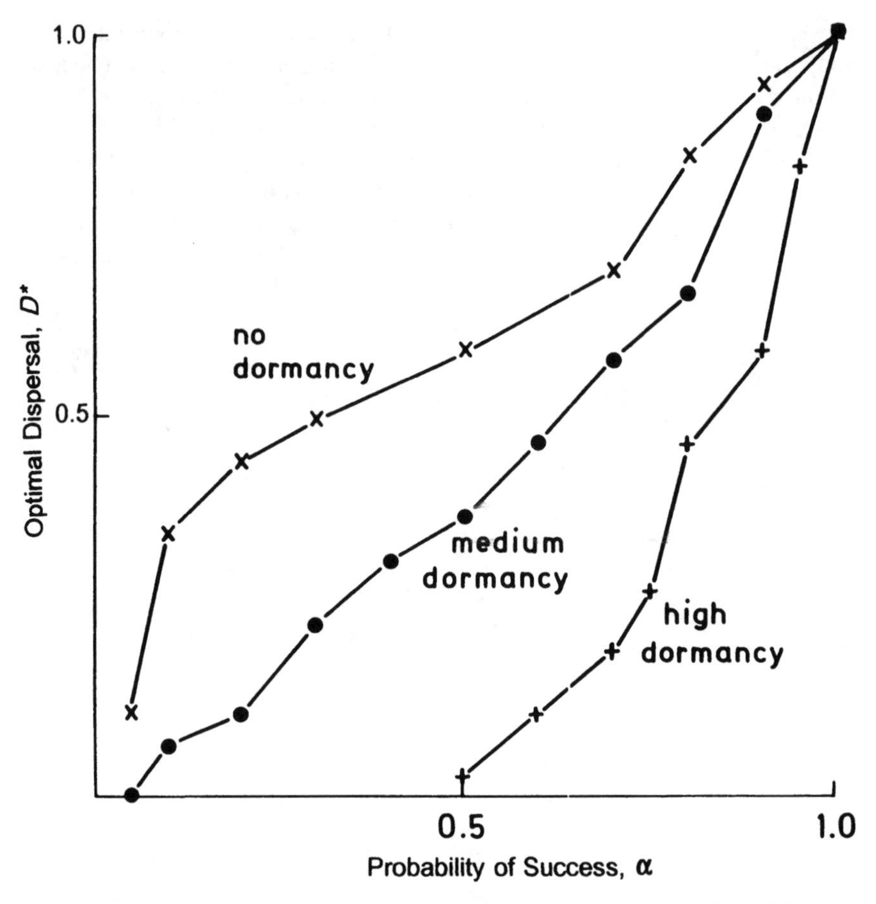

Figure 14-5. Effects of dormancy on evolutionarily stable dispersal fractions (from Levin et al. 1984). α = probability of success in finding a new site.

Such theoretical predictions are borne out by data for a range of plant species (Werner 1979; Venable and Lawlor 1980).

In describing natural phenomena, we typically invoke a similar approach. At very fine spatial and temporal scales, stochastic phenomena (or deterministically driven chaos) may make the systems of interest unpredictable. Thus we focus attention on larger spatial regions, longer time scales, or statistical ensembles, for which macroscopic statistical behaviors are more regular. This is the principal technique of scientific inquiry: by changing the scale of description, we move from unpredictable, unrepeatable individual cases to collections of cases whose behavior is regular enough to allow generalizations to be made. In so doing, we trade off the loss of detail or heterogeneity within a group for the gain of predictability; we thereby extract and abstract those fine-scale features that have

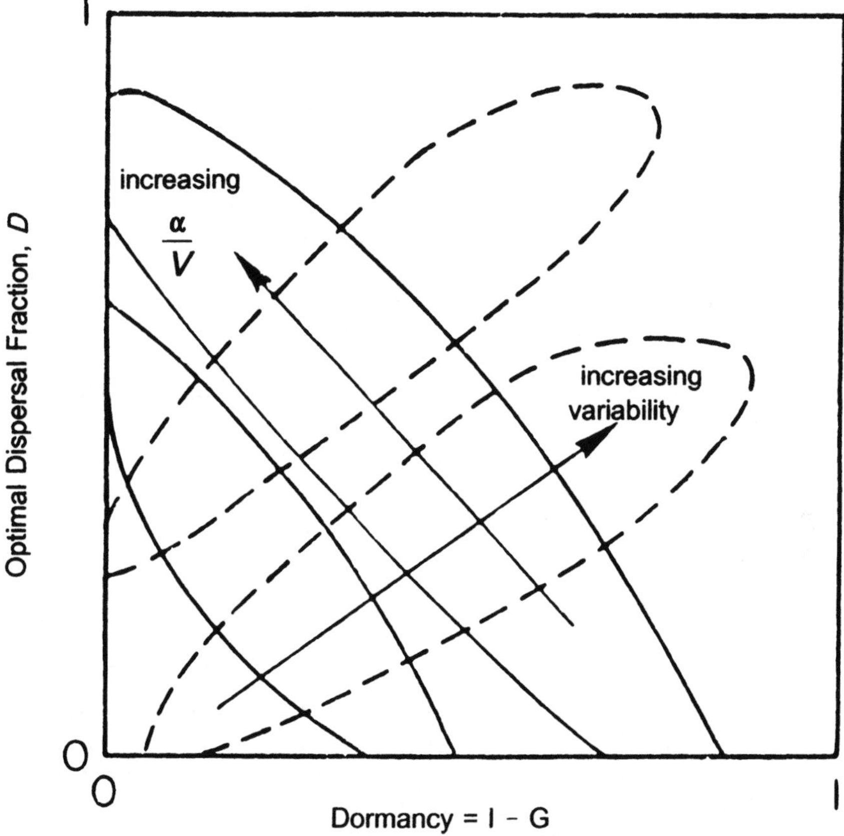

Figure 14-6. Tradeoffs between evolutionarily stable dispersal fraction D and evolutionarily stable dormancy fraction $(1 - G)$. α = probability of surviving dispersal episode, V = probability of surviving (1 yr) dormancy. Increasing variability selects for both; at given levels of variability, tradeoffs exist between dispersal and dormancy, with balance determined by α/V (from Cohen and Levin 1987).

relevance for the phenomena observed on other scales. In physics, this tradeoff is well studied, and goes to the heart of the problem of measurement (see, for example, Heisenberg 1932; Planck 1936). At fine scales, quantum mechanical laws must replace classical mechanical laws; laws become statistical in character, dealing only with probabilities of occupancy.

In population genetics, the same problem arises, and the same tradeoffs occur. Focusing on alternative definitions of fitness, Dawkins (1982) points out the tradeoffs, for example, among classical fitness, focusing on the unique properties of an individual; phenotypic measures that lump individuals together based on common phenetic traits; and genotypic measures that form even larger ensembles, for example by grouping all individuals together who carry a particular gene. As one moves up the hierarchy to larger and larger aggregates, one obtains more statistical predictability, while sweeping under the rug details of variation within an aggregate. Quantitative genetic approaches to evolution are based on a similar rationale.

The existence of these tradeoffs makes clear that there is no natural level of description. However one defines classes, there will be differential evolution among classes, and differential evolution within. Similar comments apply to the subdivision of a population for theoretical analysis, say, into trait groups (Wilson 1983) or epidemiological risk groups (Castillo-Chavez et al. 1989), or to any conceptualization of the population as a metapopulation (Wright 1977, 1978; Gilpin and Hanski 1991). The interplay among different levels of selection presents one of the key conceptual problems in evolutionary biology (Eldredge 1985). Here, as in more strictly ecological settings (O'Neill et al. 1986), the problem is not to choose the correct scale of description, but rather to recognize that change is taking place on many scales at the same time, and that it is the interaction among phenomena on different scales that must occupy our attention.

Pattern Formation

The concepts of scale and pattern are ineluctably intertwined (Hutchinson 1953). The description of pattern is the description of variation, and the quantification of variation requires the determination of scales. Thus, the identification of pattern is an entrée into the identification of scales (Denman and Powell 1984; Powell 1989 and Chapter 8 of this volume).

Our efforts to develop theories of the way ecosystems or communities are organized must revolve around attempts to discover patterns that can be quantified within systems, and compared across systems. Thus, there has been considerable attention directed to techniques for the description of ecological or population pattern (Burrough 1981; Sokal et al. 1989; Gardner et al. 1987; Milne 1988; Sugihara et al. 1990). Once patterns are detected and described, we can seek to discover the determinants of pattern, and the mechanisms that generate and maintain those patterns. With understanding of mechanisms, one has predictive capacity that is impossible with correlations alone.

In developmental biology, considerable theoretical interest has been directed to the problem of how the predictable morphologies that are the defining characteristics of organisms develop from initially undifferentiated eggs. Early models, involving gradients of chemicals (morphogens) that carry the information guiding development, were shown by Turing (1952) to lead to symmetry breaking, provided there were present both activator and inhibitor species, and provided the diffusion of the inhibitor was much more rapid than that of the activator. Under suitable conditions (Segel and Levin 1976), the destabilized uniform distributions give way to stable nonuniform patterns, which can provide the local information that specifies patterns of differentiation. The strengths of this model are that it requires no genetically determined blueprint and through purely local interactions can give rise to almost every conceivable observed pattern (Meinhardt 1982); of particular interest (Murray 1988a, 1989) has been the application of these ideas to coat or wing patterns in animals.

However, the demonstration that a specific mechanism can in theory give rise to a range of observed patterns is not proof that that mechanism is indeed responsible for those patterns. In the absence of strong evidence for the existence of morphogens, or for their universal explanatory power, attention has turned to the search for alternative explanations. In recent years, for example, another class of models, involving mechanochemical interactions, has been shown to be equally feasible from a theoretical perspective, and equally flexible in its ability to give rise to patterns (Murray and Oster 1984). There are many roads to Rome; and in general, there will be many conceivable mechanisms that could give rise to any set of patterns. All that theory alone can do is to create a catalogue of possible mechanisms; experiments are then needed to distinguish among the candidate mechanisms. This is a lesson that must be borne in mind also as we consider the problem of pattern formation in ecology.

The problem of ecological pattern is inseparable from the problem of the generation and maintenance of diversity (Levin 1981). Not only is the heterogeneity of the environment often essential to the coexistence of species, but the very description of the spatial and temporal distributions of species is a description of patterns of diversity. Thus, an understanding of pattern, its causes and its consequences, is central to understanding evolutionary processes such as speciation as well as ecological processes such as succession, community development, and the spread and persistence of species.

In the case of ecological systems, a range of mechanisms exists for generating pattern. Pattern is in part extrinsically determined, and a first step is to identify and factor out such external influences (Denman and Powell 1984). Statistical similarities in the distribution of variables provide a natural place to start, and correlations provide stronger evidence. They do, however, provide only a starting point, and other approaches must be involved for the examination of causation.

For the krill populations of the Southern Ocean, spatial distributions have been shown to be patchy on almost every scale of description. This has fundamental importance both for the dynamics of krill, and for their predator species. Various studies have characterized the Fourier spectrum of variability for krill, and shown that variance decreases with scale (Figure 14-7). However, substantial differences exist between the spectra for krill and those for temperature and fluorescence, and these differences are borne out by analysis of the cross correlations (Weber et al. 1986; Levin et al. 1989b). On broad scales, temperature (a passive marker of water movements), fluorescence (a measure of phytoplankton activity), and krill all have spectra that are consistent with the theoretical predictions of the

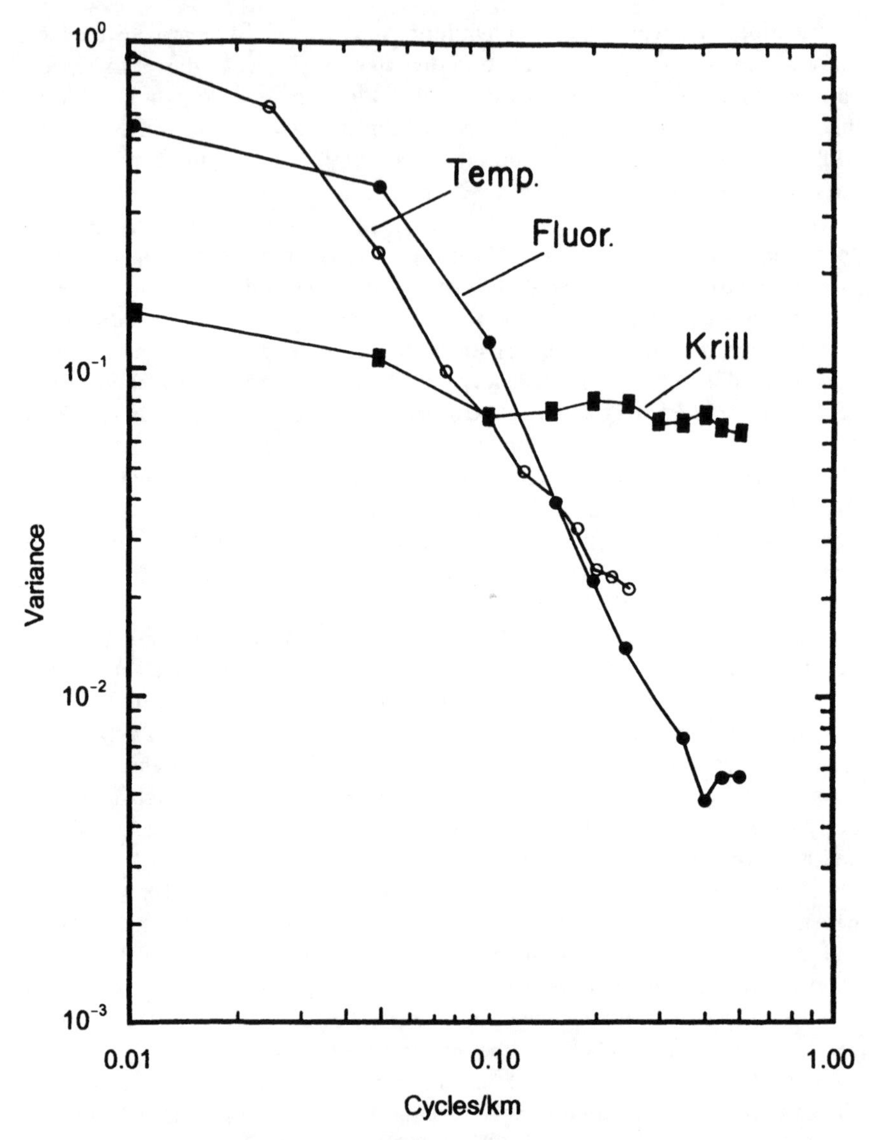

Figure 14-7. Variance spectra for temperature, fluorescence, and krill in the Scotia Sea (from Weber et al. 1986).

Kolmogorov turbulence calculations (Tennekes and Lumley 1972). On fine scales, however, the krill spectrum is noticeably flatter, suggesting that krill are much more patchily distributed than their resource, or than can be explained by water movements alone. This observation is consistent with data of Mackas (1977), who showed more generally that zooplankton populations can exhibit much more fine-scale variation than phytoplankton (Figure 14-8). For the krill, this patchiness appears to break down on the finest scales (Levin et al. 1989b).

The interpretation of these data is that large patches of krill and phytoplankton are being moved about by water movements, but that on fine scales some other mechanism must be invoked to explain pattern. Thus a two-scale model is needed, and a general lesson learned: no single mechanism explains pattern on all scales.

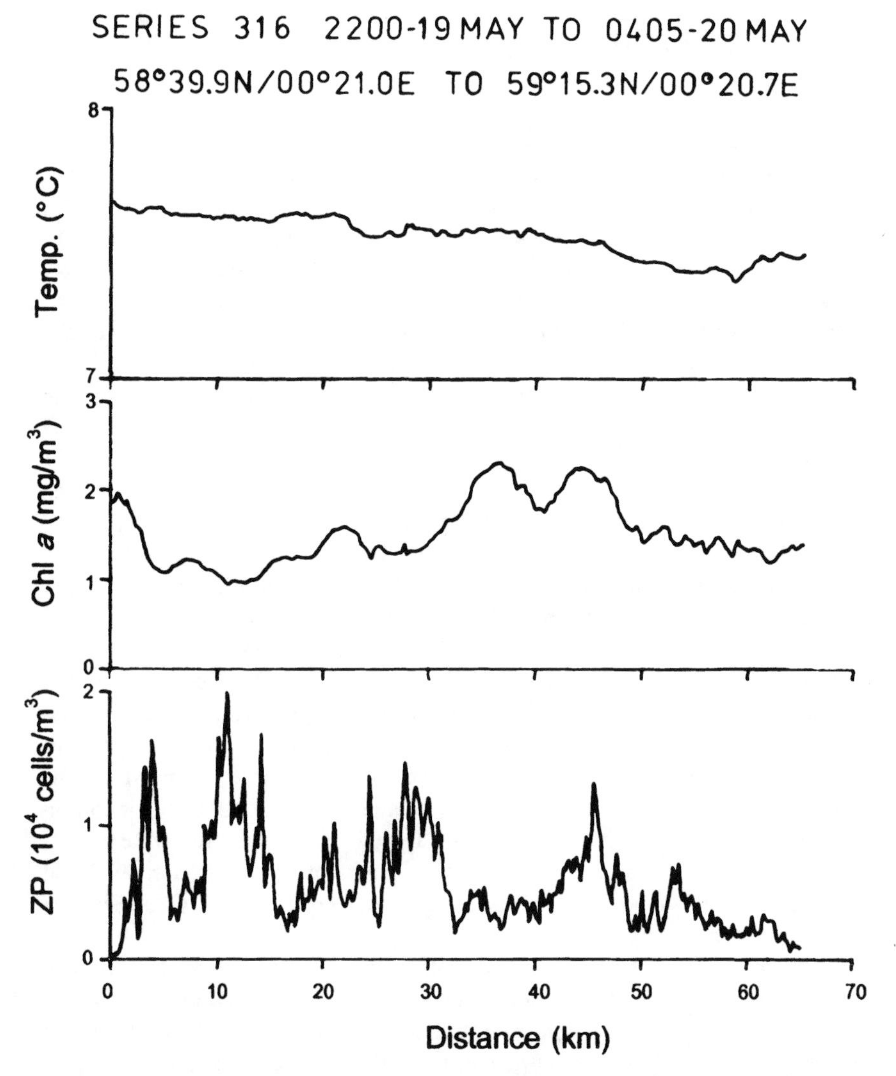

Figure 14-8. Temperature, chlorophyll *a* concentration, and zooplankton abundance in relation to latitude, for spring transect in North Sea (from Mackas 1977).

In this case, pattern seems extrinsically driven on broad scales, and autonomously generated on fine scales. The explanation lies in the swimming behavior of krill, a topic to which I shall return.

Similarly, Dayton and Tegner (1984) and Menge and Olson (1990) have discussed the range of scales at which the dynamics of communities are mediated, from biotic processes at the scale of meters to eddies and warm water intrusions at the scale of tens of kilometers. Dayton and Tegner (1984) argue that "many ecologists . . . focus on their small scale questions amenable to experimental tests and remain oblivious to the larger scale processes which may largely account for the patterns they study." One can equally point to other examples, however, such as models of general circulation, in which large-scale processes are overemphasized, at the expense of the fine scale. Rothschild and Osborn (1988) show how small-scale turbulence can have effects on predator-prey contact rates, and thus on broader scale dynamics. Butman (1987) discusses elegantly the interplay between small-scale processes such as active habitat selection and large-scale processes such as passive deposition in determining the settling of invertebrate larvae. In general, one must recognize that different processes are likely to be important on different scales, and find ways to achieve their integration (Mackas et al. 1985; Levin et al. 1989b; Denman and Powell 1984; Menge and Olson 1990). To date, this program of research has been more successfully carried out in marine than in terrestrial systems, but the situation is changing (Schimel et al. 1990).

For the krill populations of the Antarctic, once extrinsic influences have been subtracted, one still must find ways to explain what maintains fine-scale pattern. A natural first step is to consult the catalogue of ways that pattern can be created and maintained in aquatic systems. A prime candidate is the diffusive instability referred to earlier, in which phytoplankton serve as the activator species, and krill as the inhibitor (see, for example, Levin and Segel 1976). In contrast to other ways in which pattern can be generated, the notion of diffusive instability is well suited to continua such as the open ocean. However, the mechanism relies on the inhibitor (herbivore) species' being more diffusive than the activator (resource) species, which will inevitably produce distributions in which the resource, rather than the herbivore, is more patchily distributed; this is inconsistent with observations. Similar objections apply to various alternative explanations, such as those that rely on favorable patches of nutrients (Kierstead and Slobodkin 1953; Okubo 1978, 1980) or physical features such as convective cells or warm core rings.

Fortunately, considerable life history information is available for krill, which are known to aggregate actively into swarms, and indeed into schools (Hamner et al. 1983, Hamner 1984). Using this information, Grünbaum (1992) has developed individual-based models of krill populations, whose collective behavior can give rise to the formation and maintenance of aggregations consistent with those observed. His model, building on earlier work of Okubo (1972) and Sakai (1973), considers both the random and directed forces imposed by the physical and chemical environment, and the behavioral responses of individuals to other individuals. Thus, on the fine scale, the explanation of krill distributions is in the ensemble behavior of individuals acting on even finer scales; on the broad scale, the explanation is in terms of oceanographic processes acting over even broader scales (Hofmann 1988). By interfacing individual-based models with fluid dynamic models, therefore, one seeks to interrelate phenomena acting on different scales; this approach must guide us in dealing with ever more complicated problems, involving wider ranges of scales. The challenges may be more difficult, but the principles are the same.

Patterns of Spread

The effort to explain the distribution of populations in terms of the movements of individuals is an extension of one of the most successful applications of mathematics to ecological phenomena, the use of random walk and diffusion models to describe dispersal. Diffusion approximations to the description of individual movements (the continuum limits of random walk models, with only first- and second-order terms retained) have been employed by biologists for nearly a century and have received considerable attention from population geneticists (Fisher 1937; Haldane 1948; Dobzhansky and Wright 1943, 1947). The basic idea is that, although organisms do not move randomly, the collective behavior of large numbers of such individuals may be indistinguishable (at the scale of the population) from what would result if they did. For example, even when individuals respond deterministically to chemicals or other cues, the presentation of those cues may be effectively random, at least on the perceptual scales of the external observer. Indeed, the philosophy behind the application of models is not that the finer detail does not exist, but that it is irrelevant for producing the observed patterns.

The same rationale is used, for example, to justify the application of the diffusion approach to the flow of heat, or of chemicals, and with the same limitations: the predictions of the models are excellent close to a source, but their validity diminishes as distance from the source increases. Furthermore, diffusion models are based on an assumption of random collisions of molecules, which clearly is not technically valid. However, this suppression of detail is the strength rather than the weakness of the approach, because it allows the demonstration that the observed ensemble behavior can be explained entirely without reference to the extra detail. This is the kind of simplification that we must achieve more generally in learning how to connect phenomena on different scales.

The key to understanding how information is transferred across scales is to determine what information is preserved and what information is lost as one moves from one scale to the other. In the case of the diffusion approximation, the notion is that only certain macroscopic statistics of the distribution of individual movements are relevant; more generally, the goal of research into scaling is to discover what the most relevant macroscopic statistics are that inform the higher levels about lower-level behaviors.

In the case of the application of random walk models in ecology, the last 40 years have seen tremendous activity. The most important paper undoubtedly was that of Skellam (1951), who synthesized the general theory and anticipated a range of applications that have since occupied the attention of researchers: the spread of pest species, and of colonizing species following climate change; the patchiness of species distributions; and geographical clines. Okubo (1980), in a fascinating book, provides the most complete treatment of the application of diffusion models in ecology, introducing a range of novel applications.

It is perhaps not surprising that generalized diffusion models, incorporating advective movements due to winds, currents, and gravitational forces, work well to describe the passive spread of seeds and pollen (Okubo and Levin 1989; Liddle et al. 1987), or of invertebrate larvae (Hofmann 1988). It may be more surprising, however, that they also work well for organisms that can use detailed environmental cues to direct their movements. In the first rigorous test of the use of such models, Kareiva (1983) released flea beetles on collard plants in a one-dimensional habitat, and tested their movements against the predictions of a diffusion model. He later expanded this approach to other phytophagous insects, relying on studies reported in the literature. His conclusions were that the diffusion models provided remarkably good agreement with the observed data in 7 cases out of 11, and were a reasonable first approximation in the other cases. For the latter, a habitat-dependent movement model provided the necessary extra detail. Grünbaum's individual-based models of krill population dynamics are a further application of this approach. Again, the general philosophy is to incorporate a minimum of necessary detail, complicating the model only when necessary. In general, adding more parameters to a model may be expected to give a better fit to observed data, but may result in a less reliable model for prediction (see, for example, Ludwig and Walters, 1985).

One of the most powerful applications of the diffusion model has been in dealing with the spread of introduced species. Indeed, this was the application that first captured the attention of population geneticists such as Fisher (1937), interested in the spread of advantageous alleles. By adding a growth process to the diffusion model, Fisher used intuitive arguments to conclude that there would be an asymptotic speed of propagation, equal to twice the square root of the product of the intrinsic rate of increase at low density and the diffusion coefficient; this brilliant insight was confirmed formally by Kolmogorov et al. (1937), and extensions have occupied some of the best efforts of mathematicians since (e.g., Bramson 1983). Skellam (1951) applied the approach to the spread of various species, including oaks and muskrats, and the last decade has seen an increasing attention to the problem (e.g., Murray 1988b; Lubina and Levin 1988; Andow et al. 1990) for applications ranging from spores and viruses to invertebrate agricultural pests to vertebrate species such as muskrats and otters. In some cases (Figure 14-9), this simple model provides excellent agreement with observed data; in other situations, it can underestimate rates of spread by an order of magnitude. The problem is again one of scale: the spread of organisms is a multistage process involving the establishment of new centers, and the spread from those centers (Figure 14-10). Our scale of observation is such that the diffusion model may work very well on the fine scales but be unable to deal with the broader scales (or at least require a separate parameterization there). The establishment of secondary foci, involving "great leaps forward," requires an extension of the diffusion approximation to include other factors, e.g., higher-order moments (Mollison 1977) or Rvachev's intercity models of influenza transmission (Rvachev and Longini 1985).

I have described diffusion models in such detail not because they are to be taken as gospel, but because they provide a clear example of a general approach.

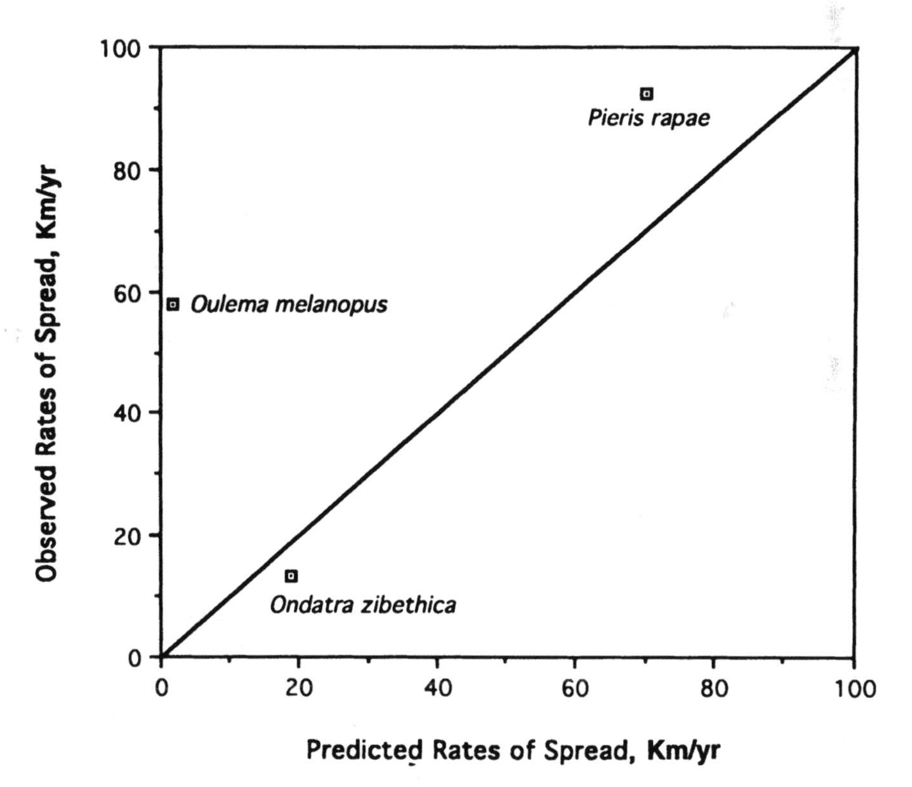

Figure 14-9. Predicted and observed rates of spread for three introduced species (data from Andow et al. 1990).

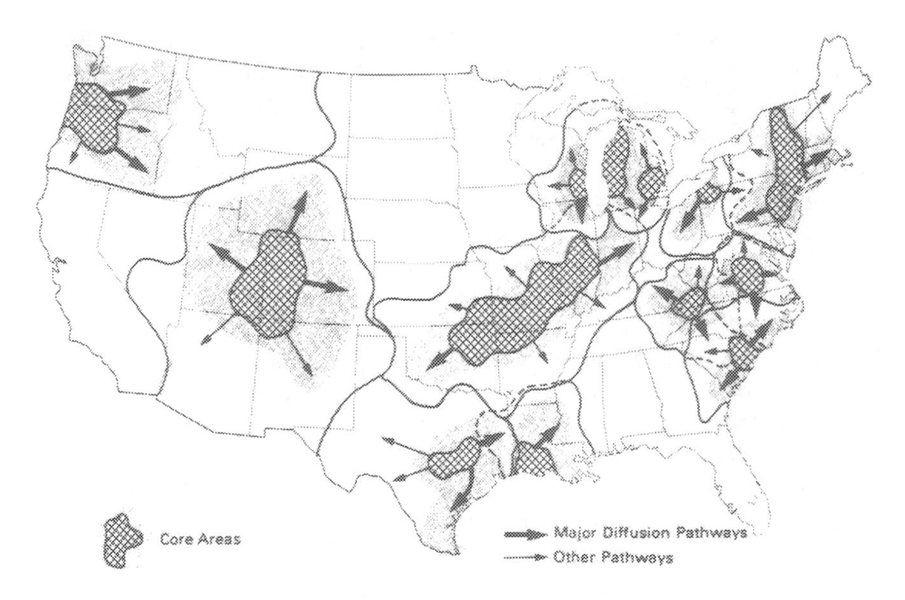

Figure 14-10. Patterns of two-level spread of influenza during start of 1967–68 season (from Pyle 1982).

Various alternatives to diffusion models have been utilized; for example, percolation models to describe the spread (and persistence) of species in fragmented habitats (Durrett 1988; Gardner et al. 1992), or correlated random walks to describe the movement of insects (Kareiva and Shigesada 1983), or the clonal growth of branching organisms (Cain 1990, 1991). All these alternatives attempt to understand behavior at one level in terms of ensembles of units at lower levels. This is the same approach I will take in this chapter as I discuss the dynamics of communities and ecosystems, and potential interactions with global climate systems.

Pattern and Scale in Terrestrial Systems

The approach described earlier for aquatic systems can be applied equally to terrestrial systems. Techniques such as remote sensing, in concert with spatial statistics, can be used to describe the broad-scale distributions of plant species and the factors that might influence them. Theoretical studies, including computer simulations, can be used to bridge to smaller-scale experimental studies, to generate and test hypotheses concerning how fine-scale processes interact with those on other scales to produce observed pattern. In this program, it is helpful to have available a suite of models of increasing complexity and detail; in this way, one can strip away detail and explore how much finer-scale information is relevant to describing observed pattern at broader scales. At one extreme are models that attempt to capture as much detail as possible, within the constraints imposed by parameter estimation problems. At the other extreme are grossly oversimplified models that retain as few features of the real system as possible, in order to explore systematically and in isolation the influence of particular factors. Combining those simple models to reassemble the ecosystem is the next step, and it is analogous to the problem of trying to understand the evolution of populations in terms of the dynamics of individual loci.

A case in point is the quantification of plant competition in terrestrial communities. In a very important series of papers, Pacala and Silander (1985) (see also Pacala 1989) have developed neighborhood models, which quantify the competitive influence of plants on each other as a function of the distance between them. That detail certainly determines the microdistribution of plants (see, for example, Lechowicz and Bell 1991); but how much of the detail is relevant to the distribution of species even at the field level, much less at the landscape level? Theoretical explorations of models that incorporate the detail can be helpful. In such work, increasing levels of aggregation are applied while model outputs are compared with field data, to determine what information (if any) at the broad scale is lost as a result of aggregation.

A similar approach is called for when one is dealing with systems in which asynchronized random disturbances play an important role. These clearly include gap phase systems, discussed in detail by A. S. Watt in his presidential address to the British Ecological Society (Watt 1947), and later by Levin and Paine (Levin and Paine 1974; Paine and Levin 1981) for a range of systems, including especially the rocky intertidal system. The notions of patch dynamic phenomena introduced in those works have been expanded, developed, and improved in a virtual explosion of studies (see especially Pickett and White 1985; Forman and Godron 1981, 1986), and are closely related to metapopulation ideas that have become so important in evolutionary theory and population biology (e.g., Gilpin and Hanski 1991; Harrison and Quinn 1989; Hastings and Wolin 1989; Fahrig and Paloheimo 1988a,b). In a novel application of these ideas, McEvoy et al. (1993) have shown how the balance among disturbance, colonization, and successional development is critical to the biological control of ragwort.

In systems in which localized disturbances play an important role, the local dynamics are unpredictable, except in terms of statistical averages over longer time scales. The local unpredictability and variability present opportunities for species that would be eliminated competitively in constant environments, and greatly increase diversity at intermediate levels of disturbance. Indeed, for many species, local unpredictability is globally the most predictable feature of the system (Levin and Paine 1974). As the scale of description is increased beyond the scale of individual disturbances, variability declines, and predictability correspondingly increases.

Levin and Paine (1974), taking the view that the rocky intertidal community can best be understood as a metapopulation of patches, focused on two scales of dynamics. On the large scale, attention was directed to the demography of patches; within patches, a more rapid and smaller-scale successional dynamic dominated. Through a modeling approach that explicitly recognized these scales, Paine and Levin (1981) were able to examine the role of disturbance as a structuring agent. Other investigators (e.g., Coffin and Lauenroth 1988; Green 1989; Clark 1991a) have subsequently quantified the dynamics of patches in a range of systems and demonstrated the validity of this approach and the importance of disturbance in maintaining the character of those systems. A similar approach can be taken for the colonization of host patches by parasites, for example, plant pathogens or animal parasites (e.g., Holmes 1983).

As I have already discussed, not all systems allow subdivision neatly into hierarchical scales of organization. In the case of a host-parasite system, the units are fairly discrete and distinct; in the case of terrestrial vegetation, the gradations are more like those already discussed for Antarctic krill: patches do not have a narrow range of sizes, but are found across a broad spectrum. That is, patchiness is found on almost every scale of observation. In this case, there is no unique natural scale, even though some scales may provide more natural biological vantage points (e.g., O'Neill et al. 1991a,b). One must recognize that the description of the system will vary with the choice of scales; that each species, including the human species, will sample and experience the environment on a unique range of scales; and that, rather than trying to determine the correct scale, we must understand simply how the system description changes across scales. This explains why there has been so much fascination in ecology, as in other fields, with the theory of fractals (Mandelbrot 1977; Sugihara and May 1990), which emphasizes both the scale dependence of data and descriptions of phenomena, and the more hopeful note that there may be scaling laws.

To begin to address the question of how system description changes with scale for disturbance-mediated systems, Levin and Buttel (1987) used landscape models for a grid in which local growth simulators were linked together through common disturbance, inter-patch dispersal, and competition. Earlier models (Levin 1974) assumed local dynamics that were deterministic, in the tradition of Lotka and Volterra, and linked cells together through an interaction matrix describing movement of propagules or individuals. For disturbance-mediated applications, however, local dynamics must incorporate stochastic elements (representing disturbance and colonization events), which may be spatially correlated. The simplest description was through an interacting particle model in which each cell was in one of a small number of successional states. A local state could be altered through invasion (and capture) by species with later successional characteristics, or by disturbance, which reset cells to the initial state. Dispersal ability was taken to be inversely correlated with competitive ability, and disturbance intensity and frequency were tied to the present state of a cell (Levin and Buttel 1987). In particular, later successional stages were susceptible to higher disturbance rates, and those disturbances were allowed to radiate outward to neighboring cells. Spatial correlations arise in such models due to this latter phenomenon, which resets groups of contiguous cells to the initial successional stage; spatial correlations can also arise from dispersal among neighboring cells, and from other influences such as competition.

In homogeneous environments, spatial and temporal variability will be a function of the size of the window used to view the world; as window size is increased, variability will decay. The exact relationship between variability and window size is difficult to predict (see empirical methods, for example, in Greig-Smith 1964; Kershaw 1957; Cain and Castro 1959; Mead 1974), but will be determined by the way spatial correlations fall off with distance (see, for example, Hubbell and Foster 1983; Robertson et al. 1988; Carlisle et al. 1989). In general, the relationship will follow a power law within the correlation length of the system (which is determined by such influences as the disturbance size distribution and the dispersal curve), and then will fall off asymptotically as the inverse of the number of cells in the window. On a log-log plot, this yields a straight line of characteristic slope (usually between 0 and -1) within the correlation length, asymptoting to a slope of minus one (Figure 14-11) for very large window sizes; such relationships have been described both for model output (Levin and Buttell 1987) and for data from terrestrial systems (Moloney et al. 1992).

The evidence presented in Figure 14-11 supports the view that there is no correct scale for describing the system, that the description of variability is contingent upon the window through which the system is viewed, but that there may be scaling laws that allow one to make comparisons among studies carried out on different scales. These results are reminiscent of similar observations that form the basis of the theory of fractals, and the power law relationships are identical to what is seen in statistical physics, in the study of critical phenomena (Wilson 1983).

The slope of the line in Figure 14-11, within the correlation length, is determined by the spatial correlations. With no correlations, one expects the variance to fall off as the inverse of the number of cells, as it does well beyond the correlation length; this would give a slope of minus one in Figure 14-11. With perfect spatial correlation, the slope would be zero. Observed slopes typically lie somewhere in between (Levin and Buttel 1987), reflecting the degree of correlation.

Simulation models allow easy exploration of the influence of factors such as dispersal and disturbance in determining the slope, and of biological consequences of disturbance, such as persistence or extinction. The technique is not perfect. As with any such method, it is limited at the fine scale by the minimum size of a cell, and at the broad scale by the extent of the grid.

Theoretical explorations can carry one just so far, and are meant primarily to guide empirical studies. The next stage is to search for such patterns in natural systems and to use models to explore causes. In the serpentine grassland of

Figure 14-11. Theoretical relationship between variance and window size. Within correlation length, characteristic slope reflects rate of decay of correlation with distance; beyond correlation length, −1 slope reflects absence of correlation.

Jasper Ridge, California, we (S. A. Levin, with Kirk Moloney and Linda Buttel), in collaboration with Hal Mooney, Nona Chiariello, and others, have sought to understand the dynamics of the vegetation as mediated by soil factors and by the activities of gophers and ants, which disturb the soil and the vegetation (Hobbs and Hobbs 1987; Hobbs and Mooney 1991; Levin et al. 1989a; Moloney et al. 1992). Related models have been developed by Coffin and Lauenroth (1989) for the dynamics of the shortgrass prairie. The random disturbance model described above must be modified here, because both the soil factors and the disturbing forces show similar evidence of strong spatial correlations and the same characteristic curves for the relationship between variance and window: on a log-log plot, variance falls off linearly, with a characteristic slope, within the correlation length, and with asymptotic slope equal to minus one. These correlating influences add another set to the list of those that can be structuring the vegetation, and the goal of research is to sort out the various influences. In a modeling approach that is an extension of that introduced by Levin and Buttel

(1987), we (Moloney et al. 1992, Levin et al. 1989a) have focused on four annual grasses in the serpentine, and on the influence of gopher disturbances on their dynamics (Figure 14-12). Timing and magnitude of disturbance are both of importance. It is easy to see from such models how some disturbance regions lead to the extinction of particular species, while others maintain coexistence through ergodic spatio-temporal mosaics (Figure 14-13). Thus, the simulation model becomes a powerful complement to other experimental techniques.

Similar models are being developed for a forested watershed in northwestern Connecticut, where disturbances create forest gaps. In collaboration with Steve

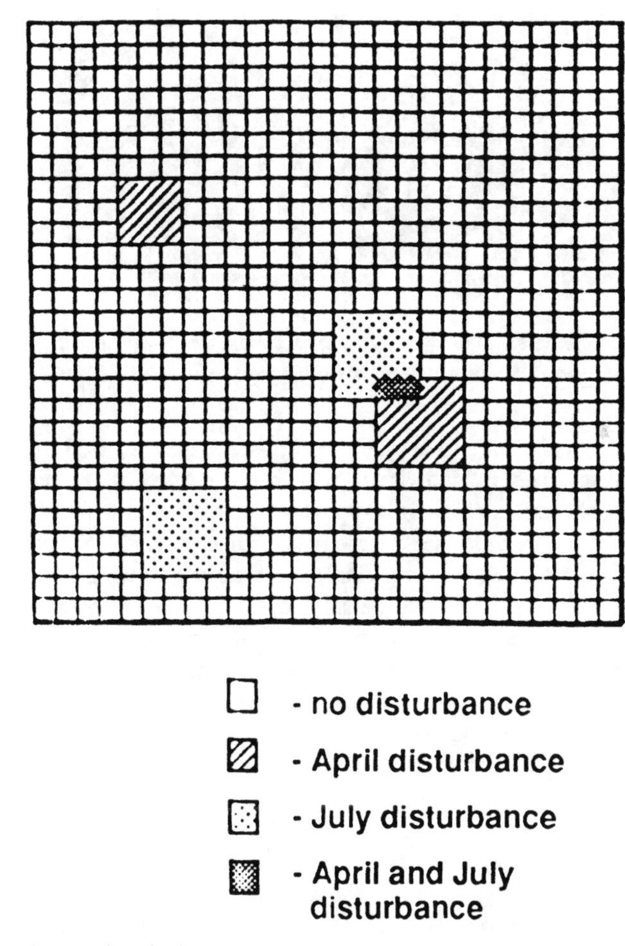

Figure 14-12. Basic grid for grassland simulation model (from Levin et al. 1989a).

Figure 14-13. Sample output from grassland simulation; snapshot of distribution of *Plantago erecta*. Shadings denote population levels (see key).

Pacala, John Silander, and Charles Canham, we are utilizing both raster- (grid-) based models and vector- (individual tree-) based models to explore the consequences of local detail concerning competition regimes. Local growth simulators, extensions of those in wide use in ecology (Botkin et al. 1972; Shugart and West 1981; Shugart 1984; Horn et al. 1989), provide the starting points, and are complemented with models of dispersal and disturbance. In both cases, model output is being interfaced with broader scale information derived from remote sensing and other studies, in work with William Philpot, Kyu-Sung Lee, and David Weinstein. Existing patterns of topography, rainfall, temperature, and soil factors will be analyzed and compared with distributions in vegetation patterns. Modeling and empirical studies will then be used to explain the variation in vegetation patterns not accounted for by physical factors.

The localized effects of tree gaps, small fires, wave disturbance, or gopher and ant mounds are special cases of situations where systems may be viewed as spatio-temporal mosaics, variable and unpredictable on the fine scale but increasingly predictable on large scales (Levin and Paine 1974; Watt 1947; Urban et al. 1987; Clark 1989, 1991a,b; Minnich 1983; Wright 1974). For such systems, transient dynamics are often ignored, with steady-state properties treated as the objects of interest. Yet some events have effects on too broad a spatial scale to permit such a perspective; the affected systems do not achieve statistical equilibrium over realistic temporal and spatial scales. In such situations, modeling approaches still can be useful in explaining how patterns of spread are mediated at the local level. Models of this sort have been discussed earlier for the spread of introduced species and also have been utilized to study the dynamics of epidemics and forest fires (Cox and Durrett 1988; von Niessen and Blumen 1988).

Metapopulation Models

One of the most important theoretical contributions of Robert MacArthur was his work with E. O. Wilson in developing the theory of island biogeography (MacArthur and Wilson 1967). Though that work, which focused on the turnover rates in island faunas, was directed to islands that were sustained by a mainland source, the ideas could be extended to archipelagoes in which the dynamics of the system became closed. In population genetics, similar considerations led to pool or stepping-stone models (Levene 1953; Kimura and Weiss 1964) of genetic correlation.

The view of systems as mosaics of islands has taken a number of interesting directions. The concept of patch dynamics (Watt 1947; Levin and Paine 1974; Paine and Levin 1981; Pickett and White 1985; Levin et al. 1993) has become a popular theme in both the terrestrial and marine literatures, and has led to new views of community structure. Metapopulation models, in which systems are viewed as composed of interacting populations of local demes, have been shown to be of importance in conservation biology (Armstrong 1988; Fahrig and Paloheimo 1988a,b; Burkey 1989; Gilpin and Hanski 1991; Nuernberger 1991), evolutionary theory (Levene 1953), and epidemiology (Levin and Pimentel 1981), and have become the focus of considerable theoretical effort (e.g., Nee and May 1992), especially in terms of the role of the metapopulation structure in facilitating coexistence of species.

The metapopulation perspective involves an explicit recognition of scales, and an explicit separation of within-patch and among-patch dynamics. In the intertidal zone, wave damage creates gaps whose demographic changes occur on a time scale of years and a spatial scale orders of magnitude larger than the typical size of a patch; within a patch, successional dynamics occur on a somewhat more rapid scale. Similar separation of scales has proved useful in viewing forest gap systems (Pickett and White 1985; Runkle 1982; Clark 1990, 1991b; Canham 1988; Grubb 1977) and gopher-mediated grassland systems (Hobbs and Hobbs 1987; Hobbs and Mooney 1991). Host-parasite systems provide yet another important application, in which the individual host is the patch (Gilbert 1977;

Anderson and May 1981; Denno and McClure 1983); the interplay between within-patch and among-patch evolutionary changes explains the evolution of intermediate levels of virulence, for example in the myxoma-*Oryctalagus* system (Fenner and Ratcliffe 1965; Levin and Pimentel 1981; Lewontin 1970) discussed earlier. Similar multiscale phenomena are important within our own genomes, in which the evolution of selfish DNA results from a conflict between the individual benefits to the segment of DNA and the costs to its host organism; or in bacteria, in which the evolutionary forces on plasmid-borne DNA must be distinguished from those on the bacteria. Finally, host plants form islands that can be colonized by insect pests (Gilbert 1977; Denno and McClure 1983; Harrison and Thomas 1991; McEvoy et al. 1993), and that recognition has led to important insights into the dynamics of these systems.

Food Webs

One of the most natural ways to describe a community or an ecosystem is in terms of the trophic relationships among species, and the tangled web that results (Elton 1958; Paine 1966, 1980; Odum 1983; Levin et al. 1977). Considerable theoretical interest has been directed to regularities that can be detected in the topological structure of such webs (Cohen 1977, 1989; Sugihara 1982; Pimm 1982; Yodzis 1989). This seems all the more remarkable given the fact that such patterns seem to hold true regardless of the criteria used to define the elements of a web, or the criteria for deciding that a link exists between two species (but see Cohen 1989; Schoener 1989). Indeed, there clearly is no unequivocal way to characterize a web. Is a taxonomic subdivision most appropriate, or would a functional one serve better? Should subdivision stop at the species level, consider different demographic classes, be partitioned according to genotype, etc.? However a class were defined, one could partition it further according to a variety of kinds of criteria, reducing variability within a class while sacrificing the predictability that can be achieved for larger assemblages. This is the same kind of problem confronted when one deals with spatial and temporal scale, but with added layers of complexity.

In an important study, Sugihara et al. (1989) examine the robustness of observed food web properties to aggregation of trophic groupings. Specifically, they look at such properties as chain length, existence of rigid circuits, and particular trophic ratios, and show that these are roughly invariant under lumping. The degree to which such regularities represent deep biological truths versus statistical anomalies is still to be resolved. What is important about their paper is the explicit recognition that one must take into consideration the biases attributable to the investigator's choice of scale and examine specifically how system description changes with scale.

A critical question regarding food web structure is whether there is a natural hierarchical decomposition of webs, or whether the particular filter imposed by a given aggregation scheme is just an arbitrary point on a continuum. Paine (1980), in his insightful Tansley lecture, introduced the notions of interaction strength, and strong and weak linkages, to suggest a fruitful way to dissect food web structure; similar ideas have proved very powerful in the sector decomposition literature in economics (Simon and Ando 1961; Iwasa et al. 1987, 1989). The general question remains unresolved, however, and a rich area for research.

Global Climate Change and Ecological Models

Global climate change, and changes in the concentrations of greenhouse gases, will have major effects on the vegetational patterns at local and regional scales (MacArthur 1972; Clark 1985); in turn, changes that occur at very fine scales, such as alterations in rates of stomatal opening and closing, ultimately will have impacts at much broader scales (Jarvis and McNaughton 1986). General circulation models (GCMs), which form the basis of climate prediction, operate on scales of hundreds of kilometers on a side, treating as homogeneous all of the ecological detail within (Hansen et al. 1987; Schneider 1989). On the other hand (Figure 14-14), most ecological studies are carried out on scales of meters or tens of meters (Kareiva and Anderson 1988); and even ecosystem studies are at scales several orders of magnitude less than those relevant to GCMs. Thus, a fundamental problem in relating the large-scale predictions of the climate models to processes at the scale of ecological information is to understand how information is transferred across scales (Jarvis and McNaughton 1986; Levin 1993).

As in the research described previously, there is a need both for statistical and correlational studies, and for modeling designed to elucidate mechanisms. A useful place to begin is in the quantification of spatial and temporal variability as a function of scale (e.g., Kratz et al. 1987; McGowan 1990). Long temporal and spatial series can be used to examine similar patterns in the variation of climate and ecosystem components; where scales of variation match, there is at least the basis for investigation of mechanistic relationships. An example is provided by the continuous plankton recorder surveys of the North Atlantic, providing data on spatial variations in the distributions of phytoplankton and zooplankton over half a century (see, for example, Colebrook 1982; McGowan 1990). The evidence from these studies has been that the larger spatial and temporal scales show the greatest variations, and that these correlate well with large-scale climatic variations (Dickson et al. 1988; McGowan 1990). The approach taken (Radach 1984), similar to that described earlier, is first to ask how much of the variation can be explained by variation in the physical environment, and then to look to autonomous biological factors to account for the balance. This mode of attack, which is the conventional one and the one that I have relied

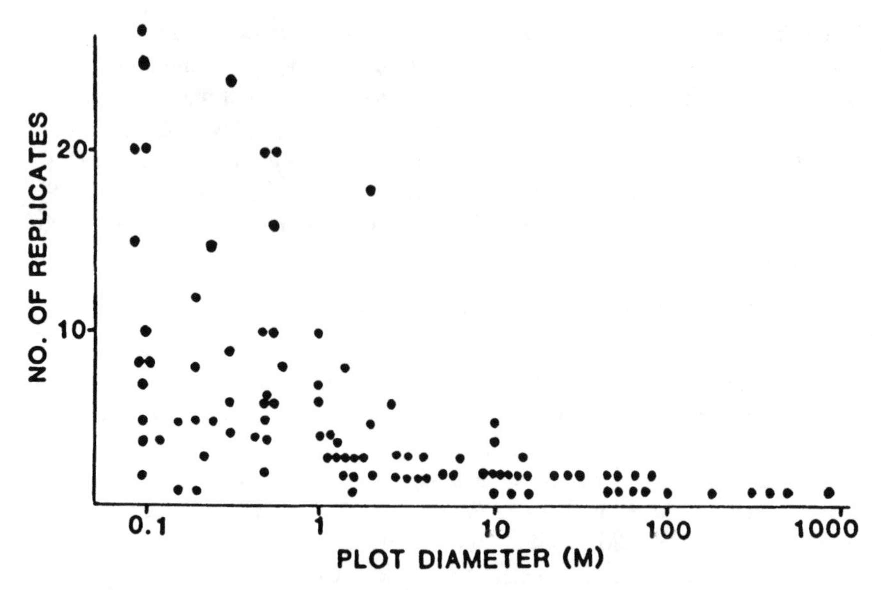

Figure 14-14. Size and number of replicates in experimental community ecology. Each point is taken from a different paper published in *Ecology*. Every paper from January 1980 to August 1986 was included if it involved experimental manipulation of resources or populations "in order to learn about community dynamics or community properties." (From P. M. Kareiva and M. Anderson, 1988, "Spatial Aspects of Species Interactions: The Wedding of Models and Experiments," in A. Hastings (ed.), *Community Ecology*. Lecture Notes in Biomathematics 77, Springer-Verlag, Berlin, Heidelberg. Reprinted by permission of Springer-Verlag.)

on in this chapter, perhaps requires scrutiny, given the possibility that intrinsic biotic factors might account for some climatic variation. Mechanistic approaches that examine how effects are scaled are the only way to address this puzzle.

Ågren et al. (1991) review models of the linkage of production and decomposition and discuss as well the linkages of process at different scales. The problem of scaling from the leaf to the ecosystem and beyond poses fundamental challenges in predicting the effects of global change (Norman 1980; Ehleringer and Field 1993). Ågren et al. point out that ecosystem models that operate at only one level of integration are not likely to incorporate mechanisms properly, and that it is essential to develop methods for integrating from finer scales; this reiterates the central theme of this chapter. A related problem is the need to interface processes that are operating at different levels of integration, as, for example, in the linkages between grassland biogeochemistry and atmospheric processes (Parton et al. 1989; Schimel et al. 1990).

It is worth noting (Holling, pers. comm.) that separating climatic and biotic influences upon changing ecosystem patterns can be extremely problematic.

Extrinsic influences can serve to trigger qualitative changes in systems dynamics (Levin 1978); cases in point may involve fires or outbreaks that are triggered by climate change but show very little correlation with it (e.g., Holling 1992b).

Patterns of Evolution at the Community Level

One of the greatest barriers to the development of interfaces between population biology and ecosystems science is the perceived scale mismatch, especially as regards evolutionary processes. To some extent, this is a misconception; for example, the evolution of responses to strong selective pressures, such as heavy metals in the environment, or antibiotics, can occur in ecological time, whereas ecosystem scientists concerned with global change must place great importance upon an understanding of the evolutionary record. Similarly, much of the study of macroevolutionary processes, such as speciation, must involve spatial scales comparable to those of relevance in ecosystem or landscape studies. Thus, there is considerable overlap between the spatial and temporal scales of interest to the population biologist and the ecosystem scientist.

Organizational complexity is another matter, by definition. Population biologists are concerned with changes taking place among and within populations; where coevolution has been documented, it is usually tight coevolution, such as in parasite-host systems, in which the fates of a small number of species (usually two), are intimately linked. Furthermore, the occasional tendency in ecosystems science to explain the evolution of ecosystem-level phenomena in terms of what is good for the ecosystem suggests to the population biologist a lack of sensitivity to the mechanisms of evolution, which they see as operating primarily at the individual level. On the other hand, to some ecosystem scientists, such attention to detail seems to miss the forest for the trees. It is hardly surprising that in some bastions of population biology, ecosystems science is scarcely tolerated, whereas in some schools of ecosystems science, population biology is seen as an academic enterprise of limited relevance to ecosystem-level problems.

A growing number of ecologists and evolutionary biologists, however, recognize the folly of failing to address the issues of mutual importance to the two communities (Orians and Paine 1983; Roughgarden 1976b; Rummel and Roughgarden 1985; Schneider and Louder 1984; Ehrlich 1991; papers in Levin 1976). As mentioned earlier, ecotoxicology and conservation biology represent two applications where systematics, population dynamics, evolutionary biology, and ecosystems science have built partnerships. Similar progress has been made in understanding the dynamics of aquatic systems, and the linkages between community structure and ecosystem processes (Bormann and Likens 1979; Carpenter and Kitchell 1988; Howarth 1991). For example, Carpenter and Kitchell (1988), in discussing the results of large-scale experimental manipulations of lakes, demonstrate the temporal and spatial dependence of the results and attribute the differences to the evolutionary history of the trophic interactions experienced; whole-lake experiments elicit processes over long time scales that involve diffuse interactions with large numbers of species. Yet central issues, such as the explanation of the evolution of community and ecosystem-level patterns in terms of forces acting on individual species, remain virtually untouched. Though various models attempt to explain how species subdivide the resources in a community, given their particular demographic characteristics (MarArthur 1970; May and MacArthur 1972), few address the question of how those demographic characteristics evolve in relation to each other (but see Lawlor and Maynard Smith 1976). And, whereas cross-system comparisons of different ecosystems may show regularities and allow functional classifications in terms of successional patterns (Connell and Slatyer 1977), trophic organization (Pimm 1982; Cohen 1989), diversity (Levin 1981; May 1978, 1981), or ecosystem processes (Ulanowicz 1986), few if any studies attempt to account for the evolution of those patterns in terms of the forces acting on individuals. The challenge is the familiar one of trying to understand patterns observed at one level of detail in terms of mechanisms that are operating on other scales. The study of such processes provides a challenging collection of problems of surpassing importance for understanding, for example, how ecosystems might be expected to change over long time scales in response to anthropogenic stresses.

As an example, consider the description of the patterns of chemical defenses and detoxification mechanisms found in communities in plants and their herbivores. Such patterns are important in defining the chemical context for introduced species, and for a variety of other problems; indeed, the study of patterns of chemical interactions in plant-herbivore communities is surely one of the most intellectually exciting in ecology (Feeny 1976; Futuyma 1983). In some cases, this evolution has been in response to tight and therefore strong interactions between pairs of species; but more often than not, the evolutionary pressures have come from diffuse influences, as have acted for example to shape the vertebrate immune system. How, then, do we develop a theoretical framework to address such problems?

The problem, actually, is not much different than is faced even when attention is on individual populations and on the evolution of ecological parameters. Classical theories have developed a rich literature for dealing with single loci, and for tight interactions among pairs of loci. Yet many and probably most traits of ecological importance are quantitative, controlled at many loci, each of which contributes in a small way. To address such problems, the theory of quantitative genetics was developed (Falconer 1960; Lande 1976). In quantitative genetics, one deals with broad phenotypic classes, suppressing variation within those classes, and focusing on macroscopic parameters such as population means and variances. One of the fundamental debates in that subject involves the appropriateness of ignoring certain levels of detail and heterogeneity, and the correct way to derive ensemble properties from information on individual dynamics.

The same problem must be confronted in developing quantitative descriptions of diffuse coevolution. Phenotypic classes must be broader, lumping individuals together not only within a species, but also across species, and phenomenological descriptions of the dynamics of macroscopic statistics must be achieved by aggregating lower-level phenomena. When this is done (Levin et al. 1990; Levin and Segel 1982), one can derive quantitative descriptions of changes at the community level, analogous to the equations of quantitative genetics. In this way, for example, one can derive equations for the rate of change of the mean level of resistance to herbivory within a community in terms of the variance and the statistics of the distribution of detoxification chemicals in the herbivore community (Levin et al. 1990).

Undoubtedly, the same objections will be raised against quantitative approaches at the community level as are lodged against the use of quantitative genetics, albeit more forcefully. After all, the approach is simply quantitative genetics carried a step further; by lumping together individuals across species, quantitative approaches would ignore the barriers to interbreeding that define species boundaries. Such objections should insert a note of caution and a reminder to be sensitive to how aggregation and simplification are done; they must not, however, prevent the development of such theories, which seem one of the most hopeful ways to bridge a yawning chasm. It cannot be the case that every detail of interspecific evolutionary interactions is essential for understanding patterns at the community level, any more than it is necessary (or useful) to account for every species in models of ecosystems dynamics. To be effective at developing appropriate descriptions at any level in terms of lower-level phenomena, it is as important to know what detail to ignore as it is to know what detail to include (Ludwig and Walters 1985).

Evolutionary theory has made great advances, and efforts over the next decade in molecular evolution will greatly expand our understanding of the mechanisms underlying evolutionary change. Yet unless we can find ways to relate detailed information at the molecular level to patterns of change at the level of the individual, the population, and the community, we will not have advanced our understanding of the evolution of the biosphere. Indeed, the disruptive selection regime, by which population genetics becomes more molecular while ecosystems science becomes more global, may enhance the speciation of the two communities of scientists unless novel efforts are made to relate the phenomena that transpire on the disparate scales of interest. Now, more than ever, we need to develop mechanistic evolutionary theories of how ecosystem patterns arise and are maintained.

Conclusion

Two fundamental and interconnected themes in ecology are the development and maintenance of spatial and temporal pattern, and the consequences of that pattern for the dynamics of populations and ecosystems. Central to these questions is the issue of how the scale of observation influences the description of pattern; each individual and each species experiences the environment on a unique range of scales, and thus responds to variability individualistically. Thus, any description of the variability and predictability of the environment makes sense only with reference to the particular range of scales that are relevant to the organisms or processes being examined.

Such issues are most clear for spatial and temporal scales, but apply as well to organizational complexity. The recognition in marine fisheries that total yield in multispecies fisheries remains fairly constant over long periods of time, though the species composition may change dramatically (May 1984), is a consequence of broadening the scale of description. Similarly, a claim that microbial communities are stable to perturbations, such as introductions of genetically engineered organisms, results from the application of a taxonomically broad filter, perhaps because only a fraction of the microbial community can be identified. In ecosystems research, one is likely to be concerned with a functional guild of microorganisms that perform a particular service to the ecosystem, and to refer to functional redundancy to explain why it is acceptable to ignore changes within a guild. This is the key to scaling and interrelating phenomena at different scales: knowing what fine detail is relevant at the higher levels, and what is noise.

There are several stages in the examination of the problem of pattern and scale. First, one must have measures to describe pattern (Gardner et al. 1987; Milne 1988), so that criteria can be established for relating that pattern to its causes and consequences. Cross-correlational analyses can provide initial suggestions as to mechanisms, but may miss emergent phenomena that arise from the collective behavior of smaller-scale processes. Theoretical investigations of the various mechanisms through which pattern can arise provide a catalogue of possibilities and may suggest relevant experiments to distinguish among hypothesized mechanisms.

All ecological systems exhibit heterogeneity and patchiness on a broad range of scales, and this patchiness is fundamental to population dynamics (Levin 1974; Roughgarden 1976a), community organization and stability (Holling 1986; Kareiva 1987), and element cycling (Bormann and Likens 1979). Patchiness is a concept that cuts across terrestrial and marine systems and provides a common ground for population biologists and ecosystem scientists. Patchiness, and the role of humans in fragmenting habitats, are key to the persistence of rare species and the spread of pest species. The level of species diversity represents a balance between regional processes, such as dispersal and species formation, and local processes, such as biotic interactions and stochasticity (Ricklefs 1987).

The consequences of spatial pattern and patchiness for the biota are many. Patchiness in the distribution of resources is fundamental to the way organisms exploit their environment (Schoener 1971; Wiens 1976; Mangel and Clark 1986; Pulliam 1989). Environmental heterogeneity provides a diversity of resources that can lead to coexistence among competitors that could not coexist in homogeneous environments (Horn and MacArthur 1972; Levin 1970, 1974), but the problem of how to count the number of resources is a vexing one. Trivially, no environment will be completely homogeneous, but how different must resources be to support different species? This question, which has been one of the central ones in community ecology (MacArthur 1970; May and MacArthur 1972; Whittaker and Levin 1977), goes to the heart of the problem of scale. Species can subdivide the environment spatially, concentrating on different parts of the same plant (Broadhead and Wapshere 1966), different layers of vegetation (MacArthur et al. 1966), or different microenvironments; or temporally, partitioning a successional gradient (Levin and Paine 1974) or a seasonal one. Thus, resource partitioning can result in temporally constant, spatially nonuniform patterns, or in spatially constant, temporally nonuniform ones, or in spatiotemporal mosaics (Levin and Paine 1974; Paine and Levin 1981; Whittaker and Levin 1977; Tilman 1988).

Because the variability and patchiness of the environment affects persistence and coexistence, it also affects species' evolutionary responses. Differential persistence occurs both among species and within species, and the latter will result in evolutionary changes that alter the species' response to environment, and hence the species' perception of the environment. Because the environment of a species is made up to a large extent of other species, the evolutionary response of one must affect the environmental variability experienced by others; thus ensues a coevolutionary dynamic that shapes the observed patterns of biotic and environmental variability.

All of this reinforces the recognition that there is no single correct scale at which to view ecosystems; the individualistic nature of responses to environment means that what we call a community or ecosystem is really just an arbitrary subdivision of a continuous gradation of local species assemblages (Whittaker 1975). It also carries important implications for predicting the responses of the biota to global change. Communities are not well integrated units that move en masse. They are collections of organisms and species that will respond individualistically to temporal variation, as they do to spatial variation. This is also true, of course, with regard to the evolutionary responses of populations. Thus, if there are predictable patterns that may be observed in what we define as communities and ecosystems, they have arisen through the individualistic ecological and evolutionary responses of their components, rather than through some higher-level evolution at the ecosystem level, Gaia notwithstanding (Lovelock 1972; see also Schneider and Boston 1991 for a wide range of views).

That there is no single correct scale or level at which to describe a system does not mean that all scales serve equally well or that there are not scaling laws. This is the major lesson of the theory of fractals (Mandelbrot 1977; Milne 1988; Sugihara and May 1990). The power of methods of spatial statistics, such as fractals, nested quadrat analysis (Greig-Smith 1964; Oosting 1956), semivariograms or correlograms (Sokal and Oden 1978; Burrough 1981, 1983a,b;

Sokal et al. 1989), or spectral analysis (Chatfield 1984), or of allometry (Calder 1984; Platt 1985; Brown and Nicoletto 1991; Harvey and Pagel 1991), is in their capability to describe how patterns change across scales. Thus, such methods have been used in ecology to quantify change in soils and in ecosystem properties at sub-field levels (Robertson et al. 1988) or landscape levels (Krummel et al. 1987), and in marine systems to quantify the distribution of physical factors, primary producers, and consumers (Steele 1978, 1991 and Chapter 1 of this volume; Haury et al. 1978; Weber et al. 1986; Levin et al. 1989b).

The simple statistical description of patterns is a starting point; but correlations are no substitute for mechanistic understanding (Lehman 1986). Modeling can play a powerful role in suggesting possible mechanisms and experiments, in exploring the possible consequences of individual factors that cannot be separated easily experimentally, and in relating fine-scale data to broad-scale patterns.

Because there is no single scale at which ecosystems should be described, there is no single scale at which models should be constructed. Methods from statistics and dynamical systems theory can play an important part in helping to determine the dimensionality of underlying mechanisms, and of appropriate models (Takens 1981; Schaffer 1981; Schaffer and Kot 1985; Sugihara et al. 1990). We need to have available a suite of models of different levels of complexity, and to understand the consequences of suppressing or incorporating detail. Models that are insufficiently detailed may ignore critical internal heterogeneity, such as that which is responsible for maintaining species diversity (Holling 1986). It is clear, for example, that the broad brush of the general circulation models ignores detail that is relevant for understanding biotic influences on climate systems, and vice versa. On the other hand, overly detailed models provide little understanding of what the essential forces are, will have more parameters and functional forms to estimate than the available data justify, will admit multiple basins of attraction, and are more prone to erratic dynamics that hamper prediction and parameter estimation. Just as we would not seek to build a model of human behavior by describing what every cell is doing, we cannot expect to model the dynamics of ecosystems by accounting for every individual, or for every species (Ludwig et al. 1978). We must determine what the appropriate levels of aggregation and simplification are for the problem at hand.

In an extremely instructive study, Ludwig and Walters (1985) have shown clearly that in some cases aggregated models can serve as better management tools than highly detailed models, even when the data used to fit the parameters of the model have been generated by the detailed model; in retrospect this should accord well with intuition. The problem of aggregation and simplification is the problem of determining minimal sufficient detail (Levin 1992; Rastetter et al. 1992).

Classical ecological models (Scudo and Ziegler 1978) treated communities as closed, integrated, deterministic, and homogeneous. Such models are simplifications of real systems, and provide a place to begin analysis. However, each of these assumptions must be relaxed if we are to understand the factors governing the diversity and dynamics of ecosystems. Virtually every population will exhibit patchiness and variability on a range of spatial and temporal scales, so that the definition of commonness or rarity is scale dependent (Schoener 1987). Virtually every ecosystem will exhibit patchiness and variability on a range of spatial, temporal, and organizational scales, with substantial interaction with other systems and influence of local stochastic events. These phenomena are critical for the maintenance of most species, which are locally ephemeral and competitively inferior, and which depend upon the continual local renewal of resources and mechanisms such as dispersal to find those opportunities. Fragmentation, local disturbance, and variability also can have major consequences for patterns of nutrient cycling (Bormann and Likens 1979), persistence (Pimm and Gilpin 1989), and patterns of spread of introduced species (Durrett 1988; Mooney and Drake 1986). The key is to separate the components of variability into those that inhibit persistence and coexistence, those that promote these, and those that are noise (Chesson 1986).

To address such phenomena, we must find ways to quantify patterns of variability in space and time, to understand how patterns change with scale (e.g., Steele 1978, 1989; Dagan 1986), and to understand the causes and consequences of pattern (Levin 1989, Wiens 1989). This is a daunting task that must involve remote sensing, spatial statistics, and other methods to quantify pattern at broad scales; theoretical work to suggest mechanisms and explore relationships; and experimental work, carried out both at fine scales and through whole-system manipulations, to test hypotheses. Together, these can provide insights as to how information is transferred across scales, and hence how to simplify and aggregate models.

The problem of relating phenomena across scales is the central problem in biology and in all of science. Cross-scale studies are critical to complement more traditional studies carried out on narrow single scales of space, time, and organizational complexity (Levin 1988, 1989; Meetenmeyer and Box 1987; Steele 1978, 1989; Holling 1992b), just as measures of β-diversity are needed to complement within-community measures of α-diversity (Whittaker 1975). By addressing this challenge, using the insights gained from similar studies in other sciences and the unique approaches that must be developed for ecological systems, we can greatly enhance our understanding of the dynamics of ecosystems and develop the theoretical basis necessary to manage them.

Acknowledgments

I gratefully acknowledge valuable comments on an earlier version of this chapter, and suggestions by Buzz Holling, Henry Horn, Bob May, Lee Miller, Bob O'Neill, and Zack Powell. It is a pleasure to acknowledge support of the National

Science Foundation under Grants BSR-8806202 and DMS-9108195; of the Andrew W. Mellon Foundation; of the Department of Energy under Grant DE-FG02-90ER60933; of the National Aeronautics and Space Administration under Grant DAGW-2085; of the U.S. Environmental Protection Agency under Cooperative Agreement CR-812685; and of McIntire-Stennis and Hatch Grants NYC-183550 and NYC-183430, respectively. Most importantly, I thank Colleen Martin for her indispensable help with this paper, and for 10 years of collegial interaction.

References

Ågren, G. I., R. E. McMurtrie, W. J. Parton, J. Pastor, and H. H. Shugart. 1991. State-of-the-art of models of production-decomposition linkages in conifer and grassland ecosystems. *Ecological Applications* **1:**118–138.

Allen, T. F. H., and T. B. Starr. 1982. *Hierarchy: Perspectives for Ecological Diversity*. University of Chicago Press, Chicago.

Anderson, R. M., and R. M. May. 1981. The population dynamics of microparasites and their invertebrate hosts. *Philosophical Transactions of the Royal Society B* **291:**451–524.

Andow, D. A., P. M. Kareiva, S. A. Levin, and A. Okubo. 1990. Spread of invading organisms. *Landscape Ecology* **4:**177–188.

Armstrong, R. A. 1988. The effects of disturbance patch size on species coexistence. *Journal of Theoretical Biology* **133:**169–184.

Bormann, F. H., and G. E. Likens. 1979. *Pattern and Process in a Forested Ecosystem*. Springer-Verlag, New York.

Botkin, D. B., J. F. Janak, and J. R. Wallis. 1972. Some ecological consequences of a computer model of forest growth. *Journal of Ecology* **60:**849–873.

Bramson, M. 1983. Convergence of solutions of the Kolmogorov equation to travelling waves. *Memoirs of the American Mathematical Society* **44:**1–190.

Broadhead, E., and A. J. Wapshere. 1966. Mesopsocus populations on larch in England—the distribution and dynamics of two closely related coexisting species of psocoptera sharing the same food resource. *Ecological Monographs* **36:**328–383.

Brown, J. H., and P. F. Nicoletto. 1991. Spatial scaling of species composition: Body masses of North American land mammals. *American Naturalist* **138:**1478–1512.

Burkey, T. V. 1989. Extinction in nature reserves: The effect of fragmentation and the importance of migration between reserve fragments. *Oikos* **55:**75–81.

Burrough, P. A. 1981. Fractal dimensions of landscapes and other environmental data. *Nature* **294:**240–242.

———. 1983a. Multiscale sources of spatial variation in soil. I. The application of fractal concepts to nested levels of soil variation. *Journal of Soil Science* **34:**577–597.

———. 1983b. Multiscale sources of spatial variation in soil. II. A non-Brownian fractal model and its applications in soil survey. *Journal of Soil Science* **34:**599–620.

Butman, C. A. 1987. Larval settlement of soft-sediment invertebrates: The spatial scales of pattern explained by active habitat selection and the emerging role of hydrodynamical processes. *Oceanography and Marine Biology Annual Review* **25:**113–165.

Cain, M. L. 1990. Models of clonal growth in *Solidago altissima*. *Journal of Ecology* **78:**27–46.

———. 1991. When do treatment differences in movement behaviors produce observable differences in long-term displacements? *Ecology* **72:**2137–2142.

Cain, S. A., and G. M. Castro. 1959. *Manual of Vegetation Analysis*. Harper & Row, New York.

Calder, W. A. III. 1984. *Size, Function, and Life History*. Harvard University Press, Cambridge, MA.

Canham, C. D. 1988. Growth and canopy architecture of shade-tolerant trees: Responses to forest gaps. *Ecology* **69:**789–795.

Carlile, D. W., J. R. Skalski, J. E. Barker, J. M. Thomas, and V. I. Cullinan. 1989. Determination of ecological scale. *Landscape Ecology* **2:**203–213.

Carpenter, S. R., and J. F. Kitchell. 1988. Consumer control of lake productivity. *BioScience* **38:**764–769.

Castillo-Chavez, C., K. Cooke, W. Huang, and S. A. Levin. 1989. "The Role of Long Periods of Infectiousness in the Dynamics of Acquired Immunodeficiency Syndrome (AIDS)." In C. Castillo-Chavez, S. A. Levin, and C. Shoemaker (eds.), *Mathematical Approaches to Problems in Resource Management and Epidemiology*. Lecture Notes in Biomathematics 81. Springer-Verlag, Berlin, pp. 177–189.

Chatfield, C. 1984. *The Analysis of Time Series: An Introduction*. Third edition. Chapman & Hall, London.

Chesson, P. 1986. "Environmental Variation and the Coexistence of Species." In J. Diamond and T. J. Case (eds.), *Community Ecology*. Harper & Row, New York, pp. 240–256.

Clark, J. S. 1989. Ecological disturbance as a renewal process: Theory and application to forest history. *Oikos* **56:**17–30.

———. 1990. Integration of ecological levels: Individual plant growth, population mortality, and ecosystem process. *Journal of Ecology* **78:**275–299.

———. 1991a. Disturbance and population structure on the shifting mosaic landscape. *Ecology* **72:**1119–1137.

———. 1991b. Disturbance and tree life history on the shifting mosaic landscape. *Ecology* **72:**1102–1118.

Clark, W. C. 1985. Scales of climate impacts. *Climatic Change* **7:**5–27.

Coffin, D. P., and W. K. Lauenroth. 1988. The effects of disturbance size and frequency on a shortgrass plant community. *Ecology* **69:**1609–1617.

———. 1989. Disturbance and gap dynamics in a semiarid grassland: A landscape-level approach. *Landscape Ecology* **3:**19–27.

Cohen, J. E. 1977. Ratio of prey to predators in community food webs. *Nature* **270:**165–177.

———. 1989. "Food Webs and Community Structure." In J. Roughgarden, R. M. May, and S. Levin (eds.), *Perspectives in Theoretical Ecology*. Princeton University Press, Princeton, NJ, pp. 181–202.

Cohen, D., and S. A. Levin. 1987. "The Interaction Between Dispersal and Dormancy Strategies in Varying and Heterogeneous Environments." In E. Teramoto and M. Yamaguti (eds.), *Mathematical Topics in Population Biology, Morphogenesis and Neurosciences,* Proceedings Kyoto 1985. Springer-Verlag, Berlin, pp. 110–122.

Colebrook, J. M. 1982. Continuous plankton records: Seasonal variations in the distribution and abundance of plankton in the North Atlantic Ocean and the North Sea (*Calanus finmarchicus*). *Journal of Plankton Research* **4:**435–462.

Connell, J. H., and R. O. Slatyer. 1977. Mechanisms of succession in natural communities and their role in community stability and organization. *American Naturalist* **111:**1119–1144.

Cox, J. T., and R. Durrett. 1988. Limit theorems for the spread of epidemics and forest fires. *Stochastic Processes and Applications* **30:**171–191.

Dagan, G. 1986. Statistical theory of groundwater flow and transport: Pore to laboratory, laboratory to formation, and formation to regional scale. *Water Resources Research* **22:**120S–134S.

Dawkins, R. 1982. *The Extended Phenotype: The Gene as the Unit of Selection*. W. H. Freeman, Oxford, England.

Dayton, P., and M. J. Tegner. 1984. "The Importance of Scale in Community Ecology: A Kelp Forest Example with Terrestrial Analogs." In P. W. Price, C. N. Slobodchikoff, and W. C. Gaud (eds.), *A New Ecology: Novel Approaches to Interactive Systems*. Wiley, New York, pp. 457–481.

den Boer, P. J. 1971. "Stabilization of Animal Numbers and the Heterogeneity of the Environment: The Problem of the Persistence of Sparse Populations." In *Dynamics of Populations,* Proceedings of the Advanced Study Institute on Dynamics of Numbers in Populations, Oosterbeek, the Netherlands. Pudoc, Wageningen, the Netherlands, pp. 77–97.

Denman, K. L., and T. M. Powell. 1984. Effects of physical processes on planktonic ecosystems in the coastal ocean. *Oceanography and Marine Biology Annual Review* **22:**125–168.

Denno, R. F., and M. S. McClure (eds.). 1983. *Variable Plants and Herbivores in Natural and Managed Systems*. Academic Press, New York.

Dickson, R. R., P. M. Kelly, J. M. Colebrook, W. S. Wooster, and D. H. Cushing. 1988. North winds and production in the eastern North Atlantic. *Journal of Plankton Research* **10:**151–169.

Dobzhansky, T., and S. Wright. 1943. Genetics of natural populations. X. Dispersion rates of *Drosophila pseudoobscura*. *Genetics* **28:**304–340.

———. 1947. Genetics of natural populations. XV. Rate of diffusion of a mutant gene through a population of *Drosophila pseudoobscura*. *Genetics* **31:**303–324.

Durrett, R. 1988. Crabgrass, measles, and gypsy moths: An introduction to interacting particle systems. *Mathematical Intelligencer* **10:**37–47.

Ehleringer, J., and C. Field (eds.) 1993. *Scaling Physiological Processes: Leaf to Globe*. Academic Press, San Diego, CA.

Ehrlich, P. R. 1991. "Coevolution and its Applicability to the Gaia Hypothesis." In S. H. Schneider and P. J. Boston (eds.), *Scientists on Gaia*. MIT Press, Cambridge, MA, pp. 19–22.

Ehrlich, P. R., and P. H. Raven 1964. Butterflies and plants: A study in coevolution. *Evolution* **18:**586–608.

Eldredge, N. 1985. *Unfinished Synthesis: Biological Hierarchies and Modern Evolutionary Thought*. Oxford University Press, New York.

Ellner, S. 1985a. ESS germination fractions in randomly varying environments. I. Logistic-type models. *Theoretical Population Biology* **28:**50–79.

———. 1985b. ESS germination fractions in randomly varying environments. II. Reciprocal yield law models. *Theoretical Population Biology* **28:**80–115.

Elton, C. S. 1958. *The Ecology of Invasions by Animals and Plants*. Methuen, London.

Fahrig, L., and J. Paloheimo. 1988a. Determinants of local population size in patchy habitat. *Theoretical Population Biology* **34:**194–213.

———. 1988b. Effect of spatial arrangement of habitat patches on local population size. *Ecology* **69:**468–475.

Falconer, D. S. 1960. *Introduction to Quantitative Genetics*. Ronald Press, New York.

Feeny, P. 1976. "Plant Apparency and Chemical Defense." In J. Wallace and R. Mansell (eds.), *Biochemical Interaction Between Plants and Insects*. Annual Review of Phytochemistry 10, pp. 1–40.

———. 1983. "Coevolution of Plants and Insects." In D. L. Whitehead and W. S. Bowers (eds.), *Natural Products for Innovative Pest Management*. Pergamon Press, Oxford, U.K., pp. 167–185.

Fenner, F., and F. N. Ratcliffe. 1965. *Myxomatosis*. Cambridge University Press, London.

Fisher, R. A. 1937. The wave of advance of advantageous genes. *Annals of Eugenics* **7:**355–369.

Forman, R. T. T., and M. Godron. 1981. Patches and structural components for a landscape ecology. *BioScience* **31:**733–740.

———. 1986. *Landscape Ecology*. Wiley, New York.

Futuyma, D. J. 1983. "Evolutionary Interactions Among Herbivorous Insects and Plants." In D. J. Futuyma and M. Slatkin (eds.), *Coevolution*. Sinauer, Sunderland, MA, pp. 207–231.

Gardner, R. H., B. T. Milne, M. G. Turner, and R. V. O'Neill. 1987. Neutral models for the analysis of broad landscape pattern. *Landscape Ecology* **1:**19–28.

Gardner, R. H., V. H. Dale, R. V. O'Neill, and M. G. Turner. 1992. "A Percolation Model of Ecological Flows." In F. di Castri and A. J. Hansen (eds.), *Landscape Boundaries: Consequences for Biotic Diversity and Ecological Flows*. Springer-Verlag, New York.

Gilbert, L. E. 1977. Development of theory in the analysis of insect-plant interactions. In D. J. Horn, R. D. Mitchell, and G. R. Stairs (eds.), *Analysis of Ecological Systems*. Ohio State University Press, Columbus, OH, pp. 117–154.

Gilpin, M. E., and I. Hanski (eds.). 1991. *Metapopulation Dynamics*. Academic Press, London.

Green, D. G. 1989. Simulated effects of fire, dispersal and spatial pattern on competition within forest mosaics. *Vegetatio* **82:**139–153.

Greig-Smith, P. 1964. *Quantitative Plant Ecology*. Second edition. Butterworth, London.

Grubb, P. J. 1977. The maintenance of species richness in plant communities. The importance of the regeneration niche. *Biology Review* **52:**107–145.

Grünbaum, D. 1992. Biomathematical models of krill aggregations, bryozoan feeding coordination, and undulatory propulsion. Ph.D. dissertation, Cornell University, Ithaca, NY.

Haldane, J. B. S. 1948. The theory of a cline. *Journal of Genetics* **48:**277–284.

Hamilton, W. D., and R. M. May. 1977. Dispersal in stable habitats. *Nature* **269:**578–581.

Hamner, W. M. 1984. Aspects of schooling in *Euphausia superba*. *Journal of Crustacean Biology* 4 (Spec. No. 1):67–74.

Hamner, W. M., P. P. Hamner, S. W. Strand, and R. W. Gilmer. 1983. Behavior of Antarctic krill, *Euphausia superba:* Chemoreception, feeding, schooling, and molting. *Science* **220:**433–435.

Hansen, J., I. Fung, A. Lacis, S. Lebedeff, D. Rind, B. Ruedy, G. Russell, and P. Stone. 1987. "Prediction of Near-Term Climate Evolution: What Can We Tell Decision-Makers Now?" In *Preparing for Climate Change, Proceedings of the First North American Conference on Preparing for Climate Changes,* October 27–29, 1987, pp. 35–47. Government Institutes, Inc., Washington, D.C.

Harrison, S., and J. F. Quinn. 1989. Correlated environments and the persistence of metapopulations. *Oikos* **56:**293–298.

Harrison, S., and C. D. Thomas. 1991. Patchiness and spatial pattern in the insect community on ragwort *Senecio jacobaea*. *Oikos* **62:**5–12.

Harvey, P. H., and M. D. Pagel. 1991. *The Comparative Method in Evolutionary Biology*. Oxford University Press, Oxford. U.K.

Hastings, A., and C. L. Wolin. 1989. Within-patch dynamics in a metapopulation. *Ecology* **70:**1261–1266.

Haury, L. R., J. A. McGowan, and P. H. Wiebe. 1978. "Patterns and Processes in the Time-Space Scales of Plankton Distributions." In J. H. Steele (ed.), NATO Conference Series IV, Vol. 3. *Spatial Pattern in Plankton Communities*. Plenum, New York, pp. 277–327.

Heisenberg, W. 1932. Nobel Prize in Physics Award Address. Nobel Foundation and Elsevier, Amsterdam.

Hobbs, R. J., and V. J. Hobbs. 1987. Gophers and grassland: A model of vegetation response to patchy soil disturbance. *Vegetatio* **69:**141–146.

Hobbs, R. J., and H. A. Mooney. 1991. Effects of rainfall variability and gopher disturbance on serpentine annual grassland dynamics. *Ecology* **72:**59–68.

Hofmann, E. E. 1988. Plankton dynamics on the outer southeastern U.S. Continental Shelf. Part III: A coupled physical-biological model. *Journal of Marine Research* **46:**919–946.

Holling, C. S. 1986. "The Resilience of Terrestrial Ecosystems: Local Surprise and Global Change." In W. C. Clark and R. E. Munn (eds.), *Sustainable Development of the Biosphere*. Cambridge University Press, London, pp. 292–317.

———. 1992a. Cross-scale morphology, geometry and dynamics of ecosystems. *Ecological Monographs* **62:**447–502.

———. 1992b. "The Role of Forest Insects in Structuring the Boreal Landscape." In H. H. Shugart (ed.), *A Systems Analysis of the Global Boreal Forest*. Cambridge University Press, London.

Holmes, J. C. 1983. "Evolutionary Relationships Between Parasitic Helminths and Their Hosts." In D. J. Futuyma and M. Slatkin (eds.), *Coevolution*. Sinauer, Sunderland, MA, pp. 161–185.

Horn, H. S., and R. H. MacArthur. 1972. Competition among fugitive species in a harlequin environment. *Ecology* **53:**749–752.

Horn, H. S., H. H. Shugart, and D. L. Urban. 1989. "Simulators as Models of Forest Dynamics." In J. Roughgarden, R. M. May, and S. A. Levin (eds.), *Perspectives in Theoretical Ecology*. Princeton University Press, Princeton, NJ, pp. 256–267.

Howarth, R. W. 1991. "Comparative responses of aquatic ecosystems to toxic chemical stress." In J. Cole, G. Lovett, and S. Findlay (eds.), *Comparative Analyses of Ecosystems: Patterns, Mechanisms, and Theories*. Springer-Verlag, New York, pp. 169–195.

Hubbell, S. P., and R. B. Foster. 1983. "Diversity of Canopy Trees in a Neotropical Forest and Implications for Conservation." In S. L. Sutton, T. C. Whitmore, and A. C. Chadwick (eds.), *Tropical Rain Forest: Ecology and Management*. Blackwell, Oxford, U.K., pp. 25–41.

Hutchinson, G. E. 1953. The concept of pattern in ecology. *Proceedings of the National Academy of Sciences USA* **105:**1–12.

Iwasa, Y., V. Andreasen, and S. Levin. 1987. Aggregation in model ecosystems. I. Perfect aggregation. *Ecological Modelling* **37:**287–302.

Iwasa, Y., S. A. Levin, and V. Andreasen. 1989. Aggregation in model ecosystems. II. Approximate aggregation. *IMA Journal of Mathematics Applied in Medicine and Biology* **6:**1–23.

Jarvis, P. G., and K. G. McNaughton. 1986. Stomatal control of transpiration: Scaling up from leaf to region. *Advances in Ecological Research* **15:**1–49.

Kareiva, P. M. 1983. Local movement in herbivorous insects: Applying a passive diffusion model to mark-recapture field experiments. *Oecologia* (Berlin) **57:**322–327.

———. 1987. Habitat fragmentation and the stability of predator-prey interactions. *Nature* **321:**388–391.

Kareiva, P. M., and M. Anderson. 1988. "Spatial Aspects of Species Interactions: The

Wedding of Models and Experiments." In A. Hastings (ed.), *Community Ecology*. Lecture Notes in Biomathematics 77. Springer-Verlag, Berlin, pp. 35–50.

Kareiva, P. M., and N. Shigesada. 1983. Analyzing insect movement as a correlated random walk. *Oecologia* (Berlin) **56:**234–238.

Kershaw, K. A. 1957. The use of cover and frequency in the detection of pattern in plant communities. *Ecology* **38:**291–299.

Kierstead, H., and L. B. Slobodkin. 1953. The size of water masses containing plankton bloom. *Journal of Marine Research* **12:**141–147.

Kimura, M., and G. H. Weiss. 1964. The stepping stone model of population structure and the decrease of genetic correlation with distance. *Genetics* **49:**561–576.

Kolmogorov, A., I. Petrovsky, and N. Piscunov. 1937. Étude de l'équation de la diffusion avec croissance de la quantité de matière et son application à un problème biologique. *Moscow University Bulletin Series International Sec. A.* **1:**1–25.

Kratz, T. K., T. M. Frost, and J. J. Magnuson. 1987. Inferences from spatial and temporal variability in ecosystems: Long-term zooplankton data from lakes. *American Naturalist* **129:**830–846.

Krummel, J. R., R. H. Gardner, G. Sugihara, R. V. O'Neill, and P. R. Coleman. 1987. Landscape patterns in a disturbed environment. *Oikos* **48:**321–324.

Lande, R. 1976. Natural selection and random genetic drift in phenotypic evolution. *Evolution* **30:**314–334.

Lawlor, L. R., and J. Maynard Smith. 1976. The coevolution and stability of competing species. *American Naturalist* **110:**79–99.

Lechowicz, M. J., and G. Bell. 1991. The ecology and genetics of fitness in Impatiens. II. Microscale heterogeneity of the edaphic environment. *Journal of Ecology* **79:**687–696.

Lehman, J. T. 1986. The goal of understanding in limnology. *Limnology and Oceanography* **31:**1160–1166.

Levene, H. 1953. Genetic equilibrium when more than one ecological niche is available. *American Naturalist* **87:**331–333.

Levin, S. A. 1970. Community equilibria and stability, and an extension of the competitive exclusion principle. *American Naturalist* **104:**413–423.

———. 1974. Dispersion and population interactions. *American Naturalist* **108:**207–228.

———. (Ed.). 1976. *Ecological Theory and Ecosystem Models*. The Institute of Ecology, University of Wisconsin, Madison.

———. 1978. "Pattern Formation in Ecological Communities." In J. H. Steele (ed.), *Spatial Pattern in Plankton Communities*. Plenum, New York, pp. 433–466.

———. 1981. "Mechanisms for the Generation and Maintenance of Diversity." In R. W. Hiorns and D. Cooke (eds.), *The Mathematical Theory of the Dynamics of Biological Populations*. Academic Press, London, pp. 173–194.

———. 1988. "Pattern, Scale and Variability: An Ecological Perspective." In A. Hastings (ed.), *Community Ecology*. Lecture Notes in Biomathematics 77. Springer-Verlag, Heidelberg, pp. 1–12.

———. 1989. "Challenges in the Development of a Theory of Community and Ecosystem Structure and Function." In J. Roughgarden, R. M. May, and S. A. Levin (eds.), *Perspectives in Ecological Theory*. Princeton University Press, Princeton, NJ, pp. 242–255.

———. 1992. "The Problem of Relevant Detail." In S. Busenberg and M. Martelli (eds.), *Differential Equations Models in Biology, Epidemiology and Ecology*, Proceedings, Claremont 1990. Lecture Notes in Biomathematics 92, Springer-Verlag, Berlin, pp. 9–15.

———. 1993. "Concepts of Scale at the Local Level." In J. R. Ehleringer and C. B. Field (eds.), *Scaling Physiological Processes: Leaf to Globe*. Academic Press, San Diego, CA.

Levin, S. A., and L. Buttel. 1987. Measures of Patchiness in Ecological Systems. Ecosystems Research Center Report No. ERC-130, Cornell University, Ithaca, NY.

Levin, S. A., D. Cohen, and A. Hastings. 1984. Dispersal strategies in patchy environments. *Theoretical Population Biology* **26(2)**:165–191.

Levin, S. A., J. E. Levin, and R. T. Paine. 1977. Snowy owl predation on short-eared owls. *The Condor* **79**:395.

Levin, S. A., K. Moloney, L. Buttel, and C. Castillo-Chavez. 1989a. Dynamical models of ecosystems and epidemics. *Future Generation Computer Systems* **5**:265–274.

Levin, S. A., A. Morin, and T. H. Powell. 1989b. "Patterns and Processes in the Distribution and Dynamics of Antarctic Krill." In *Scientific Committee for the Conservation of Antarctic Marine Living Resources, Selected Scientific Papers Part 1*, SC-CAMLR-SSP/5, CCAMLR, Hobart, Tasmania, Australia, pp. 281–299.

Levin, S. A., and R. T. Paine. 1974. Disturbance, patch formation, and community structure. *Proceedings of the National Academy of Sciences USA* **71**:2744–2747.

Levin, S. A., and D. Pimentel. 1981. Selection of intermediate rates of increase in parasite-host systems. *American Naturalist* **117**:308–315.

Levin, S. A., T. M. Powell, and J. H. Steele (eds.). 1993. *Patch Dynamics*. Lecture Notes in Biomathematics 96. Springer-Verlag, Berlin, Heidelberg, New York.

Levin, S. A., and L. A. Segel. 1976. An hypothesis for the origin of planktonic patchiness. *Nature* **259(5545)**:659.

———. 1982. Models of the influence of predation on aspect diversity in prey populations. *Journal of Mathematical Biology* **14**:253–284.

Levin, S. A., L. A. Segel, and F. Adler. 1990. Diffuse coevolution in plant-herbivore communities. *Theoretical Population Biology* **37**:171–191.

Lewontin, R. C. 1970. The units of selection. *Annual Review of Ecology and Systematics* **1**:1–18.

Liddle, M. J., J.-Y. Parlange, and A. Bulow-Olsen. 1987. A simple method for measuring diffusion rates and predation of seed on the soil surface. *Journal of Ecology* **75**:1–8.

Lovelock, J. E. 1972. Gaia as seen through the atmosphere. *Atmospheric Environment* **6**:579–580.

Lubina, J. A., and S. A. Levin. 1988. The spread of a reinvading species: Range expansion in the California sea otter. *American Naturalist* **131**:526–543.

Ludwig, D., D. D. Jones, and C. S. Holling. 1978. Qualitative analysis of insect outbreak systems: The spruce budworm and forest. *Journal of Animal Ecology* **44**:315–332.

Ludwig, D., and C. J. Walters. 1985. Are age-structured models appropriate for catch-effort data? *Canadian Journal of Fisheries and Aquatic Sciences* **42**:1066–1072.

MacArthur, R. H. 1968. "The Theory of the Niche." In R. C. Lewontin (ed.), *Population Biology and Evolution*. Syracuse University Press, Syracuse, NY, pp. 159–176.

———. 1970. Species packing and competitive equilibrium among many species. *Theoretical Population Biology* **1**:1–11.

———. 1972. *Geographical Ecology*. Harper & Row, New York.

MacArthur, R. H., H. Recher, and M. Cody. 1966. On the relation between habitat selection and species diversity. *American Naturalist* **100**:319–332.

MacArthur, R. H., and E. O. Wilson. 1967. *The Theory of Island Biogeography*. Princeton University Press, Princeton, NJ.

Mackas, D. L. 1977. Horizontal spatial variability and covariability of marine phytoplankton and zooplankton. Ph.D. dissertation. Dalhousie University, Halifax, Nova Scotia, Canada.

Mackas, D. L., K. L. Denman, and M. R. Abbott. 1985. Plankton patchiness: Biology in the physical vernacular. *Bulletin of Marine Sciences* **37**:652–674.

Mandelbrot, B. B. 1977. *Fractals: Form, Chance, and Dimension*. Freeman, San Francisco, CA.

Mangel, M., and C. W. Clark. 1986. Towards a unified foraging theory. *Ecology* **67**:1127–1138.

May, R. M. 1978. "The Dynamics and Diversity of Insect Faunas." In L. A. Mound and N. Waloff (eds.), *Diversity of Insect Faunas*. Blackwell Scientific, Oxford, U.K., pp. 188–204.

———. 1981. "Patterns in Multi-Species Communities." In R. M. May (ed.), *Theoretical Ecology*. Blackwell Scientific, Oxford, U.K., pp. 197–227.

———. (ed.). 1984. *Exploitation of Marine Communities*. Springer-Verlag, Berlin.

May, R. M., and R. H. MacArthur. 1972. Niche overlap as a function of environmental variability. *Proceedings of the National Academy of Sciences* **69**:1109–1113.

McEvoy, P. B., C. S. Cox, J. VanSickle, and N. T. Rudd. 1993. Disturbance, competition, herbivory effects on ragwort's rate of increase: A biological control case study. *Ecological Monographs* **63**:55–75.

McGowan, J. A. 1990. Climate and change in oceanic ecosystems: The value of time-series data. *Trends in Ecology and Evolution* **5**:293–299.

Mead, R. 1974. A test for spatial pattern at several scales using data from a grid of contiguous quadrats. *Biometrics* **30**:295–307.

Meentenmeyer, V., and E. O. Box. 1987. "Scale Effects in Landscape Studies." In M. G. Turner (ed.), *Landscape Heterogeneity and Disturbance*. Springer-Verlag, New York, pp. 15–34.

Meinhardt, H. 1982. *Models of Biological Pattern Formation*. Academic Press, London.

Menge, B. A., and A. M. Olson. 1990. Role of scale and environmental factors in the regulation of community structure. *Trends in Ecology and Evolution* **5**:52–57.

Milne, B. T. 1988. Measuring the fractal geometry of landscapes. *Applied Mathematics and Computation* **27**:67–79.

Minnich, R. A. 1983. Five mosaics in southern California and northern Baja California. *Science* **219**:1287–1294.

Mollison, D. 1977. Spatial contact models for ecological and epidemic spread. *Journal of the Royal Statistical Society B* **39**:283–326.

Moloney, K. A., S. A. Levin, N. R. Chiariello, and L. Buttel. 1992. Pattern and scale in a serpentine grassland. *Theoretical Population Biology* **41**:257–276.

Mooney, H. A., and J. A. Drake. 1986. *Ecology of Biological Invasions of North America and Hawaii*. Springer-Verlag, New York.

Murray, J. D. 1988a. How the leopard gets its spots. *Scientific American* **258**:80–87.

———. 1988b. Spatial dispersal of species. *Trends in Ecology and Evolution* **3**:307–309.

———. 1989. *Mathematical Biology*. Biomathematics 19. Springer-Verlag, Heidelberg.

Murray, J. D., and G. F. Oster. 1984. Cell traction models for generating pattern and form in morphogenesis. *Journal of Mathematical Biology* **19**:265–279.

Nee, S., and R. M. May. 1992. Dynamics of metapopulations: Habitat destruction and competitive coexistence. *Journal of Animal Ecology* **61(1)**:37–40.

Norman, J. M. 1980. "Interfacing Leaf and Canopy Light Interception Models." In J. D. Hesketh and J. W. Jones (eds.), *Predicting Photosynthesis for Ecosystem Models*. Volume II. CRC, Boca Raton, FL, pp. 49–67.

Nuernberger, B. D. 1991. Population structure of *Dineutus assimilis* in a patchy environment: Dispersal, gene flow, and persistence. Ph.D. dissertation, Cornell University, Ithaca, NY.

Odum, H. 1983. *Systems Ecology: An Introduction*. John Wiley, New York.

Okubo, A. 1972. A note on small organism diffusion around an attractive center: A mathematical model. *Journal of the Oceanographic Society of Japan* **28**:1–7.

———. 1978. "Horizontal Dispersion and Critical Scales for Phytoplankton Patches." In J. H. Steele (ed.), *Spatial Pattern in Plankton Communities*. Plenum, New York, pp. 21–42.

———. 1980. *Diffusion and Ecological Problems: Mathematical Models*. Biomathematics 10. Springer-Verlag, Berlin.

Okubo, A., and S. A. Levin, 1989. A theoretical framework for the analysis of data on the wind dispersal of seeds and pollen. *Ecology* **70**:329–338.

O'Neill, R. V., D. L. DeAngelis, J. B. Waide, and T. F. H. Allen. 1986. *A Hierarchical Concept of Ecosystems*. Princeton University Press, Princeton, NJ.

O'Neill, R. V., R. Gardner, B. T. Milne, M. G. Turner, and B. Jackson. 1991a. "Heterogeneity and Spatial Hierarchies." In J. Kolasa and S. T. A. Pickett (eds.), *Ecological Heterogeneity*. Springer-Verlag, New York, pp. 85–95.

O'Neill, R. V., S. J. Turner, V. I. Cullinan, D. P. Coffin, T. Cook, W. Conley, J. Brunt, J. M. Thomas, M. R. Conley, and J. Gosz. 1991b. Multiple landscape scales: An intersite comparison. *Landscape Ecology* **5**:137–144.

Oosting, H. J. 1956. *The Study of Plant Communities*. Second edition. W. H. Freeman, San Francisco, CA.

Orians, G. H., and R. T. Paine. 1983. Convergent Evolution at the Community Level." In D. J. Futuyma and M. Slatkin (eds.), *Coevolution*. Sinauer, Sutherland, MA, pp. 431–458.

Pacala, S. W. 1989. "Plant Population Dynamics Theory." In J. Roughgarden, R. M. May, and S. A. Levin (eds.), *Perspectives in Theoretical Ecology*. Princeton University Press, Princeton, NJ, pp. 54–67.

Pacala, S. W., and J. A. Silander. 1985. Neighborhood models of plant population dynamics. I. Single-species models of annuals. *American Naturalist* **125**:385–411.

Paine, R. T. 1966. Food web complexity and species diversity. *American Naturalist* **100**:65–75.

———. 1980. Food webs: Linkage, interaction strength and community infrastructure. The Third Tansley Lecture. *Journal of Animal Ecology* **49**:667–685.

Paine, R. T., and S. A. Levin. 1981. Intertidal landscapes: Disturbance and the dynamics of pattern. *Ecological Monographs* **51(2)**:145–178.

Parton, W. J., C. V. Cole, J. W. B. Stewart, D. S. Ojima, and D. S. Schimel. 1989. "Simulating Regional Patterns of Soil C, N, and P Dynamics in the U.S. Central Grassland Region." In L. Bergstrom and M. Charholm (eds.), *Ecology of Arable Land*. Kluwer Academic, Dordrecht, the Netherlands, pp. 99–108.

Pickett, S. T. A., and P. S. White (eds.). 1985. *The Ecology of Natural Disturbance and Patch Dynamics*. Academic Press, Orlando, FL.

Pimm, S. L. 1982. *Food Webs*. Population and Community Biology Series. Chapman & Hall, New York.

Pimm, S. L., and M. E. Gilpin. 1989. "Theoretical Issues in Conservation Biology." In J. Roughgarden, R. M. May, and S. A. Levin (eds.), *Perspectives in Theoretical Ecology*. Princeton University Press, Princeton, NJ, pp. 287–305.

Planck, M. 1936. *The Philosophy of Physics*. Translated by W. H. Johnston. Allen and Unwin, London.

Platt, T. 1985. Structure of the marine ecosystem: Its allometric basis. In R. E. Ulanowicz and T. Platt (eds.), Ecosystem Theory for Biological Oceanography. *Canadian Bulletin of Fisheries and Aquatic Sciences* **213**:55–75.

Powell, T. M. 1989. "Physical and Biological Scales of Variability in Lakes, Estuaries, and the Coastal Ocean." In J. Roughgarden, R. M. May, and S. A. Levin (eds.),

Perspectives in Theoretical Ecology. Princeton University Press, Princeton, NJ, pp. 157–180.

Pulliam, H. R. 1989. "Individual Behavior and the Procurement of Essential Resources." In J. Roughgarden, R. M. May, and S. A. Levin (eds.), *Perspectives in Theoretical Ecology*. Princeton University Press, Princeton, NJ, pp. 25–38.

Pyle, G. F. 1982. "Some Observations on the Geography of Influenza Diffusion." In P. Selby (ed.), *Influenza Models*. MTP Press, Lancaster, U.K., pp. 213–223.

Radach, G. 1984. Variations in the plankton in relation to climate. *Rapports et Procès-Verbaux des Réunions, Conseil International pour l'Exploration de la Mer* (ICES) **185:**234–254.

Rastetter, E. B., A. W. King, B. J. Cosby, G. M. Hornberger, R. V. O'Neill, and J. E. Hobbie. 1992. Aggregating fine-scale ecological knowledge to model coarser-scale attributes of ecosystems. *Ecological Applications* **2:**55–70.

Reddingius, J. 1971. "Gambling for Existence. A Discussion of Some Theoretical Problems in Animal Population Ecology." In *Acta Theoretica Supplementum Climum* (added to Acta Biotheoretica 20, 1972). Brill, Leiden, the Netherlands.

Ricklefs, R. E. 1987. Community diversity: Relative roles of local and regional processes. *Science* **235:**167–171.

Robertson, G. P., M. A. Huston, F. C. Evans, and J. M. Tiedje. 1988. Spatial patterns in a successional plant community: Patterns of nitrogen availability. *Ecology* **69:**1517–1524.

Rothschild, B. J., and T. R. Osborn. 1988. Small-scale turbulence and plankton contact rates. *Journal of Plankton Research* **10:**465–474.

Roughgarden, J. 1976a. Influence of competition on patchiness in a random environment. *Theoretical Population Biology* **14:**185–203.

———. 1976b. Resource partitioning among competing species—a coevolutionary approach. *Theoretical Population Biology* **9:**388–424.

———. 1989. "The Structure and Assembly of Communities." In J. Roughgarden, R. M. May, and S. A. Levin (eds.), *Perspectives in Theoretical Ecology*. Princeton University Press, Princeton, NJ, pp. 203–226.

Rummel, J., and J. Roughgarden. 1985. A theory of faunal buildup for competition communities. *Evolution* **39:**1009–1033.

Runkle, J. R. 1982. Patterns of disturbance in some old-growth mesic forests of eastern North America. *Ecology* **63:**1533–1586.

Rvachev, L. A., and I. M. Longini. 1985. A mathematical model for the global spread of influenza. *Mathematical Biosciences* **75:**3–22.

Sakai, S. 1973. A model for group structure and its behavior. *Biophysics* (Japan) **13:**82–90.

Schaffer, W. M. 1974. Selection for optimal life histories: The effects of age structure. *Ecology* **55:**291–303.

———. 1981. Ecological abstraction: The consequences of reduced dimensionality in ecological models. *Ecological Monographs* **51:**383–401.

Schaffer, W. M., and M. Kot. 1985. Nearly one dimensional dynamics in an epidemic. *Journal of Theoretical Biology* **112:**403–427.

Schimel, D. S., W. J. Parton, T. G. F. Kittel, D. S. Ojima, and C. V. Cole. 1990. Grassland biogeochemistry: Links to atmospheric processes. *Climatic Change* **17:**13–25.

Schneider, S. H. 1989. The greenhouse effect: Science and policy. *Science* **243:**771–781.

Schneider, S. H., and P. J. Boston (eds.). 1991. *Scientists on Gaia*. MIT Press, Cambridge, MA.

Schneider, S. H., and R. Louder. 1984. *The Coevolution of Climate and Life*. Sierra Books, San Francisco, CA.

Schoener, T. W. 1971. Theory of feeding strategies. *Annual Review of Ecology and Systematics* **2:**369–404.

———. 1987. The geographical distribution of rarity. *Oecologia* (Berlin) **74:**161–173.

———. 1989. Food webs from the small to the large. *Ecology* **70:**1559–1589.

Scudo, F. M., and J. R. Ziegler. 1978. *The Golden Age of Theoretical Ecology: 1923–1940*. Lecture Notes in Biomathematics 22. Springer-Verlag, Berlin.

Segel, L. A., and S. A. Levin. 1976. "Applications of Nonlinear Stability Theory to the Study of the Effects of Dispersion on Predator-Prey Interactions." In R. Piccirelli (ed.), *Selected Topics in Statistical Mechanics and Biophysics*. Conference Proceedings Number 27. American Institute of Physics, New York, pp. 123–152.

Sherry, T. W., and R. T. Holmes. 1988. Habitat selection by breeding American Redstarts in response to a dominant competitor, the Least Flycatcher. *The Auk* **105:**350–364.

Shugart, H. H. 1984. *A Theory of Forest Dynamics: The Ecological Implications of Forest Succession Models*. Springer-Verlag, New York.

Shugart, H. H., and D. C. West. 1981. Long-term dynamics of forest ecosystems. *American Scientist* **69:**647–652.

Simon, H. A., and A. Ando. 1961. Aggregation of variables in dynamic systems. *Econometrica* **29:**111–138.

Skellam, J. G. 1951. Random dispersal in theoretical populations. *Biometrika* **38:**196–218.

Sokal, R. R., and N. L. Oden. 1978. Spatial autocorrelation in biology 2: Some biological implications and four examples of evolutionary and ecological interest. *Biological Journal of the Linnean Society* **10:**229–249.

Sokal, R. R., G. M. Jacquez, and M. C. Wooten. 1989. Spatial autocorrelation analysis of migration and selection. *Genetics* **121:**845–856.

Steele, J. H. 1978. "Some Comments on Plankton Patches." In J. H. Steele (ed.), *Spatial Pattern in Plankton Communities*. NATO Conference Series IV, Vol. 3, Plenum, New York, pp. 1–20.

———. 1989. "Discussion: Scale and Coupling in Ecological Systems." In J. Roughgarden, R. M. May, and S. Levin (eds.), *Perspectives in Theoretical Ecology*. Princeton University Press, Princeton, NJ, pp. 177–180.

———. 1991. Can ecological theory cross the land-sea boundary? *Journal of Theoretical Biology* **153:**425–436.

Stommel, H. 1963. Varieties of oceanographic experience. *Science* **139:**572–576.

Strathmann, R. 1974. The spread of sibling larvae of sedentary marine invertebrates. *American Naturalist* **108:**29–44.

Sugihara, G. 1982. Niche hierarchy: Structure, organization, and assembly in natural communities. Ph.D. dissertation. Princeton University, Princeton, NJ.

Sugihara, G., B. Grenfell, and R. M. May. 1990. Distinguishing error from chaos in ecological time series. *Philosophical Transactions of the Royal Society of London B* **330:**235–251.

Sugihara, G., and R. M. May. 1990. Applications of fractals in ecology. *Trends in Ecology and Evolution* **5:**79–86.

Sugihara, G., K. Schoenly, and A. Trombla. 1989. Scale invariance in food web properties. *Science* **245:**48–52.

Takens, F. 1981. "Detecting Strange Attractors in Turbulence." In D. A. Rand and L. S. Young (eds.), *Dynamical Systems and Turbulence: Warwick 1980*. Lecture Notes in Mathematics 898. Springer-Verlag, Berlin, pp. 366–381.

Tennekes, H., and J. Lumley. 1972. *A First Course in Turbulence*. MIT Press, Cambridge, MA.

Tilman, D. 1988. *Plant Strategies and the Dynamics and Structure of Plant Communities*. Princeton University Press, Princeton, NJ.

Turing, A. M. 1952. The chemical basis of morphogenesis. *Philosophical Transactions of the Royal Society of London B* **237:**37–72.

Ulanowicz, R. E. 1986. *Growth and Development: Ecosystems Phenomenology*. Springer-Verlag, New York.

Urban, D. L., R. V. O'Neill, and H. J. Shugart, Jr. 1987. Landscape ecology. *BioScience* **37:**119–127.

Venable, D. L., and L. Lawlor. 1980. Delayed germination and dispersal in space and time: Escape in space and time. *Oecologia* (Berlin) **46:**272–282.

von Niessen, W., and A. Blumen. 1988. Dynamic simulation of forest fires. *Canadian Journal of Forest Research* **18:**805–812.

Watt, A. S. 1947. Pattern and process in the plant community. *Journal of Ecology* **35:**1–22.

Weber, L. H., S. Z. El-Sayed, and I. Hampton. 1986. The variance spectra of phytoplankton, krill and water temperature in the Antarctic Ocean south of Africa. *Deep-Sea Research* **33:**1327–1343.

Werner, P. A. 1979. "Competition and Coexistence of Similar Species." In O. T. Solbrig, S. Jain, G. Johnson, and P. Raven (eds.), *Topics in Plant Population Biology*. Columbia University Press, New York, pp. 287–310.

Whittaker, R. H. 1975. *Communities and Ecosystems*. Macmillan, New York.

Whittaker, R. H., and S. A. Levin. 1977. The role of mosaic phenomena in natural communities. *Theoretical Population Biology* **12:**117–139.

Wiens, J. A. 1976. Population responses to patchy environments. *Annual Review of Ecology and Systematics* **7**:81–120.

———. 1986. "Spatial Scale and Temporal Variation in Studies of Shrubsteppe Birds." In J. Diamond and T. J. Case (eds.), *Community Ecology*. Harper & Row, New York, pp. 154–172.

———. 1989. Spatial scaling in ecology. *Functional Ecology* **3**:385–397.

Wiens, J. A., J. T. Rotenberry, and B. Van Horne. 1986. A lesson in the limitations of field experiments: Shrubsteppe birds and habitat alteration. *Ecology* **67**:365–376.

Wilson, K. G. 1983. The renormalization group and critical phenomena. *Reviews of Modern Physics* **55**:583–600.

Wright, H. E. 1974. Landscape development, forest fires, and wilderness management. *Science* **186**:487–495.

Wright, S. 1977. *Evolution and the Genetics of Populations. Volume 3. Experimental Results and Evolutionary Deductions*. University of Chicago Press, Chicago.

———. 1978. *Evolution and the Genetics of Populations. Volume 4. Variability Within and Among Natural Populations*. University of Chicago Press, Chicago.

Yodzis, P. 1989. *Introduction to Theoretical Ecology*. Harper & Row, New York.

15

Long-Term Environmental Change

Patricia F. McDowell, Thompson Webb III, and Patrick J. Bartlein

Time series for the Quaternary, which covers the past 1.6 million years, reveal large, continuous, and sometimes abrupt ecological and environmental changes on time scales of decades to thousands of years. For biological variables, the time series can involve data from single localities or multiple localities over a spectrum of areas, and therefore can appear as single time series, multiple time series, or time series of maps for regions, continents, and the globe. These figures can show data from single organisms (e.g., tree rings), multiple organisms, individual taxa, or multiple taxa. The biological data and time series can be viewed in the context of time series for environmental systems such as climate, soils, and geomorphologic features. The techniques for studying these long ecological time series are many and include statistical analysis, numerical modeling, and simulation. The selection of techniques depends on the problem being investigated and the types of data available.

Quaternary time series such as oxygen isotope data from foraminifera in deep sea cores record changes in global ice volume. Carbon dioxide and methane measurements from ice cores monitor global signals. Thus a single time series can be used to represent global-scale changes. Most time series, however, represent local to regional changes at best and require maps to show how the variations at the sampling site fit into the global picture. In many of these time series of 5000 years or longer, the abundance of selected species may vary for only short intervals, so records from multiple species are required to show the full complement of variations at that site or for that region. The analysis of spatial-temporal data sets is needed if the full spectrum of biological variations is to be studied.

This chapter summarizes some of what is known about the "natural" (i.e., non-

---

human-induced) temporal and spatial variations for climatic, vegetational, and geomorphic systems during the past million years and the past 20,000 years. Originally prepared for a jointly edited book on *The Earth as Transformed by Human Action,* it served as a background for the discussion in the other chapters, inasmuch as the assessment of human impact requires some measure of the magnitude, scale, and cause of natural variations in the environment and biosphere.

The chapter focuses on temporal scales from the last million years down to decades and on spatial scales from the global to the local. It describes the systems in terms of forcing-response models, although, where possible, we have noted feedbacks among components in these systems. This approach led to the discussion of possible controls for the major responses of the climate, vegetation, and geomorphologic systems. Each of these systems responds according to its characteristics and the different time constraints within it. For example, the climate system is multilayered, and for some forcings and time scales only the atmosphere is involved, whereas for others the lithosphere, cryosphere, and hydrosphere are also involved. For the biosphere, the vegetation is just one component, and its variations span local succession with one set of time constants to continental-scale migrations with a different set of controls and time constants.

The unifying figure in the chapter is the log-scaled space-time scatter diagram, which helped in organizing the phenomena observed in each system (e.g., hurricanes, glacial advances, and mountain uplift for the climate system) in terms of the spectrum of temporal and spatial scales within which they occur. The figure for climatic phenomena shows the multiple time scales over which climate varies. Climate is commonly defined as average weather, and the nature of climate variation will depend upon the temporal and spatial interval used for averaging the weather and other phenomena like ocean currents and ice sheet movements. Because most biological systems respond to local or regional climate variations, it is important to understand how regional climates respond to global-scale variations. Recent modeling experiments with global climate models are helpful in providing this kind of information.

The chapter presents a speculative space-time figure for certain biotic systems and attempts to show that the life spans for ecological units like communities differ from those for evolutionary units like species. The current view is that most tree species have survived the multiple glacial and hydrological cycles of the Quaternary. If so, members of these species have been forced to live in a continuously changing panorama of plant associations and, hence, competitive situations. Webb (1987), Huntley and Webb (1988a), and Bennett (1990) have discussed some of the implications of this situation for speciation and evolution, and Hunter et al. (1988) have looked at some of the implications for conservation biology.

The Quaternary record is a rich source of ecological time series. Webb and Bartlein (1992) recently summarized the climate and biological time series for three time scales: the last 20,000 years, the last 175,000 years, and the last 3 million years. Their review complements the one presented here and shows in time series form some of the many variations illustrated in the time-space scale diagrams. Imbrie et al. (1992) have reviewed the diverse time series for the past 400,000 years and described their linear response to orbital forcing. Webb (1993) and MacDonald (1993) have recently reviewed the temporal, spatial, and taxonomic resolution of pollen data, an important source of quantitative information on past vegetational variations, and MacDonald (1993) has reviewed the use of these data in reconstructing plant invasions, one type of temporal and spatial response of plants to changing environments. For reviews of temporally and spatially precise time series of past vegetation, see chapters in Chambers (1993) and Huntley and Webb (1988a), and for a general review of North American vegetation history, see Ritchie (1987).

Ecologists often ask: Does history matter? The data and discussion in the following chapter offer a resounding Yes. In fact, landscapes contain geomorphologic features (e.g., moraines or dried-out lake beds) that tell of very different past environments. Any explanation of landscape pattern requires knowledge of its history over several time scales. Similarly, but to a lesser extent, the modern vegetation can record patterns of past land-use or environments. The climate system is one that varies in this respect, because the atmosphere has a very short time constant and hence little memory of its previous states, whereas the lithosphere is still responding, via glacial rebound, to the last period of maximum glaciation. Such different time constants of response and memory are present in ecological time series and add to the challenge of their analysis.

—T. W. III, 1993

The Importance of Long-Term Environmental Change to Understanding Human Transformation of the Earth

The interrelated systems of climate, vegetation, and the physical landscape (landforms and soils) are both the major determinants of the human environment and the major subjects of human environmental exploitation. Even without human perturbations, these systems are highly dynamic on time scales ranging from hours to millions of years. Each of these environmental systems changes naturally when influenced by external forcing factors and internal dynamics. Positive and negative feedbacks within these systems are common. Environmental systems exhibit complex behavior, because their responses to a particular type of stimulus can differ at different temporal or spatial scales.[1]

This environmental dynamism and complexity has long hampered attempts to

---

[1]Our focus on climate, vegetation, and physical landscapes in not meant to imply that other systems vary insignificantly. Fluxes in marine biota present one of many examples of other natural systems experiencing major changes.

develop some unifying theory to explain general patterns of long-term environmental change. Many early students of human impact ignored natural dynamics altogether; George Perkins Marsh attributed to nature an almost "unchanging permanence of form, outline, and proportion," and "a condition of equilibrium . . . which, without the action of man, would remain with little fluctuation for countless ages" (Marsh 1885, p. 26). At a 1955 conference (Thomas 1956), several authors discussed climatic and landscape change, but the emphasis was on description of the symptoms of environmental change and on review of new, potentially synthetic approaches to understanding long-term environmental change. Today, we use many synthetic approaches and unifying theories that were not foreseen in 1955. The emerging picture is at once more complex and more integrated than that of 1955.

Our predecessor at the 1955 symposium observed that "[i]n a conference on "Man's Role in Changing the Face of the Earth," a summary of environmental changes through forces independent of man appears to be an assignment to the voice of a loyal opposition" (Russell 1956, p. 453). Long-term environmental change, however, provides an important background to human transformation of the earth. Members of the genus *Homo* first appeared at the beginning of the Quaternary, and many cultural activities, such as agriculture, developed within the past 20,000 years during which the present interglacial environment and landscape became established. Natural environmental changes provide a scale for judging the magnitude and some of the possible effects of human impacts on the environment. Indeed, without a knowledge of natural change, the change due to human intervention cannot be specified.

The goals of this chapter are to describe how environmental systems have changed not only during the past 20,000 years[2] but also within the 1.8-million-year long Quaternary period, and to provide an overview of various emergent unifying theories of long-term environmental change. The most important of these is the astronomical theory of climatic change (Milankovitch 1941; Imbrie and Imbrie 1979; Berger et al. 1984), which states that variations in the earth's orbit and axial tilt cause variations in solar energy input to earth's atmosphere

[2]Our summary draws heavily upon COHMAP, a Cooperative Holocene Mapping Project, which is an international paleoclimatic research group that has used data and simulation models to study the changes in climate, vegetation, lakes, and oceans that occurred during the past 20,000 years (COHMAP Members 1988; Kerr 1984b). Its main goals are to assemble global sets of paleoclimatic data, to calibrate these data in climatic terms, and to use these data to test the ability of global circulation models to simulate past climates (COHMAP Members 1988). COHMAP researchers have mapped and described the vegetation, surface oceans, and surface hydrological regimes of the past 20,000 years. COHMAP research follows the lead set by CLIMAP (Climate, Long-range Investigation, Mapping, and Prediction; CLIMAP Project Members 1976, 1981), which was a well-organized international interdisciplinary paleoclimatic research group that first mapped and modeled the full-glacial climate of 18,000 years ago (Gates 1976). We have used a summary of environmental changes and their causes by the COHMAP group (COHMAP Members 1988) to review the major environmental changes of the past 20,000 years.

and are the overriding pacemakers of the highly significant, long-term variations of global ice volume and monsoon intensity (Hays et al. 1976; Kutzbach 1981). The climatic dependence of vegetation, geomorphology, hydrology, and soils, at time scales of 100 to 100,000 years, makes an understanding of climate change key to obtaining a unifying theory for the dynamics of these systems. The environmental symptoms of these longer and shorter climatic cycles are many and widespread, affecting not only high-latitude glaciated regions, but also tropical and desert areas and the oceans.

To understand the changes of the last 20,000 years, we will examine system controls and responses at scales ranging from 100 to 1,000,000 years. It is important to specify time (and space) scales when discussing environmental change, because (1) the scale of the system's response and recovery provides a perspective by which to gauge the magnitude of human and natural impacts, and (2) systems that are controls at one temporal or spatial scale may be dependent responding systems at another scale (Saltzman 1985). Chisholm (1980) recognized that the latter point also applies to social systems.

Climate: Scale, Control, and Response

Of the three environmental systems that we describe, we view climate and its variations as setting the pace for many of the variations in the other systems. We therefore describe the scales and types of climatic variations before discussing the associated vegetational and geomorphic changes.

Global Climatic Change at Different Time Scales

The earth's climate varies at time scales ranging from very short (severe weather phenomena, $<$ 1 hour) to extremely long (geological warming or cooling trends, 10 million years). Kutzbach (1976) suggested that climate variance is greatest at short time scales (10 years to days) and at long time scales (10,000 to $>$ 100,000 years), and that it is fairly low at intermediate time scales. Climate variation at time scales of 1 month to 100 years is significant, however, because its scale is comparable to (or shorter than) those of social, political, and ecological adjustment processes such as national industrialization, national urbanization, and demographic change. Thus, even though relatively small in magnitude, climatic variations at these intermediate scales can have a disrupting effect on human subsistence (Clark 1985). They are also those most likely to be attributed incorrectly to human action.

From a long-term perspective, climatic and other environmental changes operating at time scales of 1000 to 1 million years are likely to have a significant influence on human cultural and biological evolution (Huntley and Webb 1988a). Furthermore, the physical landscape that humans occupy is largely the result of climatic, marine, and hydrologic processes operating at these time scales. The distribution of natural biotic resources varies significantly at these time scales. For these reasons, so-called long-term climatic changes, which we examine here, have great relevance for the human environment.

The spatial scales of short and intermediate-term climatic variations were compiled by Clark (1985), and his scheme is extended here (Figure 15-1). We have divided the temporal range of global climatic variations into five time-dependent systems, each of which includes processes and events occurring at various spatial scales. Within each system, smaller phenomena generally have shorter response times and operate over shorter periods. The positive slopes for the phenomena in each system reflect this trend. Our five systems include (1) weather phenomena, (2) quasi-biennial to short-term climatic fluctuations, (3) 100- to 1000-year neoglacial variations, (4) the orbitally driven glacial-interglacial and monsoon variations, and (5) the long-term, tectonically driven climatic variations over millions of years.

The general climatic pattern of the past 800,000 years and much of the Quaternary period is one of large-amplitude, glacial-interglacial cycles about 100,000

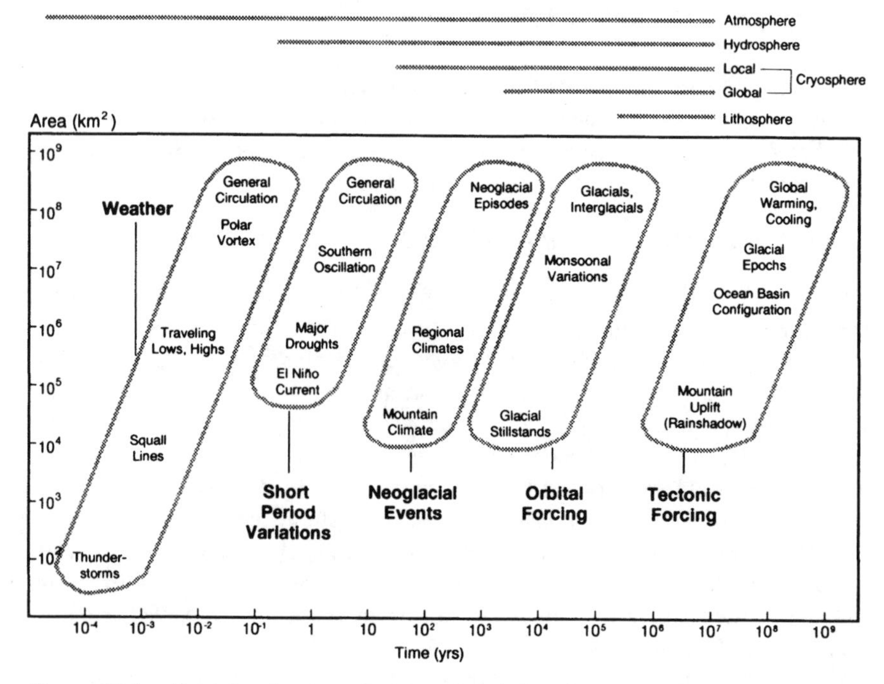

Figure 15-1. Spatial and temporal scales of variations in weather and climate. Short-term variations are limited primarily to the atmosphere, but longer-term variations involve progressively more earth systems. Each bubble encloses a group of related types of variations where the shorter-term, smaller-area features are part of the longer-term, larger-area features. (Figure by Montine Jordan.)

years long. These involve maximum extension and retreat of large continental ice sheets. Superimposed on these long-term cycles are smaller-amplitude fluctuations at several shorter time scales, including the long-term regular variations in monsoon intensity in the tropics. The longest of long-term climatic variations, such as the shift into the present glacial epoch, glacial-interglacial cycles, and stadial-interstadial cycles, affect the entire earth, although the effects are not necessarily synchronous or similar in magnitude and direction everywhere. At a somewhat shorter temporal scale are fluctuations such as the Little Ice Age, a period of expanded alpine glaciers, cooler temperatures, and expanded sea ice in North America, Europe, and the North Atlantic, that occurred about 600 to 100 years ago (Lamb 1982; Grove 1988). This is the latest and best-documented in a series of episodes of alpine glacier expansion that have occurred during the past 12,000 years (Grove 1979). Outside the mid-latitudes, climate fluctuations at a similar temporal and spatial scale probably also occurred in subtropical areas and in arid regions (e.g., Street-Perrott 1981; Mehringer 1986), but they are less well documented. These changes are subhemispheric, not global, in scale. Climatic variations of intermediate time scale include the historical warming that has taken place since the late 1800s, major droughts such as the North American Dust Bowl, and El Niño-Southern Oscillation (ENSO) events. The best-studied and most regular intermediate-scale changes are weather and sea-surface variations associated with the ENSO. Formation of a warm southerly flowing current off Peru is called El Niño, because it forms in December. The Southern Oscillation is a quasi-biennial east–west change in atmospheric pressure across the tropical Pacific. In western South America and the adjacent Pacific Ocean, these effects are always similar, but the effects in other regions can vary from one ENSO event to the next. At the smallest temporal and spatial scales are weather phenomena, including major storm systems.

Controls of Climatic Variations

Controls of long- and short-term climatic variation are not fully understood. Many possible controls exist (Kutzbach 1976; Mitchell 1976; Saltzman 1985), but their relative significance is unclear, and in general, the ultimate controls for many known patterns of climatic variation have not been identified. Major controls, including those external and those internal to the climate system, are listed in Table 15-1. Some of these controls operate very slowly (continental drift) or in a regular, predictable fashion (orbital variations), but many occur as sudden, random perturbations (volcanic activity, asteroid impact). The global climate system is complex, consisting of four major linked subsystems, the atmosphere, oceans, biosphere, and cryosphere (and, for changes of 10,000 to 100 million years, even variations in the lithosphere are involved). The linkages among the subsystems occur through the exchange of heat, and to a lesser extent through the exchange of water vapor and CO_2. The atmosphere, the fastest-

Table 15-1. Controls of long-term climatic variability

Control	Effect	Status of control at time scales of (years)			
		10^7	10^5	10^3	$\leqslant 10^1$
External:					
Shift of continents to higher latitudes	—increased pole-equator temperature gradient —cooling of high latitudes	I	NA	NA	NA
Increased land area by tectonic processes	—increased albedo, cooling —increased monsoon intensity	I	NA	NA	NA
Changing orientation of ocean basins	—changes in circulation controls and monsoons	I	NA	NA	NA
Mountain-building and increased continental elevation	—changes in circulation controls, monsoon intensity, and albedo	I	I	NA	NA
Closing (tectonic and eustatic) of gateways between ocean basins:					
zonal gateways	—increased meridional transport of warm water —high-latitude warming —increased precipitation on high-latitude continents	I	I	NA	NA
meridional gateways	—isolation and cooling of high-latitude oceans				
Volcanic activity	—increased dust and SO_2 in atmosphere —increased back-scattering of solar radiation, cooling	X	X	I	I
Asteroid impact	—increased dust in atmosphere —increased back-scattering of solar radiation, cooling	I	I	I	I
Changes in the earth's orbital parameters:					
Increased eccentricity of orbit	—greater seasonality for one hemisphere	X	I	I	NA
Increased obliquity of axis	—greater seasonality for both hemispheres	X	I	I	NA
Precession of solstices to coincide with peri/aphelion	—greater seasonality for hemisphere with summer solstice at perihelion	X	I	I	NA

Internal:					
Decreased salinity of oceans	—surface stratification of oceans, decrease in heat storage	I	X	D	I
Growth of ice sheets	—increased albedo, cooling —shifts in circulation features	D	D	I	I
Growth of sea ice	—increased albedo, cooling —shifts in circulation features	D	D	I	D
Eustatic sea level fall	—increased albedo, cooling	D	D	I	I
Cooling of sea surface temperatures	—decreased transfer of water vapor to atmosphere, starvation of ice sheets	D	D	D	I
Change in terrestrial biomass	—changes in CO_2 in ocean and atmosphere	X	X	D	I
Change in aerosols	—warming or cooling, depending on aerosol height and composition	X	X	D	D

Note: I = independent; characteristic that operates independently of, and to some extent controls, climatic variation, D = dependent; characteristic that is determined by climatic conditions. X = indeterminate; characteristic that is too variable to be reconstructed at the time scale. NA = not applicable; characteristic that is not controlled by climate, or that varies too slowly to be significant. For brevity, most controls and corresponding effects are chosen to represent a shift toward glacial conditions. During a shift toward interglacial conditions, the direction of the controls and effects is reversed.

responding and most variable of these subsystems, is most directly responsible for climate conditions, but atmospheric conditions are partially dependent on the temperature and circulation patterns of the ocean surfaces, and on the extent and height of major ice sheets (Saltzman 1985; Kutzbach and Guetter 1986). The oceans, and particularly the cryosphere, vary more slowly than the atmosphere and therefore have a damping effect upon it. In addition, moisture, temperature, and albedo characteristics of the continental surfaces influence the atmosphere, at least on the short-term scale.

Numerous positive and negative feedbacks arise among these four subsystems, and some create resonances in climatic variation (Saltzman 1985). Feedbacks may be nonlinear, resulting in a distinctive pattern of climatic variation without any external cause, termed autovariation (Kutzbach 1976). At time scales shorter than 1000 years, the regional climatic signal can be of high amplitude and noisy. This fact makes it difficult to extrapolate long-term trends in historical weather records (Bradley et al. 1987).

Climatic Variations During the Cenozoic: Geological Control

During the past 100 million years (since the late Cretaceous), the earth's climate has cooled (Lloyd 1984). This long-term trend is most likely related to the changing position of the continents (in particular the shift of continental masses into higher latitudes), and accompanying changes of sea level, topography, and albedo, although changes in atmospheric composition—particularly CO_2—may also be important (Barron and Washington 1984; Kerr 1984a).

Several key stages appear in the long-term cooling trend [extensive reviews are given by Lloyd (1984) and Crowley (1983)]. Between 30 and 40 million years ago, Antarctica separated from Australia and South America, thus thermally isolating Antarctica and initiating glaciation there. Continuous glaciation was established in Antarctica by 15 million years ago, initiating the current ice age (i.e., a period during which continental ice sheets are continuously present somewhere on the earth's surface). The Arctic Ocean became ice covered by 5 million years ago, and continental glaciation in the Northern Hemisphere began 2.5 million years ago (Shackleton et al. 1984), ushering in the ice-age climates of the Quaternary. During this late-Cenozoic ice age, many glacial/interglacial episodes have occurred.

Climatic Variations During the Quaternary: The Astronomical Theory of Climatic Variations

Superimposed on the long-term trend are intervals of ice sheet growth and decay, as well as regular fluctuations in the intensity of the tropical monsoons (Kutzbach 1981; Crowley et al. 1986; Prell and Kutzbach 1987). These range from stadial-interstadial or monsoon-intensity fluctuations lasting on the order of 20,000 years, to glacial-interglacial fluctuations every 100,000 years, to longer-term fluctuations several million years in duration (Miller et al. 1987). These variations result mainly from changes in the earth's orbital elements (Hays et al. 1976; Imbrie et al. 1984; Milankovitch 1941), with amplification from atmospheric chemistry (CO_2 content), land-sea contrast, ice cover, and resulting changes in albedo and sea level.

Three elements of the earth's orbit that vary systematically are the eccentricity of the earth's orbit around the sun, the obliquity of the earth's axis relative to the ecliptic, and the precession of the equinoxes relative to perihelion and aphelion (Figure 15-2). These regular variations in orbital geometry control the amount and seasonal distribution of solar radiation intercepted at the top of the earth's atmosphere. For example, 10,000 years ago, the tilt of the earth's axis was about 24.2 degrees (as opposed to the present 23.4 degrees), and perihelion (time when the earth is closest to the sun) occurred in July (as opposed to January, at present). Consequently, the northern mid-latitudes received about 8% more radiation in July than at present, and about 10% less in January (Figure 15-3). The general circulation of the atmosphere and oceans must respond to such changes in energy input. The general circulation is maintained by the imbalance in energy received between high and low latitudes, and the atmosphere and ocean transport energy and moisture to redistribute the energy surplus of the tropics. Regional climatic conditions, in turn, are controlled by atmospheric circulation, which determines storm tracks and the movement of different air masses, and by the direct effects of the insolation changes on heating and evapotranspiration.

The signal of the orbital variations is clearly seen in the Quaternary climatic record reconstructed from deep-sea sediment cores, in particular the oxygen isotope records from these cores (Broecker and van Donk 1970; Hays et al. 1976). The relative proportion of heavy (^{18}O) to light (^{16}O) oxygen isotopes in carbonate skeletons of plankton (which are preserved in deep sea sediments) varies in response to long-term climatic variations, with heavy oxygen being relatively concentrated in the oceans during intervals when ice sheets are larger and oceans cooler. Continuous records of oxygen-isotope variations over the last 500,000 to 1,000,000 years have been reconstructed from cores from all of the major ocean basins, including both high-latitude and low-latitude sites (Figure 15-3). These records present a remarkably coherent and detailed record of ice-age climatic variations, consisting of alternating cold and warm periods, with warm, relatively ice-free interglacial periods occurring every 100,000 years or so, and lasting about 10,000 years (Kukla and Matthews 1972).

Glacial periods are initiated by reduced winter-summer contrast (due to low obliquity and/or a precessional position such that perihelion coincides with an equinox). The resulting mild winters enhance the delivery of winter precipitation (snow) in the Northern Hemisphere high latitudes; the cool summers suppress snow melt in the same regions. The result is glacierization (Ruddiman and McIntyre 1981). Interglacial and interstadial periods are initiated by the opposite

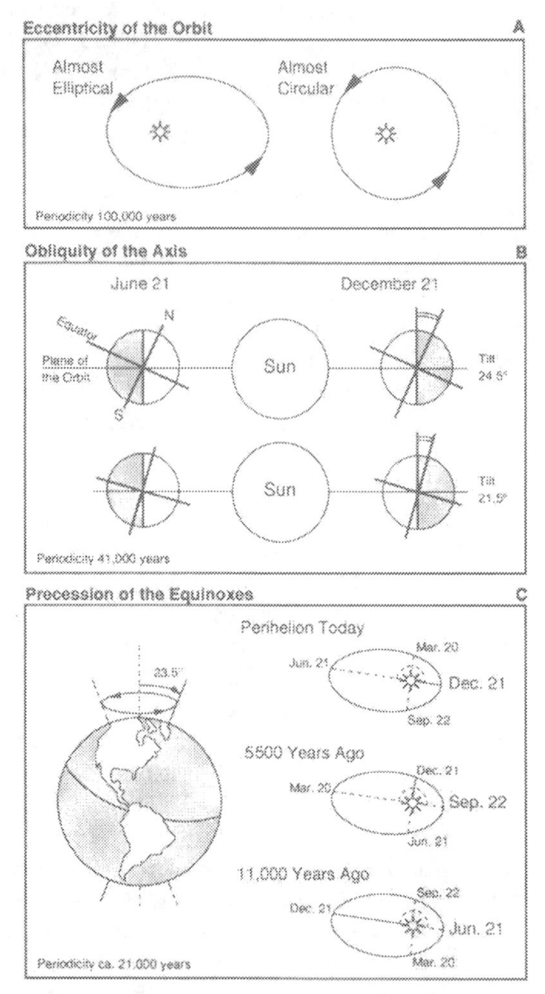

Figure 15-2. Optical variations and their effects on long-term climatic change. (A) Eccentricity of the orbit. When the earth's orbit is nearly circular, solar radiation receipt on earth (summed over both Northern and Southern Hemispheres) is nearly equal throughout the year, but as the orbit becomes more elliptical, intra-annual variation in radiation receipt increases. (B) Obliquity of the earth's axis. Increased tilt of the earth's axis increases each hemisphere's difference between summer and winter solar input, and therefore enhances seasonality. (C) Precession of the equinoxes. Precession, which is due to "wobble" of the earth's axis (change in the orientation of the axis of rotation), causes differences in the degree of seasonality between the Northern and Southern Hemispheres. When perihelion occurs near December 21, the Northern Hemisphere experiences reduced seasonality and the Southern Hemisphere experiences increased seasonality; when perihelion occurs nears June 21, the reverse is true. (Based on Imbrie and Imbrie 1979; Lowe and Walker 1984; figure by Montine Jordan.)

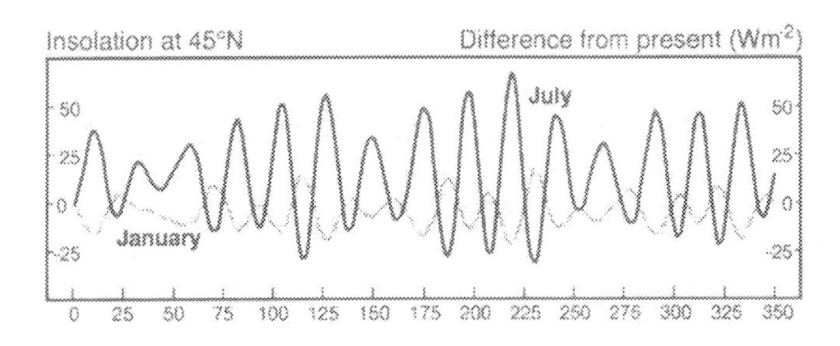

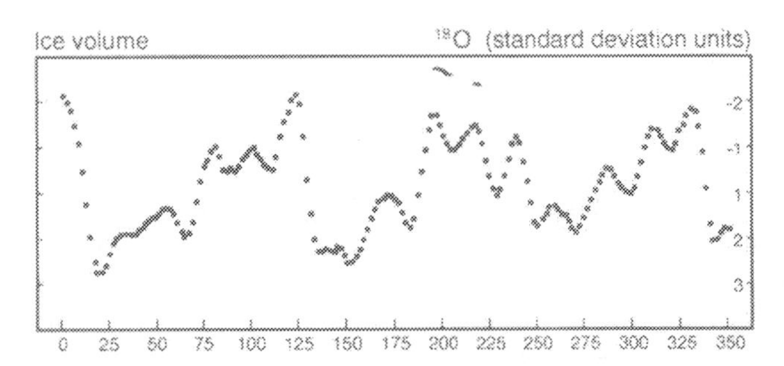

Figure 15-3. Insolation variation from the present-day level, at 45° N, and the composite oxygen isotope record from deep sea cores. ^{18}O is expressed as standard deviations. The data are plotted so that interglacial conditions (less ice, warmer) plot to the right of the graph, and glacial conditions to the left. (Sources: Berger 1978; Imbrie et al. 1984; figure by Montine Jordan.)

conditions: cold winters and warm summers. For example, the last deglaciation, 18,000 to 6000 years ago, coincided with a precessional position (between 13,000 and 7000 years ago) in which perihelion occurred during Northern Hemisphere summer. Northern Hemisphere summers were therefore generally warmer than those of today.

Climatic Variations During the Past 20,000 Years: Control by Surface Boundary Conditions and Insolation

During the past 20,000 years, climate changed from the extreme of a glacial interval to an interglacial one (see Fig. 2 in Jäger and Barry 1990). CLIMAP Project Members (1976, 1981) summarized the climate of the last glacial maximum, about 18,000 years ago. At this time, the ice sheets were close to their maximum extent, sea level was lower, and portions of the northern oceans were more than 10° C cooler than at present.

During this time span, the ice sheets assumed the role of an independent variable in the climate system, owing to their large thermal inertia. Similarly, the oceans, which on the Quaternary time scale warm and cool in response to the orbitally controlled insolation variations, changed rather slowly over this time period. Together with the insolation variations (Figure 15-3), the ice sheets and oceans have provided a set of changing boundary conditions during the past 20,000 years (Figure 15-4). Kutzbach and Guetter (1986) used these changing boundary conditions to carry out a set of "natural experiments" with a climate simulation model. They used the National Center for Atmospheric Research Community Climate Model (NCAR CCM) to simulate climate at 3000-year intervals from 18,000 years ago to the present. The NCAR CCM is a general circulation model and provides a simulated "snapshot" of regional and global climatic patterns consistent with a specified set of boundary conditions.

Geologic evidence supports two important results of these simulations: (1) the Laurentide ice sheet had a major impact on the atmospheric circulation of the Northern Hemisphere when it was large, and (2) the amplification of the seasonal cycle of insolation around 10,000 years ago resulted in strengthened monsoonal circulation and heating of the Northern Hemisphere continental interiors (COHMAP Members 1988). Both results are of some importance in understanding the impact of climatic variations on human activities during the past 10,000 years, because they contributed to regional climatic variations in Africa and Eurasia.

Between 9000 and 6000 years ago, lakes in the broad region extending across Africa north of the equator to the Indian subcontinent had water levels greater than at either present or earlier times (Street-Perrott and Harrison 1985). These

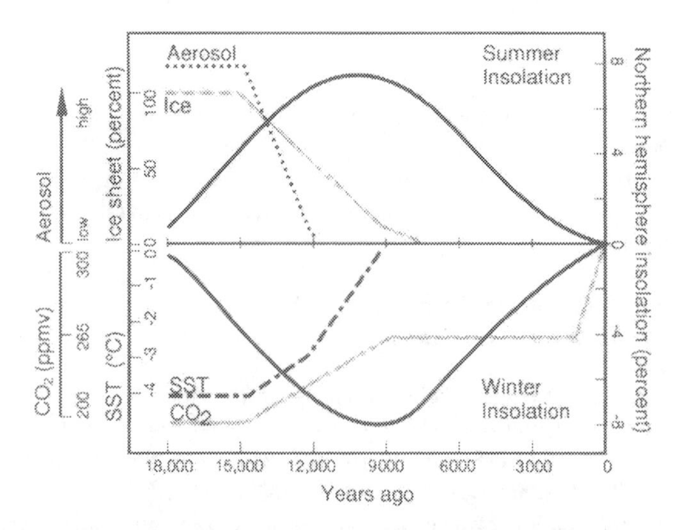

Figure 15-4. Changing boundary conditions during the past 18,000 years. Aerosol is a qualitative estimate of atmospheric aerosol and dust loading, ice is a reconstruction of global ice volume, SST is sea-surface temperature, and CO_2 is a qualitative estimate of concentration of carbon dioxide in the atmosphere. (Source: Kutzbach and Guetter 1986; figure by Montine Jordan.)

elevated levels were the result of greater effective moisture produced by the stronger monsoonal circulation that operated when Northern Hemisphere summer insolation was greater than at present. In North America (Bartlein and Webb 1985) and in Europe (Huntley and Prentice 1988), pollen evidence suggests that these regions were characterized by warmer summers 6000 years ago than at present, again in response to elevated summertime insolation.

A striking feature of the maps of past climates (for example, CLIMAP Project Members 1981; Wahl and Lawson 1970, Figs. 1 and 7; Street-Perrott and Harrison 1985, Figs. 11 to 14; Bartlein and Webb 1985, Fig. 9) is that the spatial anomaly patterns (the differences between the past and present) are not uniform. Rather, the climatic changes vary spatially, which is a necessary result of the circulation changes in both the atmosphere and ocean that occur in response to global-scale changes in boundary conditions (Figure 15-4). The spatial patterns are strong enough that even though the global mean temperature 18,000 years ago may have been 4° to 6° C lower than it is today (Gates 1976; Kutzbach and Guetter 1986), some regions of the ocean had sea-surface temperatures higher than those today (CLIMAP Project Members 1976, 1981). The existence of spatial pattern in climate anomalies is a general feature of climatic variations at all time scales, and requires caution in the use of global climatic changes to predict changes in regional or local climates.

Climatic Variations on the Decadal to Millennial Time Scale: Sunspots, Volcanoes, Carbon Dioxide, and Autovariations

Despite the greater abundance of paleoclimatic evidence and even instrumental data for the decadal to millennial time scale, the causes of the climatic variations that occur on this time scale are perhaps less well understood at present than are the causes of the longer-term variations. Several external factors, such as volcanic eruptions, may account for the observed record, but internal variations of the climate system itself (i.e., autovariation) may also be important on this time scale. Variations at this time scale may be controlled by volcanic eruptions (e.g., Bryson and Goodman 1980; Hammer et al. 1980; Porter 1981), by solar variability (in particular, sunspot frequency) (Harvey 1980; Stuiver 1980; Labitzke and van Loon 1988; van Loon and Labitzke 1988), or by combinations of these controls and variations in the concentration of CO_2 in the atmosphere (Hansen et al. 1981). Because a given climatic change might be attributable to several different causes, acting separately or interactively, short-term climatic variations may never be assigned to one specific cause. For instance, the extreme El Niño event of 1982 was accompanied by the eruption of the Mexican volcano El Chichón, and the resulting unusual weather patterns were induced by both "causes" (Schneider 1983).

During the past century, global, as well as hemispheric, mean temperatures have increased (Jones et al. 1986; see Fig. 3 in Jäger and Barry 1989); but again, the pattern of temperature change is not spatially uniform (for example, see Wahl and Lawson 1970, Figs. 1, 7). Much concern has focused on this recent warming because of its possible linkage to the increase in the concentration of CO_2 and other trace gases in the atmosphere, which has resulted from human activities (MacCracken and Luther 1985). The global change in CO_2 concentration will also result in spatial patterns of change, but the nature of these patterns is still uncertain. It is ironic that, as we are achieving a basic understanding of how the climate system works, the nature of the system itself is being fundamentally changed by human activity.

Vegetation: Scale, Control, and Response

Vegetational Change at Different Time and Space Scales

Vegetation is dynamic and varies on all time scales. It is responsive to several frequencies of climate variation as well as to variations in soils, grazing, fire frequency, disease, disturbance, and human impact. The factors that appear to influence vegetation patterns vary according to the spatial extent and resolution of the data used. A zoom lens view of the mapped patterns of modern vegetation at different spatial scales illustrates this point (Figure 15-5). The 10 million km^2 of eastern North America contains seven major vegetation regions (called plant formations), each about 1 million km^2, ranging from tundra and boreal forest in the north to pine-oak forest in the south and prairie in the west. These are aligned along temperature and moisture gradients. The 300,000 km^2 of Wisconsin and Michigan contains seven forest types, each about 40,000 km^2 and differing between the north and south because of climate, but grouped into specific areas by soil texture contrasts (fine-grained tills versus sandy outwash and glaciolacustrine deposits). Within the 1000 km^2 of Menominee County (in northeastern Wisconsin), the eight small forest types of about 100 km^2 grow on soils of different texture and also in regions with different degrees of human or natural disturbance. Climate variation is minimal within the county. The southwest of Menominee County is an outwash plain supporting Hill's oak and jack pine, the northern hardwoods grow on fine-textured soils, but the large area of aspen forest reflects disturbance by logging and fire. This series of vegetation maps shows how (1) spatial variations in macroclimate control the continental-scale patterns of plant formations, (2) soil variations control the location of forest types within a region, and (3) contrasts in soils and human disturbance control vegetation patterns within a county.

This scale dependence in the pattern and controls of the spatial variations in vegetation suggests that a scale dependence may also exist for the patterns and controls of temporal variations. Delcourt et al. (1983) published a preliminary time-space diagram for vegetation that resembles in form Figure 15-1 and the standard textbook scale-diagrams for atmospheric turbulence (e.g., Anthes et al.

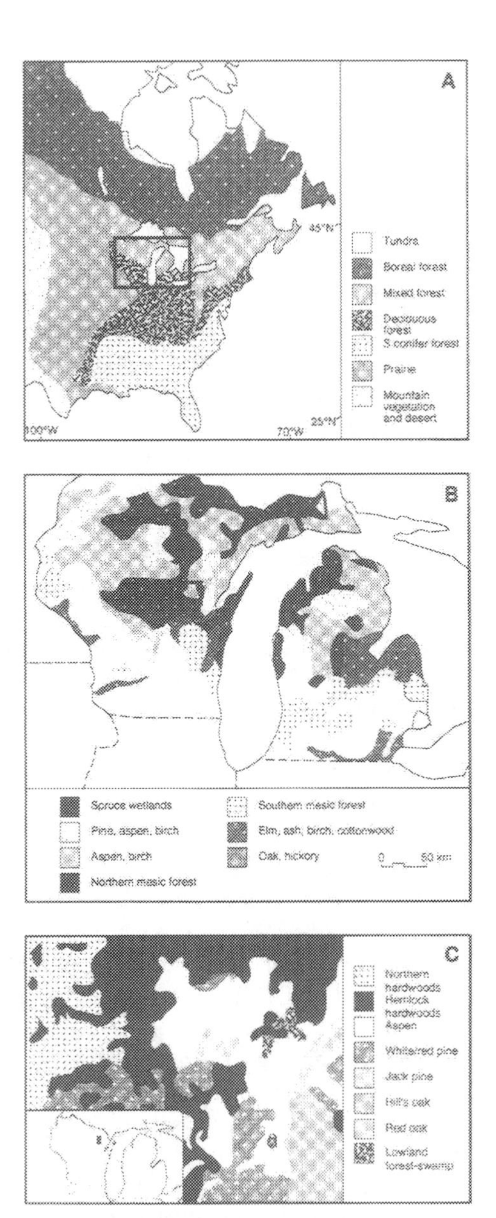

Figure 15-5. Modern vegetation patterns at decreasing spatial scales. (A) Generalized vegetation map for eastern North America. (B) Major forest types in Michigan and Wisconsin. Spruce wetlands, pine-aspen-birch, aspen-birch, and northern mesic forest are forest types within the larger mixed forest region of A. Southern mesic forest, elm-ash-birch-cottonwood, and oak-hickory are forest types within the larger deciduous forest region. (C) Major forest types within the Menominee Indian Reservation (MIR) of northeastern Wisconsin (see inset). Within the MIR, the northern mesic forest of B is subdivided into hemlock hardwoods and northern hardwoods. Part of the aspen forest is included in the aspen-birch type of B, and the rest is included within the pine-aspen-birch type of B. (Source: Bradshaw and Webb 1985; figure by Montine Jordan.)

1978) or oceanic circulation (Haury et al. 1978). Recent studies of vegetational dynamics, however, suggest that certain modifications are needed in this vegetation diagram. Webb (1987) has used temporal sequences of pollen maps for the past 18,000 years (Figure 15-6; see Huntley and Webb 1988a, for a general review of methods and studies concerned with vegetation history) to illustrate the changing associations among several major plant taxa. Each taxon has had a unique trajectory, and therefore major plant formations and ecosystems have appeared and disappeared over this time span. Webb et al. (1983b) showed similar types of variations for forest types in southern Quebec, and many pollen diagrams illustrate such variation in local to regional vegetation. These data provide evidence that individual plant formations and forest types have relatively short lifetimes.

Between 18,000 and 12,000 years ago, the patterns and spatial gradients of vegetation in eastern North America differed markedly from those of the present (Figure 15-6), with the modern patterns of spatial gradients appearing about 10,000 years ago. These modern patterns could emerge only after the ice sheet had shrunk below a critical size and associated changes in atmospheric circulation had occurred (Kutzbach and Guetter 1986; Webb et al. 1987). During the past 10,000 years, smaller but still significant changes have occurred, including a shift of the prairie/forest border first to the east and then to the west (Webb et al. 1983a). Comparisons of past climatic conditions reconstructed from pollen data and model simulations of past climates indicate that climate played a major role in producing the pattern and timing of vegetation changes during the past 18,000 years (Webb et al. 1987).

Because plant taxa change independently of each other in space and time, taxon associations at all spatial scales have changed through time (Figure 15-7). The different types of vegetational associations (from formations down to forest types in counties) all have specific sizes and lifetimes. So, too, do individual trees at 10 m^2 and 10 to 100 years (when younger than 10 years, trees are only seedlings or saplings), as well as plant communities at the spatial scale at which two or more trees interact and directly compete. For the latter, the frequency of disturbance-induced succession sets their average lifetime. This series from individual trees up to large vegetation regions has a steep slope in Figure 15-7, and the duration of the orbitally induced climate variations at periodicities of 10,000 to 100,000 years sets an upper temporal bound for these (ecological) units (Jacobson et al. 1987).

The sequences from individual trees up to plant formations illustrate the spatial and temporal characteristics of vegetational phenomena that result from a variety of ecological controls. A comparable sequence can be constructed for taxonomic units that result from evolutionary processes (e.g., changes in gene frequency, extinction, and speciation). Estimates for the average temporal and spatial limits of various taxonomic units from individuals up to the level of class (e.g., angiosperma or flowering plants) determine the slope on the log-log time-space plot

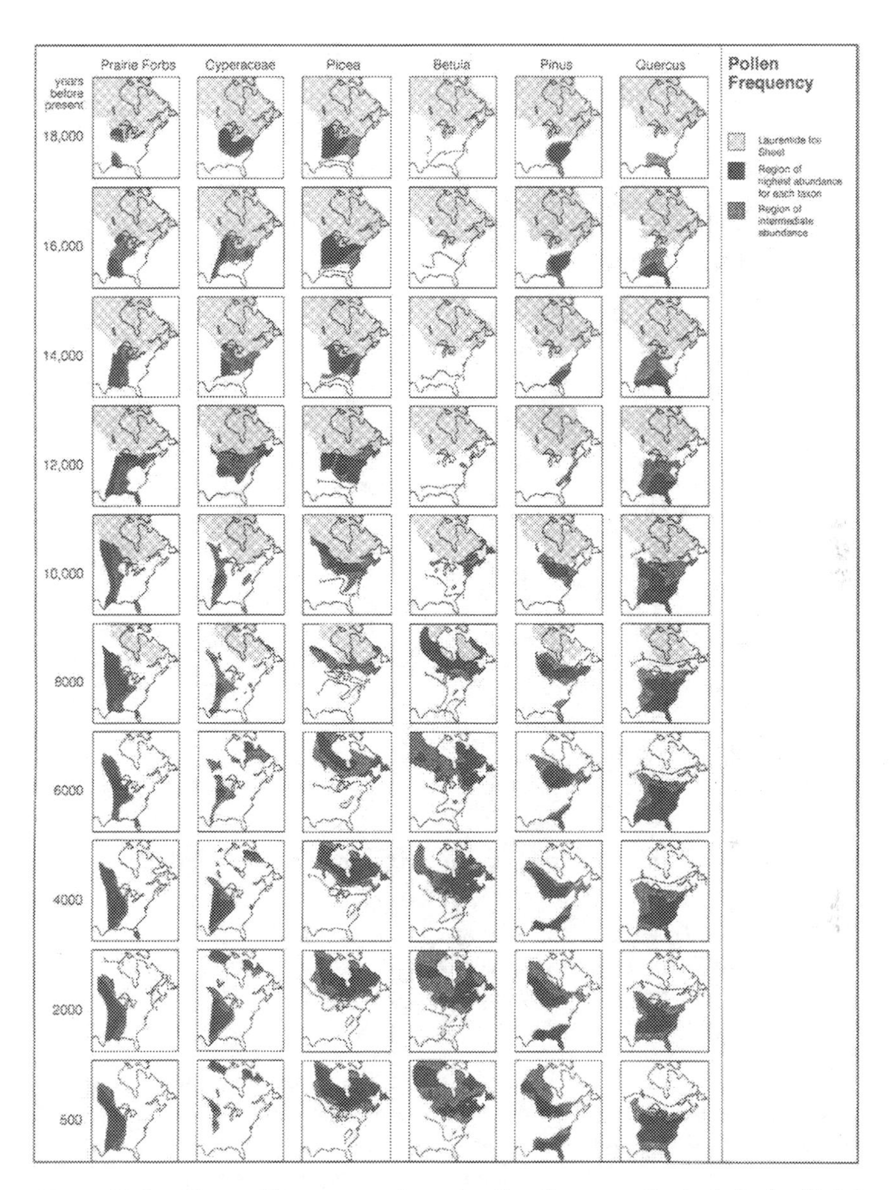

Figure 15-6. Maps with contours of equal pollen frequency for leafy herbs (forbs), sedge (Cyperaceae), spruce (*Picea*), birch (*Betula*), pine (*Pinus*), and oak (*Quercus*). For each taxon, the map series shows spatial changes at 2000-year intervals from 18,000 years ago to present (500 years ago). Light-shaded areas indicate the Laurentide ice sheet. Dark-stippled areas are the region of highest abundance for each taxon. (Source: Webb 1988; figure by Montine Jordan.)

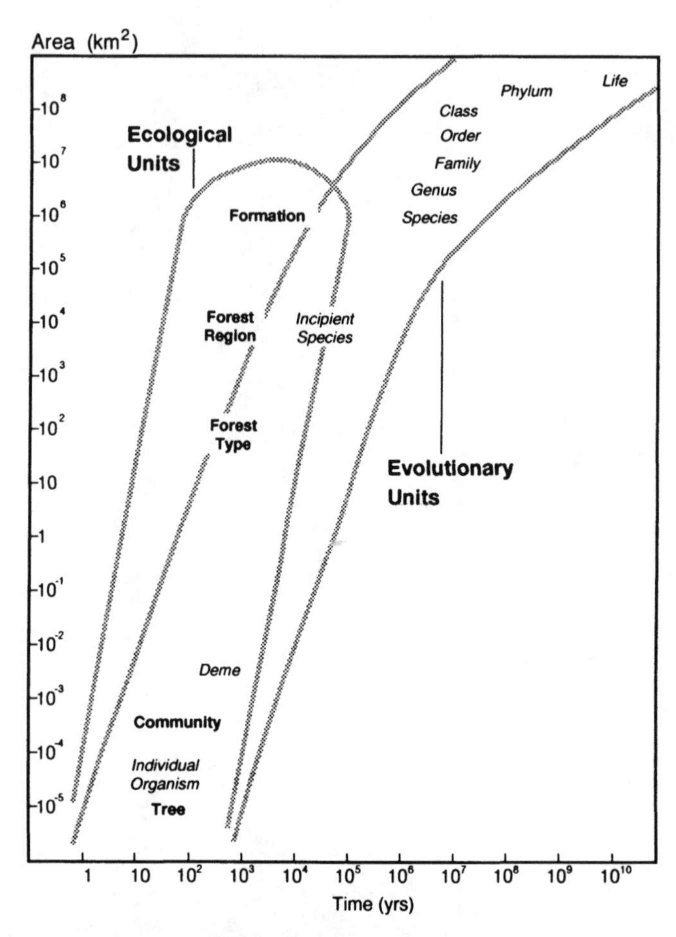

Figure 15-7. Spatial and temporal scales of biologic phenomena. The scales shown here apply to wind-pollinated trees and may not be appropriate for other organisms, such as microorganisms, for example. Evolutionary units are shown in italics, and ecological units are shown in boldface. (Figure by Montine Jordan.)

for these units (Figure 15-7). The key observations are the time and space scales for individual trees, species, and the class of flowering plants. The latter evolved 100 million years ago and today grow over much of the earth's land surface. Stanley (1985) and Tiffney (pers. comm.) estimate the average lifetime of species to be 1 to 10 million years, and the area of coverage for most species is within an order of magnitude of 100,000 km^2. These genetic and taxonomic units plot along a less steep slope on the log-log time-space plot than do the vegetational units (Figure 15-7). Taxonomic units tend to have a longer lifetime than vegetational units for a similar area of coverage. This difference in slope implies that species can survive the continuous appearance and disappearance of vegetational units (associations) at all spatial scales. Tectonically induced changes in climate, mountain barriers, and continent alignments are probably a key factor leading to extinctions and speciation, but meteorite impacts may be another key factor.

What the future holds, however, is difficult to tell. The predicted increased rate for CO_2-induced climatic change during the next century could significantly increase rates of species extinction. The ability of species to survive large past changes in climate depended in part on the rate of major climate change being slow enough that individual species could track the environmental conditions favorable for their growth (Webb 1986). A human-caused climate change in the next 100 years, equal in magnitude to past changes that previously took 1000 to 10,000 years, may be too fast for certain taxa to adjust and may result in their extinction (Davis 1989). The location, size, and composition of various scales of vegetational regions will also change. These changes will alter the competitive relationships among taxa and thus influence the future course of evolution.

Controls of Vegetational Change

Certain aspects of vegetational change are closely linked to climatic change, but other processes can also be important, such as long-term geological processes, short-term fluctuations in weather, and various environmental disturbances (such as floods, volcanic eruptions, windstorms, and fires) (Table 15-2). One reason for plotting vegetational and taxonomic units on the log-log space-time plots is to use the characteristic temporal and spatial scales of each unit to help in identifying the various physical and biological processes that act to control the unit. Mathematical theory for the general laws governing ecological and evolutionary change is not as well advanced as the theory used by meteorologists to understand and model weather and climatic variations. Empirical studies are therefore valuable in helping to show the relative importance of various processes in determining key spatial and temporal variations in the biosphere (Delcourt et al. 1983; Webb et al. 1987). A full understanding of cause and effect between biospheric and environmental phenomena is complicated because feedbacks and linkages in biological systems can blur the distinction between control and response and because processes that are controls of biological change at one time scale may be dependent on biological conditions at another time scale. For example, fire frequency is partly dependent (at short time scales and small spatial scales) on vegetation successional status for fuel build-up, but it is an independent control of vegetation patterns at intermediate time scales (Delcourt et al. 1983). Linkages and feedbacks in the vegetation cause soil conditions, fire regime, disease, micrometeorological conditions, and even human activities to vary as the biological system varies over time scales of < 1 to 1000 years. Our assignment of controls (Table 15-2) is therefore tentative, and we will welcome future research that improves on our summary.

Table 15-2. Controls of biological change, with emphasis on vegetation

Control	Effect	Status of control at time scales of (years) 10^7	10^5	10^3	$\leq 10^1$
Plate tectonics					
continental drift	—divides (joins) continents	I	NA	NA	NA
	—separates (joins) floras and faunas				
	—moves continents from climates favorable to life to climates largely devoid of life				
mountain building	—creates barriers to migration, interaction	I	I	NA	NA
	—creates new habitats (rainshadows, deserts)				
	—controls atmospheric circulation features				
Catastrophic events*	—increased extinction rates or mass extinctions	I	NA	NA	NA
Climate	—biomes and ecotones appear and disappear	X	I	I	NA
orbital variations in seasonal insolation	—taxa migrate				
	—marine and terrestrial taxa change in range, area, and abundance				
neoglacial events (e.g., Younger Dryas, Little Ice Age)	—alpine glaciers advance and retreat	X	X	I	NA
	—forest types change in composition				
	—ecotones move short distances				
	—alpine tree line fluctuates				
droughts (Sahel, Dust Bowl, ENSO)	—composition and dominance changes in biomes, communities	X	X	X	I
	—local extinctions				
	—annual growth variations in trees				
seasonai cycle	—birds and other animals migrate	NA	NA	NA	I
	—phenological events (leafing out, flowering, etc.)				
	—animals hibernate, perennials vernalize				
	—leaf loss in the face of colder temperatures or dry seasons				
	—plankton blooms (abundance variations)				
	—cycle for annual plants or insects				
	—waxing and waning of secondary tree growth				

diurnal cycle	—photosynthesis turned on and off —diurnal behavior in animals	NA	NA	NA	I
Soil soil development (profile differentiation)	—changes in soil fertility, tilth that alter biomass and composition of vegetation	X	I	I	NA
soil degradation (e.g., salinization, laterization)	—can limit diversity of plant taxa	X	I	I	I
Disturbance regimes disease	—selective death of most individuals of one or a few taxa throughout range of taxa	X	X	I	I
	—cyclic impact on abundance of individuals within a taxon	X	X	X	I
fire	—selection for or against certain taxa —initiates secondary succession	X	X	I	D
blowdowns	—initiate secondary succession	X	X	X	I
glacial fluctuations	—bulldoze vegetation or make land available for vegetation	X	I	I	NA
human activities	—alteration or elimination of habitats	X	D	I	I
	—selective removal of taxa (trees, mammals, birds, smallpox)	X	D	I	I
	—extinctions on continents	X	D	I	NA
	—extinctions on islands	X	X	X	I
	—spread of exotic plants and animals	X	X	I	I
	—domestication	X	X	I	I

Note: I = independent; characteristic that operates independently of, and to some extent controls, biologic variation. D = dependent; characteristic that is determined by biologic conditions. X = indeterminate; characteristic that is too variable to be reconstructed at the time scale. NA = not applicable; characteristic that is not controlled by biologic conditions, or that varies too slowly to be significant. *Catastrophic events (e.g., episodes of increased dustiness due to increased frequency of meteor impacts or explosive volcanism) are potentially regular features on the 10^7 to 10^8 time scale, and are unlikely for most periods shorter than 10^7 years. When they occur, their impact is greatest at time scales of 10^{-1} to 10^5 years.

Landscape Change: Scale, Control, and Response

Dynamics and Inheritance in the Physical Landscape

In contrast to elements of the climatic and vegetational systems, elements (individual landforms, soil bodies) within the physical landscape have long life spans. A climatic regime is made up of numerous transient weather events. The life span of an individual tree is 10 to 1,000 years, generally shorter than the life span of the community of which it is a part. A cliff, glacial moraine, or river floodplain, on the other hand, may outlive the geomorphic regime under which it formed. An episode of landscape adjustment involving erosion of pre-existing landforms and construction of new ones usually does not rework the entire older landscape; remnants of it are preserved. The result is that most landscapes contain landform assemblages dating from several generations of landscape development.

Inherited elements in a landscape may have been formed by processes similar to those operating today, but some landscapes are polygenetic. In polygenetic landscapes, elements formed by a particular geomorphic process, such as wind erosion, may persist long after that process has ceased to operate on the landscape and another process has replaced it as the dominant process. The best-known examples of polygenetic landscapes are the mid-latitude regions that were glaciated during the most recent ice age, but now are completely free of glacial and even periglacial activity. In the midwestern and northeastern United States, and in central and northern Europe, glacial landforms are being reshaped by weathering, soil development, slope erosion, mass movement, and fluvial activity, but present-day landscape characteristics such as slope, regolith depth (i.e., depth of loose surficial material over bedrock), and drainage patterns are mainly inherited features from glaciation.

Polygenetic landscapes are widespread, and they certainly are not limited to mid-latitude glaciated regions. In deserts, relict elements from past humid phases, such as ancient river systems, old archaeological sites, fossil karst phenomena, strandlines of expanded lakes, and deep weathering profiles, have long been recognized (Goudie 1985, 1986). Relicts of past arid phases, such as stabilized fossil dune fields, are also present on desert margins in Africa (Grove and Warren 1968; Thomas and Goudie 1984) and Asia (Allchin et al. 1978). Modern coastlines are the result of (1) long-term, climatically controlled processes (post-glacial eustatic and glacio-eustatic sea-level rise), (2) types and rates of coastal erosional and depositional processes operating under present climate, and (3) inherited Quaternary landforms and deposits (such as shore platforms of former interglacials or beach ridges) (Hopley 1985). Present-day coastal features, including beaches, estuaries, lagoons, barrier islands, coastal dune sheets, and near-shore zones, all have developed since the last glacial-eustatic sea level minimum at about 18,000 years ago, when the ice sheets were at their maximum extent. Many have developed only within the past 5000 years. Landscapes similar to today's may have been present during the previous interglacial at about 125,000 years ago, when sea levels were at a maximum, but they were not necessarily at the same locations (Bloom 1983). Even in the humid tropics, long thought to be the most stable morphoclimatic zone, the physical landscape is polygenetic and many features are relict. For example, fluvial dissection occurred in the central Amazon Basin (Tricart 1985) and in Sierra Leone (Thomas and Thorp 1985) under dry conditions at about 20,000 to 12,500 years ago.

Past changes in environmental controls, therefore, have resulted not only in changes in rates of geomorphic activity, but in shifts in dominant geomorphic process, for example, from glacial to fluvial processes or from eolian to fluvial processes. King (1980) pointed out that a shift in the dominant geomorphic process results in a landscape that is necessarily out of equilibrium with the current processes. The result is a period of intense geomorphic activity, in which reshaping of hillslopes and reworking of soil profiles occurs, in addition to major changes in the sediment storage zones such as valley bottoms, etc. Even if climatic change does not result in a shift in the dominant process, it will change the rates of processes. The time of change in rates is a time of disequilibrium and increased geomorphic activity (Figure 15-8).

Theories of Geomorphic Change

Long-term geomorphic change has no unifying theory equivalent to the astronomical theory for long-term climatic changes. In the late nineteenth and early twentieth centuries, W. M. Davis' (1899) theory of the geographical cycle served as a unifying theory. It was a theory of gradualism, with a cycle of landscape evolution being initiated by a regional fall in relative base level, followed by a long period of unidirectional landscape evolution and adjustment to the base level. By the 1960s, new attention to the physics of geomorphic processes and a growing body of empirical data on geomorphic processes (in short-term operation) led to abandonment of Davisian theory and a focus of attention on time-independent processes. Research on Quaternary geomorphic change has focused on regional studies, with climatic and/or base-level changes as the inferred causes. In the nineteenth century, endogenic forces (earth's internal forces, including tectonics and volcanism) were considered the most important external control of geomorphic processes, but today climate and adjustments internal to the geomorphic system are recognized as equally important. Most geomorphologists working at the Quaternary and historical time scales operate from a loosely defined paradigm consisting of concepts of thresholds, feedbacks, complex response, and episodic activity (Chorley et al. 1984, p. 1–42). Theories that specify how climate change affects geomorphic systems are not well developed or widely accepted (see Schumm and Parker 1973; Cooke and Reeves 1976, p. 1–23). In process studies, landscape units commonly are analyzed in a framework of a system

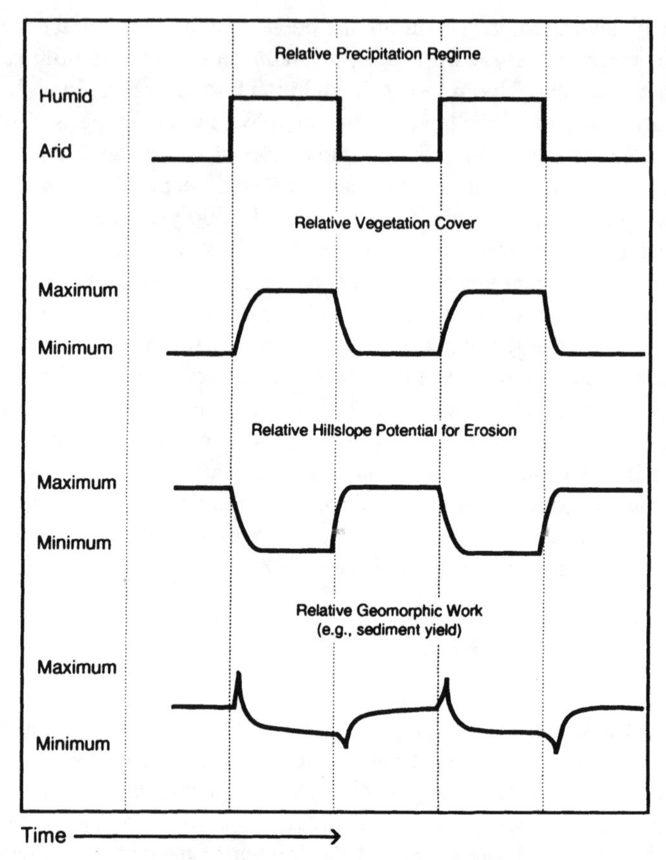

Figure 15-8. Suggested biogeomorphic responses to a series of climatic intervals with abrupt transitions. Curves apply to fluvial systems in mid-latitude climates with 25–150 cm mean annual precipitation. Each humid or arid climate period is several hundred to several thousand years long. Geomorphic work, integrating changes in the driving force of precipitation and the resisting force of vegetation cover, is concentrated at the times of change rather than throughout humid or arid periods. (Source: Knox 1972; figure by Montine Jordan.)

budget, which divides the landscape into areas of sediment source, sediment transport, and sediment storage, linked by geomorphic processes.

The concepts of long-term geomorphic change have many similarities with those of long-term behavior of the climate system. Both systems are viewed as complex groups of linked subsystems, with multiple feedback links, both positive and negative. These feedback links, as in climatology, make it difficult to separate controls from responses. In river systems, for example, channel gradient is both a control (of sediment transport rates) and a response (to net aggradation or degradation). As in the climate system, responses are frequently nonlinear.

Threshold response occurs when a continuous or gradually changing control process triggers a sudden change in geomorphic form or rate of process. Thresholds cause geomorphic activity to be episodic rather than steady through time, certainly on time scales up to 1000 years. Threshold response also makes identification of a particular controlling factor difficult, because control and response do not necessarily occur closely in time.

Schumm (1973, 1980, 1981) called attention to the existence of intrinsic geomorphic thresholds—sudden changes in system behavior due to gradual morphological changes that are inherent in the system. For example, in arid watersheds, sediment delivered to the channel system tends to be stored there over the short term (10 to 1000 years), leading to a steeper channel gradient and ultimately to gullying. The shift from an aggrading to an incising channel is typically abrupt in both space and time. This kind of change, with no obvious external trigger, can significantly affect the morphology of the system. Adjacent watersheds with similar geology, climate, and base level may have quite different morphologies at a certain time, because they stand in different positions in relation to their intrinsic geomorphic thresholds. Schumm and his students also developed the concept of complex response, and they documented experimental and empirical cases of it. Complex response refers to a series of oscillations in the direction of geomorphic change, in response to a single perturbation of the system. The best empirical example is that of Douglas Creek in western Colorado, which developed four cut-and-fill terraces in 80 years, in response to historical grazing disturbance (Womack and Schumm 1977). Davis (1976) documented a similar response to agricultural conversion of a midwestern U.S. watershed. Complex response presents obvious problems for the identification of causes of geomorphic change and for prediction of geomorphic change.

Despite the complexity of geomorphic change suggested by this modern paradigm of geomorphic system behavior, consistent patterns of geomorphic behavior are evident for time scales ranging from 1000 to 10,000 years. The effects of threshold response and complex response apparently are concentrated at time scales of less than 1000 years. In environments as diverse as Texas (Baker and Penteado-Orellana 1977), Wisconsin (McDowell 1983), and Poland (Kozarski 1983), episodes of river terracing attributed to climatic change have occurred at intervals of 2000 to 10,000 years during the past 20,000 years. These episodes of landscape development lasting 1000 to 5000 years appear to be largely controlled by system response and adjustment to changes in climatic, eustatic, or tectonic conditions.

Physical Landscape Change at Different Time and Space Scales

Like climatic processes, geomorphic and pedologic processes operate over a large range of temporal and spatial scales (Figure 15-9). The temporal and spatial scales of landscape features are related; large features change and persist over much longer time scales than do small features (Baker 1986).

At the largest spatial scale are tectonic units, ranging from physiographic provinces to individual fault blocks. The dominant controls of landscape development at this scale are tectonic processes and denudation. This is the classical cyclic time scale of W. M. Davis. Development of plate tectonics theory has provided insight into the temporal variability of these processes. In tectonically stable regions, large segments of paleolandscapes may persist for 100 million years or longer, such as the Gondwana and post-Gondwana planation surfaces

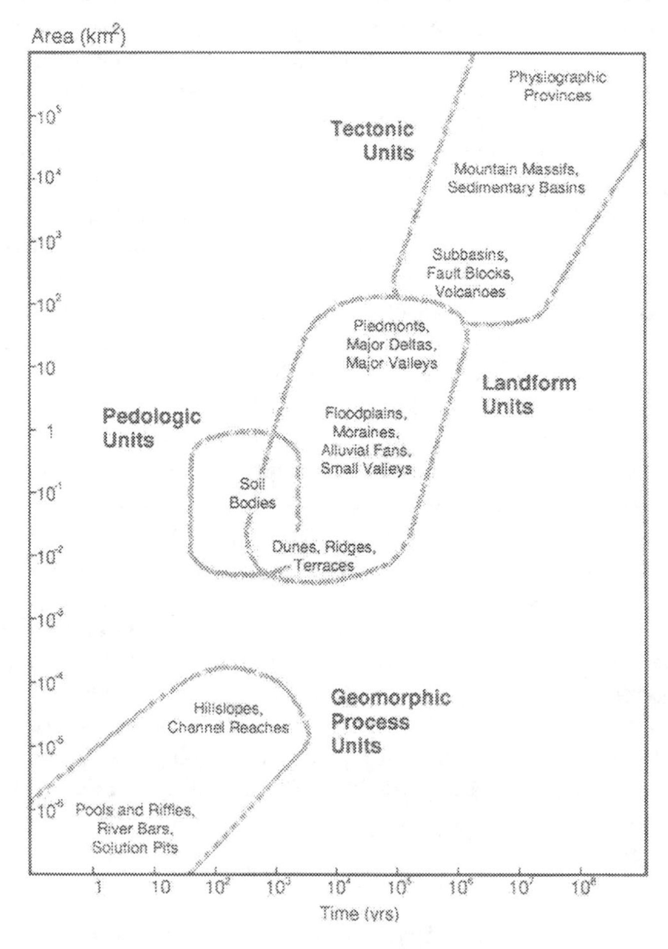

Figure 15-9. Spatial and temporal scales of landscape features. Bubbles enclose the typical time-space scales of each of the four types of landscape features. Some specific examples for each type are shown within the bubbles. (Based partly on Baker 1986; figure by Montine Jordan.)

of the Southern Hemisphere (King 1983). In contrast, in tectonically active areas such as the Pacific Rim, some mountain ranges may be less than 1 million years old. In both cases, however, these features show significant changes only over 1000 years or longer, much longer than the scale of human social and political adjustment processes.

Features of intermediate temporal and spatial scale are termed erosional-depositional units by Baker (1986). They include landforms deposited or eroded by the fluid geomorphic forces—running water, wind, waves, and glaciers—and by mass-movement processes. Causal explanations of intermediate-scale landscape features are complex, because three types of controlling factors (climate, tectonic control of base levels, and human impacts on hydrology and vegetation cover) can influence geomorphic and pedologic processes at this scale. These controls may operate individually or interactively. Intermediate-scale landscape features are most significant to human occupation of the environment, but they vary more slowly than do social and political adjustment processes.

Development of soil bodies is controlled primarily by climatic conditions (and the resultant vegetation cover) and by geomorphic processes that supply parent material and determine relief. Soils show significant variation at equivalent or slightly shorter time scales than the landforms on which they develop. Soil A horizons develop to a steady-state level of maturity in about 100 years, while argillic B horizons need at least 1000 years to develop and do not reach a steady state for 100,000 years (Birkeland 1984).

Small geomorphic process units are not significant to human societies, but they may be significant to individual farms, families, and villages. Their variation at time scales of 1 to 1000 years is controlled largely by climatic conditions and internal dynamics of, for example, the river system or weathering mantle.

Controls of Physical Landscape Change

Schumm and Lichty (1963) first pointed out that geomorphic characteristics that are dependent variables at one time scale may be independent variables at shorter time scales. Table 15-3 illustrates such functional changes in system variables, with an emphasis on denudation and fluvial geomorphic systems. The three decreasing spatial scales of fluvial geomorphic systems shown—physiographic province, drainage basin, and river channel—are examples of tectonic units, landform units, and geomorphic process units, respectively (from Figure 15-9). Similar tables could be constructed for glacial, eolian, coastal, and mass-movement processes, and for pedologic processes. At time scales most relevant to human activities, 10 to 1000 years, climate and the shortest-scale manifestations of tectonic processes are the most important variables controlling landscape development. For example, movement on faults and earthquakes can effect minor reshaping of the landscape on time scales of 1000 years, in regions that are tectonically active. More important, most regional landscapes are affected by

Table 15-3. Controls of geomorphic processes

Control	Effect	Status of control at time scales of (years)			
		10^7	10^5	10^3	$\leq 10^1$
Physiographic province:					
Megatectonic cycle	—rifting, sea floor spreading, subduction, continental collision, orogeny	I	I	NA	NA
Tectono-eustasy	—base-level change	I	I	NA	NA
Neotectonic pulses	—uplift, subsidence, faulting	X	X	I	I
Earthquakes	—uplift, subsidence, faulting	X	X	X	I
	—mass movements				
	—drainage changes				
Climatic change	—glaciation	X	I	I	I
	—hydrologic cycle changes				
Change in vegetation cover	—changes in rates of sediment yield and/or runoff	X	I	I	I
Glacio-isostasy	—base-level change	X	I	I	NA
Glacio-eustasy	—base-level change	X	X	I	NA
Drainage basin:					
Geology (lithology, structure)	—controls drainage pattern, slope morphology, sediment type	I	I	NA	NA
Climate	—influences type and rate of weathering, hydrologic regime, vegetation cover	X	I	I	I
Relief	—controls slope morphology, erosion potential	D	D	I	I
Vegetation cover	—changes in erosion rates	D	D	D	I
Human impacts	—changes in land cover, hydrologic system, erosion rates	X	D	I	I
Drainage network and morphology	—influences delivery of water and sediment	D	D	I	I
Hillslope morphology	—influences erosion rates, water delivery to channel, mass-movement rates	D	D	I	I

River channel:					
Geology	—influences valley slope, sediment type	I	I	NA	NA
Climate	—influences runoff into channel, vegetation	X	I	I	I
Vegetation	—influences bank stability, roughness	D	D	I	I
Human impacts	—dams, channel modifications, water diversions	X	D	I	I
Valley slope	—influences channel slope, channel pattern	D	D	D	I
Channel morphology (slope, sinuosity, shape, etc.)	—influences velocity, flooding regime, sediment transport	X	X	D	I

Note: I = independent; characteristic that operates independently of, and to some extent controls, geomorphic variation. D = dependent; characteristic that is determined by geomorphic conditions. X = indeterminate; characteristic that is too variable to be reconstructed at the time scale. NA = not applicable; characteristic that is not controlled by geomorphology, or that varies too slowly to be significant.

climatic change at this time scale. At the spatial scale of a watershed, geology (lithology and structure) is the dominant variable controlling the physical landscape at 100,000 to 10 million years, but it is virtually unvarying at 1000 years or less. Climate and its effects on hydrology, weathering, vegetation, and soil development are the most important landscape controls at 100 to 1000 years. River channels, the smallest and most dynamic landform scale included in Table 15-3, are truly human scaled and are significant features to social units such as farms, villages, and cities. For river channels, climate is the most important controlling variable at 1000 years.

Glacial-interglacial cycles probably are the most effective type of climatic variation in reshaping landscapes in the high- and mid-latitudes, dry subtropics, and perhaps even the humid tropics (where geomorphic history is poorly known at present). Mid-latitude geomorphic conditions during full- and late-glacial times were characterized by glaciation, and outside the glacial limit by intensified fluvial activity (Schumm and Brakenridge 1987), periglacial processes (Péwé 1983), eolian activity (Osterkamp et al. 1987), and pluvial lakes (Smith and Street-Perrott 1983)—all absent or greatly reduced today. Eustatic sea level was about 100 to 120 m lower than at present, exposing extensive areas of continental shelves. The glacial-time landscapes of the high- and mid-latitudes were in general geomorphically active, with intensified sediment transfer processes and diminished soil development, compared to Holocene landscapes, characterized by more geomorphic stability and intensified soil development.

In many regional geomorphic systems, such as the glaciated areas, areas of periglacial regime, and pluvial lakes regions, the major shift in geomorphic regimes occurred rapidly at the end of full-glacial climatic conditions at about 12,000 to 10,000 years ago. In other geomorphic systems, the shift from full-glacial conditions to post-glacial conditions was prolonged and complex. Coastlines, for example, retreated continuously as eustatic sea level rose between 18,000 and 6,000 years ago, and coastal landform assemblages were not established in their modern positions until 5000 years ago or later (Bloom 1983; McDowell 1987).

The full-glacial to post-glacial shift in geomorphic regimes was generally greater than changes during the last 10,000 years, but changes during the latter period have been significant, particularly in fluvial and eolian systems. Sand dune fields in the central and western United States have been reactivated intermittently during dry periods of the Holocene (Osterkamp et al. 1987; Mehringer 1986). In many river systems, the last 10,000 years have been characterized by alternating periods of fluvial stability and instability, of 500 to 2000 years' duration, caused primarily by climatic change (Knox 1983).

Human Impacts in the Context of Long-Term Environmental Change

Natural environmental changes provide a yardstick for judging the magnitude of human impacts on the environment. For example, the impact of European settlement on the North American pollen record between 1700 and today has been large in local areas (Van Zant et al. 1979; McAndrews 1988). Conversion of some areas of the eastern deciduous forest to an artificial savanna of cropland, pastureland, and woodlots was an event comparable in magnitude to climatic-induced vegetation changes over the last 6000 years (Bernabo and Webb 1977). The human impact on the regional or national scale, however, appears less significant when compared to vegetation changes documented in the pollen records for the last 4000 years.

Desertification of the Sahel is one of the most significant human environmental impacts of this century. Yet even this event is minor compared to lake-level fluctuations, river system changes, and vegetation changes that occurred during the last 10,000 years. Deforestation of the Amazon Basin, in contrast, is similar in scale to natural vegetational variations during the past 10,000 years (Dickinson 1987).

Pollen records of human impact in Europe date back to paleolithic times—about 300,000 years ago (West 1956). Impact at a regional scale dates from Mesolithic times (10,000 to 5300 years ago) (Edwards and Ralston 1984; Simmons and Innes 1988), and many paleoecologists have interpreted the widespread synchronous (± 500 years) decline in elm pollen abundance at 5000 (radiocarbon) years ago as resulting from human activity (see discussion in Huntley and Birks 1983). Although disease rather than selective forest clearance might be the factor that killed the elms, human activities may have helped spread the disease. Pollen evidence of Neolithic agricultural activity is widespread in Europe after 5000 years ago. Human impact in Europe before A.D. 1700 contrasts markedly with that in North America, where only nine out of over 300 radiocarbon-dated sites with pollen data record any evidence of human activity (McAndrews 1988).

The major controls of long-term change (1000 to 10,000 years) in geomorphic systems are climatic change, sea level, and tectonic changes (Table 15-3). One important question is whether human impacts are significant compared to natural changes, at these time scales. Although good evidence exists for significant human impact of geomorphic systems during recent historic times, evidence of significant prehistoric human impact is less clear.

One of the best-studied examples of probable prehistoric human impact on geomorphic systems is the case of accelerated hillslope erosion and resulting valley alluviation in Greece and elsewhere in the Mediterranean region, in Classical (3500 to 1800 years ago) and post-Classical times. These alluvial episodes have been attributed to deforestation and intensive land use under Greek and Roman cultures (Butzer 1980), climatic change (Vita-Finzi 1969), and a combination of the two factors (Bintliff 1982). Evidence tends to support the hypothesis of human cause for erosion in the first and second millenia B.C. (2000 to 4000 years ago), but this evidence is fragmentary (Davidson 1980).

In central Europe, alluviation during and after the late Bronze Age has also been attributed to human impact through deforestation, expansion of agriculture, and grazing. Some significant alluvial episodes predate major human impact on vegetation cover and soils. Butzer (1980) argued, however, that after 3500 years ago, fluvial adjustments took place in the context of human disturbance of the landscape, and that the role of this disturbance in causing fluvial adjustments is not understood and is probably underestimated. Bell (1982) examined evidence from archaeological sites of erosion, alluviation, and colluviation in Britain and northwestern Europe. He concluded that times of geomorphic activity were asynchronous, suggesting that local factors, particularly prehistoric human land use, were the primary controls rather than climatic change.

Prehistoric arroyo cutting in the valleys of the southwestern United States has also been cited as an example of early human impact on fluvial systems. Bryan (1954) suggested that during development of the Pueblo culture 1500 to 600 years ago (A.C. 500–1350), the inhabitants practiced floodwater agriculture on unincised valley floors. He hypothesized that a combination of human impacts and drought decreased the watershed vegetation cover, and that channel incision occurred at the end of the drought. Channel incision lowered water tables on the canyon floor, resulting in the decline of agriculture and abandonment of the pueblos. But more recent studies (Love 1979) have shown that arroyo incision did not follow pueblo settlement, and that arroyo filling actually occurred toward the end of the pueblo occupation.

These examples suggest that the impact of prehistoric human activity on geomorphic systems, although real in some instances, was minor and highly localized. Human impacts since the Industrial Revolution, on the other hand, have been major and are well documented (Trimble 1974; Knox 1977; Trimble and Lund 1982; Graf 1985; Goudie 1986). In some local areas, these recent human impacts have resulted in alluviation or erosion, concentrated within a few decades, that is equal in magnitude to natural changes on the 1000-year time scale.

Conclusions

An understanding of natural change in the human environment is essential to an understanding of human impacts. Because of the inherent dynamics of natural processes, it is difficult to identify with confidence many of the changes that may be due to human forces. Some transformations—deforestation, for instance—are easily identified as anthropogenic, but others—human impact on global climate, for instance—remain equivocal so long as natural processes of change and internal system dynamics are poorly understood. The study of natural changes, moreover, provides a means for assessing the relative significance of anthropogenic perturbations, as compared with natural variations. Our knowledge of the response of environmental systems to natural stimuli may also assist in predicting their response to human forces that are comparable in type, rate, and magnitude.

Climatic change is one of the most important controls of long-term change in the natural environment, and at some time scales it is the dominant control. An understanding of climatic variations is therefore basic to understanding and assessing environmental change. Our knowledge of climatic variations has advanced significantly in the past 30 years, but the picture remains complex. Numerous controls of long-term climatic change have been identified. These controls depend on the length and resolution of the time scale. Astronomical factors, which dominate at time scales of 1000 to 100,000 years, are predictable, but climatic change is not yet predictable at shorter time scales.

Controls of temporal change in biological systems and in the physical landscape are also scale dependent. Climatic variations influence the variations in vegetation patterns and associations at periods greater than 1000 years; at shorter time scales, other factors, including natural biological processes and human activities, become important. In some areas, for example Europe and parts of Asia, human activities have influenced the vegetation for thousands of years. Landform development is controlled by climatic and geologic factors at temporal scales greater than 1000 years, and by internal processes, short-term climatic variation, and human actions at shorter time scales. The physical landscape has a longer "memory" than the climate system and the biosphere, and therefore major human impacts on the physical landscape may be more serious and long-lived than major human impacts on the atmosphere and biosphere. But many uncertainties still exist concerning the impacts of and recovery from current human actions, such as those affecting the vegetation in tropical regions and the atmospheric concentrations of greenhouse gases like carbon dioxide and methane.

Acknowledgments

This work was partially supported by grants NSF/ATM-8713981 and DOE FG02-85ER60304 (Webb) and NSF/ATM-8713980 (Bartlein). We thank K. Anderson, K. Wolter, K. Edwards, and I. C. Prentice for their input.

References

Allchin, B., A. Goudie, and K. Hedge. 1978. *The Prehistory and Palaeogeography of the Great Indian Desert*. Academic Press, London.

Anthes, R. A., H. A. Panofsky, J. J. Cahir, and A. Rango. 1978. *The Atmosphere*. Second edition. Merrill, Columbus, OH.

Baker, V. R. 1986. "Introduction: Regional Landforms Analysis." In N. M. Short and R. W. Blair, Jr. (eds.), *Geomorphology From Space, A Global Overview of Regional Landforms*. National Aeronautics and Space Administration, SP-486. U.S. Government Printing Office, Washington, D.C., pp. 1–26.

Baker, V. R., and N. M. Penteado-Orellana. 1977. Adjustment to climatic change of the Colorado River in central Texas. *Journal of Geology* **85**:395–422.

Barron, E. J., and W. M. Washington. 1984. The role of geographic variation in explaining paleoclimates, results from Cretaceous model sensitivity studies. *Journal of Geophysical Research* **89:**1267–1279.

Bartlein, P. J., and I. C. Prentice. 1988. Orbital variations, climate and paleoecology. *Trends in Ecology and Evolution* **4:**195–199.

Bartlein, P. J., and T. Webb III. 1985. Mean July temperature at 6000 yr B.P. in Eastern North America: Regression equations for estimates from fossil-pollen data. In C. R. Harrington (ed.), Climatic Change in Canada 5: Critical Periods in the Quaternary Climatic History of Northern North America. *Syllogeus* **55:**301–342.

Bell, M. 1982. "The Effects of Land-Use and Climate on Valley Sedimentation." In A. Harding (ed.), *Climatic Change in Later Prehistory.* Edinburgh University Press, Edinburgh, Scotland, pp. 127–142.

Bennett, K. D. 1990. Milankovitch cycles and their effects on species in ecological and evolutionary time. *Paleobiology* **16(1):**11–21.

Benson, L., and R. S. Thompson. 1987. "The Physical Record of Lakes in the Great Basin." In W. F. Ruddiman and H. E. Wright, Jr. (eds.), *North America and Adjacent Oceans During the Last Deglaciation,* The Geology of North America Series, Vol. K-3. The Geological Society of America, Boulder, CO., pp. 241–260.

Berger, A. L. 1978. Long-term variations of caloric insolation resulting from the earth's orbital elements. *Quaternary Research* **9:**139–167.

Berger, A., J. Imbrie, J. Hays, G. Kukla, and B. Saltzman. 1984. *Milankovitch and Climate.* Riedel, Dordrecht, the Netherlands.

Bernabo, J. C., and T. Webb III. 1977. Changing patterns in the Holocene pollen record from northeastern North America: A mapped summary. *Quaternary Research* **8:**64–96.

Bintliff, J. L. 1982. "Climatic Change, Archaeology, and Quaternary Science in the Eastern Mediterranean Region." In A. Harding (ed.), *Climatic Change in Later Prehistory.* Edinburgh University Press, Edinburgh, Scotland.

Birkeland, P. W. 1984. *Soils and Geomorphology.* Oxford University Press, New York.

Bloom, A. L. 1983. "Sea Level and Coastal Changes." In H. E. Wright, Jr. (ed.), *Late-Quaternary Environments of the United States,* Vol. 2, *The Holocene.* University of Minnesota Press, Minneapolis, pp. 42–51.

Bradley, R. S., H. F. Diaz, J. K. Eischeid, P. D. Jones, P. M. Kelley, and C. M. Goodess. 1987. Precipitation fluctuations over Northern Hemisphere land areas since the mid-19th century. *Science* **237:**171–175.

Bradshaw, R. H. W., and T. Webb III. 1985. Relationships between contemporary pollen and vegetation data from Wisconsin and Michigan, U.S.A. *Ecology* **66:**721–737.

Brakenridge, G. R. 1987. "Fluvial Systems in the Appalachians." In L. W. Graf (ed.), *Geomorphic Systems of North America,* Centennial Special Volume 2, The Geological Society of America, Boulder, CO, pp. 37–44.

Broecker, W. S., and J. van Donk. 1970. Insolation changes, ice volume, and the O^{18} record in deep sea cores. *Review of Geophysics and Space Physics* **8:**169–198.

Bryan, K. 1954. The geology of Chaco Canyon, New Mexico, in relation to the life and remains of the historic peoples of Pueblo Bonito. Smithsonian Misc. Collections, v. 122, pub. 4141, pp. 1–65.

Bryson, R. A., and B. M. Goodman. 1980. Volcanic activity and climatic changes. *Science* **207:**1041–1044.

Butzer, K. W. 1980. "Holocene Alluvial Sequences: Problems of Dating and Correlation." In R. A. Cullingford, D. A. Davidson, and J. Lewin (eds.), *Timescales in Geomorphology*. John Wiley & Sons, Chichester, U.K.

Chambers, R. M. (ed.). 1993. *Climate Change and Human Impact on the Landscape*. Chapman & Hall, London.

Chisholm, M. 1980. The wealth of nations. *Transactions of the Institute of British Geographers*. New Series **5:**255–276.

Chorley, R. J., S. A. Schumm, and D. E. Sugden. 1984. *Geomorphology*. Methuen, London.

Clark. W. C. 1985. Scales of climatic impact. *Climatic Change* **7:**5–27.

CLIMAP Project Members. 1976. The surface of the ice-age earth. *Science* **191:**1131–1137.

———. 1981. Seasonal reconstructions of the earth's surface at the last glacial maximum. *Geological Society of America Map and Chart Series,* **MC-36:**1–18.

COHMAP Members 1988. Climatic changes of the last 18,000 years: Observations and model simulations. *Science* **241:**1043–1052.

Cooke, R. U., and R. W. Reeves. 1976. *Arroyos and Environmental Change in the American South-West*. Clarendon Press, Oxford, U.K.

Crowley, T. J. 1983. The geologic record of climatic change. *Reviews of Geophysics and Space Physics* **21:**828–877.

Crowley, T. J., D. A. Short, D. G. Mengel, and G. K. North. 1986. Role of seasonality in the evolution of climate during the last 100 million years. *Science* **231:**579–584.

Davidson, D. A. 1980. "Erosion in Greece During the First and Second Millennia B.C." In R. A. Cullingford and J. Lewin (eds.), *Timescales in Geomorphology*. John Wiley & Sons, Chichester, U.K., pp. 143–158.

Davis, M. B. 1976. Erosion rates and land-use history in southern Michigan. *Environmental Conservation* **3:**139–148.

———. 1989. Lags in vegetation response to greenhouse warming. *Climatic Change* **15:**75–82.

Davis, W. M. 1899. The geographical cycle. *Geographical Journal* **14:**481–504.

Delcourt, H. R., P. A. Delcourt, and T. Webb III. 1983. Dynamic plant ecology: The spectrum of vegetational change in space and time. *Quaternary Science Review* **1:**153–175.

Dickinson, R. E. 1987. *The Geophysiology of Amazonia*. John Wiley, New York.

Edwards, K. J., and I. Ralston. 1984. Postglacial hunter-gatherers and vegetational history in Scotland. *Proceedings of the Society of Antiquaries of Scotland* **114:**15–34.

Gates, W. L. 1976. Modeling the ice-age climate. *Science* **191:**1138–1144.

Gaudreau, D. C., and T. Webb III. 1985. "Late-Quaternary Pollen Stratigraphy and Isochrone Maps for the Northeastern United States." In V. M. Bryant and R. G. Holloway (eds.), *Pollen Records of Late-Quaternary North American Sediments*. American Association of Stratigraphic Palynologists, Dallas, TX, pp. 247–280.

Goudie, A. 1985. "Themes in Desert Geomorphology." In A. Pitty (ed.), *Geomorphology: Themes and Trends*. Barnes & Noble, Totowa, NJ, pp. 56–71.

———. 1986. *Environmental Change*. Second edition. Clarendon, Oxford, U.K.

Graf, W. L. 1985. *The Colorado River, Instability and Basin Management*. Resource Pub. in Geog. no. 1984/2. Association of American Geographers, Washington, D.C.

Grove, J. M. 1979. The glacial history of the Holocene. *Programs in Physical Geography* **3:**1–53.

———. 1988. *The Little Ice Age*. Methuen, London.

Grove, A. T., and A. Warren. 1968. Quaternary landforms and climate on the south side of the Sahara. *Geographic Journal* **134:**194–208.

Hammer, C. U., H. B. Clausen, and W. Dansgaard. 1980. Greenland ice sheet evidence of post-glacial volcanism and its climatic impact. *Nature* **288:**230–235.

Hansen, J., D. Johnson, A. Lacis, S. Lebedeff, P. Lee, D. Rind, and G. Russell. 1981. Climatic impacts of increasing atmospheric CO_2. *Science* **213:**957–961.

Harvey, L. D. D. 1980. Solar variability as a contributing factor to Holocene climatic change. *Programs in Physical Geography* **4:**487–530.

Haury, L. R., J. A. McGowan, and P. H. Wiebe. 1978. "Patterns and Processes in the Time-Space Scales of Plankton Distributions." In J. H. Steele (ed.), *Spatial Patterns in Plankton Communities*. Plenum, New York, pp. 277–327.

Hays, J. D., J. Imbrie, and N. Shackleton. 1976. Variations in the earth's orbit: Pacemaker of the ice age. *Science* **194:**1121–1132.

Hopley, D. 1985. "Geomorphological Development of Modern Coastlines: A Review." In A. Pitty (ed.), *Geomorphology: Themes and Trends*. Barnes & Noble, Totowa, NJ, pp. 56–71.

Hunter, M. L., Jr., G. L. Jacobson, and T. Webb III. 1988. Paleoecology and the coarse-filter approach to maintaining biological diversity. *Conservation Biology* **2(4):**375–385.

Huntley, B. 1988. "Glacial and Holocene Vegetation History: Europe." In B. Huntley and T. Webb III (eds.), *Vegetation History*. Kluwer Academic, Dordrecht, the Netherlands, pp. 341–383.

Huntley, B., and H. J. B. Birks. 1983. An atlas of past and present pollen maps for Europe: 0–13000 years ago. Cambridge University Press, London.

Huntley, B., and I. C. Prentice. 1988. July temperatures in Europe from pollen data 6000 years before present. *Science* **241:**687–690.

Huntley, B., and T. Webb III. 1988a. *Vegetation History*. Kluwer Academic, Dordrecht, the Netherlands.

———. 1988b. Migration: Species' response to climatic variations caused by changes in the earth's orbit. *Journal of Biogeography* **16(1)**:5–19.

Imbrie, J., E. A. Boyle, S. C. Clemens, A. Duffy, W. R. Howard, G. Kukla, J. E. Kutzbach, D. G. Martinson, A. McIntyre, A. C. Mix, B. Molfino, J. J. Morley, L. C. Peterson, N. G. Pisias, W. L. Prell, M. E. Raymo, N. J. Shackleton, and J. R. Toggweiler, 1992. On the structure and origin of major glaciation cycles. 1. Linear responses to Milankovitch forcing. *Paleoceanography* **7(6)**:701–738.

Imbrie, J., J. D. Hays, D. G. Martinson, A. McIntyre, A. C. Mix, J. J. Morley, N. G. Pisias, W. L. Prell, and N. J. Shackleton. 1984. "The Orbital Theory of Pleistocene Climate: Support from a Revised Chronology of the Marine O18 Record." In A. Berger, J. Imbrie, J. Hays, G. Kukla, and B. Saltzman (eds.), *Milankovitch and Climate*. Reidel, Dordrecht, the Netherlands, pp. 269–305.

Imbrie, J., and K. P. Imbrie. 1979. *Ice Ages: Solving the Mystery*. Enslow, Short Hills, NJ.

Jacobson, G. L., T. Webb III, and E. C. Grimm. 1987. "Patterns and Rates of Vegetation Change During the Deglaciation of Eastern North America." In W. F. Ruddiman and H. E. Wright, Jr. (eds.), *North America and Adjacent Oceans During the Last Deglaciation,* The Geology of North America series, vol. K–3. The Geological Society of America, Boulder, CO, pp. 277–288.

Jäger, J., and R. G. Barry. 1990. "Climate." In B. L. Turner II et al. (eds.), *The Earth as Transformed by Human Action*. Cambridge University Press, New York, pp. 335–351.

Jones, P. D., T. M. L. Wigley, and P. G. Wright. 1986. Global temperature variations between 1861 and 1984. *Nature* **322**:430–434.

Kerr, R. A. 1984a. How to make a warm Cretaceous climate. *Science* **223**:677–678.

———. 1984b. Climate since the ice began to melt. *Science* **226**:326–327.

King, C. A. M. 1980. "Thresholds in Glacial Geomorphology." In D. R. Coates and J. D. Vitek (eds.), *Thresholds in Geomorphology*. George Allen and Unwin, London, pp. 297–321.

King, L. C. 1983. *Wandering Continents and Spreading Sea Floors on an Expanding Earth*. Wiley-Interscience, New York.

Knox, J. C. 1972. Valley alluviation in southwestern Wisconsin. *Annals of the Association of American Geographers* **62**:401–410.

———. 1977. Human impacts on Wisconsin stream channels. *Annals of the Association of American Geographers* **67**:323–342.

———. 1983. "Responses of River Systems to Holocene Climates." In H. E. Wright, Jr. (ed.), *Late Quaternary Environments of the United States,* Vol. 2, *The Holocene*. University of Minnesota Press, Minneapolis, pp. 26–41.

Kozarski, S. 1983. River channel changes in the middle reach of the Warta valley, Great Poland Lowland. *Quaternary Studies in Poland* **4**:159–169.

Kukla, G. J., and R. K. Matthews. 1972. When will the present interglacial end? *Science* **178**:190–191.

Kutzbach, J. E. 1976. The nature of climate and climatic variations. *Quaternary Research* **6**:471–480.

———. 1981. Monsoon climate of the early Holocene: Climate experiment with the earth's orbital parameters for 9000 years ago. *Science* **214**:59–61.

Kutzbach, J. E., and P. J. Guetter. 1986. The influence of changing orbital patterns and surface boundary conditions on climate simulations for the past 18,000 years. *Journal of the Atmospheric Sciences* **43**:1726–1759.

Kutzbach, J. E., and F. A. Street-Perrott. 1985. Milankovitch forcing of fluctuations in the level of tropical lakes from 18 to 0 kyr B.P. *Nature* **317**:130–134.

Labitzke, K., and H. van Loon. 1988. Associations between the 11-year solar cycle, the QBO and the atmosphere. Pt. I: The troposphere and stratosphere in the Northern Hemisphere in winter. *Journal of Atmospheric and Terrestrial Physics* **50**:197–206.

Lamb, H. H. 1982. *Climate, History and the Modern World*. Methuen, London.

Lloyd, C. R. 1984. Pre-Pleistocene paleoclimates: The geological and paleontological evidence; modeling strategies, boundary conditions, and some preliminary results. *Advances in Geophysics* **26**:35–140.

Love, D. W. 1979. "Quaternary Fluvial Geomorphic Adjustments in Chaco Canyon, New Mexico." In D. D. Rhodes and G. P. Williams, (eds.), *Adjustments of the Fluvial System*. Kendall-Hunt, Dubuque, IA, pp. 277–308.

Lowe, J. J., and M. J. C. Walker. 1984. *Reconstructing Quaternary Environments*. Longman, London.

MacCracken, M. C., and F. M. Luther. 1985. *Detecting the Climatic Effects of Increasing Carbon Dioxide*. DOE/ER-0235. U.S. Department of Energy, Washington, D.C.

MacDonald, G. M. 1993. Fossil pollen analysis and the reconstruction of plant invasions. *Advances in Ecological Research* **24**:67–110.

Marsh, G. P. 1885. *The Earth as Modified by Human Action*. Charles Scribner's Sons, New York.

McAndrews, J. H. 1988. "Human Disturbance of North American Forests and Grasslands: The Fossil Pollen Record." In B. Huntley and T. Webb III. *Vegetation History*. Kluwer Academic Publishers, Dordrecht, the Netherlands, pp. 673–697.

McDowell, P. F. 1983. Evidence of stream response to Holocene climatic change in a small Wisconsin watershed. *Quaternary Research* **19**:100–116.

———. 1987. "Geomorphic Processes in the Pacific Coast and Mountain System of Oregon and Washington." In W. L. Graf (ed.), *Geomorphic Systems of North America*. Centennial Special Volume 2. The Geological Society of America, Boulder, CO.

Mehringer, P. J., Jr. 1986. "Prehistoric Environments." In W. L. D'Azevedo (ed.), *Handbook of North American Indians*, Vol. 11, *Great Basin*. Smithsonian Institution, Washington, D.C., pp. 31–50.

Milankovitch, M. 1941. *Kanon der Erdbestrahlung und Seine Anwendung auf das Eiszeitenproblem*, Special Publication 133. Royal Serbian Academy, Belgrade. (English translation published in 1969 by Israel Program for Scientific Translations).

Miller, K. G., R. G. Fairbanks, and G. S. Mountain. 1987. Tertiary oxygen isotope synthesis, sea level history, and continental margin erosion. *Paleoceanography* **2:**1–19.

Mitchell, J. M., Jr. 1976. An overview of climatic variability and its causal mechanisms. *Quaternary Research* **6:**481–493.

Osterkamp, W. L., M. M. Fenton, T. C. Gustavson, R. F. Hadley, V. T. Holliday, R. B. Morrison, and T. J. Toy. 1987. "Great Plains." In W. L. Graf (ed.), *Geomorphic Systems of North America,* Centennial Special Volume 2. The Geological Society of America, Boulder, CO, pp. 163–210.

Péwé, T. L. 1983. "The Periglacial Environment in North America During Wisconsin Time." In S. C. Porter (ed.), *Late Quaternary Environments of the United States,* Vol. 1, *The Late Pleistocene.* University of Minnesota Press, Minneapolis, pp. 157–189.

Porter, S. C. 1981. Recent glacier variations and volcanic eruptions. *Nature* **291:**139–142.

Prell, W. L., and J. E. Kutzbach. 1987. Monsoon variability over the past 150,000 years. *Journal of Geophysical Research* **92:**8411–8425.

Prentice, I. C. 1986. Vegetation responses to past climatic changes. *Vegetatio* **67:**131–141.

Ritchie, J. C. 1987. *Postglacial Vegation of Canada.* Cambridge University Press, London.

Ruddiman, W. F. 1985. "Climate Studies in Ocean Cores." In A. D. Hecht (ed.), *Paleoclimate Analysis and Modeling.* John Wiley & Sons, New York, pp. 197–257.

Ruddiman, W. F., and A. McIntyre. 1981. Oceanic mechanisms for amplification of the 23,000 year ice volume cycle. *Science* **212:**617–627.

Russell, R. J. 1956. "Environmental Change Through Forces Independent of Man." In W. L. Thomas, Jr. (ed.), *Man's Role in Changing the Face of the Earth.* University of Chicago Press, Chicago, pp. 453–470.

Saltzman, B. 1985. "Paleoclimatic Modeling." In A. D. Hecht (ed.), *Paleoclimate Analysis and Modeling.* John Wiley & Sons, New York, pp. 341–396.

Sarntheim, M. 1978. Sand deserts during glacial maximum and climatic optimum. *Nature* **272:**43–46.

Schneider, S. H. 1983. Volcanic dust veils and climate: How clear is the connection?—An editorial. *Climatic Change* **5:**111–113.

Schumm, S. A. 1973. "Geomorphic Thresholds and Complex Response of Drainage Systems." In M. Morisawa (ed.), *Fluvial Geomorphology,* Proceedings of the Fourth Annual Binghamton Geomorphology Symposium, Binghamton, NY. Publications in Geomorphology, State University of New York, Binghamton, pp. 299–310.

———. 1980. Geomorphic thresholds: The concept and its applications. *Transactions of the Institute of British Geographers,* New Series, **15:**485–515.

———. 1981. "Some Applications of the Concept of Geomorphic Thresholds." In D. R. Coates and J. D. Vitek (eds.), *Thresholds in Geomorphology.* George Allen and Unwin, London, pp. 473–485.

Schumm, S. A., and G. R. Brakenridge. 1987. "River Response." In W. F. Ruddiman and H. E. Wright, Jr. (eds.), *North America and Adjacent Oceans During the Last Deglaciation,* The Geology of North America series, vol. K–3. Geological Society of America, Boulder, CO, pp. 221–240.

Schumm, S. A., and R. W. Lichty. 1963. Time, space and causality in geomorphology. *American Journal of Science* **263:**110–119.

Schumm, S. A., and R. S. Parker. 1973. Implications of complex response of drainage systems for Quaternary alluvial stratigraphy. *Nature* **243:**99–100.

Shackleton, N. J., J. Backman, H. Zimmerman, D. V. Kent, M. A. Hall, D. G. Roberts, D. Schnitker, J. G. Baldauf, A. Desprairies, R. Honrighausen, P. Huddlestun, J. B. Keene, A. J. Kaltenback, K. A. O. Krumsiek, A. C. Morton, J. W. Murray, and J. Westburg-Smith. 1984. Oxygen isotope calibration of the onset of ice-rafting and history of glaciation in the North Atlantic region. *Nature* **307:**620–623.

Shackleton, N. J., and N. G. Pisias. 1985. "Atmospheric Carbon Dioxide, Orbital Forcing, and Climate." In E. T. Sundquist and W. S. Broecker (eds.), *The Carbon Cycle and Atmospheric* CO_2 *: Natural Variations Archean to Present,* Geophysical Monograph, vol. 32. American Geophysical Union, Washington, D.C., pp. 303–317.

Simmons, I. G., and J. B. Innes. 1988. Late Quaternary vegetational history of the North York Moors, VIII–X. *Journal of Biogeography* **15:**249–272.

Smith, G. I., and F. A. Street-Perrott. 1983. "Pluvial Lakes of the Western United States." In S. C. Porter (ed.), *Late Quaternary Environments of the United States,* Vol. 1, *The Late Pleistocene*. University of Minnesota Press, Minneapolis, pp. 190–212.

Stanley, S. M. 1985. Rates of evolution. *Paleobiology* **11:**13–26.

Street-Perrott, F. A. 1981. Tropical paleoenvironments. *Programs in Physical Geography* **5:**157–185.

Street-Perrott, F. A., and S. P. Harrison. 1985. "Lake Levels and Climate Reconstruction." In A. D. Hecht (ed.), *Paleoclimate Analysis and Modeling*. John Wiley & Sons, New York, pp. 291–340.

Stuiver, M. 1980. Solar variability and climatic change during the current millennium. *Nature* **286:**868–871.

Thomas, D. S. G., and A. Goudie. 1984. "Ancient Ergs of the Southern Hemisphere." In J. C. Vogel (ed.), *Late Cenozoic Palaeoclimates of the Southern Hemisphere*. Balkema, Rotterdam, pp. 407–418.

Thomas, M. F., and M. B. Thorp. 1985. "Environmental Change and Episodic Etchplantation in the Humid Tropics of Sierra Leone." In I. Douglas and T. Spencer (eds.), *Environmental Change and Tropical Geomorphology*. George Allen and Unwin, London, pp. 239–267.

Thomas, W. L., Jr. 1956. *Man's Role in Changing the Face of the Earth*. University of Chicago Press, Chicago.

Tricart, J. 1985. "Evidence of Upper Pleistocene Dry Climates in Northern South America." In I. Douglas and T. Spencer (eds.), *Environmental Change and Tropical Geomorphology*. George Allen and Unwin, London, pp. 197–217.

Trimble, S. W. 1974. Man-induced soil erosion on the southern Piedmont, 1700–1970. Soil Conservation Society of America, Ankeny, IA.

Trimble, S. W., and S. W. Lund. 1982. Soil conservation and the reduction of erosion and sedimentation in the Coon Creek Basin, Wisconsin. U.S. Geological Survey Professional Paper no. 1234. U.S. Government Printing Office, Washington, D.C.

van Loon, H., and K. Labitzke. 1988. Associations between the 11-year solar cycle, the QBO and the atmosphere. Pt. II: surface and 700mb in the Northern Hemisphere in winter. *Journal of Climate* **1(9)**:905–920.

Van Zant, K. L., T. Webb III, G. M. Peterson, and R. G. Baker. 1979. Increased Cannabis/Humulus pollen, an indicator of European agriculture in Iowa. *Palynology* **3**:229–233.

Vita-Finzi, C. 1969. *The Mediterranean Valleys*. Cambridge University Press, London.

Wahl, E. W., and T. L. Lawson. 1970. The climate of the midnineteenth century United States compared to current normals. *Monthly Weather Review* **98**:259–265.

Webb, T., III. 1986. Is vegetation in equilibrium with climate? How to interpret late-Quaternary pollen data. *Vegetatio* **67**:75–91.

———. 1987. The appearance and disappearance of major vegetational assemblages: Long-term vegetational dynamics in eastern North America. *Vegetatio* **69**:177–187.

———. 1988. "Eastern North America." In B. Huntley and T. Webb III (eds.), *Vegetation History*. Kluwer Academic, Dordrecht, the Netherlands, pp. 385–414.

———. 1993. "Constructing the Past from late-Quaternary Pollen Data: Temporal Resolution and a Zoom Lens Space-Time Perspective." In S. M. Kidwell and A. K. Behrensmeyer (eds.), *Taphonomic Approaches to Time Resolution in Fossil Assemblages*. Paleontological Society Short Courses in Paelontology No. 6. University of Tennessee, Knoxville, pp. 79–101.

Webb, T., III, and P. J. Bartlein. 1992. Global changes during the last 3 million years: Climatic controls and biotic responses. *Annual Review of Ecology and Systematics* **23**:141–173.

Webb, T., III, P. J. Bartlein, and J. E. Kutzbach. 1987. "Climatic Change in Eastern North America During the Past 18,000 Years: Comparison of Pollen Data with Model Results." In W. F. Ruddiman and H. E. Wright, Jr. (eds.), *North America and Adjacent Oceans During the Last Deglaciation,* The Geology of North American series, vol. K–3. The Geological Society of America, Boulder, CO, pp. 447–462.

Webb, T., III, E. J. Cushing, and H. E. Wright, Jr. 1983a. "Holocene Changes in the Vegetation of the Midwest." In H. E. Wright, Jr. (ed.), *Late-Quaternary Environments of the United States*, vol. 2, *The Holocene*. University of Minnesota Press, Minneapolis, pp. 142–165.

Webb, T., III, P. J. Richard, and R. J. Mott. 1983b. A mapped history of Holocene vegetation in southern Quebec. *Syllogeus* **49**:273–336.

West, R. G. 1956. The Quaternary deposits at Hoxne, Suffolk. *Philosophical Transactions of the Royal Society of London* B **239**:265–356.

Womack, W. R., and S. A. Schumm. 1977. Terraces of Douglas Creek, northwestern Colorado: An example of episodic erosion. *Geology* **5**:72–76.

Wright, H. E., Jr., J. C. Almendinger, and J. Gruger. 1985. Pollen diagram from the Nebraska Sandhills, and the age of the dunes. *Quaternary Research* **24**:115–120.

16

Monitoring Ocean Productivity by Assimilating Satellite Chlorophyll into Ecosystem Models

Robert A. Armstrong, Jorge L. Sarmiento, and Richard D. Slater

Satellite color imagery is currently the only data source that can be used to monitor long-term, global-scale changes in ocean biology. The ability to monitor ocean ecosystem processes is important not only because oceanic biological resources have direct value to mankind, but also because ocean biology plays a major role in the global carbon cycle (Siegenthaler and Sarmiento 1993). The effects of biological processes on seasonal and interannual changes in surface ocean pCO_2 are very large and are extremely difficult to monitor with *in situ* measurements, since shipboard measurements can provide only limited coverage of the world ocean. In addition, long-term changes in ocean circulation may occur in response to greenhouse warming (Manabe and Stouffer 1993); and it is likely that such changes will have a significant impact on biological systems, and hence on the ocean-atmosphere CO_2 balance (e.g., Sarmiento and Orr 1991).

Satellite observations of ocean color have been used with considerable success to produce estimates of phytoplankton pigment concentrations in the upper few meters of the water column (e.g., Esaias et al. 1986). In contrast, using these chlorophyll estimates to predict biologically important quantities, such as productivity, has to date met with more limited success. For example, Balch et al. (1992) reviewed several methods for estimating integral primary productivity (total productivity in the water column, as measured from ships) from satellite estimates of surface chlorophyll. The best of these productivity models accounted for only 12% of the variance in productivity, which is only a modest improvement over the 6% variance explained when productivity is simply regressed on satellite surface chlorophyll. Perhaps the most surprising result of this study was that one highly sophisticated procedure (Platt and Sathyendranath 1988) did even worse than the simple regression, suggesting that detailed, physiology-based models may be more sensitive than simpler models to choice of parameter values (Platt and Sathyendranath 1993a,b; Balch 1993).

This current inability to extract reliable productivity estimates from satellite data lies squarely in the way of our ability to monitor other important ocean processes. There is evidence, for example, that the quantity of most interest for the carbon cycle is "new production," the fraction of total production that is fueled by nitrogen that has not been recycled within the photic zone (Dugdale and Goering 1967; Platt et al. 1992a). A need therefore exists for a way to link total production to new production. Platt and Sathyendranath (1988) suggest using measurement-based estimates of the ratio of new production to primary production (the "*f*-ratio"; Eppley and Peterson 1979) to estimate new production. Simple empirical relationships (e.g., Eppley and Peterson 1979; Harrison et al. 1987) have been applied to obtain estimates of new production from satellite data (e.g., Dugdale et al. 1989). However, we do not yet have enough information on *f*-ratios or new production to know how successful this approach will be. For example, when nutrients are added in pulses, a correlation is induced between total production and the *f*-ratio, making determination of average values for the *f*-ratio difficult (Platt and Sathyendranath 1988).

A second major problem involves monitoring High Nutrient Low Chlorophyll (HNLC) areas. These are areas of the world ocean (most notably the equatorial Pacific, the subarctic Pacific, and the Southern Ocean) where macronutrients (nitrate, phosphate) are not used completely at any time during the year (Figure 16-1). Sarmiento and Orr (1991) have shown that changes in the productivity of these regions, particularly the Southern Ocean, could potentially have large impacts on the ability of the oceans to store carbon. Although large increases in the productivity of such areas would not be sufficient to negate the effects of emissions of carbon from fossil fuels (Sarmiento and Orr 1991), monitoring of the spatial extent and intensity of these areas would seem prudent, lest unexpected changes in their productivity lead to rapid increases or decreases in pCO_2 similar

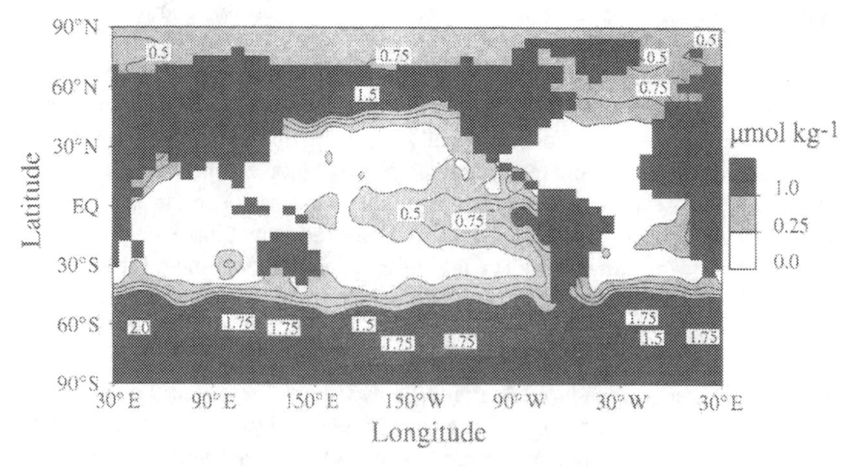

Figure 16-1. Observed annual mean surface phosphate as mapped by S. Levitus and R. G. Najjar (pers. comm.); contour interval is 0.25 µmol/kg.

to those that appear to have occurred in association with the eruption of Mount Pinatubo (Sarmiento 1993).

Using satellite data to estimate quantities of ecological significance is not easy, both because the satellite data themselves are noisy and because by themselves these data are insufficient for calculating these quantities. First, the satellite predicts chlorophyll only with a factor of two (Balch et al. 1992), because of atmospheric interference with the water-leaving signal and also because "chlorophyll" seen by the satellite is actually a mixture of chlorophyll and phaeopigments (chlorophyll breakdown products); the latter can range from almost all (96%) to almost none (3%) of the "chlorophyll" signal (Gordon and Clark 1980; Balch et al. 1992). Second, the satellite "sees" only about one optical depth (where the light is reduced to 1/e of its value at the surface; Gordon and McCluney 1975; Gordon et al. 1982), whereas the photic zone extends about 5.6 optical depths (to the depth at which light is reduced to 1% of its surface value, the nominal "compensation point" below which respiration exceeds photosynthesis). Thus, chlorophyll concentration at the surface may not be representative of its concentration at depth (Balch et al. 1992). Finally, the satellite sees only chlorophyll (plus phaeopigments), not nutrients or other quantities of ecological importance. Thus, for example, satellite images cannot be used to estimate the values of important parameters linking chlorophyll to productivity, most notably the maximum rate of photosynthesis per unit chlorophyll P^b_{max} (Balch et al. 1992), which can depend on nutrients as well as light (Platt et al. 1992b).

Platt and Sathyendranath (1988) proposed solving this last problem by dividing the ocean into a set of "biogeographic provinces," each of which would be assigned a characteristic depth profile and photosynthetic parameters (see also Mueller and Lange 1989; Sathyendranath et al. 1991). Although this idea seems reasonable, the recent work by Balch et al. (1992) suggests that their problem may be difficult to implement in practice. An additional consideration is that the biogeographic provinces defined in this approach would be fixed for all time, making the model useful only for monitoring relatively small changes in the productivity of the oceans. Although both the characteristic values of parameters and the boundaries between biogeographic provinces may be allowed to change seasonally, this level of flexibility may not be adequate for predicting carbon cycle responses to climate changes large enough to induce large-scale changes in ocean circulation (Manabe and Stouffer 1993).

We propose instead to use a simple model of the pelagic ecosystem embedded in an ocean General Circulation Model (GCM) to link satellite chlorophyll estimates to net primary productivity and other important ecological quantities. Satellite observations are used to constrain surface chlorophyll concentrations to their observed values by "assimilating" these data into the model. The ecosystem model then evolves dynamically, subject to these boundary conditions, to produce the required predictions.

We have made considerable progress in understanding the properties that ecosystem models and assimilation schemes must possess if they are to be used for this purpose. Our first efforts at constraining the surface chlorophyll to observed values failed because the one-phytoplankton–one-zooplankton ecosystem model we used (Fasham et al. 1990; Sarmiento et al. 1993) could not be forced to assume high chlorophyll values without producing pathological behaviors (most notably, ammonium values that were at least an order of magnitude too high). In response, we developed a model based on multiple phytoplankton and zooplankton size classes (Moloney and Field 1991; Armstrong in press), so that when one phytoplankton size class reached its maximum value, others would be able to assimilate the additional chlorophyll. In all other respects the model was designed to be as close to the Fasham et al. (1990) model as possible, so that the effects of multiple size classes *per se* would not be entangled with other factors. Our experience with this model has been quite encouraging: the model can be forced toward observed surface chlorophyll concentrations without producing anomalies in any variable.

Assimilating Satellite Data into Ecosystem Models

We began by assimilating monthly average Coastal Zone Color Scanner (CZCS) chlorophyll estimates into an ocean GCM containing a simple model of the pelagic ecosystem (Fasham et al. 1993; Sarmiento et al. 1993). The simplest ecosystem model (Fasham et al. 1990) has seven compartments (Figure 16-2): phytoplankton (P), zooplankton (Z), nitrate (NO_3), ammonium (NH_4), detritus (D), bacteria (B), and dissolved organic nitrogen (DON). The reader is referred to papers by Fasham et al. (1993) and Sarmiento et al. (1993) for a full description of the model and to Fasham et al. (1990) for a justification of the functional forms used in the model. Only those details of model structure that are directly related to the present study will be discussed below.

We used a seasonal ocean GCM of the Atlantic from 30° S to 68° N with 2° horizontal resolution and 25 vertical levels (Sarmiento 1986). Six of these layers lie in the upper 123 m, where the ecosystem equations are solved; the bottoms of these six layers are at 10 m, 23 m, 40 m, 61 m, 88 m, and 123 m. The ocean circulation is determined by solving the equations of motion and state and the conservation of temperature and salinity as described by Sarmiento (1986). Convective overturning is simulated by homogenizing adjacent layers when they are unstable with respect to each other. The model is forced at the surface with the monthly climatological winds of Hellerman and Rosenstein (1983) and the monthly averaged temperatures and seasonally averaged salinities of Levitus (1982). It is run for 1900 years before being used for the ecosystem simulation. In the simulation runs, the ecosystem equations are solved simultaneously with the GCM equations in the top six boxes (0–123 m). For more details, see Sarmiento (1986) and Sarmiento et al. (1993).

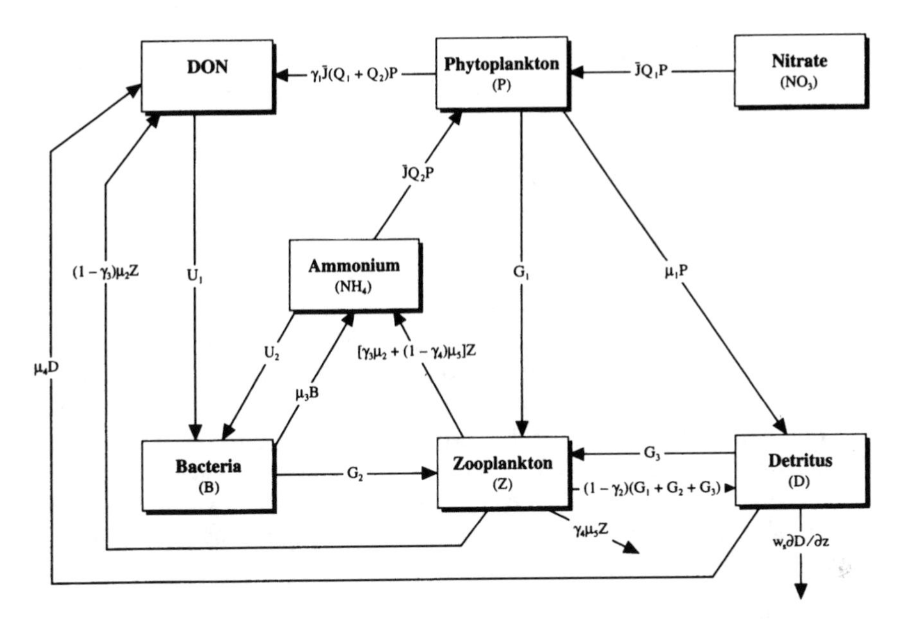

Figure 16-2. The Fasham et al. (1990) ecosystem model.

Although there exist many sophisticated schemes for assimilating data into models (e.g., Derber and Rosati 1989; Mellor and Ezer 1991; Harms et al. 1992), we chose the simplest method that was consistent with our goals. Model predictions of phytoplankton nitrogen concentrations in the top box of the GCM (0–10 m depth) were converted to chlorophyll concentrations using a nominal chlorophyll:nitrogen ratio of 1.59 mg Chl *a* $[\text{mmol N}]^{-1}$, which corresponds to a chlorophyll:carbon ratio of 0.02 mg Chl *a* $[\text{mg C}]^{-1}$ and a carbon:nitrogen ratio of 6.625 mmol C $[\text{mmol N}]^{-1}$ (Sarmiento et al. 1993). Daily estimates of CZCS chlorophyll P_{sat} were generated by linear interpolation between monthly means, and model chlorophyll (nitrogen) values P_{model} in the top box (top 10 m) of the GCM were forced toward the satellite values by adding a fraction of $\gamma\Delta t$ of the difference between the two values to the model value; that is, at each time step a quantity

$$\delta P = \gamma \Delta t \times (P_{sat} - P_{model}) \tag{1}$$

was added to the model value to force it toward the observed value $P_{sat.}$ The parameter γ defines the rate at which the model is forced toward the data, so that $\gamma\Delta t$ is the fraction by which chlorophyll will be forced toward the satellite values per time step Δt (= 1 hr in the current model). Values of $\gamma = 1\ \text{d}^{-1}$ and $\gamma = 5\ \text{d}^{-1}$ were used in this experiment. A value of $\gamma = 24\ \text{d}^{-1}$, corresponding to complete replacement of the model data by the satellite data (cf. Ishizaka 1990), was not tried, as the main patterns in model responses emerged clearly at the more gentle forcings.

We adopted this simple "nudging" of the data toward the satellite observations (Harms et al. 1992) because it possesses two desirable properties. First, by nudging only the top box, we correct only what we know: chlorophyll concentrations at the top of the photic zone; we make no assumptions about chlorophyll concentrations in the deeper boxes, which are left to evolve to appropriate values by themselves. Second, by nudging only the chlorophyll, we avoid having to make assumptions about what should happen to other ecosystem components in response to the forcing; again, the ecosystem model itself does the adjusting. Note that this procedure does not conserve total nitrogen, since adding phytoplankton also adds nitrogen; for the purpose of identifying model deficiencies, this drawback was considered unimportant. To achieve the longer-term goal of using satellite data to monitor ocean biology, however, we may well need to develop alternative assimilation procedures that do conserve total nitrogen.

Experiments Using the Fasham Ecosystem Model

Figure 16-3 allows comparison of satellite data (Figure 16-3D) to three separate model experiments using the Fasham (1990) model. Figure 16-3A shows model results without data assimilation. The general impression is that yearly average chlorophyll levels are too high in the equatorial region and too low at high latitudes. Forcing at $\gamma = 5\ d^{-1}$ (Figure 16-3C) makes a substantial improvement over most of the ocean, while making predictions somewhat worse at high latitudes. The less vigorous forcing of $\gamma = 1\ d^{-1}$ (Figure 16-3B) makes more modest improvements over most of the ocean, but still makes predictions worse at high latitudes. Plots of ecosystem properties (Figure 16-4) at the location of Ocean Weather Station (OWS) India (59° N 19° W, between Ireland and Iceland) show that the model can produce unrealistically high ammonium levels (cf. Fasham et al. 1993) under data assimilation: $> 20\ mmol/m^3$ at a forcing of $\gamma = 5\ d^{-1}$ (Figure 16-4F) and $> 5\ mmol/m^3$ at a forcing of $\gamma = 1\ d^{-1}$ (not shown).

These results point to significant differences in the ability of the ecosystem model to assimilate data under different circumstances. In areas where the model overpredicts phytoplankton density (the equatorial region) and in areas where phytoplankton density is too low to support zooplankton (the subtropical gyre), the model performs well if forced hard enough (Figure 16-3C). However, where the model underpredicts phytoplankton densities and zooplankton are present, the model performs poorly, indicating the need to improve model performance under these circumstances. We can gain insight into the model's performance by considering how zooplankton grazing is modeled. The model assumes that zooplankton can be represented by a single "composite" type that changes its

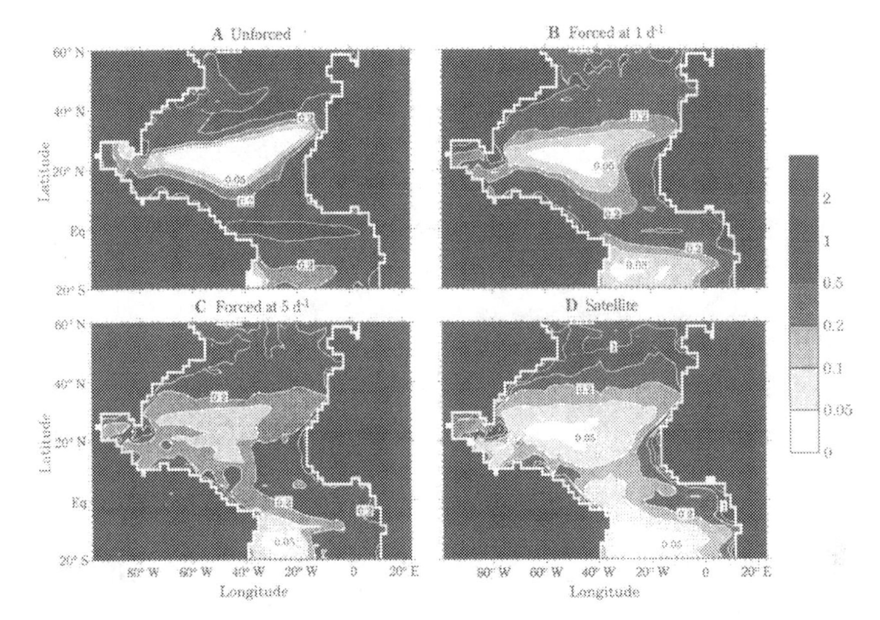

Figure 16-3. Annual mean chlorophyll concentrations (mg Chl/m^3) at 5 meters depth predicted by the Fasham model: (A) the free-running (unforced) model; (B) the model forced with satellite data at $\gamma = 1\ d^{-1}$; (C) the model forced at $\gamma = 5\ d^{-1}$; (D) satellite observations.

preference for its various food types (phytoplankton P, detritus D, and bacterial B) according to their densities. The zooplankton growth equation can be written

$$\frac{dZ}{dt} = Z[\gamma_2 G(P,B,D) - (\mu_2 + \mu_5)], \tag{2}$$

where G is the total rate of grazing on the three food types (d^{-1}), μ_2, and μ_5 are loss rates to excretion and mortality, respectively (d^{-1}), and γ_2 is a feeding efficiency (dimensionless). At steady state ($dZ/dt = 0$) and in the absence of physical transport, the value of G is determined by

$$G(P,B,D) = (\mu_2 + \mu_5)/\gamma_2. \tag{3}$$

Equation (3) and the function G determine all possible values of the food triplet (P,B,D) when zooplankton are present; in particular, they define the maximum possible steady state values of P, B, and D. For example, if the total grazing rate G were to depend on a linear combination $\alpha_P + \alpha_B B + \alpha_D D$ of its foods, then at steady state we would require that $\alpha_P P + \alpha_B B + \alpha_D D = G^{-1}[(\mu_2 + \mu_5)/\gamma_2]$. It then follows that the steady state value of P in the presence of zooplankton and in the absence of physical transport must lie between 0 (when

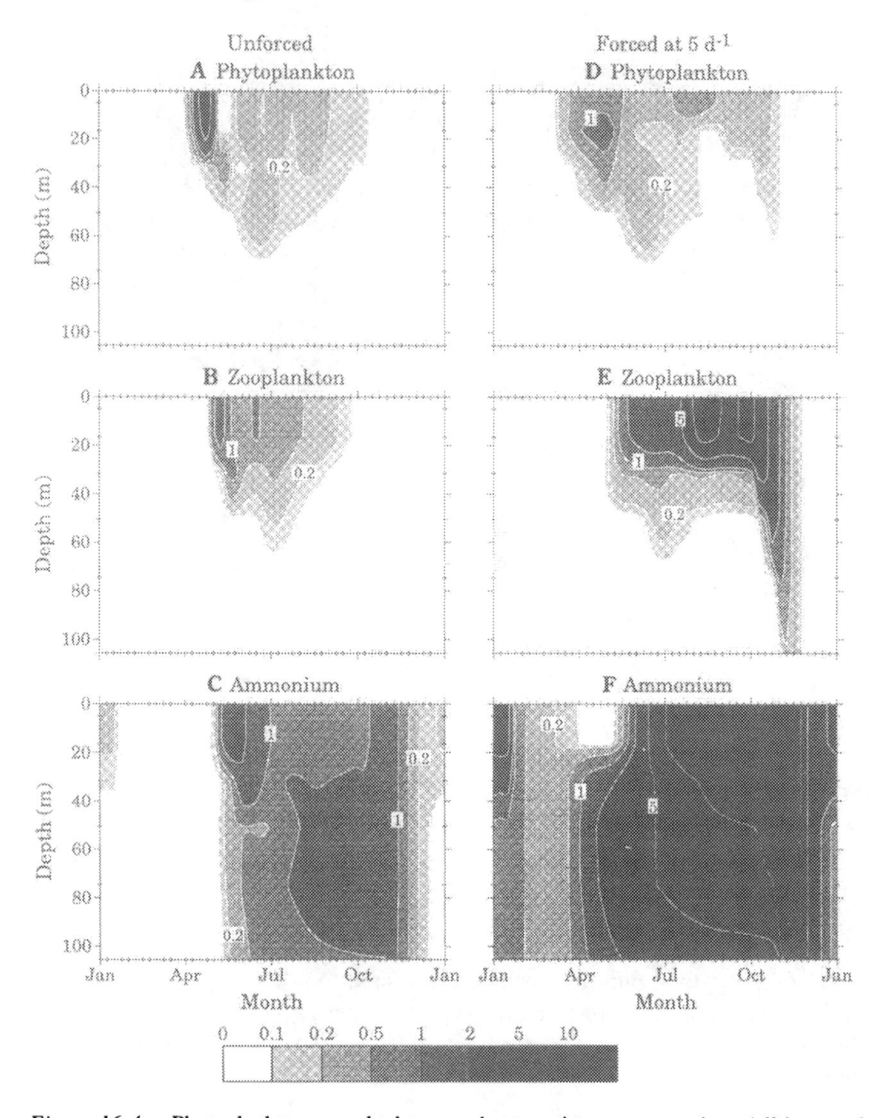

Figure 16-4. Phytoplankton, zooplankton, and ammonium concentrations (all in mmol N/m^3) at Ocean Weather Station India as predicted by the Fasham model. (A)–(C) concentrations predicted by the unforced (free-running) model; (D)–(F) concentrations predicted at $\gamma = 5d^{-1}$.

the zooplankton survive only on bacteria and detritus) and $P = G^{-1}[(\mu_2 + \mu_5)/\gamma_2]/\alpha_P$ (when $B = D = 0$). In contrast, the functional form of G specified by Fasham et al. (1990), after Hutson (1984) and used in Sarmiento et al. (1993) is not linear in food types; it is instead a complicated function that allows zooplankton to feed preferentially on food types that are more abundant. For the particular functional form of G and the particular parameter values used in their model, and in the absence of physical transport, Sarmiento et al. (1993, p. 441) showed that steady state phytoplankton densities in the presence of zooplankton must always be less than 0.342 mmol N/m^3. It is this property that produces the pathological behavior when we assimilate satellite data into this model.

When the model is near steady state and zooplankton are present (conditions that obtain over the whole year in the equatorial zone and in summer and autumn in high latitudes), zooplankton grazing constrains phytoplankton densities to be less than 0.342 mmol N/m^3 (0.55 mg Chl a/m^3), as noted above. If satellite chlorophyll estimates are above this range, as they are at OWS India, forcing model chlorophyll values toward satellite values causes zooplankton densities to increase (Figure 16-4E versus Figure 16-4B) to the point where they are able to eat all the added phytoplankton, forcing phytoplankton densities back to their steady state range (Figure 16-4D versus Figure 16-4A) in spite of the assimilation. This additional grazing produces additional ammonium. Finally, because the phytoplankton are constrained to a small population size by zooplankton grazing, they cannot consume all the additional ammonium, allowing ammonium to build up to unrealistic levels (Figure 16-4F versus Figure 16-4C).

A Multiple-Chains Fasham Model

To address this problem, we replaced the single phytoplankton-zooplankton food chain of the Fasham (1990) model with a more complex structure, so that when one phytoplankton size class reaches its limit, others will be present to assimilate the additional chlorophyll. This change has ample biological justification. In a review of the effects of size on algal physiology and phytoplankton ecology, Chisholm (1992, p. 222) noted that the total amount of chlorophyll in each of several successively larger size fractions seems to have an upper limit, and that "beyond certain thresholds, chlorophyll can only be added by adding a larger size class of cells." Raimbault et al. (1988), for example, noted that there appears to be a maximum of roughly 0.8 mg Chl a/m^3 in the 1–3-μm phytoplankton size class and 0.7 mg Chl a/m^3 in the 3–10-μm size class. These size classes are of approximately equal width on a logarithmic scale, suggesting roughly equal maximum chlorophyll per logarithmic increment in size. Smaller cells also tend to dominate in HNLC areas (Chavez 1989; Banse 1992; Frost and Franzen 1992), with larger cells appearing only in response to the addition of micronutrients, notably iron (Martin et al. 1991; Morel et al. 1991; Price et al. 1991).

These patterns can be produced quite naturally by models with multiple food chains, where the steady state phytoplankton concentration in each size class is set by the herbivorous predator on that size class, and where parameters are chosen to allow food chains based on smaller phytoplankton to dominate under oligotrophic conditions, with chains based on larger phytoplankton appearing only at increased nutrient loadings (Armstrong in press).

The simplest way to create a model with multiple size classes is to define it in terms of modules and to choose the nominal sizes of individuals in each module to be a certain fraction of their counterparts in the next larger module. The parameters of each food chain can then be related to those of other chains using allometric (power law) relations (Moloney and Field 1989, 1991; Armstrong in press). In particular, the rate constant R_i for the ith food chain is related to the corresponding rate constant R_1 for the smallest food chain according to the relationship

$$R_i = R_1(L_i/L_1)^{\beta_R}. \tag{4}$$

In equation (4), β_R is the allometric constant for rate constant R, and L_i and L_1 are the characteristic lengths of phytoplankton at the bases of food chain i and food chain 1, respectively. The use of allometric relationships in an oceanographic context has been extensively justified on empirical grounds by Moloney and Field (1989, 1991; see also Chisholm 1992); from a modeling perspective, using such relationships is an excellent way to add more complexity (in this case, more food chains) to the model with minimal addition of new parameters (Armstrong in press).

These simple considerations lead to the following extension of the Fasham et al. (1990) model (Figure 16-5), where we have simply replicated the P-Z-D part of the original model three times for successive size classes. The smallest zooplankton size class feeds on its own phytoplankton size class, its own detrital size class, and bacteria, with relative preferences of 0.50, 0.25, and 0.25 for these three food types (Fasham et al. 1990; Sarmiento et al. 1993). The next larger zooplankton size class also feeds on its own phytoplankton size class and its own detrital size class; but instead of feeding on bacteria, it feeds on the smallest zooplankton size class, retaining the same relative preferences (0.50, 0.25, and 0.25) for these three food types. The third zooplankton size class is constructed in like manner, and additional size classes could be added as needed. Although this structure, where each zooplankton size class feeds only on its own phytoplankton size class and its own detrital size class, is unrealistic, it provides the simplest possible extension of the Fasham et al. (1990) model to test whether a model with multiple food chains can be used to assimilate satellite data.

Again, to make comparison with previous results (Fasham et al. 1993; Sarmiento et al. 1993) as clean as possible, all parameter values for the smallest food chain (P_1, Z_1, D_1) and the nutrient recycling part of the model (NO_3, NH_4, B, DON) were chosen to be identical to those used in these previous studies. Parameters for other chains were then constructed relative to those of the first chain. We first assumed that adjacent size classes differed in length by a factor of 5. Allometric coefficients other than zero were then assigned to four key rate constants. Phytoplankton maximum growth rate was assumed to decline with size ($\beta_{V_P} = -0.4$) according to the empirically established relationship for diatoms (Chisholm 1992; see also Armstrong in press). Zooplankton maximum growth

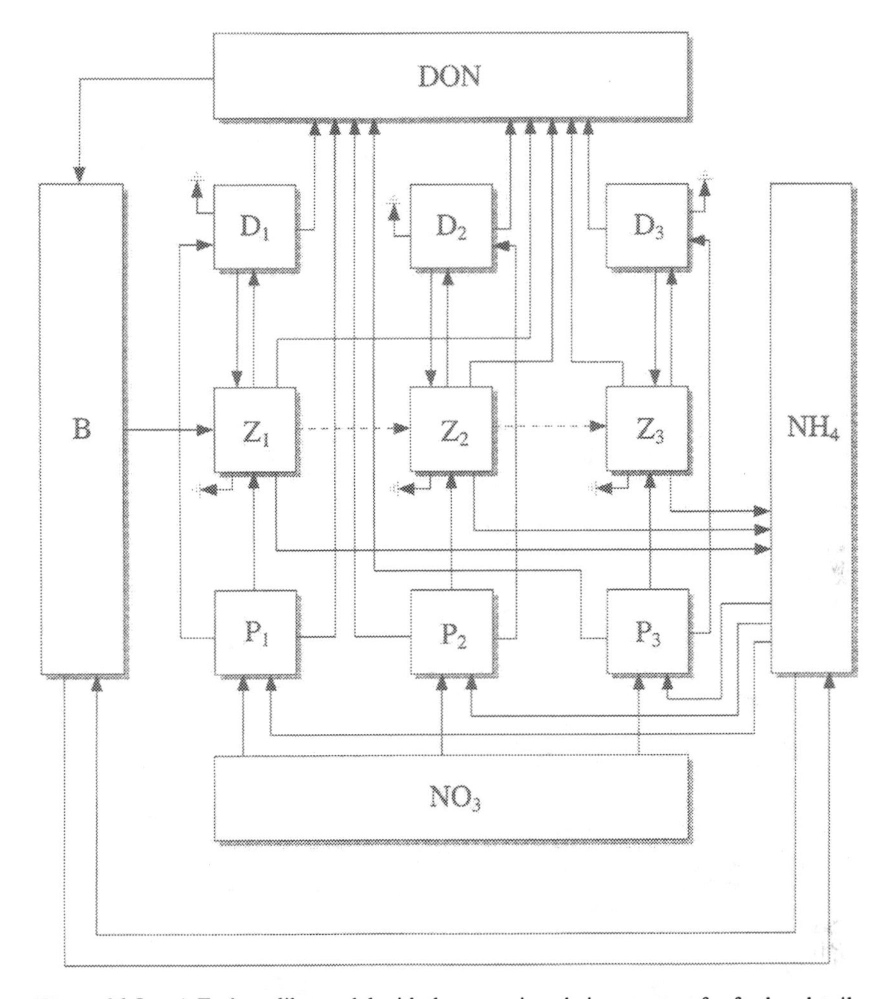

Figure 16-5. A Fasham-like model with three grazing chains; see text for further details.

rate ($\beta_g = -0.75$) and zooplankton excretion rate ($\beta_{\mu_2} = -0.75$) were also assumed to decline with length according to the relationship of Moloney and Field (1989), which was derived using data on zooplankton growth and respiration. Finally, detrital sinking rate was assumed to increase with size ($\beta_V = 1.2$), again from data fits performed by Moloney and Field (1991).

Results show that the multiple-chains model achieved its objectives. Figure 16-6A shows that the unforced multiple-chains model produces higher chlorophyll in high latitudes (closer to the satellite image) and also higher chlorophyll at the equator (even farther than the original model from the satellite image). Figures 16-6B and 16-6C show the runs with data assimilation. Note that Figure 16-6B ($\gamma = 1\ d^{-1}$) is in considerably better agreement with satellite data than is Figure

16-3B, and that Figure 16-6C ($\gamma = 5\ d^{-1}$) reproduces the satellite data almost exactly. Figure 16-7F shows that ammonium is confined to reasonable levels (< 1 mmol/m^3) in the $\gamma = 5\ d^{-1}$ run; the result at $\gamma = 1\ d^{-1}$ (not shown) is similar.

In Figure 16-8, zonal mean ratios of model chlorophyll to satellite chlorophyll are plotted against time. This plot shows that assimilation at $\gamma = 5\ d^{-1}$ (Figure 16-8B) is quite effective at forcing model chlorophyll toward satellite chlorophyll at virtually all latitudes and seasons; forcing at $\gamma = 1\ d^{-1}$ (Figure 16-8A) is considerably less effective at the equator.

Productivities

Figure 16-9 shows productivities estimated from the free-running Fasham et al. (1990) model (Figure 16-9A), from the free-running multiple-chains model (Figure 16-9B), and from the multiple-chains model into which the averaged CZCS data have been assimilated (Figure 16-9C,D). Comparison of these figures shows that data assimilation decreases model estimates of yearly productivity at the equator (relative to the free-running models) by a factor of two or more. Figure 16-10 shows that in contrast to these twofold changes in yearly average productivity, zonal average productivities can at certain times be changed by factors of four or more from those produced by the free-running simulation, implying that

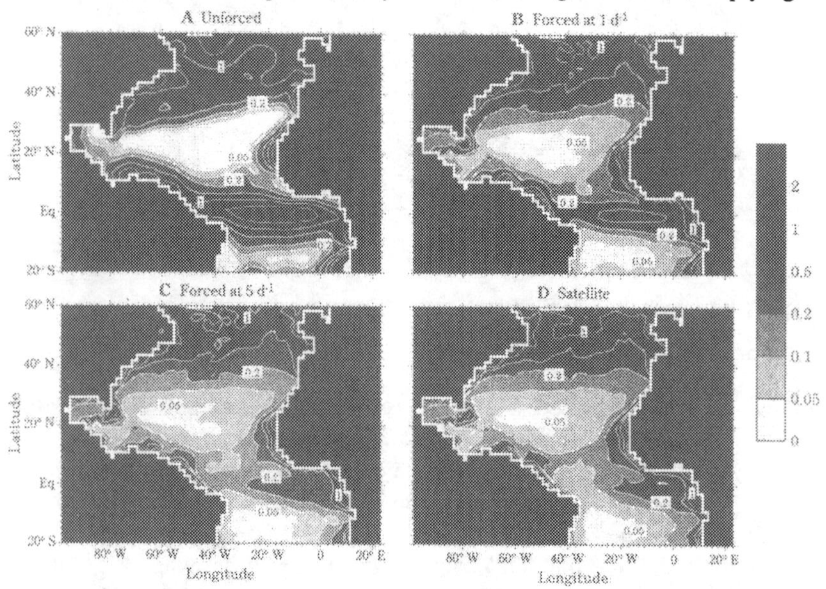

Figure 16-6. Annual mean chlorophyll concentrations (mg Chl/m^3) at 5 meters depth predicted by the multiple-chains Fasham model. When forced with satellite data (B,C), total chlorophyll to be assimilated δP at a given time step was partitioned among the three algal size classes in proportion to their current concentrations. (A) the free-running (unforced) model; (B) the model forced at $\gamma = 1\ d^{-1}$; (C) the model forced at $\gamma = 5\ d^{-1}$; (D) satellite observations.

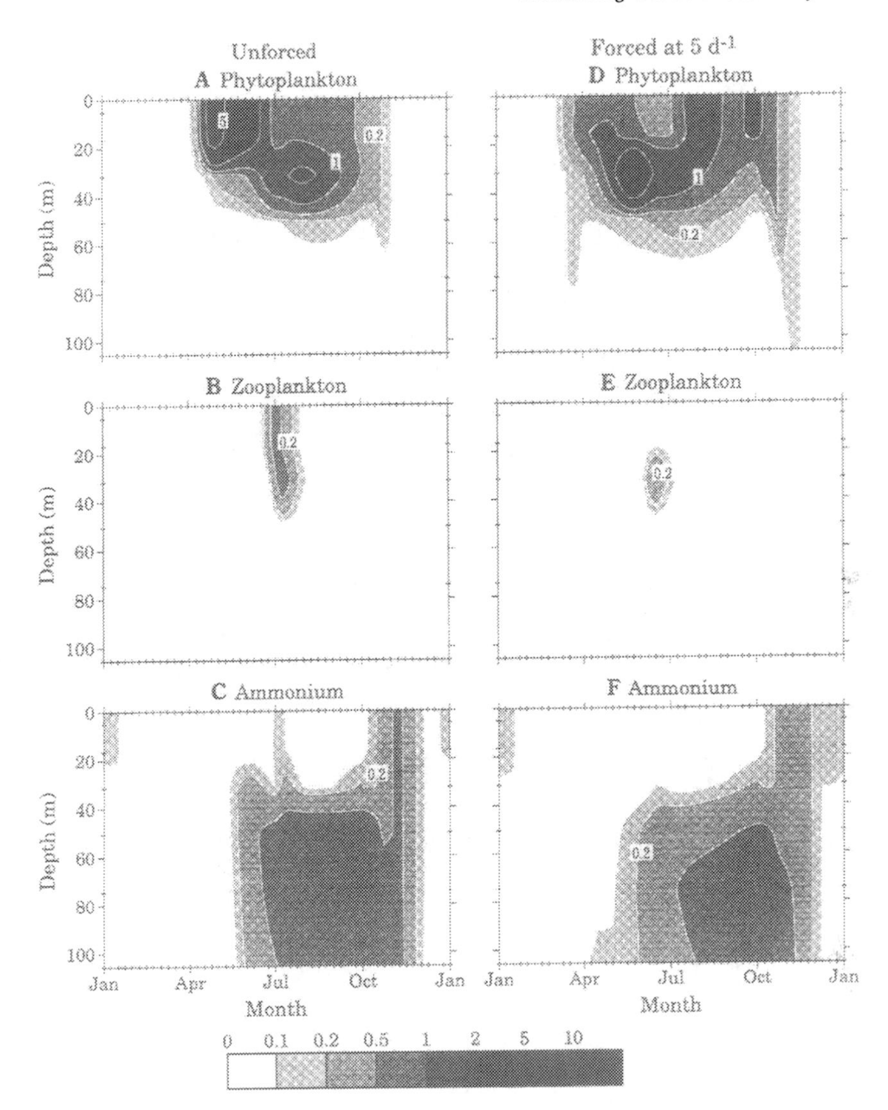

Figure 16-7. Phytoplankton, zooplankton, and ammonium concentrations (all in mmol/m^3) at Ocean Weather Station India as predicted by the multiple-chains Fasham model. (A)–(C) concentrations predicted by the unforced (free-running) model; (D)–(F) concentrations predicted at $\gamma = 5\ d^{-1}$.

the assimilation procedure can produce dramatic changes in individual productivities.

Discussion

We have made considerable progress in defining the characteristics that ecosystem models must possess if they are to be used for assimilating satellite data to make

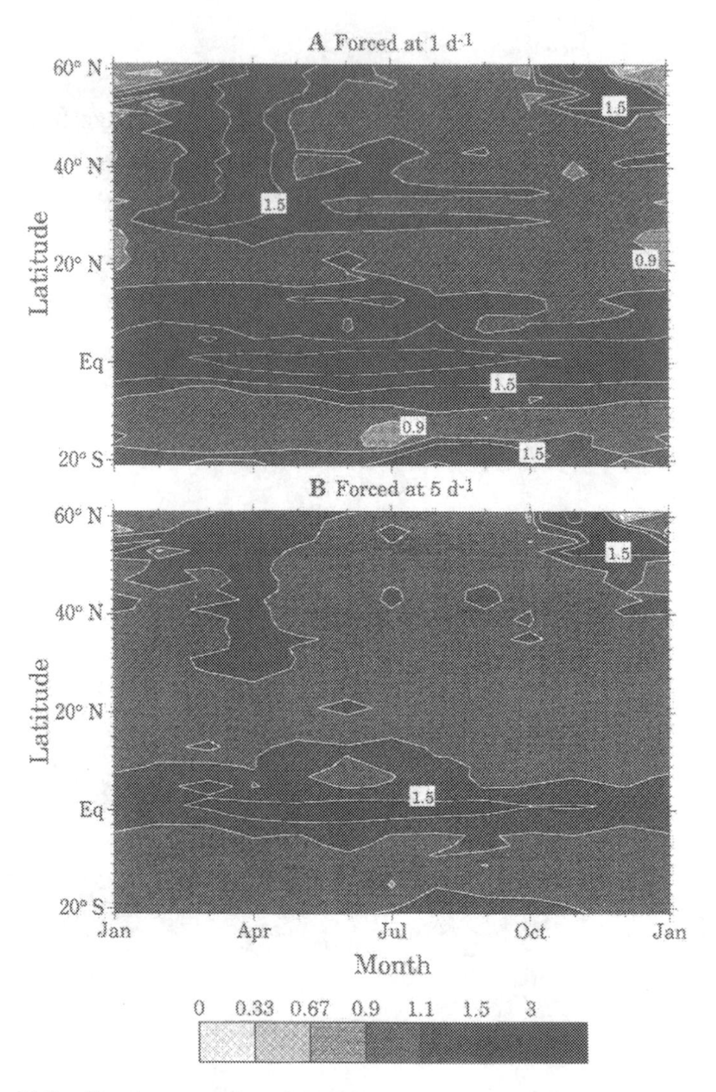

Figure 16-8. Zonal mean ratios of model chlorophyll to satellite chlorophyll for the multiple-chains model: (A) model forced at $\gamma = 1\ d^{-1}$; (B) model forced at $\gamma = 5\ d^{-1}$. Note that in (B) most of the plot is covered by ratios in the range of 0.9–1.1, indicating excellent agreement between model chlorophyll and satellite chlorophyll.

ecological predictions. Further progress must be made in several areas, however, before the promise of our approach can be fully realized. We plan an iterative approach, with successive rounds of improvement to the models, to attain our goals.

First, we must critically assess the ability of our approach to predict productivity

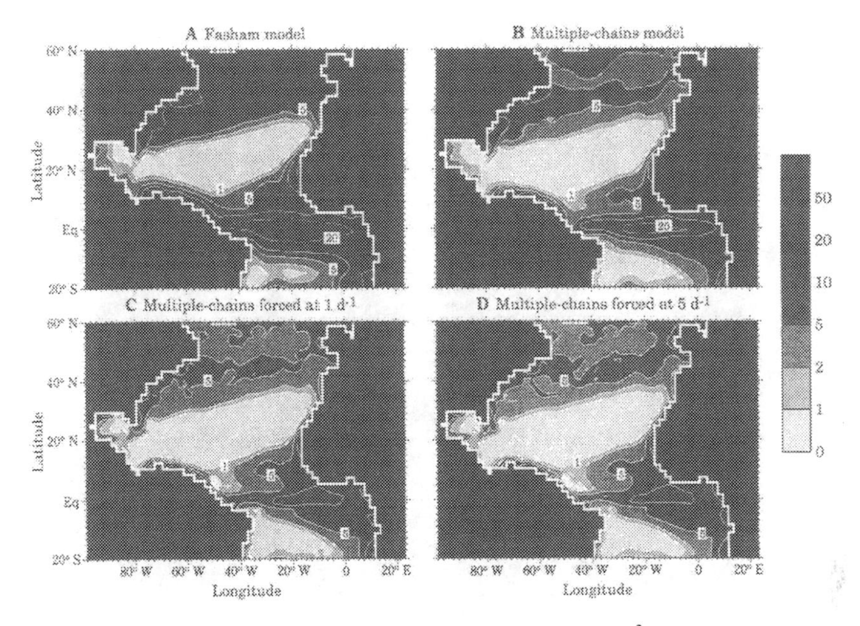

Figure 16-9. Annual total primary productivity (mmol N/m^3, yr) (vertical mean 0-123m): (A) the free-running Fasham model; (B) the free-running multiple-chains model; (C) the multiple-chains model forced at $\gamma = 1\ d^{-1}$; (D) the multiple-chains model forced at $\gamma = 5\ d^{-1}$.

from satellite chlorophyll estimates. For this test we must simultaneously evaluate shipboard measurements of productivity and satellite estimates of surface chlorophyll, as in the dataset of Balch et al. (1992). To make use of these data we must ensure that the surface chlorophyll values of GCM grid cells containing productivity measurements for given dates are forced toward the satellite chlorophyll estimates (for those dates) both long enough and hard enough to let the ecosystem model converge to the appropriate state. This can be accomplished by constructing from the satellite data appropriate "analysis fields," sets of spatial and temporal values [the P_{sat}'s of equation (1)] toward which model surface chlorophyll values must be forced (Harms et al. 1992). We must assess where (both spatially and temporally) the model performs well and where it performs poorly; this analysis should allow us to differentiate problems with the physical model from problems with the ecosystem model, allowing us to pinpoint deficiencies before successive rounds of testing.

Second, the multiple-chains Fasham model (Figure 16-5) is cumbersome at best, and we continue to seek simpler models that contain only those attributes that are necessary for successful assimilation of satellite data. For example, rather than modeling multiple size classes explicitly, it would be advantageous to parameterize the effects of size structure on algal growth rates and on algal and detrital sinking rates.

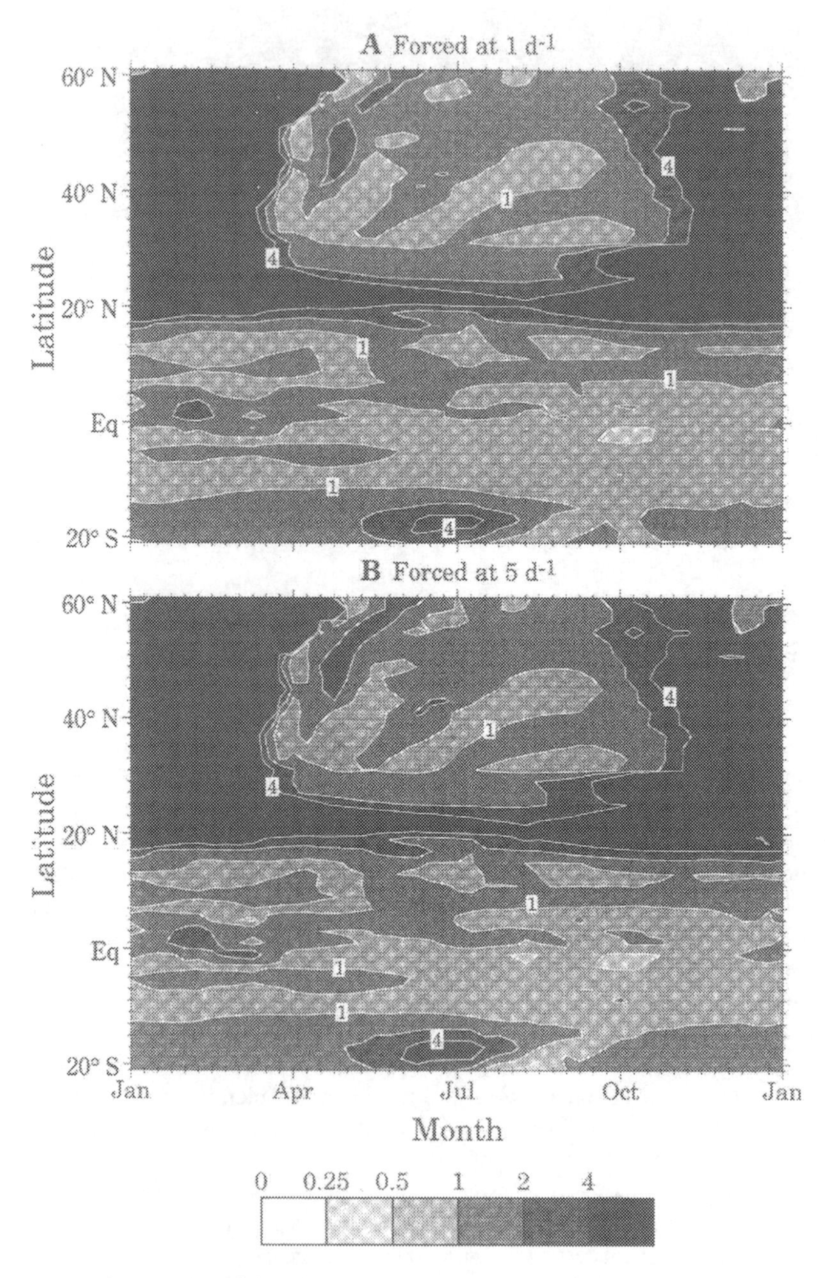

Figure 16-10. Ratios (forced:unforced) of primary production from multiple chains model runs: (A) the run with assimilation at $\gamma = 1\ d^{-1}$; (B) the run with assimilation at $\gamma = 5\ d^{-1}$.

Third, our models are nitrogen based, and we have used a constant chlorophyll:nitrogen (Chl:N) ratio to convert our nitrogen-based results to chlorophyll units. However, laboratory studies on nutrient-limited cultures of *Skeletonema costatum* (Sakshaug et al. 1989) showed more than tenfold variability of Chl:N ratios over a light gradient. We have recently modified the approach of Kiefer and Atkinson (1984; see also Kiefer and Mitchell 1983) to modeling variable Chl:N ratios. Adding these algorithms to a simplified model of the pelagic ecosystem dramatically improved the model's ability to predict simultaneously chlorophyll concentrations and productivities in Bermuda Atlantic Time Series (BATS) data; adding variable Chl:N ratios to the multiple-chains model should therefore increase considerably the predictive power of our method.

Progress on other fronts will also help in developing productivity time series. The physical model can be improved, and techniques for assimilating satellite data into the GCM can be implemented (e.g., Mellor and Ezer 1991). Data assimilation techniques can also be improved, especially in the area of deriving appropriate analysis fields (Gandin 1963; Derber and Rosati 1989) from satellite data, where coverage is often spotty because of cloud cover. Studies fitting time series data to the ecosystem model will help ensure that the level of recycling performed by the ecosystem model is appropriate, so that we may calculate appropriate *f*-ratios (Eppley and Peterson 1979) and new productions (Dugdale and Goering 1967) for use in monitoring pCO_2.

Finally, when the physical model and the ecosystem model have been developed to the point where we can make reliable predictions over most of the world ocean, there may still remain areas where the model must be forced hard toward lower phytoplankton concentrations. If such anomalies correspond spatially to current HNLC areas, their presence will indicate that processes not included in our ecosystem model (e.g., iron limitation) are occurring in these areas; we can then use the assimilation procedure to monitor the status of these areas.

Acknowledgments

We thank Geoff Evans for many helpful discussions. Support was provided by NASA grant No. NAGW-3137 and NSF grant No. OCE-9012333 to JLS. This paper is funded in part by Grant No. NA26RG0102-01 from the National Oceanic and Atmospheric Administration. The views expressed herein are those of the authors and do not necessarily reflect the views of NOAA or any of its subagencies.

References

Armstrong, R. A. In press. Grazing limitation and nutrient limitation in marine ecosystems: Steady-state solutions of an ecosystem model with multiple food chains. *Limnology and Oceanography*.

Balch, W. M. 1993. Reply. *Journal of Geophysical Research* **98:**16,585–16,587.

Balch, W., R. Evans, J. Brown, G. Feldman, C. McClain, and W. Esaias. 1992. The remote sensing of ocean primary productivity: Use of a new data compilation to test satellite algorithms. *Journal of Geophysical Research* **97:**2279–2293.

Banse, K. 1992. "Grazing, Temporal Changes in Phytoplankton Concentrations, and the Microbial Loop in the Open Sea." In P. G. Falkowski and A. D. Woodhead (eds.), *Primary Productivity and Biogeochemical Cycles in the Sea*. Plenum, New York, pp. 409–440.

Chavez, F. P. 1989. Size distribution of phytoplankton in the central and eastern tropical Pacific. *Global Biogeochemical Cycles* **3:**27–35.

Chisholm, S. W. 1992. "Phytoplankton Size." In P. G. Falkowski and A. D. Woodhead (eds.), *Primary Productivity and Biogeochemical Cycles in the Sea*. Plenum, New York, pp. 213–237.

Derber, J., and A. Rosati. 1989. A global oceanic data assimilation system. *Journal of Physical Oceanography* **19:**1333–1347.

Dugdale, R. C., and J. J. Goering. 1967. Uptake of new and regenerated forms of nitrogen in primary productivity. *Limnology and Oceanography* **12:**196–206.

Dugdale, R. C., A. Morel, A. Bricaud, and F. P. Wilkerson. 1989. Modeling new production in upwelling centers: A case study of modeling new production from remotely sensed temperature and color. *Journal of Geophysical Research* **94:**18,119–18,132.

Eppley, R. W., and B. J. Peterson. 1979. Particulate organic matter flux and planktonic new production in the deep ocean. *Nature* **282:**677–680.

Esaias, W. E., G. C. Feldman, C. R. McClain, and J. A. Elrod. 1986. Monthly satellite-derived phytoplankton pigment distribution for the North Atlantic Ocean basin. *EOS* (Transcripts AGU) **67:**835–837.

Fasham, M. J. R., H. W. Ducklow, and S. M. McKelvie. 1990. A nitrogen-based model of plankton dynamics in the oceanic mixed layer. *Journal of Marine Research* **48:**591–639.

Fasham, M. J. R., J. L. Sarmiento, R. D. Slater, H. W. Ducklow, and R. Williams. 1993. Ecosystem behavior at Bermuda Station "S" and Ocean Weather Station "India": A general circulation model and observational analysis. *Global Biogeochemical Cycles* **7:**379–415.

Frost, B. W., and N. C. Franzen. 1992. Grazing and iron limitation in the control of phytoplankton stock and nutrient concentration: A chemostat analog of the Pacific equatorial upwelling zone. *Marine Ecology Progress Series* **83:**291–303.

Gandin, L. S. 1963. Objective analysis of meteorological fields.

Gidrometeorologicheskoe Izdatel'stvo, Leningrad, USSR. (Israel Program for Scientific Translations, 1965).

Gordon, H. R., and D. K. Clark. 1980. Remote sensing optical properties of a stratified ocean: An improved interpretation. *Applied Optics* **19:**3428–3430.

Gordon, H. R., and W. R. McCluney. 1975. Estimation of sunlight penetration in the sea for remote sensing. *Applied Optics* **14:**413–416.

Gordon, H. R., D. K. Clark, J. W. Brown, O. B. Brown, and R. H. Evans. 1982. Satellite measurements of the phytoplankton pigment concentration in the surface waters of a warm core Gulf Stream ring. *Journal of Marine Research* **40:**491–502.

Harms, D. E., S. Raman, and R. V. Madala. 1992. An examination of four-dimensional data-assimilation techniques for numerical weather prediction. *Bulletin of the American Meteorological Society* **73:**425–440.

Harrison, W. G., T. Platt, and M. R. Lewis. 1987. f-ratio and its relationship to ambient nitrate concentration in coastal waters. *Journal of Plankton Research* **9:**235–248.

Hellerman, S., and M. Rosenstein. 1983. Normal monthly wind stress over the world ocean with error estimates. *Journal of Physical Oceanography* **13:**1093–1104.

Hutson, V. 1984. Predator mediated coexistence with a switching predator. *Mathematical Biosciences* **68:**233–246.

Ishizaka, J. 1990. Coupling of coastal zone color scanner data to a physical-biological model of the Southeastern U.S. continental shelf ecosystem. 3. Nutrient and phytoplankton fluxes and CZCS data assimilation. *Journal of Geophysical Research* **95:**20,201–20,212.

Kiefer, D. A., and C. A. Atkinson. 1984. Cycling of nitrogen by plankton: A hypothetical description based upon efficiency of energy conversion. *Journal of Marine Research* **42:**655–676.

Kiefer, D. A., and B. G. Mitchell. 1983. A simple, steady state description of phytoplankton growth based on absorption cross-section and quantum efficiency. *Limnology and Oceanography* **28:**770–776.

Levitus, S. 1982. *Climatological Atlas of the World Ocean*. NOAA Professional Papers 13. U.S. Government Printing Office, Washington, D.C.

Manabe, S., and R. J. Stouffer. 1993. Century-scale effects of increased atmospheric CO_2 on the ocean-atmosphere system. *Nature* **364:**215–218.

Martin, J. H., R. M. Gordon, and S. E. Fitzwater. 1991. The case for iron. *Limnology and Oceanography* **36:**1793–1802.

Mellor, G.L., and T. Ezer. 1991. A Gulf Stream model and an altimetry assimilation scheme. *Journal of Geophysical Research* **96:**8779–8795.

Moloney, C. L., and J. G. Field. 1989. General allometric equations for rates of nutrient uptake, ingestion, and respiration in plankton organisms. *Limnology and Oceanography* **34:**1290–1299.

———. 1991. The size-based dynamics of plankton food webs. I. A simulation model of carbon and nitrogen flows. *Journal of Plankton Research* **13:**1003–1038.

Morel, F. M. M., R. J. M. Hudson, and N. M. Price. 1991. Limitation of productivity by trace metals in the sea. *Limnology and Oceanography* **36:**1742–1755.

Mueller, J. L., and R. E. Lange. 1989. Bio-optical provinces of the Northeast Pacific Ocean: A provisional analysis. *Limnology and Oceanography* **34:**1572–1586.

Platt, T., and S. Sathyendranath. 1988. Oceanic primary production: Estimation by remote sensing at local and regional scales. *Science* **241:**1613–1620.

———. 1991. Biological production models as elements of coupled, atmosphere-ocean models for climate research. *Journal of Geophysical Research* **96:**2585–2592.

———. 1993a. Comment on "The remote sensing of ocean primary productivity: Use of a new data compilation to test satellite algorithms" by William Balch et al. *Journal of Geophysical Research* **98:**16,583–16,584.

———. 1993b. Estimators of primary production for interpretation of remotely sensed data on ocean color. *Journal of Geophysical Research* **98:**14,561–14,576.

Platt, T., P. Jauhari, and S. Sathyendranath. 1992a. "The Importance and Measurement of New Production." In P. G. Falkowski and A. D. Woodhead (eds.), *Primary Productivity and Biogeochemical Cycles in the Sea*. Plenum, New York, pp. 273–284.

Platt, T., S. Sathyendranath, O. Ulloa, W. G. Harrison, N. Hoepffner, and J. Goes. 1992b. Nutrient control of phytoplankton photosynthesis in the Western North Atlantic. *Nature* **356:**229–231.

Price, N. M., L. F. Andersen, and F. M. M. Morel. 1991. Iron and nitrogen nutrition on equatorial Pacific plankton. *Deep-Sea Research* **38:**1361–1378.

Raimbault, P., M. Rodier, and I. Taupier-Letage. 1988. Size fraction of phytoplankton in the Ligurian Sea and the Algerian Basin (Mediterranean Sea): Size distribution versus total concentration. *Marine Microbial Food Webs* **3:**1–7.

Sakshaug, E., K. Andresen, and D. A. Kiefer. 1989. A steady state description of growth and light absorption in the marine planktonic diatom *Skeletonema costatum*. *Limnology and Oceanography* **34:**198–205.

Sarmiento, J. L. 1986. On the North and Tropical Atlantic heat balance. *Journal of Geophysical Research* **91:**11,677–11,689.

———. 1993. Atmospheric CO_2 stalled. *Nature* **365:**697–698.

Sarmiento, J. L., and J. C. Orr. 1991. Three-dimensional simulations of the impact of Southern Ocean nutrient depletion on atmospheric CO_2 and ocean chemistry. *Limnology and Oceanography* **36:**1928–1950.

Sarmiento, J. L., R. D. Slater, M. J. R. Fasham, H. W. Ducklow, J. R. Toggweiler, and G. T. Evans. 1993. A seasonal three-dimensional ecosystem model of nitrogen cycling in the North Atlantic photic zone. *Global Biogeochemical Cycles* **7:**417–450.

Sathyendranath, S., T. Platt, E. P. W. Horne, W. G. Harrison, O. Ulloa, R. Outerbridge, and N. Hoepffner. 1991. Estimation of new production in the ocean by compound remote sensing. *Science* **353:**129–133.

Siegenthaler, U., and J. L. Sarmiento. 1993. Atmospheric carbon dioxide and the ocean. *Nature* **365:**119–125.

17

Long-Term Measurements at the Arctic LTER Site

John E. Hobbie, Linda A. Deegan, Bruce J. Peterson, Edward B. Rastetter, Gaius R. Shaver, George W. Kling, W. John O'Brien, F. S. Terry Chapin, Michael C. Miller, George W. Kipphut, William B. Bowden, Anne E. Hershey, and Michael E. McDonald

Ecologists are collecting a number of long-term datasets at the widespread sites of the National Science Foundation's (NSF) Long-Term Ecological Research (LTER) Program. The goals of the LTER are to carry out basic ecological research on questions best studied over years and decades and over large areas. In this chapter, we illustrate the decisions taken at one site, the Arctic LTER, about the questions appropriate for long-term study, and we show the multitude of ways in which long-term datasets have contributed to our understanding of the Arctic.

The LTER network began in 1981 and now has 18 sites in various ecosystems of the United States and the Antarctic. In addition to the site-based projects, LTER includes multisite collaborations in data collection and sharing, experimentation, and synthesis. Networking is greatly enhanced through an electronic communication system and development of standards for collection and management of data.

The research at the Arctic LTER site is funded by several projects from the NSF, the Department of Energy (DOE), and the Environmental Protection Agency; the core LTER funding provides long-term stability, maintains a number of large experiments, and collects monitoring data. The Arctic LTER site in the Toolik Lake region encompasses rolling tundra-covered hills, lakes, and streams in the foothills of the North Slope of Alaska along the Dalton Highway, which parallels the oil pipeline (Figure 17-1). The entire watersheds of Toolik Lake and the upper Kuparuk River are included in the site. Most of the LTER site falls within the 10-mile-wide pipeline corridor controlled by the Bureau of Land Management (BLM). A long-term research agreement with the BLM provides site protection, which is the first requirement for a long-term research site.

Each LTER site has its own goals and questions. The Arctic LTER poses three questions that apply across tundra, stream, and lake ecosystems.

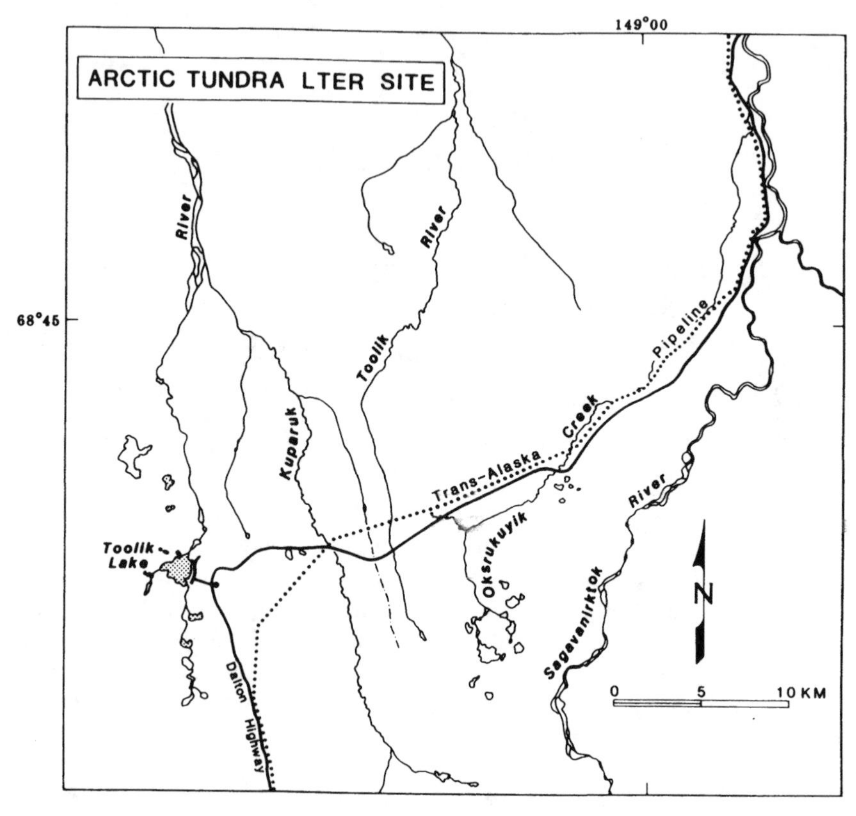

Figure 17-1. Location of Toolik Lake and the Kuparuk River research sites in northern Alaska.

(1) What are the long-term changes occurring at the LTER site?

(2) What is the extent of top-down control (by predators and grazers) versus bottom-up control (by resources such as nutrients for plants and algal mass for grazers) in arctic ecosystems?

(3) What are the rates and controls of the transfer of nutrients and organic matter from the land to the water in arctic ecosystems?

These questions are addressed by measurements of both natural and manipulated systems. That is, we gain understanding of rates and controls by observations of the existing ecosystems and by experimentally changing those factors or species that observations suggest may be important controllers. Observations on Toolik Lake and other small lakes in the area began in 1975, when the Dalton Highway opened. Stream and terrestrial data collection began soon thereafter.

Climate data have been collected continuously only since 1985, when a DOE-sponsored project began nearby; the LTER project began to collect climate and other long-term data in 1987. By 1993, the data collected included climate at several sites; lake and stream chemistry including nutrients, lake temperatures, and light penetration; biomass production of tussock tundra, streams, and lakes; and numbers of consumers including insects, zooplankton, and fish in streams and lakes. A large part of the Arctic LTER effort goes into long-term experimental manipulations such as changes in nutrients, temperature, and light on the tussock tundra and changes in nutrients and fish populations in lakes and streams.

The overall approach of the Arctic LTER research is question driven and experimentally oriented. This approach presented a dilemma during the planning of the monitoring for long-term changes: should the monitoring be planned to answer a question, or should everything possible be measured with the expectation that patterns would emerge with time? A further complication was the general belief of the researchers that no detectable climate change was expected to occur for decades in the Arctic. For this reason, the monitoring results would not be expected to reveal any actual changes, but instead would serve as a baseline for estimating changes in the distant future.

Another complication was that funds were limited, so only a few measurements could be made for each of the habitats. We therefore had to set priorities with the aid of conceptual models of how the arctic ecosystem works. We used the models to develop a list of physical drivers and biotic parameters that would describe both the current state and the most likely changes to the ecosystem. The conceptual model served as a linked series of hypotheses that have been used to guide our research. For example, one hypothesis was that differences in species response to temperature would result in changes in the species composition of tundra in response to climate change. One problem with setting priorities is that unanticipated changes might be missed. We were confident, however, that our choice of parameters to measure, such as the species composition, productivity, and nutrients, included most of the important biotic indicators of change of ecosystem structure and function.

A second type of long-term dataset monitors responses to experimental manipulations. These experiments are designed to change a physical, chemical, or biologic control of an ecosystem. Many experiments are carried out in one or several days and need no long-term datasets to determine the effects. Other experiments lead to changes in ecosystem properties or species that may take years to unfold.

The monitoring for both responses to climate change and the experimental manipulations produced results that added to our understanding of these arctic ecosystems. The results illustrated in the following sections of this chapter show the absolute necessity of a long-term approach and of long-term datasets to the LTER program and to a thorough ecological understanding.

Long-Term Datasets on Measurements of Natural Ecosystems

Detection of Climate Changes

All ecosystems vary naturally from year to year. The problem is to establish the long-term trends in these highly variable arctic systems. In view of the General Circulation Model (GCM) predictions for a doubled CO_2 atmosphere (e.g., Maxwell 1992)—that northern Alaska will have a 2–5° C increase in summer temperatures—the climate is an obvious driver to monitor; the LTER project has maintained a year-round climate station at Toolik Lake since 1987. This record is too short to have any meaning at this point, but Weather Bureau data from two North Slope weather stations, Barrow and Prudhoe Bay, provide a longer record of regional change. Both stations are on the coastal plain, at the edge of the Arctic Ocean. Prudhoe Bay is 200 km directly north of Toolik Lake, and Barrow lies to the northwest. The data for the summer months (Figure 17-2A) show that in recent years the summers at both stations have been much warmer than the long-term mean.

Air temperature is related to many aspects of the arctic system, including precipitation, depth of the thawed layer, cloudiness, and length of the growing season. One direct link between climate and the ecology of lakes is water temperature, which is controlled by air temperature and solar radiation. Lake temperature integrates both radiation and air temperature and may serve as a useful indicator of long-term trends. One of the many ways in which water temperature is important in arctic lakes is its effect on the growth of fish. Lake trout, a dominant predator in Toolik Lake, are food limited and can grow only when water temperatures are cool enough so that the fish's metabolic activity is relatively low.

The Toolik Lake record of summer temperatures in the epilimnion, the mixed layer of the lake above a summer thermocline, indicates warmer water temperatures in recent years (Figure 17-2B), similar to the air temperature trend at Prudhoe Bay and Barrow weather stations. It is impossible, of course, to say that this warming is caused by long-term climate change, but it is a regional phenomenon.

As demonstrated above, one value of long-term datasets is finding a change in an ecologically important factor. In this case, lake temperature changed over the course of our research projects and tracked regional warming of the North Slope of Alaska. We can now continue to search for the ecological changes this physical change has induced.

Detection of Correlations—Climate and Plant Flowering

Climate strongly limits the distribution of plants in the Arctic, but there are also strong year-to-year variations in plant growth and reproduction that may be related to variations in weather. Shaver et al. (1986) analyzed 34 sites, including

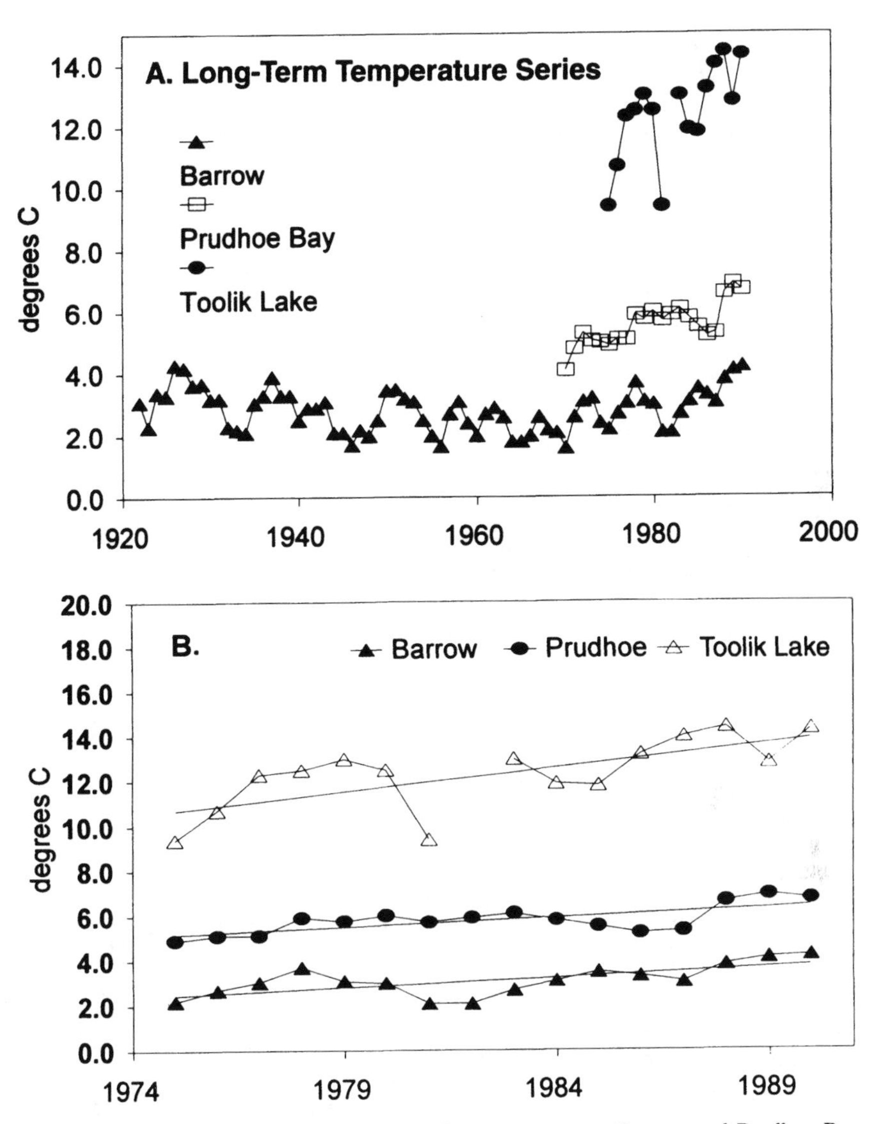

Figure 17-2. (A) July and August mean air temperatures at Barrow and Prudhoe Bay and the mean July temperature of the epilimnion of Toolik Lake in recent years (data from Oechel et al. 1993 for Barrow and Prudhoe Bay and from M. C. Miller for Toolik Lake). (B) Linear regression fits of the temperature records. The r^2 values are 0.38, 0.42, and 0.48 for Barrow, Prudhoe Bay, and Toolik Lake, respectively.

one at Toolik Lake, in an 800-km south-to-north transect along the Elliott and Dalton Highways in northern Alaska from Fairbanks to Prudhoe Bay. The authors observed growth, flowering, and nutrient content of one plant species, *Eriophorum vaginatum*, or arctic cotton grass, over four years. Years of high flower density were simultaneous across the entire range of the transect (Figure 17-3). Although this regional uniformity has been observed previously in an anecdotal way in the Arctic, it has not been quantified.

The ecological question is: what controls such clear, uniform annual variations in an important plant process over an entire region? Weather is the most likely control. However, there is no correlation between flowering and weather for the present year or for the previous year. For example, 1981 was a very good year for flowering (Figure 17-3); this was an exceptionally cold year that was preceded by several relatively warm years (Figure 17-2). Yet 1983 was also a year of high

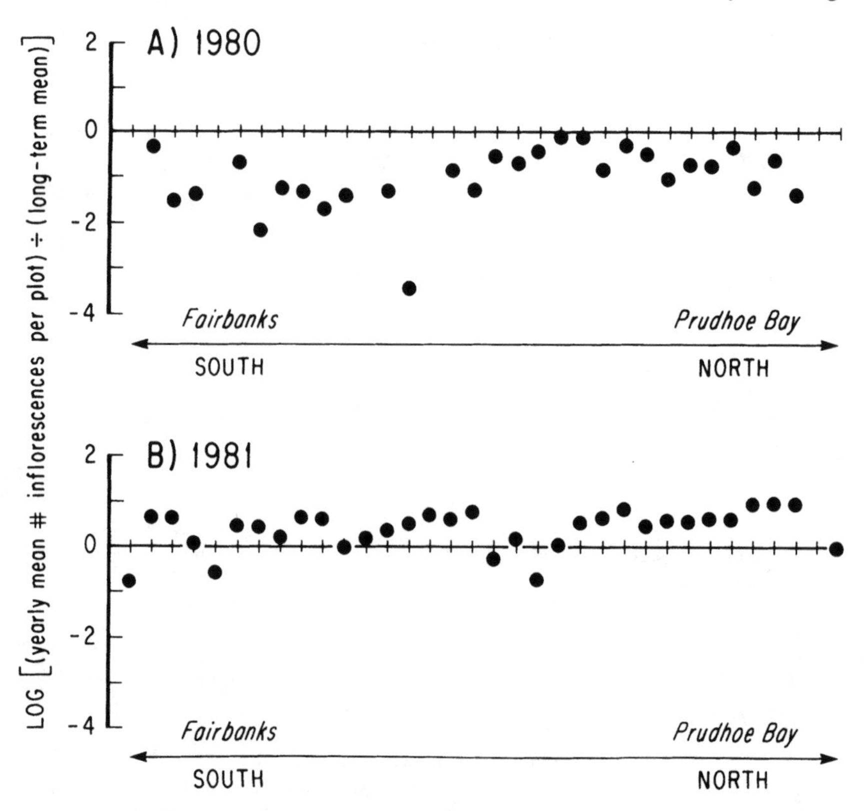

Figure 17-3. The yearly mean inflorescence density relative to the long-term mean for 34 sites along the Dalton Highway from Fairbanks to Prudhoe Bay. Data for an above-average-flowering (1981) and a below-average-flowering (1980) year are calculated from the natural logarithm of the yearly mean divided by the long-term mean (data from Shaver et al. 1986).

flowering (data not shown), but it was close to average in its summer temperatures and was preceded by two to three cold years (Shaver et al. 1986). We know that in this species, flower buds are actually formed one or two years before they emerge, and therefore the climate in the years before flowering must be important. However, this is not a complete enough answer to be ecologically useful. One possibility is that the effect could occur through the temperature mediation of decomposition in the soil and the subsequent release of nutrients. Answers to this question are currently being sought by maintaining our long-term monitoring of *Eriophorum vaginatum* flowering and climate.

Detection of Plant Community Change

The interactions among tundra plants will also change as a result of different responses to changed temperatures, nutrient availability, and growing season length. The outcome of this competition among plants for nutrients and light may well change the structure or composition of the plant community. At first, the same plants will be present, but their proportional composition in the community will change; later, certain species of plants may drop out of the community altogether.

At the LTER site, Donald and Marilyn Walker of the University of Colorado have established a detailed hierarchical database of vegetation that will allow changes to be monitored. The LTER project has contributed funds for their work, but most of their funds have come from DOE's "R4D" project. The basis for their vegetation mapping is a carefully surveyed 1-km × 1-km grid of permanent stakes (every 100 m) at known latitude and longitude. At the finest resolution of < 1 cm, the Walkers have constructed maps of the current vegetation inside a 1-m × 1-m plot close to each of the 100 stakes. They have also used aerial photos to construct vegetation maps of the entire 1-km × 1-km grid and of a 5-km × 5-km grid (1 m and 10 m resolution).

Community composition is also being monitored at other sites located on different types of terrain. At a tussock tundra site near Toolik Lake, a multiyear dataset reveals that there has been a significant change in the composition of the vegetation. The total vascular plant biomass (graminoids, evergreens, deciduous plants) has remained nearly constant over the period of observation, whereas the graminoid biomass has dropped by two-thirds (e.g., *Eriophorum vaginatum* in Figure 17-4). The difference has been made up by the biomass of deciduous shrubs, especially of *Betula nana,* the dwarf birch. Given that the data cover only a few years, we do not know how long this shift will continue or if shifts like these are transient.

These results lead to the hypothesis that the total biomass of the plant community is controlled by the nutrient supply rate. This supply rate is well buffered against change, perhaps over decades. Year-to-year variations do occur, leading to changes in the allocation of the nutrient supply and resulting small changes

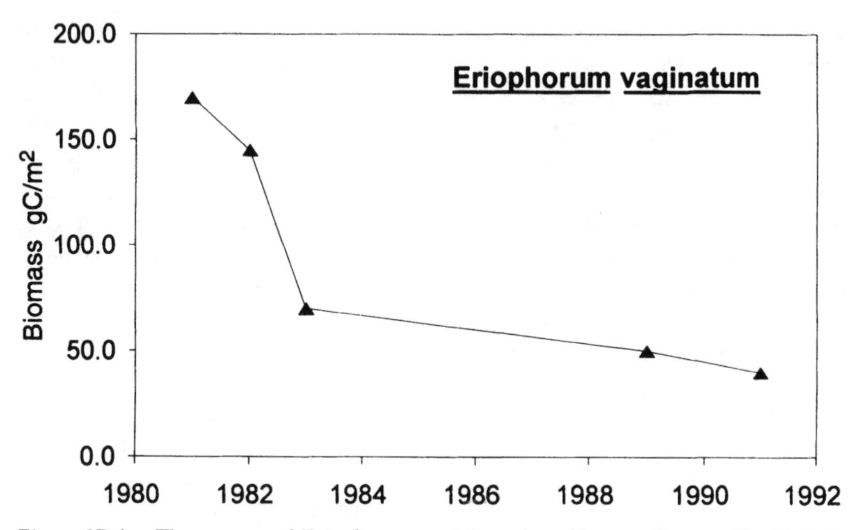

Figure 17-4. The amount of *Eriophorum vaginatum* in a ridgetop plot near Toolik Lake (results of F. S. Chapin and G. R. Shaver).

in abundance of extant species. These changes in the species composition have potential ramifications for the population dynamics of caribou (bottom-up control). The flowering heads of *Eriophorum vaginatum* are an important energy and protein source for caribou in early spring, when calving occurs.

Detection of Correlations—Stream Flow and Fish Production

The variability of components of natural systems often provides the opportunity to develop ecological correlations. Sometimes these correlations are robust enough to be used to predict events. For example, an extensive dataset on the positive correlation between the annual rate of input of phosphorus to lakes and the amount of algal growth (Persson et al. 1975) shows that Toolik Lake, despite its northern location, responds to phosphorus addition in the same way as temperate lakes. Usually the correlation information is not sufficient to draw cause-and-effect conclusions but does provide a valuable indication of factors and relationships to be examined experimentally.

One correlation that has arisen from the long-term dataset is the direct relationship between annual stream flow and the growth of fish in the upper Kuparuk River (Figure 17-5). Arctic grayling, the sole species of fish in the river, feed primarily on insects that drift in the current. Because the headwater reaches of this river freeze completely and the river stops flowing during the winter, all of the grayling in the upper 50 km of the river must migrate south to a single lake to survive (small lake close to the bottom of Figure 17-1). We have tagged almost half the adult population of these migratory grayling and catch and weigh

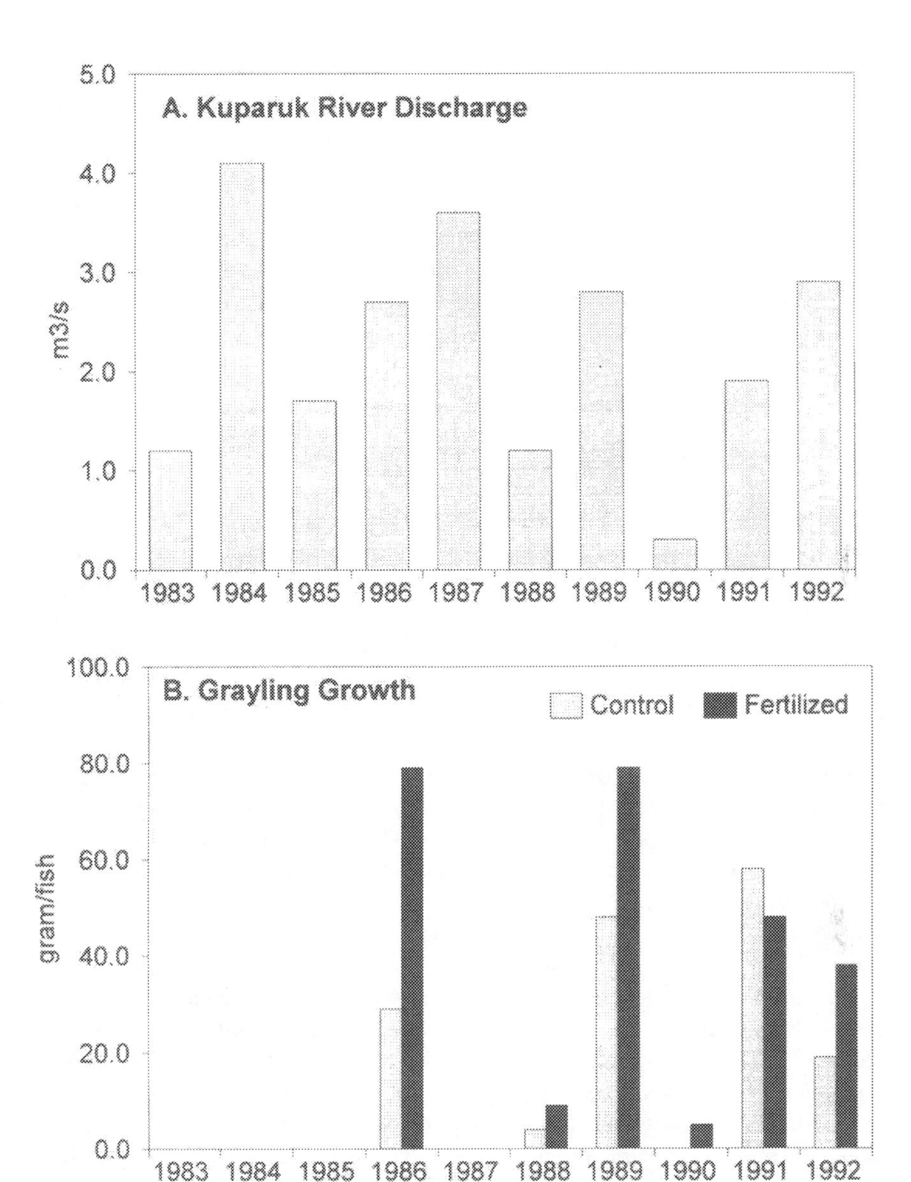

Figure 17-5. (A) Mean discharge in the upper Kuparuk River for the period of phosphorus fertilization each summer (1 July–20 August). (B) The summer growth of arctic grayling (*Thymallus arcticus*) in the Kuparuk River for a number of years. Growth in the control reach is shown by the shaded bars; growth in the fertilized reach is shown as solid bars (1986–1991 data from Deegan and Peterson 1992).

them every few years when they return to the lake. Growth is estimated through mark-and-recapture in the study reaches of the river.

Over a number of years, the growth of adult grayling is greatest at high mean summer flows and least at low mean flows (Figure 17-5). Although correlation does not give us the ecological understanding we desire, it does indicate a profitable area for future research. Future efforts will investigate the effects of temperature and water flow on the drifting insects, the main food of the grayling, and the effects of warm temperatures on the metabolic requirements of the fish. Without the long-term approach and the long-term datasets, these correlations would not have been observed and the questions would not have arisen.

Long-Term Datasets on Measurements of Manipulated Ecosystems

Effects of Lake Predator Experiments

From the beginning of observations on Toolik Lake in 1975, the plankton community intrigued scientists with its combination of small and large species of zooplankton and of predaceous fish. A number of studies of temperate lakes suggested that the fish should eliminate the large zooplankters. Several species found in the 1970s in Toolik Lake, for example, *Daphnia middendorffiana* and *Holopedium gibberum,* attain 2 mm in length; these are among the giants of the freshwater zooplankton world. We concluded that the paradox of the coexistence of these large zooplankters and fish could be explained by the rarity of planktivorous fish. It became apparent over the next decade that something was happening to the zooplankton, however, for the abundance of these two large zooplankton species dropped a hundredfold (Figure 17-6). By 1993 these species were not detectable in Toolik Lake. The virtual extinction is likely caused by biological interactions in this one lake; the populations of *Daphnia middendorffiana* have not changed over the 18 years of observations in unfished lakes near Toolik Lake.

Our current hypothesis is that we have been observing an unplanned experiment in Toolik Lake as a result of human removal of larger lake trout (McDonald and Hershey 1989). In the undisturbed lake, the predation of the large lake trout controlled the abundance and size of all the fish species in the lake. This resulted in the smaller zooplanktivorous fishes' being forced into near-shore, rocky refuges. When the pipeline construction began and a 400-person camp was built in 1970, followed by a road in 1975, recreational fishing began to remove the large lake trout, a process that still continues. Grayling and small lake trout, which are the predators of the large zooplankton, have become more abundant and are responsible for the decline of the *Daphnia* and *Holopedium*. Evidence about the changes in fish populations is difficult to obtain in this very oligotrophic lake, because a detailed study would likely destroy most of the population. Two surveys of the length and weight of lake trout in Toolik Lake have been made, one in 1977

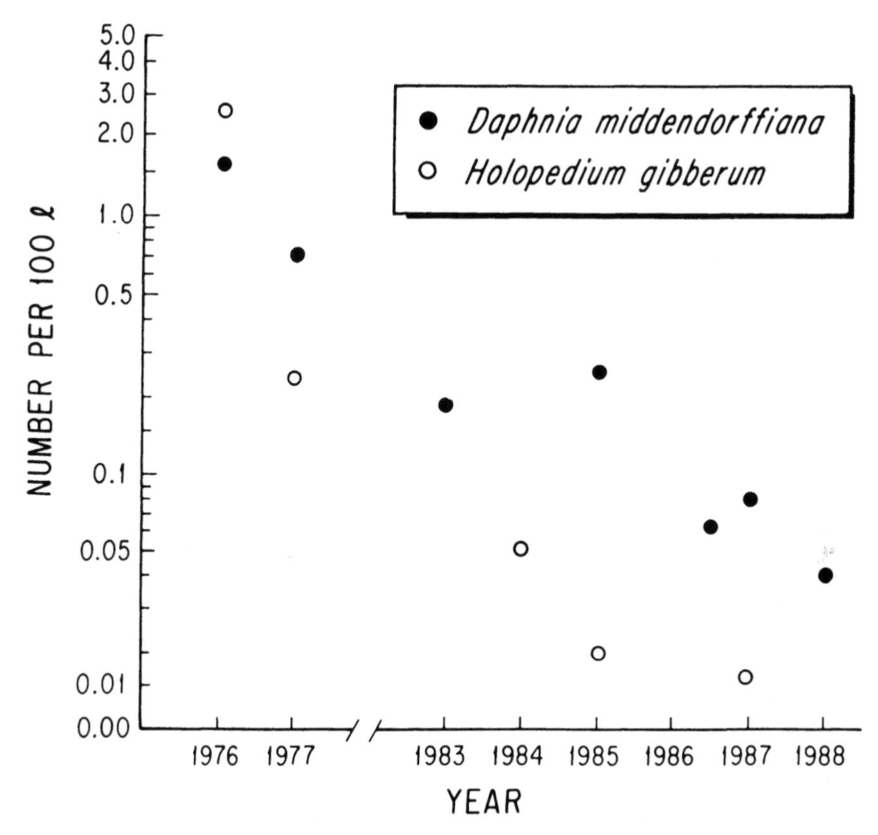

Figure 17-6. Large-bodied zooplankton abundance in Toolik Lake, 1976–1988 (data from W. J. O'Brien).

and one in 1987 (reported in McDonald and Hershey 1989). The median fork length declined from 388 to 313 mm in this decade, while the median weight declined from 578 to 318 g. Additional support for the hypothesis comes from changes within the lake of the distribution of the populations of a prey fish, the slimy sculpin (McDonald and Hershey 1992). Between 1978 and 1987, the percentage of the population caught over soft sediments increased; the authors interpreted this and other evidence to indicate that predation risk from lake trout had decreased over this period.

To test this hypothesis, we are currently carrying out an experiment in two large lakes near Toolik Lake. After several years of observations to establish a baseline dataset on zooplankton and fishes, we removed lake trout from one lake and added lake trout to another lake where they had been previously absent. The long-term dataset on zooplankton in Toolik Lake has enabled us to formulate an hypothesis and develop an experimental test.

The Effect of Phosphorus Addition to a River

One scenario for the future is that changes in land use and in climate in the Arctic will result in additional nutrients' reaching streams, rivers, and lakes. We have carried out nutrient addition experiments in the Kuparuk River since 1983 to determine the response of this stream ecosystem to added nutrients, as well as to find out more about the pathways of nutrient cycling (see Peterson et al. 1993). For four to six weeks each summer for the last 10 years, phosphate has been continuously added to the river as phosphoric acid at a rate designed to raise the concentration in the stream by 0.3 μM P (or 10 μg PO_4−P/liter). The section of stream immediately upstream from the point of addition serves as the experimental control and reference. From these and other experiments we know that phosphorus is the limiting nutrient for algae in the river; however, nitrate is present in near-limiting quantities as well.

The algae attached to the rocks in the riffles of the river greatly increased their photosynthesis and their biomass (measured as chlorophyll) in the first two years of phosphorus addition (Figure 17-7). This build-up of algal biomass was expected, but the low level of algae present in the fertilized reach in subsequent years was a surprise. We hypothesized that the insect larvae in the stream responded after a two-year lag period and were sufficiently abundant after the first two years of the experiment to control algal growth by grazing. Gibeau and Miller (1989) tested this hypothesis in 1987 by means of stream incubations of

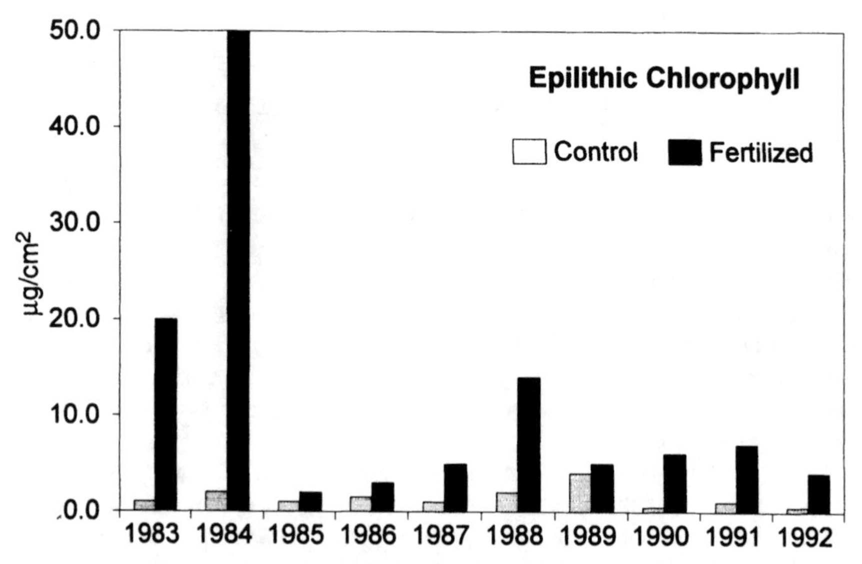

Figure 17-7. The mean concentration of epilithic chlorophyll on riffle rocks of the upstream control (shaded) and downstream fertilized (solid) reaches after four to six weeks of continuous phosphorus addition each year (1983–1986 data from Peterson et al. 1992).

ceramic disks attached to small vials containing various combinations of nutrients and an insecticide. Abundant algal biomass developed only on the disks with nutrients plus the insecticide, indicating that insects were likely holding down the algal biomass by grazing. A repeat of this experiment in 1988 did not yield these differences; evidently the insects, especially the chironomid larvae, are able to exert this control of the algae only when the insects are as abundant as they were in 1987.

In the past two years, after nine years of fertilization, we were surprised to observe that mosses had become a dominant organism in the fertilized reaches of the Kuparuk River (Figure 17-8; Bowden et al. in press; Finlay and Bowden in press). Mosses are always present in the river—for example, in the reach upstream from the phosphate dripper—but are not abundant. Experiments have shown that *Hygrohypnum* grows rapidly in the phosphate-enriched zone, but very slowly in the control zone. We hypothesize that *Hygrohypnum* abundance in the fertilized reach is a consequence of several years of exponential growth under nutrient-rich conditions (Figure 17-8). Again, the long-term dataset was important in the documentation of this and other changes in the experimental reach of the river.

The Effect of Nutrient Additions to a Lake

The effect of added nutrients on lake ecosystems was studied in a fertilization experiment in Lake N-2, located 0.5 km northwest of Toolik Lake. The lake was divided with a polyethylene curtain in 1985, and the outflow half of the lake was fertilized with inorganic nitrogen and phosphorus for six years. The past three years of study (1991 to present) have provided information on how fast an arctic lake can recover from such a perturbation.

As expected, the phytoplankton primary productivity and biomass responded in nearly direct proportion to the amount of nutrients added; the rate of addition per square meter of lake surface was set at five times the estimated rate for nutrients entering Toolik Lake. Overall, the phytoplankton primary productivity in the fertilized side of Lake N-2 increased about fivefold (Figure 17-9). Fertilization was discontinued after 1990, and there was a dramatic drop in primary production in 1991. The most interesting observation was how slowly primary production increased from year to year. Production doubled over the first four years (1985 to 1988) and then doubled again in the next two years (1989, 1990) despite the large amount of nutrients added to the lake each summer. We know from other observations that the explanation lies in the interaction between phosphate in the water column and the very abundant iron in the lake sediments. For most of the experimental period, the added phosphate became chemically bound to the iron in the sediments. Toward the end of the experimental period, the increased rain of plankton to the sediment covered most of the iron. During 1990, the last year of nutrient addition, we were

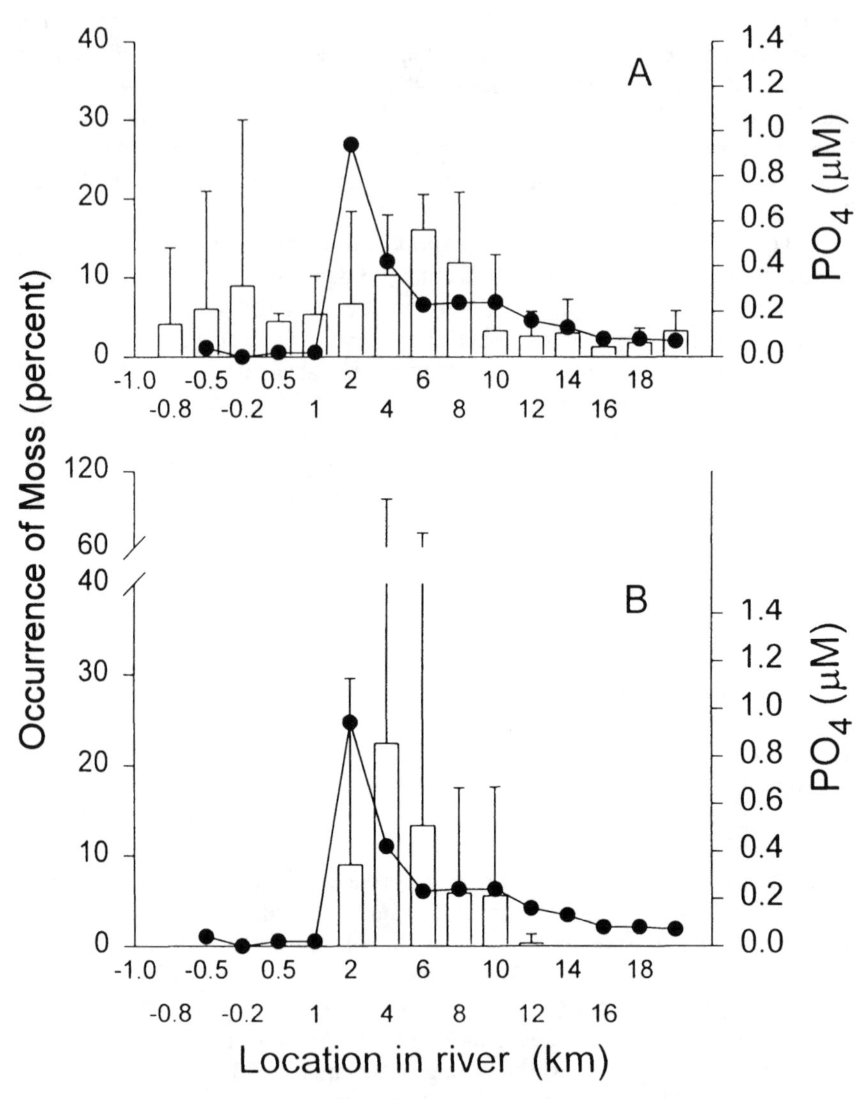

Figure 17-8. Percentage occurrence of mosses (bars) during a July and an August sampling in 1992 and the concentration of phosphate (closed circles) during late August 1992 in the Kuparuk River at various distances (km) from the phosphate dripper. The dripper was located at 0.59 km; downstream distances are positive and upstream distances are negative. The percentage occurrence was calculated as the mean percentage occurrence of moss at points along 5 or 10 transects at each sampling location. Points (at least 100 per transect) were usually spaced 10 cm apart. The standard error for n = 5 is shown for each bar. Two species are shown: *Schistidium agassizii* (A) occurs everywhere in the river, whereas *Hygrohypnum* spp (B) occurs only in the fertilized zone of the river (data from Bowden et al. in press).

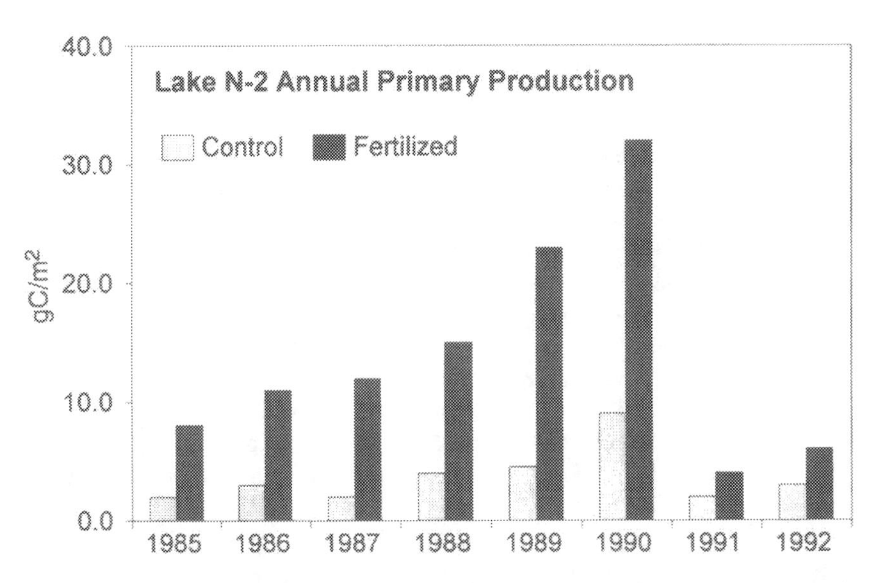

Figure 17-9. The annual primary production in the plankton of Lake N-2 measured in the control (shaded) and fertilized (solid) sides. Measurements were made each year over the period of 1 July to 15 August (data from M. C. Miller.)

able to measure the first significant phosphate flux back from the sediments to the water column. The importance of sediment chemistry in sequestering nutrients is reinforced by the dramatic drop in primary productivity after the cessation of fertilization in 1990.

Responses of Tussock Tundra to Changes in Light and Nutrients

Any changes in land use and climate are likely to alter the vegetation growth patterns and eventually the structure of the plant community. Our field experiments in tussock tundra at Toolik Lake provide a good example of how variation in the rates of response of different species can lead to different interpretations of results of the same experiment unless measurements are made over a long term. We changed air temperature on our experimental plots by placing small greenhouses over the tundra, changed light intensity by shading, and changed nitrogen and phosphorus availability by adding fertilizer. We also combined both fertilizer and a greenhouse in one treatment.

We maintained these experiments for nine consecutive summers and harvested the plants on the plots twice in that time, once after three years and once after nine years. The effects of treatment on above-ground net primary production of vascular plants were nearly identical in both years (Figure 17-10), with the greatest responses occurring on the sites that received fertilizer. The greenhouse heating alone also stimulated net primary productivity (NPP) in both years, although not significantly, and the shade treatment reduced NPP.

The similarity between the graphs in Figure 17-10 might at first lead to the conclusion that repeated observation of such an experiment was unnecessary. However, a closer look at the data shows that the compositions of the two harvests were quite different (Figure 17-11). The data from 1983 indicate that the grasses and sedges ("graminoids") responded most rapidly to fertilization, and that the deciduous shrubs were able to take advantage of fertilizer only when air temperatures were also increased. But by 1989, the main component of the increased productivity in the two fertilized plots was the dramatic increase in the slower-growing deciduous shrubs.

The ability to compare data, separated by six years, allows us much greater insight into controls over both the productivity and the composition of the productivity of this tussock tundra ecosystem. Nutrient supply now appears to be the

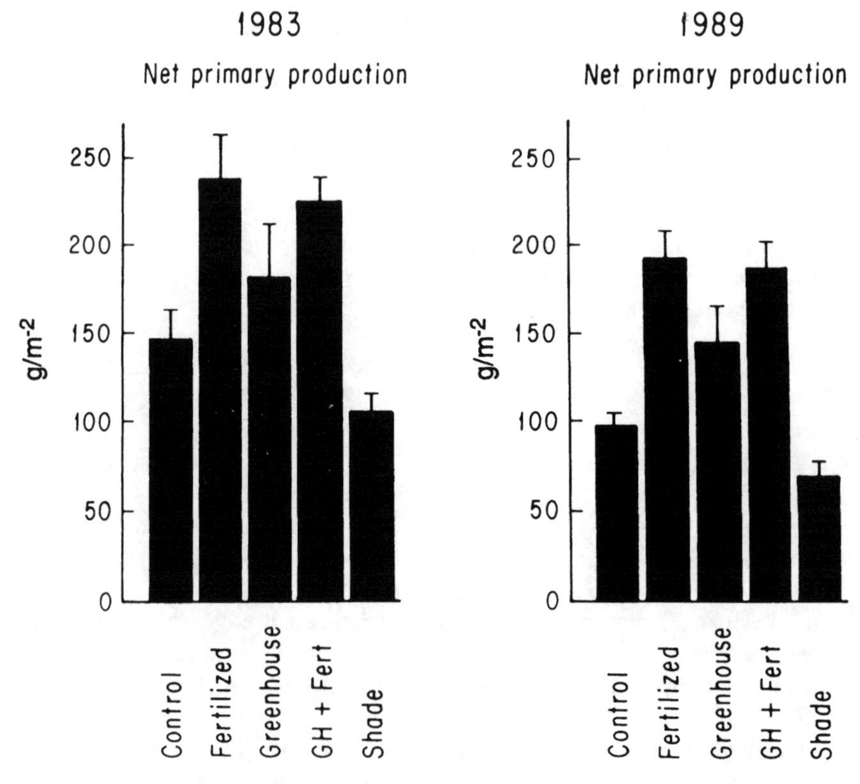

Figure 17-10. Annual above-ground net primary production, as measured by harvest of experimental plots in tussock tundra (not including root production or stem secondary growth). Plots were harvested in 1983 and again in 1989 (data from F. S. Chapin and G. R. Shaver).

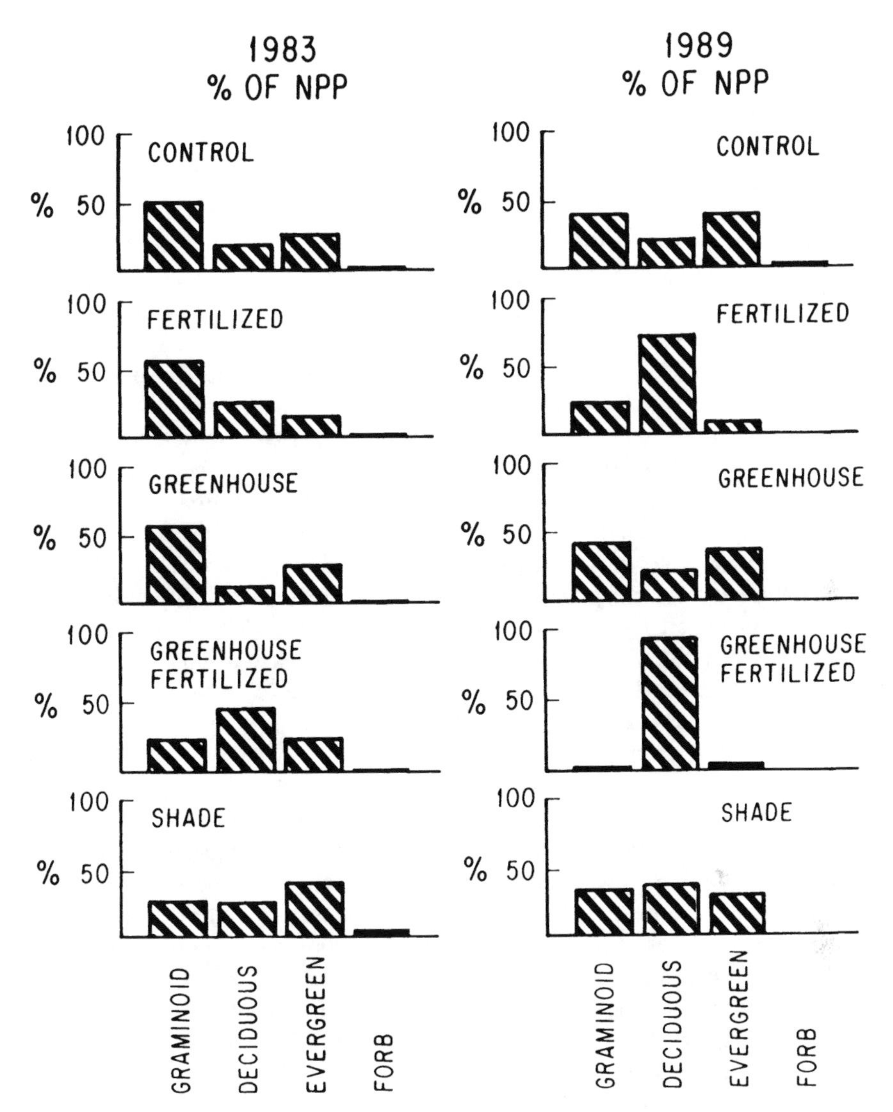

Figure 17-11. The data from the 1983 and 1989 harvests of experimental plots in tussock tundra (Figure 17-10), expressed as the percentage of net production composed of graminoid, deciduous, evergreen, and forb types of plants.

primary control over net productivity, although reduction of light below ambient levels substantially reduces productivity; air temperature has relatively little effect. The small increases in productivity in the greenhouse-only treatment probably result from increased soil temperatures and nutrient mineralization. Although nutrient supply also controls the eventual composition of the community, temperature plays a major role in controlling the rate of change in community composition. Species differ in the rate at which they respond to changes in air temperature and nutrient availability.

Conclusions

At the Arctic LTER research site, we have made extensive use of long-term datasets. As shown by the preceding examples, long-term datasets are crucial to interpretations of the current state of arctic ecosystems and our predictions of how they may change. A number of testable hypotheses have arisen from the long-term measurements of tundra, stream, and lake ecosystems. In some cases, the natural changes led to correlations among components of these systems that suggested causality. An example is the correlation between stream flow and growth of fish. But we have also made observations on a number of long-term experiments. Here, long-term datasets are vital because the first results of an experimental manipulation were often not the final results. For example, the conclusions we would have reached after three years of the tussock tundra light, temperature, and nutrient manipulation would have been quite different from the ones we reached after nine years of the experiment.

One reason for the slowness of these arctic ecosystems to respond is the short period of activity each year. Stream insects, for example, have only a single generation per year, so it is entirely understandable that stream insect populations take a number of years to respond to an increase in productivity of the stream algae.

In this chapter, we have discussed examples of the most successful long-term datasets: those that led to hypotheses or showed unexpected results of experiments. Most of these successful datasets turn out to be measurements of biological responses, including biomass changes, community composition changes, and changes in rates of an ecosystem function such as primary production. We suggest that the biota are much more sensitive indicators of change than are physical or chemical measurements. The biota integrate small changes over long time periods and are very sensitive to small changes in nutrient supply or in predation pressure.

It is very likely that as these long-term datasets continue to be collected, they will become even more valuable. New biological insights will emerge, and the patterns of change of physical and chemical factors, such as climate and stream-water chemistry, will become evident.

Acknowledgments

The research reported here has been funded by a series of grants from the National Science Foundation. These include BSR 8205344, DEB (BSR) 9019055, DEB 9211775, and OPP 9024188.

References

Bowden, W. B., J. C. Finlay, and P. E. Maloney. In press. Long-term effects of PO_4 fertilization on the distribution of bryophytes in an arctic river. *Freshwater Biology*.

Deegan, L. A., and B. J. Peterson. 1992. Whole river fertilization stimulates fish production in an arctic tundra river. *Canadian Journal of Fisheries and Aquatic Science* **49:**1890–1901.

Finlay, J. C., and W. B. Bowden. In press. Controls on production of bryophytes in an arctic tundra river. *Freshwater Biology*.

Gibeau, G. G., and M. C. Miller. 1989. A micro-bioassay for epilithon using nutrient diffusing artificial substrata. *Journal of Freshwater Ecology* **5:**171–176.

Maxwell, B. 1992. "Arctic Climate: Potential for Change Under Global Warming." In F. S. Chapin, J. F. Reynolds, R. L. Jefferies, G. R. Shaver, J. Svoboda, and E. W. Chu (eds.), *Arctic Ecosystems in a Changing Climate*. Academic Press, New York, pp. 11–34.

McDonald, M. E., and A. E. Hershey. 1989. Size structure of a lake trout (*Salvelinus namaycush*) population in an arctic lake: Influence of angling and implications for fish community structure. *Canadian Journal of Fisheries and Aquatic Science* **46:**2153–2156.

McDonald, M. E., and A. E. Hershey. 1992. Shifts in abundance and growth of slimy sculpin in response to changes in the predator population in an arctic Alaskan lake. *Hydrobiologia* **240:**219–223.

Oechel, W. C., S. J. Hastings, G. Vourlitis, M. Jenkins, G. Riechers, and N. Grulke. 1993. Recent change of Arctic tundra ecosystems from a carbon sink to a source. *Nature* **361:**520–526.

Persson, G., S. K. Holmgren, M. Jansson, A. Lundgren, B. Nyman, C. Solander, and C. Anell. 1975. "Phosphorus and Nitrogen and the Regulation of Lake Ecosystems: Experimental Approaches in Subarctic Sweden." In *Proceedings of the Circumpolar Conference on Northern Ecology*. National Research Council, Canada, pp. III-1–19.

Peterson, B. J., L. Deegan, J. Helfrich, J. E. Hobbie, M. Hullar, B. Moller, T. E. Ford, A. Hershey, A. Hiltner, G. Kipphut, M. A. Lock, D. A. Fiebig, V. McKinley, M. C. Miller, J. R. Vestal, R. Ventullo, and G. Volk. 1993. Biological responses of a tundra river to fertilization. *Ecology* **74:**653–672.

Shaver, G. R., N. Fetcher, and F. S. Chapin III. 1986. Growth and flowering in *Eriophorum vaginatum:* Annual and latitudinal variations. *Ecology* **67:**1524–1535.

18

Basic Models in Epidemiology

Fred Brauer and Carlos Castillo-Chavez

The longstanding debate as to whether or not chaotic dynamics occur in natural populations (see Hastings et al. 1993) arises because various natural populations exhibit apparently "random" fluctuations in abundance. Furthermore, in several instances these fluctuations will look random not only to the naked eye but also through the "eyes" of some standard time series analysis (see Ellner 1989). Several simple nonlinear mathematical ecological models that exhibit chaos for apparently reasonable parameter values have been studied over the last few years. Initially chaos was found in simple discrete time models for single species (May 1974, 1976; May and Oster 1976). Chaotic dynamics have also been encountered in discrete models with age structure (Ebenman 1987; Guckenheimer et al. 1977; Levin 1981; Levin and Goodyear 1980); in discrete models with two species (Allen 1989a,b, 1990a,b, 1991; Beddington et al. 1975; Bellows and Hassell 1988; May 1974; May and Oster 1976; Neubert and Kot 1992); in simple discrete models of parasites (May 1985); in models for host-parasitoid-pathogen systems (Hochberg et al. 1990); in discrete demographic models with two sexes (Caswell and Weeks 1986); and in models that include frequency-dependent selection (Altenberg 1991; Cressman 1988; May and Anderson 1983a). Continuous-time ecological models that exhibit chaos have also been found in Lotka-Volterra-type models (Gilpin 1992; Schaffer 1985a,b; Gardini et al. 1987; Takeuchi and Adachi 1983) and in three-species food web chain models (Hastings and Powell 1991). Chaos has also been found in epidemic models (Kot et al. 1988; Olsen and Schaffer 1990; Schaffer 1985a; Schaffer and Kot 1985; Schaffer et al. 1990). Reviews that include many more models that have exhibited chaos in ecology include those of May (1987) and Hastings et al. (1993, the organizing reference for the above-cited sources). In summary, simple and strategic ecological models can exhibit chaotic dynamics for what appears to be likely regions of parameter space, and current research focuses on increasing our understanding of the mechanisms that are more likely to lead to chaos, as well as their implications in theoretical biology.

The study of chaos in ecology through the use of strategic models can be complemented by the analysis of time series because some parameter estimates will have errors, other parameters will not be measurable, and our results will be model dependent. Several time series methods have been used to detect chaos in epidemiological data, including some based on dynamical systems theory—model-free approaches (Kot et al. 1988; Olsen and Schaffer 1990; Schaffer 1985a; Schaffer and Kot 1985; Ellner 1989; Shaffer et al. 1990). Reviews of model-free and model-based approaches are discussed in Hastings et al. (1993) and Ellner (1989). Because of the extensive use of epidemiological data in combination with dynamical systems theory to study the presence of chaos in real epidemiological systems, it has become important to gain some understanding of epidemic modeling. The rest of this chapter concentrates on a brief introduction to epidemic theory.

Epidemiological models have been used to study ecological and epidemiological phenomena extensively (see Anderson 1982; Anderson and May 1991; Bailey 1975; Busenberg and Cooke 1993; Capasso 1993; Castillo-Chavez 1989; Gabriel et al. 1990; Hethcote and Yorke 1984; Hethcote and van Ark 1992; Jewell et al. 1991; Kaplan and Brandeau 1994; and references therein). Their success is based partly on the fact that they provide an ideal tool for modeling *tight* coevolutionary interactions (see Levin 1983b, Levin and Castillo-Chavez 1990, and references therein)—that is, interactions in which the fate of the host (the resource) is intimately connected to the fate of the pathogen. Implicitly, hosts are modeled as dynamic patches (e.g., with their own life history) that are subject to invasions (by pathogens). These invasions temporarily or permanently affect the quality (even the life) of the patch. Furthermore, because pathogens compete for resources (patches) with other pathogens, epidemiological models are useful in addressing questions of coexistence and stability of biological systems.

The patch dynamics usually take place on a different time scale than the pathogen's dynamics. Furthermore, natural selection usually acts faster on the evolution of a pathogen than on the evolution of the host. At this time, there is no clear mathematical framework in which these matters can be properly addressed, but epidemiological models have provided a useful framework to begin to study these evolutionary dynamics (see Levin 1983a,b; May and Anderson 1983a,b; Levin and Castillo-Chavez 1990; Anderson and May 1991). Specific coevolutionary interactions such as those experienced by the myxomatosis virus (fairly lethal) and the European rabbit have provided an excellent and challenging opportunity for the application of epidemiological models to questions in evolutionary biology (see Levin and Pimentel 1981; Levin 1983a,b; May and Anderson 1983a,b).

The goals of this chapter are limited. We provide a brief introduction to epidemiological models and in the process try to provide an introduction to epidemiological "thinking" (e.g., we discuss the thought process that goes, or that we think goes, into building models that address specific epidemiological questions). Our introduction to epidemiological thinking is handled through the insertion of remarks and comments throughout the text, many of which are obvious but nonetheless (we think) important.

The prior neglect of some of the ideas expressed in these remarks was inspired by mathematical convenience. Unfortunately, it kept mathematical epidemiology away from the field of theoretical epidemiology—a situation that has begun to change (see the recent encyclopedic volume of Anderson and May 1991).

Alphabet Models

Alphabet models—so called after the useful mnemonic description introduced by Hethcote (1976)—consider a population of susceptible individuals that is being invaded by an infectious agent. These models concentrate on the disease dynamics that result if a small number of members of the population become infected and (eventually) infectious. The population is usually divided into three epidemiological subclasses, labeled S, I, and R. $S(t)$ denotes those individuals who are susceptible to the disease at time t, $I(t)$ denotes the number of infected individuals (here assumed infectious) who are hence capable of spreading the disease at time t, and $R(t)$ denotes the number of individuals who at time t no longer contribute to the direct spread of the disease because they have been removed through isolation or immunization, or through recovery from the disease with immunity against reinfection, or through death caused by the disease. Although the characteristics of $R(t)$ are very different from an epidemiological point of view, they are equivalent in a modeling sense because these individuals do not contribute further to the spread of disease and because the models do not incorporate demographic effects. For models that also involve demographic effects, the situation can be very different. The total population involved in the disease dynamics at time t will be denoted by $N(t)$ where $N(t)$ may be equal to $S(t) + I(t)$ or $S(t) + I(t) + R(t)$.

A model was proposed by Kermack and McKendrick in 1927 to explain the rapid rise and fall of cases observed frequently in epidemics such as the Great Plague in London (1665–1666), the cholera epidemic in London (1865), and the plague in Bombay (1906). The simplest version of their model (with a trivial modification) is given by the following set of nonlinear ordinary differential equations (ODEs):

$$S' = -\beta S\, I/N$$

$$I' = \beta SI/N - \frac{1}{\tau}I,$$

$$R' = \frac{1}{\tau}I$$

where (usually) $N(t) = S(t) + I(t) + R(t)$.

This model is based on the following three assumptions:

(1) There are no deaths unrelated to the disease and no births. This assumption is reflected in the relation $(S + I + R)' = 0$ so that total population size $N(t)$ is constant. It is a reasonable assumption for epidemics of short duration. [If there are disease-related deaths, or if R is used to label individuals that are permanently removed from the population and, hence, unable to interact with S- and I-individuals, then a better definition of $N(t)$ would be $N(t) \equiv S(t) + I(t)$.]

(2) The homogeneous mixing assumption, that is, the probability that a susceptible has a contact with an infective, is I/N. The contact process for spreading the infection is based on contacts between infectives and susceptibles, and these contacts occur at random (homogeneous mixing). Here $\beta \equiv \beta(N)$ denotes the per capita *effective* contact rate, which in general is a function of the size of the active population. It is common to assume that $\beta(N) = p\phi(N)$ where p denotes the probability of transmission per contact and $\phi(N)$ denotes the per capita contact rate. Here ϕS gives the average number of contacts that the population of susceptibles experiences per unit time. These susceptibles have contacts with S-, I-, and R-individuals, but under the homogeneous mixing assumption the proportion of contacts with infectives is I/N. The total average contact rate (handshakes, kisses, etc.) is $\phi SI/N$, but not all contacts lead to infection. The *effective* contact rate or the *incidence* (new cases of infection per unit of time) is $\beta(N)SI/N$. (If N is constant, then the rate of new infection is proportional to both the number of susceptibles and infectives; this is also the case if N is not constant but $\beta(N) \equiv pN$—a linear function of N. The use of a bilinear incidence rate corresponds to the use of *the mass-action law*.)

(3) The per capita recovery rate is constant, that is, infectives are removed from the I-class into the R-class at the constant rate $1/\tau$ (per individual and per unit of time). (This assumption implies that the distribution of time spent by individuals in the I-class has been an exponential distribution with mean τ.)

A key question is whether or not a population experiences an epidemic, that is, whether or not a pathogen can successfully invade a population. Success here means that the number of patches occupied, that is, the number of infective individuals, increases temporarily over time after the initial introduction of a pathogen into a population of purely susceptible patches. Under some conditions, the pathogen succeeds and there is an epidemic (the number of infected patches rises); otherwise it fails. These conditions are usually formulated in terms of a threshold value: the basic reproductive number of the contact number. If this value exceeds a critical value (usually one), there is an epidemic; otherwise there is no epidemic. For a mathematical definition of this concept that is well grounded in the biology, the reader is referred to Diekmann et al. (1990).

From the above model, Kermack and McKendrick (1927) deduced their threshold theorem. The importance of their result is due in part to the fact that most epidemic models exhibit similar threshold phenomena. Kermack and McKendrick found a dimensionless quantity called the contact, or the basic reproductive number, whose magnitude determines whether the infection will die out (I decreases monotonically to zero) or whether there will be an epidemic (I first increases to a maximum and then decreases to zero).

Since $S + I + R$ is constant, we may either let $S + I + R = K$, the constant total population size, or we may measure S, I, and R in units of K, effectively dividing S, I, and R by K so that they now represent fractions of the population in each class with $S + I + R = 1$. In either case, we may eliminate R from the system by substituting $R = K - S - I$ or $R = 1 - S - I$. We choose to work with the original variables and obtain the two-dimensional system:

$$S' = -\beta SI/K,$$
$$I' = \beta SI/K - \frac{1}{\tau}I.$$

Here β is measured in units of contacts per unit time and τ is measured in units of time.

We observe that $S' < 0$ for all t and $I' < 0$ if and only if $S < \frac{K}{\beta\tau}$. This implies that if $S(0) < \frac{K}{\beta\tau}$, then $I' < 0$ for all t and the infection dies out, but if $S(0) > \frac{K}{\beta\tau}$, then $I' > 0$ while S decreases to $\frac{K}{\beta\tau}$, after which $I' < 0$. If we "think" of introducing a small number of infectives into a susceptible population at time zero, so that $I(0) = \epsilon > 0$, $R(0) = 0$, $S(0) = K - \epsilon \approx K$, then we have a threshold dimensionless quantity

$$\mathcal{R}_0 = \beta\tau.$$

If $\mathcal{R}_0 < 1$ the infection dies out, and if $\mathcal{R}_0 > 1$ there is an epidemic. This is the threshold theorem of Kermack and McKendrick (who used the mass-action law; that is, they implicitly assumed that $\beta \equiv pK$). To simplify the notation, we assume without loss of generality that $K = 1$ during the mathematical analysis of the Kermack and McKendrick model. (Note that this assumption does not alter the value of $\mathcal{R}_0$, as β is assumed to be a constant; however, it does make a quantitative difference if we assume, as Kermack and McKendrick probably did, that $\beta \equiv pK$; in which case β would have to be measured in different units.)

A qualitative analysis of the system of differential equations (here we take $K = 1$ to simplify the analysis) shows that equilibria are solutions (S_∞, I_∞) of $S_\infty I_\infty = 0$, $I_\infty(\beta S_\infty - 1/\tau) = 0$. Thus there is a line of equilibria $(S_\infty, 0)$, with arbitrary S_∞, $0 \leq S_\infty \leq 1$. The linearization at an equilibrium $(S_\infty, 0)$ is

$$u' = -\beta S_\infty v$$
$$v' = (\beta S_\infty - \frac{1}{\tau})v$$

with matrix

$$\begin{bmatrix} 0 & -\beta S_\infty \\ 0 & \beta S_\infty - \frac{1}{\tau} \end{bmatrix}$$

having eigenvalues 0 and $\beta S_\infty - \frac{1}{\tau}$. The zero eigenvalue means that linearization is useful for the direct mathematical analysis of the system. However, it is not difficult to show that the phase portrait is as in Figure 18-1.

Hence we observe that the limiting value $S(\infty)$ of each orbit depends on the starting point $S(0)$, but $S(\infty) > 0$ for all $S(0)$. This means that in an epidemic, not all members of the population become infected—a deduction that agrees with recorded data for many epidemics. It is not difficult to show that, for an orbit that goes from point $(S(0),0)$ to point $(S(\infty),0)$,

$$S(\infty) - \frac{1}{\beta\tau} \ln S(\infty) = S(0) - \frac{1}{\beta\tau} \ln S(0).$$

To see this, we observe that

$$\begin{aligned} \frac{d}{dt}\left[S + I - \frac{1}{\beta\tau}\ln S\right] &= S' + I' - \frac{1}{\beta\tau}\frac{S'}{S} \\ &= S'\left[1 - \frac{1}{\beta\tau S}\right] + I' \\ &= -\beta SI\left[1 - \frac{1}{\beta\tau S}\right] + I' \\ &= -\beta SI + \frac{1}{\tau}I + I' = 0. \end{aligned}$$

Thus $S(t) + I(t) - 1/\beta\tau \ln S(t)$ is a constant and

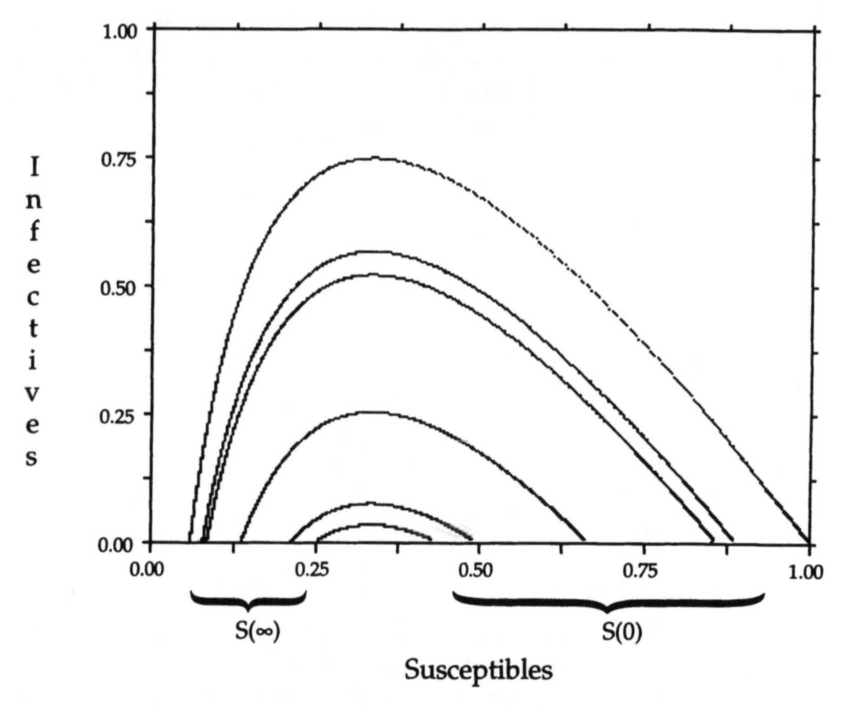

Figure 18-1. Kermack-McKendrick model; phase plot with basic reproductive number = 3 (Runge-Kutta 2, fixed time-step solution). S(0) denotes the initial proportion of susceptibles, whereas S(∞) denotes the susceptible proportion after the epizootic event.

$$S(\infty) + I(\infty) - \frac{1}{\beta\tau}\ln S(\infty) = S(0) + I(0) - \frac{1}{\beta\tau}\ln S(0).$$

Since $I(\infty) = I(0) = 0$, it follows that

$$S(\infty) - \frac{1}{\beta\tau}\ln S(\infty) = S(0) - \frac{1}{\beta\tau}\ln S(0).$$

The contact rate β in the model we are considering (assumed constant), depends on the particular disease being studied, and may also depend on social and behavioral factors. In general, it is difficult to estimate β, or equivalently $\mathcal{R}_0$ (since τ can be estimated experimentally). However, $S(0) \sim K$ as a disease is invading a purely susceptible population; and $S(0)$, that is, the proportion of individuals without antibodies, can be estimated from serological data. The above relation gives a means of calculating $\beta\tau$ for a given epidemic from knowledge of $S(0)$ and $S(\infty)$. For example, an influenza epidemic has been reported with $S(0) = 0.911$, $S(\infty) = 0.513$. From

$$S(0) - S(\infty) = \frac{1}{\beta\tau}[\ln S(0) - \ln S(\infty)] = \frac{1}{\beta\tau}\ln\frac{S(0)}{S(\infty)},$$

we derive

$$\beta\tau = \frac{\ln\dfrac{S(0)}{S(\infty)}}{S(0) - S(\infty)},$$

and the given data yield $\beta\tau = 1.44$, and since τ is about 3 days, then $\beta \sim 0.48$/day. Note that as long as $S(0) > 1/\beta\tau$, that is, as long as the *proportion* of susceptibles exceeds a critical value, an epidemic is possible (the data are from Evans 1982, as reported in Hethcote 1989).

The Kermack-McKendrick model considered above is an example of a simple model in which the population is divided into different compartments and the transitions between compartments are modeled. The understanding derived from simple models is useful in describing general properties and outlining the possible effects of control programs. It is also possible to formulate complex models with detailed assumptions incorporating more structure appropriate to a particular disease or situation. For example, compartments may be subdivided further to describe situations in which different subclasses have different transmission rates corresponding to different behavior (as in sexually transmitted diseases), age dependence, seasonal variations, or spatial dependence. Such more complex models may give more specific quantitative predictions.

The Kermack-McKendrick model is described as an S-I-R model because the transitions are from susceptible to infective to removed, with the removal coming through recovery with full immunity (as in measles) or through death (as in rabies and many other animal diseases). Another type of model is an S-I-S model, in which infectives return to the susceptible class upon recovery because the disease confers no immunity. Such models are appropriate for most diseases transmitted by bacterial or helminth agents, and for some sexually transmitted diseases including gonorrhea, but not for others like HIV or herpes, which infect individuals for life.

The simplest S-I-S model, also from Kermack and McKendrick (1932), is

$$S' = -\beta SI/N + \frac{1}{\tau}I$$

$$I' = \beta SI/N - \frac{1}{\tau}I,$$

where $S + I = N$.

The S-I-S differs from the S-I-R model because recovered members return to the class S at a rate $\frac{1}{\tau}I$ instead of passing to the class R. Again, the total population size is a constant $N = K$, and we may measure it in units of K (that is, we can assume that $K = 1$). We may use the relation $I = K - S$ to reduce the model to the single equation

$$S' = (1 - S/K)(-\beta S + \frac{K}{\tau})$$

with two equilibria,

$$S = K \text{ and } S = \frac{K}{\beta\tau}.$$

If $\frac{1}{\beta\tau} < \frac{S}{K}$, or $\beta\tau > 1$, both these equilibria are in the interval $0 < S \leq K$ but if $\frac{1}{\beta\tau} > \frac{S}{K}$ or $\beta\tau < 1$, only the equilibrium $S = K$ is relevant.

The threshold quantity $\beta\tau$ again distinguishes between two qualitatively different possible behaviors, but the possibilities are different from the S-I-R model. To simplify the analysis we take $K = 1$. Note that our modification of the incidence rate and the fact that β is a constant implies that K plays no role in the discussion.

The linearization about the equilibrium $S = 1$ (or $I = 0$), obtained with the aid of the change of variable $u = 1 - S$, is

$$u' = (\beta - \frac{1}{\tau})u.$$

Thus, the equilibrium is asymptotically stable if $\beta - 1/\tau < 0$, or $\beta\tau < 1$, and unstable if $\beta\tau > 1$. The linearization about the equilibrium $S = 1/\beta\tau$, $(K = 1)$ is of interest only if $\beta\tau > 1$, and it is obtained through the change of variable $u = S - 1/\beta\tau$. It is

$$u' = (\frac{1}{\tau} - \beta)u,$$

and thus the equilibrium is asymptotically stable if $1/\tau - \beta < 0$, or $\beta\tau > 1$. The result of this analysis is that there is always a single asymptotically stable equilibrium to which all solutions tend. If $\beta\tau < 1$, this equilibrium is $S = 1$, $I = 0$, corresponding to the disappearance of the disease. If $\beta\tau > 1$, the asymptotically stable equilibrium is $S = \frac{1}{\beta\tau}, I = 1 - \frac{1}{\beta\tau} > 0$, called an *endemic equilibrium,* in which the disease persists.

Measles is a disease for which endemic equilibria have been observed in many places, with oscillations about the equilibrium. In an attempt to formulate an S-I-R model that could give such behavior, Soper proposed a model in 1929 that assumed a constant birth rate μK in the susceptible class and a constant death rate μK in the removed class,

$$S' = \frac{-\beta S}{K} I + \mu K$$
$$I' = \frac{\beta S}{K} I - \frac{1}{\tau} I$$
$$R' = \frac{1}{\tau} I - \mu K.$$

The model is unsatisfactory because the linkage of births of susceptibles to deaths of removed members is unreasonable. It is also an improper model mathematically because if $R(0)$ and $I(0)$ are sufficiently small, then $R(t)$ will become negative. Furthermore, as discussed before, this model implicitly assumes that β is a function of K, and hence one must be careful with the units. For any disease model to be plausible, it is essential that the problem be properly posed in the sense that the number of members in each class must remain non-negative. (A full analysis of a model should include verification of this property.)

The difficulties in Soper's model were resolved by the assumption of deaths in each class proportional to the number of members in the class, keeping the total population size constant (Hethcote 1976):

$$S' = -\beta SI/K + \frac{1}{L}(K - S)$$
$$I' = \beta SI/K - \frac{1}{\tau} I - \frac{1}{L} I$$
$$R' = \frac{1}{\tau} I - \frac{1}{L} R.$$

Here L represents the average life span (to be justified later), because $1/L$ is the death rate. Because $S + I + R = K$, we can eliminate R and consider the two-dimensional system

$$S' = -\beta SI/K + \frac{1}{L}(K - S)$$
$$I' = \beta SI/K - \frac{1}{\tau} I - \frac{1}{L} I.$$

From the equilibrium conditions $\beta SI/K = \frac{1}{L}(K - S)$ and $I(\beta S/K - \frac{1}{\tau} - \frac{1}{L}) = 0$, we see that there is always an equilibrium $S = K$, $I = 0$ (vanishing of the disease), and if $I' > 0$, that is, if

$$\frac{\tau + L}{\beta\tau L} < \frac{K}{S(0)},$$

or (if everybody is susceptible, that is, if $S(0) \approx K$) if

$$\beta\tau\left(\frac{L}{L + \tau}\right) > 1,$$

there is a second equilibrium $S = \frac{\tau + L}{\beta\tau L}K$ with $I > 0$ (endemic equilibrium). Using linearization, it is not difficult to show that, if

$$\beta\tau\left(\frac{L}{L + \tau}\right) < 1,$$

the equilibrium $S = K$, $I = 0$ is asymptotically stable, whereas if

$$\beta\tau\left(\frac{L}{L + \tau}\right) > 1,$$

then it is unstable.

The threshold

$$\mathcal{R}_0 + \beta\tau\left(\frac{L}{L + \tau}\right)$$

is the analogy of the Kermack-McKendrick model, except that τ needs to be corrected to take into account the effects of natural mortality on the length of the infectious period; that is,

$$\tau\left(\frac{L}{L + \tau}\right)$$

denotes the death-adjusted mean infectious period.

The equilibrium $S = K$, $I = 0$ is unstable and the endemic equilibrium is asymptotically stable if $\mathcal{R}_0 > 1$. Thus the model allows endemic equilibria. In conclusion, the S-I-R model with births and deaths and the S-I-S model both support endemic equilibria, and hence we may also conclude that the requirement for the support of an endemic equilibrium is the flow of new susceptibles, either through births or through recovery without immunity.

In general, if $\beta(N) \equiv p\phi(N)$, then the condition $\mathcal{R}_0 > 1$ is equivalent to

$$\phi(S(0)) > \frac{\tau + L}{p\tau L}$$

or, if ϕ has an inverse, it is equivalent to the condition

$$S(0) > N_{\text{critical}} \equiv \phi^{-1}\left(\frac{\tau + L}{p\tau L}\right);$$

That is, the susceptible population must reach a critical value for an epidemic to occur. This is a useful concept; however, the estimation of the critical value of N depends on our knowledge of the functional form of $\beta(N)$. Rough values of N_{critical} have been estimated implicitly assuming that $\beta(N) = pN$, that is, N_{critical} has been computed (indirectly from data) using the formula

$$N_{\text{critical}} \equiv \frac{\tau + L}{p\tau L}$$

These estimates have led to the conclusion that influenza will persist in continental cities with at least 300,000 individuals, but it will persist in islands only with at least 500,000 individuals (for explicit details, see Anderson 1982). The values are different because the p's are probably different, as is the average age of first infection. An alternative way of reading these results (using available data) is as follows: if N is at least 300,000 (for cities), then p (the probability of transmission per contact) is high enough to have an infection that will remain endemic. The same result holds for islands as long as N is at least 500,000. To summarize these results and to make the data consistent with the same functional form for $\beta(N)$, we must accept different values of p for islands and continental cities.

Thresholds are useful, as they help us determine approaches for disease eradication or control. In order to eradicate an infection, it is necessary to reduce the contact number $\mathcal{R}_0$ below 1. This may sometimes be achieved by immunization, which has the effect of transferring members from the susceptible class to the removed class and thus reducing K (here $K = 1$). The effect of immunizing a proportion (which could be a function) q of the population is to replace K by $K(1 - q)$, and thus to replace the contact number $\mathcal{R}_0 = \beta\tau\left(\frac{L}{L + \tau}\right)$ by

$\beta\tau\left(\frac{L}{L+\tau}\right)(1-q)$. The condition $\beta\tau\left(\frac{L}{L+\tau}\right)(1-q) < 1$ gives $1-q = \frac{1}{\mathcal{R}_0}$, or $q > 1 - \frac{1}{\mathcal{R}_0}$, guarantees that successes has been achieved.

A population is said to have *herd immunity* if a large enough fraction has been immunized to ensure that the disease will not spread if an infective case is introduced. The only infection for which this has actually been achieved worldwide is smallpox. That is, through vaccination we will have eliminated a species. (There are still 600 laboratory samples in Russia and the United States that were supposed to have been destroyed by the end of 1993, but as of February 1994 they had not.) For measles, epidemiological data in the United States indicates that $\mathcal{R}_0$ for rural populations (using models for which $\beta = pN$, see later sections) ranges from 5.4 to 6.3, requiring vaccination of 81.5% to 84.1% of the population. In urban areas, $\mathcal{R}_0$ ranges from 8.3 to 13.0, requiring vaccination of 88% to 92.3% of the population. In Great Britain, $\mathcal{R}_0$ ranges from 12.5 to 16.3, requiring vaccination of 92% to 94% of the population. As vaccine efficacy for measles vaccination at age 15 months is approximately 95%, it is effectively impossible to achieve herd immunity for measles (Anderson 1982). Smallpox has a lower value of $\mathcal{R}_0$ and requires about 80% immunization. As the consequences of the disease are more serious, this is achievable.[1] The homogeneous mixing model says that it is feasible, but nearly impossible, to eradicate measles.

The S-I-R model with births and deaths predicts part of what is observed for measles but does not support the possibility of sustained oscillations about the endemic equilibrium. To explain oscillations about the endemic equilibrium as observed, we might assume seasonal variations in the contact rate β—not an unreasonable supposition for a childhood disease most commonly transmitted through school contacts, especially in winter in cold climates.

The S-I-R model with births and deaths is quite inappropriate for a disease such as rabies, from which recovery is rare. For such a disease, the class R consists of members removed by death, and the total population size is $S + 1$. Thus the number of births should be proportional to $(S + 1)$, or possibly proportional to S, to reflect the fact that such diseases are sufficiently debilitating that infective members do not reproduce. It is not possible to assume this and also keep the total population size constant in the absence of disease. A plausible model for such a disease will necessarily assume a nonlinear birth rate and will imply that the total population size must vary in time whether or not the disease is present. The formulation and mathematical analysis of general models that include these features can be found in the works of Brauer (1990, 1991) and Pugliese (1990a,b).

---

[1]The situation is not so simple, as the above models assume homogeneous mixing. In fact, the difficulties experienced in eradicating smallpox were due to heterogeneity in the contact rates. For an earlier paper on immunization, see Hethcote (1978) and references therein. Only after an intensive campaign that took into account these heterogeneities was it possible to eradicate smallpox.

Transition Rates and Waiting Times

In the models we have been considering, we have assumed rates of transition between classes that are proportional to the size of the class, such as recovery rates $\frac{1}{\tau}I$. Such assumptions do not really mean that recovery depends on the size of the infective class. A more realistic formulation is obtained if we replace the differential equation

$$I' = \beta SI/N - \frac{1}{\tau}I$$

by the integral equation

$$I(t) = \int_0^t \beta S(t)[I(x)/N(x)]e^{-\frac{1}{\tau}(t-x)}\,dx.$$

If t is large enough for members who were initially (at time $t = 0$) infective to have recovered and we assume that $N(x) = K$, a constant, then this integral equation is equivalent to the differential equation because differentiation under the integral sign gives

$$\begin{aligned} I'(t) &= \beta S(t)I(t)/K + \int_0^t \beta S(x)[I(x)/K]\frac{d}{dt}e^{-\frac{1}{\tau}(t-x)}\,dx \\ &= \beta S(t)I(t)/K - \frac{1}{\tau}\int_0^t \beta S(x)[I(x)/K]e^{-\frac{1}{\tau}(t-x)}\,dx \\ &= \beta S(t)I(t)/K - \frac{1}{\tau}I(t). \end{aligned}$$

The interpretation of the integral equation is that $e^{-\frac{1}{\tau}s}$ is the probability of remaining infective at time s after having become infective. Then

$$\beta S(x)[I(x)/K]e^{-\frac{1}{\tau}(t-x)}$$

represents the number of members who became infective at time x and remain infective $(t - x)$ time units later, and

$$\int_0^t \beta S(x)[I(x)/K]e^{-\frac{1}{\tau}(t-x)}\,dx$$

is the integral over all times $x \leq t$ of this number or the total number of infectives at time t. The mean length of the infective period is

$$\int_0^\infty e^{-\frac{1}{\tau}s}\,ds = \tau.$$

A similar interpretation applies to the constant of proportionality $\frac{1}{L}$ in the death rate in each class in the S-I-R model with births and deaths. The reinterpretation is that $e^{-\frac{1}{L}s}$ is the probability of survival to age s, and thus the average life expectancy is L.

We can also give a waiting-time interpretation to the rate of infection in the S-I-R model with births and deaths. At the endemic equilibrium (S_∞, I_∞), the rate of new infections is $\beta S_\infty[I_\infty/K]$. This means that $\beta[I_\infty/K]$ is the rate of infection per susceptible. If we assume an exponential distribution of probability of infection of an individual, the average time spent in the susceptible class before infection is $\frac{K}{\beta I_\infty}$. Thus the average age at infection, which we denote by A, is $\frac{K}{\beta I_\infty}$. As we have seen for this model, the endemic equilibrium (S_∞, I_∞) is the solution with $I_\infty > 0$ of the pair of equations

$$\beta S_\infty[I_\infty/K] = \frac{1}{L}(K - S_\infty),\ [I_\infty/K]\,[\beta S_\infty - \frac{1}{\tau} - \frac{1}{L}] = 0$$

from which we obtain

$$S_\infty = \frac{\tau + L}{\beta\tau L}K,\ I_\infty = \frac{K}{\beta L S_\infty}(K - S_\infty).$$

Also, the contact number is

$$\mathcal{R}_0 = \beta\tau\frac{L}{\tau+L} = \frac{K}{S_\infty},$$

with $\mathcal{R}_0 > 1$. From these relations, we see that

$$\frac{L}{A} = \beta L[I_\infty/K] = \frac{K}{S_\infty}(1 - [S_\infty/K]) = \frac{K}{S_\infty} - 1 = \mathcal{R}_0 - 1.$$

The relation $\mathcal{R}_0 = 1 + \frac{L}{A}$ is a useful means of estimating contact numbers, since L and A can be measured. Note that this derivation assumes the existence of an endemic state and, consequently, that $\mathcal{R}_0 > 1$. The estimates given earlier for contact numbers in various measles outbreaks were derived from this relation. One of the effects of immunization programs is to reduce the number of infectives at equilibrium, and this means that the average age at infection will increase. This could have serious consequences in childhood diseases for which the danger of complications increases with age. However, immunization programs will tend to reduce the total number of cases in older individuals even though they tend to increase the proportion of older cases.

Instead of assuming exponentially distributed recovery rates, we could assume that there is a probability $P(s)$ of remaining infective for a time s after becoming infective, where $P(s)$ is a monotone decreasing function with $P(0) = 1$. We will let

$$\tau = \int_0^\infty P(s)ds,$$

the average length of the infective period. An important special case is the choice

$$P(s) = \begin{cases} 1 & , 0 \leq s \leq \tau \\ 0 & , s > \tau \end{cases},$$

corresponding to an infective period of fixed length τ. In this case, the integral equation for $I(t)$, namely

$$I(t) = \int_0^t \beta S(x)I(x)P(t - x)dx$$

becomes

$$I(t) = \int_{t-\tau}^t \beta S(x)I(x)dx,$$

which is equivalent to the differential-difference equation

$$I'(t) = \beta S(t)I(t) - \beta S(t-\tau)I(t-\tau).$$

The choices $P(s) = e^{-\frac{1}{\tau}s}$ and $P(s) = \begin{cases} 1 & , 0 \leq s \leq \tau \\ 0 & , s > \tau \end{cases}$ represent extremes, and there is some reason to conjecture that models with different choices of the distribution function $P(s)$ with the same average infective period τ should have the same qualitative behavior. Whether this is true is of some mathematical interest, but is of epidemiological interest only if it is false and there are differences in behavior for different choices of $P(s)$. At this point, we must remark that a differential-difference model is not completely formulated until the initial conditions are given. The situation is different than for ordinary differential equations, as the proper formulation of a delay-differential equation requires the specification of initial conditions over the delayed period. That is, we must specify a real-valued function that is defined over $[-\tau, 0]$. From the theory of differential-difference equations, it is well known that solutions will exist under very general conditions for almost all initial conditions. The situation may be different for integral equations, which are, in fact, more general than differential-difference equations. The problem of existence of solutions is more delicate, and it is intimately connected to the kernel $P(s)$. Sometimes it is not possible to prescribe initial conditions, and sometimes solutions may not exist for almost all initial conditions; so when we claim that in some sense the above two choices represent two extremes, we are in fact providing a practical rule of thumb rather than a precise mathematical statement. However, this rule works quite well in some situations. The HIV/AIDS models analyzed by Castillo-Chavez et al. (1989b,c,d) provide a good example.

It is also possible to introduce an exposed period of fixed length ω, or more generally with a probability $Q(s)$ of remaining in the exposed class for a time s after becoming exposed. The analysis of disease models with exposed and infective periods of fixed length reduces to the analysis of systems of differential-difference equations with two delays, and this leads to questions of whether all roots of a transcendental characteristic equation have a negative real part. We shall describe one example, namely, an S-E-I-R model (where E denotes the exposed class, that is, individuals that have acquired the disease but are not yet infectious). The model without an exposed class is a special case with the exposed period ω equal to zero.

Example: The S-E-I-R model, without births or deaths, where $S + E + I + R \equiv K$.

The S-E-I-R model with an exposed period of fixed length ω and an infective period of fixed length τ is

$$S'(t) = -\beta S(t)\frac{I(t)}{K}$$

$$E'(t) = \beta S(t)\frac{I(t)}{K} - \beta S(t-\omega)\frac{I(t-\omega)}{K}, \text{ or } E(t) = \int_{t-\omega}^{t} \beta S(x)\frac{I(x)}{K}dx$$

$$I'(t) = \beta S(t-\omega)\frac{I(t-\omega)}{K} - \beta S(t-\tau-\omega)\frac{I(t-\tau-\omega)}{K}, \text{ or } I(t) = \int_{t-\tau-\omega}^{t-\omega} \beta S(x)\frac{I(x)}{K}dx$$

$$R'(t) = \beta S(t-\tau-\omega)\frac{I(t-\tau-\omega)}{K}.$$

Since E and R are determined when S and I are known, this can be reduced to the pair of equations

$$S'(t) = -\beta S(t)\frac{I(t)}{K},$$

$$I(t) = \int_{t-\tau-\omega}^{t-\omega} \beta S(x)\frac{I(x)}{K}dx,$$

and then to a single equation by writing

$$I(t) = \int_{t-\tau-\omega}^{t-\omega} S'(x)dx = S(t-\tau-\omega) - S(t-\omega)$$

and substituting into $S'(t) = -\beta S(t)\frac{I(t)}{K}$ to obtain

$$S'(t) = -\beta S(t)\frac{[S(t-\tau-\omega) - S(t-\omega)]}{K}.$$

To this single differential-difference equation with two delays we must add some initial data for $-(\tau + \omega) \leq t \leq 0$ to give a well-posed problem, because every constant is a solution of the differential-difference equation. It is possible to prove that $S(t)$ is nonincreasing and tends to a limit $S_\infty > 0$ as $t \rightarrow \infty$; the limiting value will depend on the initial data. The analysis of this model is considerably more difficult than the analysis of the original Kermack-McKendrick S-I-R model, but the results are qualitatively the same; that is, there is a threshold number that determines whether or not there is an epidemic.

Obviously, diseases for which variable and long infectious periods are important must be modeled using the approach of this section. An example is provided by the modeling of the sexual transmission of the HIV/AIDS virus in a homosexually active population. Castillo-Chavez et al. (1989b,c,d) have shown that the results are qualitatively the same for all incubation period distributions (see also Blythe and Anderson 1988a,b). However, the results may be different if the transmission probability per (sexual) contact, that is, p, depends on how long an individual has been infected (see Thieme and Castillo-Chavez 1989, 1994). In some sense, these are the kinds of results that one may expect from mathematical models; their use helps us focus on factors that do affect the qualitative dynamics of a disease. For HIV/AIDS in a homosexually active and homogeneously mixing population, only the *age at infection* has the potential of changing the qualitative dynamics. Of course, epidemiologists are also interested in quantitative results (predictions). These results are more difficult to obtain, as specific data are needed. Furthermore, the above discussion makes it clear that quantitative predictions under a simple epidemic model (which requires fewer parameters and hence fewer data) would be based on specific assumptions, such as homogeneous mixing (all individuals behave in the same average way) and the functional form of the contact function $\phi(N)$. The same situation is true, however, if one uses statistical models. For long-term predictions, dynamic models seem to be the best choice; whereas for short-term predictions, statistical models are sufficient. This division is somewhat simplistic, as other methods are available, such as time series analysis. Efforts to combine time series analysis, statistical methods, and dynamical systems have been the focus of intensive research in epidemiology (or, more generally, in theoretical biology). For some recent results on this active area of research, we recommend the articles by Ellner (1989) and Hastings et al. (1993), and references therein.

Models With Social Structure

Classical mathematical models in demography (Leslie 1945; Lotka 1922; McKendrick 1926) concentrate on the dynamics of birth and death processes for female populations. They ignore the mating/marriage structures and their effect on social dynamics. The incorporation of mating structures or marriage functions was pioneered by Kendall (1949) and by Keyfitz (1949). Their work was extended by Fredrickson (1971), McFarland (1972), Parlett (1972), and Pollard (1973) two decades ago, but with very limited impact. The HIV/AIDS epidemic attracted theoreticians' attention to the study of the effect of social dynamics on the spread of epidemics. Questions raised primarily by researchers interested in HIV/AIDS epidemiology have brought back interest in the modeling of marriage functions and their connection to social dynamics (Anderson et al. 1989; Blythe and Castillo-Chavez 1989, 1990; Blythe et al. 1991, 1992; Busenberg and Castillo-Chavez 1989, 1991; Castillo-Chavez 1989; Castillo-Chavez et al. 1989d, 1991; Dietz 1988; Dietz and Hadeler 1988; Gupta et al. 1989; Hadeler 1989a,b; Hadeler and Nagoma 1990; Huang 1989; Huang et al. 1992; Hyman and Stanley 1988, 1989; Jacquez et al. 1988, 1989; May and Anderson 1989; Sattenspiel 1987a,b; Sattenspeil and Simon 1988; Sattenspiel and Castillo-Chavez 1990; Waldstätter 1989). Research on the social spread of disease has grown at a very fast pace over the last decade.

The contact or social structure of a population plays a fundamental role in the transmission dynamics of diseases, cultural traits, genetic traits, and so on. In the past, it has been modeled by assuming that the rate of transmission of the trait in consideration is directly proportional to the product of those that have the trait and those that do not. That is, it is assumed that the incidence rate is a bilinear function of susceptibles and infected (see Anderson 1982; Anderson and May 1991; Bailey 1975; and references therein). The assumption that the rate of new "cases" (the incidence) is proportional to the product of "susceptibles" and "converts" (those infected)—the mass-action law—is useful, but only, as described before, in limited circumstances. In fact, it is not useful in the modeling of sexually transmitted diseases (STDs). A thorough analysis of the modeling assumptions involved in the construction of the incidence rate or "force" of infection has been carried out in a systematic fashion by Busenberg and Castillo-Chavez (1989, 1991).

The importance of the contact process for frequency-dependent systems was recognized by Ross as early as 1911 in his work on malaria (see also Lotka 1923). The contact/social structure of the population must respond to demographic/epidemiological changes in the population. A flexible framework for the modeling of population interactions is being developed because several questions of theoretical and practical importance cannot be properly studied under the existing framework. Some recent applications to this new framework include those to food web dynamics (Velasco-Hernandez and Castillo-Chavez 1994) and those to cultural dynamics (Lubkin and Castillo-Chavez 1994).

Busenberg and Castillo-Chavez (1989, 1991) defined the contact/social structures through mixing/pair-formation matrices. In addition, they provided a useful characterization of these matrices, which constitutes the basis of our further analysis. Following Castillo-Chavez and Busenberg (1991), we introduce this framework within a two-sex mixing population. We begin with some needed notation and definitions:

$p_{ij}(t)$ = probability that a male in group i mixed with a female in group j at time t, given that he mixed with somebody;

$q_{ji}(t)$ = probability that a female in group j mixed with a male in group i at time t, given that she mixed with somebody;

$T_i^m(t)$ = number of males in group i at time t;

$T_j^f(t)$ = number of females in group j at time t;

b_i^m = average (assumed constant) number of female partners per group i male per time unit
= per capita pair-formation rate for group i males;

b_j^f = average (assumed constant) number of male partners per group j female per time unit
= per capita pair-formation rate for group j females.

Definition. $(p_{ij}(t), q_{ji}(t))$ is called a mixing/pair-formation matrix if and only if it satisfies the following properties at all times:

(A1) $p_{ij}(t) \geq 0$ and $q_{ji}(t) \geq 0$ for $i = 1, \ldots, L, \quad j = 1, \ldots, N$.

(A2) $\sum_{j=1}^{N} p_{ij}(t) = 1$ for $i = 1, \ldots, L$; $\sum_{i=1}^{L} q_{ji}(t) = 1$ for $j = 1, \ldots, N$.

(A3) $c_i T_i^m(t) p_{ij}(t) \dfrac{\sum_{j=1}^{N} b_j T_j^f(t)}{\sum_{j=1}^{N} c_i T_i^m(t)} = b_j T_j^f(t) q_{ji}(t)$ for $i = 1, \ldots, L, \quad j = 1, \ldots, N$.

(A4) $P_{ij}(t) \equiv q_{ji}(t) \equiv 0$ by definition if $c_i\, b_j T_i^m(t)\, T_j^f(t) = 0$ for some i, $1 \leq i \leq L$, and/or some j, $1 \leq j \leq N$.

The fraction in Property (A3) is the total sexual-activity ratio between females and males. This ratio is obviously not equal to 1 at all times. In fact, it expresses the fact that the total average rate of pair formation between males of type i and females of type j must be equal. A possible interpretation of our *explicit* formulation of the equality of these rates is that we assume that the partnership system favors females (that is, they will tend to achieve their desired *optimal* average), whereas males will have to modify their desired optimal rates (c_i's) to take into account the total sexual-activity ratio. Or, in other words, if we assume that the average pairing rates for females are constant, then the male pairing rates cannot be arbitrary. Property (A4) asserts that individuals from populations that do not interact cannot possibly mix. An immediate consequence of the above properties is that the total effective average rates of male and female activity must agree at all times; that is,

$$\sum_{i=1}^{L} C_i T_i^m = \sum_{j=1}^{N} b_j T_j^f,$$

where

$$C_i \equiv c_i \frac{\sum_{j=1}^{N} b_j \mathrm{T}_j^f(t)}{\sum_{j=1}^{N} c_i \mathrm{T}_i^m(t)}.$$

The only separable solution to (A1)–(A4) is the Ross solution (named after Ross, in recognition of his implicit knowledge of the above axioms; see Lotka 1922) given by $((\bar{p}_j(t), \bar{q}_i(t))$, where

$$\bar{p}_j(t) = \frac{b_j^f T_j^f(t)}{\sum_{i=1} b_i^f T_i^f(t)}, \quad \bar{q}_i(\mathrm{t}) = \frac{b_i^m T_i^m(t)}{\sum_{k=1} b_j^m T_j^m(t)}.$$

Castillo-Chavez and Busenberg (1991) characterized all solutions to axioms (A1)–(A4) as multiplicative perturbations of the Ross solution. These perturbations are defined in terms of two matrices, $\mathbf{\Phi}^m = \{\phi_{ij}^m\}$ and $\mathbf{\Phi}^f = \{\phi_{ji}^f\}$. The matrices $\mathbf{\Phi}^m$ and $\mathbf{\Phi}^f$ define the preferences and/or affinities of types of individuals of one gender for other types (here, of the opposite gender), and these preferences may change with time directly or through changes in the frequency of types. We refer to these two matrices as the male and female *preference matrices*, respectively. To formulate this representation theorem, the following expressions are needed:

$$\log_i^m \equiv \sum_{j=1}^{N} \bar{p}_j \phi_{ij}^m, \qquad R_i^m \equiv 1 - \log_i^m, \qquad V^m \equiv \sum_{i=1}^{L} \bar{q}_i R_i^m, \tag{1}$$

$$\log_j^f \equiv \sum_{i=1}^{L} \bar{q}_i \phi_{ji}^f, \qquad R_j^f \equiv 1 = \log_j^f, \qquad V^f \equiv \sum_{j=1}^{N} \bar{p}_j R_j^f.$$

Theorem 1.

For each marriage function $(\mathbf{P},\mathbf{Q})$, matrices $\mathbf{\Phi}^m$ and $\mathbf{\Phi}^f$ can be found such that

$$p_{ij} = \bar{p}_j \left[\frac{R_j^f R_i^m}{V^f} + \phi_{ij}^m \right] \text{and } q_{ji} = \bar{q}_i \left[\frac{R_i^m R_j^f}{V^m} + \phi_{ji}^f \right] \tag{2}$$

with $\quad 0 \le R_i^m \le 1$ and $0 \le R_j^f \le 1$ for $i = 1, \ldots, L, j = 1, \ldots, N,$ and $\sum_{i=1}^{L} \log_i^m \bar{q}_i < 1$ and $\sum_{j=1}^{N} \log_j^f \bar{p}_j < 1$

if and only if

$$\phi_{ij}^m = \phi_{ji}^f + R_i^m R_j^f \left[\frac{1}{V^m} - \frac{1}{V^f}\right].$$

Condition (3) shows the implicit frequency- (and time-) dependent relationship forced by (A3) between the elements of $\mathbf{\Phi}^m$ and $\mathbf{\Phi}^f$. Using vector notation,

$$\overrightarrow{\overline{P}} = \begin{pmatrix} \overline{P}_1 \\ \vdots \\ \overline{P}_N \end{pmatrix} \text{and} \overrightarrow{\overline{q}} = \begin{pmatrix} \overline{q}_1 \\ \vdots \\ \overline{q}_L \end{pmatrix},$$

we can combine the constraints imposed by (3) through an implicit nonlinear relationship

$$\mathbf{\Phi}^m = \psi(\overrightarrow{\overline{p}}, \overrightarrow{\overline{q}}, \mathbf{\Phi}^f, \mathbf{\Phi}^m), \tag{4}$$

where the elements of ψ are defined component-wise by (3).

We proceed to outline some useful results (see Hsu Schmitz et al. 1993, Hsu Schmitz 1994) to give an insight into the role of $\mathbf{\Phi}^m$ and $\mathbf{\Phi}^f$:

Theorem 2.

If either $\phi_{ij}^m = \alpha, 0 \le \alpha < 1 \; \forall \, i, j$, or $\phi_{ji}^f = \beta, 0 \le \beta < 1 \; \forall \, j, i$, where α and β are constants, then $p_{ij} = \bar{p}_j$ and $q_{ij} = \bar{q}_1$. That is, equation (2) reduces to the unique separable Ross solution in axiom (A2).

It may be argued that this representation merely passes the buck by transferring the difficulties from one set of matrices, (**P**,**Q**), to another ($\mathbf{\Phi}^m$, $\mathbf{\Phi}^f$). However, the use of preference matrices ($\mathbf{\Phi}^m$,$\mathbf{\Phi}^f$) helps increase our understanding of the marriage/social structure of a population, because population matrices facilitate the modeling of specific, nontrivial mixing patterns between subpopulations. Our data (see Rubin et al. 1992; Castillo-Chavez et al. 1992; Hsu Schmitz and Castillo-Chavez 1993, 1994; Hsu Schmitz 1994) show that females mix with older males and males mix with younger females. In the past, these mixing patterns were not modeled, either because they led to intractable mathematical models or because there was no obvious way of doing it. The use of affinity matrices ($\mathbf{\Phi}^m$, $\mathbf{\Phi}^f$) facilitates the construction of these "unusual" mating/social structures. The following result (Hsu Schmitz et al. 1993, Hsu Schmitz 1994) provides a parametric family of mixing matrices that allows modeling of age-dependent mixing.

Theorem 3.

$V^f = V^m$ if and only if $\mathbf{\Phi}^m = (\mathbf{\Phi}^f)^T$, where T denotes transposition.

The above result implies that the only solutions to axioms (A1)–(A4) with frequency-independent $\mathbf{\Phi}^m$ and $\mathbf{\Phi}^f$ are those with $\mathbf{\Phi}^m = (\mathbf{\Phi}^f)^T$. This situation occurs when males and females have matching preferences that do not change with the dynamics of $T_i^m(t)$ and $T_j^f(t)$. Although the class of solutions with $\mathbf{\Phi}^m = (\mathbf{\Phi}^f)^T$ is quite restrictive, this class extends, considerably, the mixing/mating structures available in the literature. If we use constant preference matrices $\mathbf{\Phi}^m$ and $\mathbf{\Phi}^f$, then the class of parametric mixing models generated with this selection becomes quite rich and flexible. Figure 18-2 shows a real matrix, whereas Figure 18-3 shows the corresponding Ross solutions. Both were constructed using our data from a known population of undergraduate students and their partners (Rubin et al. 1992; Castillo-Chavez et al. 1992; Hsu Schmitz et al. 1993; Hsu Schmitz and Castillo-Chavez 1993, 1994; Hsu Schmitz 1994), and both can be fit to matrices that satisfy the relationship $\mathbf{\Phi}^m = (\mathbf{\Phi}^f)^T$.

Theorem 2 requires the relationship in equation (3). This relation implies the existence of a function ψ such that

$$\mathbf{\Phi}^m = \psi\,(\vec{p}, \vec{q}, \mathbf{\Phi}^f, \mathbf{\Phi}^m),$$

or, in other words, the preferences of males for females and vice versa satisfy a complex relationship. Common sense dictates that if the set of preference of one gender (e.g., $\mathbf{\Phi}^f$) is known, then so must be the preference of the other (e.g., $\mathbf{\Phi}^m$); and in fact it can be shown (see Hsu Schmitz and Castillo-Chavez 1993; Hsu Schmitz 1994) that this is the case. Hence there exists a function ψ such that

$$\mathbf{\Phi}^m = \psi(\vec{p}, \vec{q}, \mathbf{\Phi}^f).$$

This result will be referred to as the "T^3 Theorem" because it reminds us that, in all situations, it "Takes Two to Tango."[2]

In this section, we have assumed that the pair-formation rates b_i^m and b_j^f are

[2]This theorem got its colorful name from Stavros Busenberg.

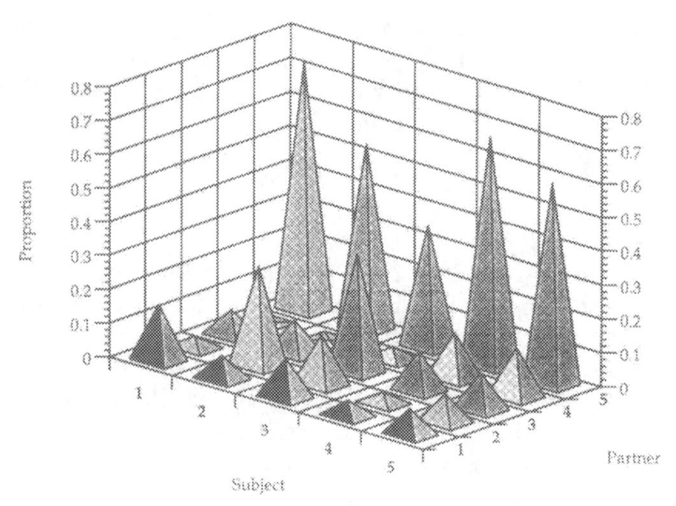

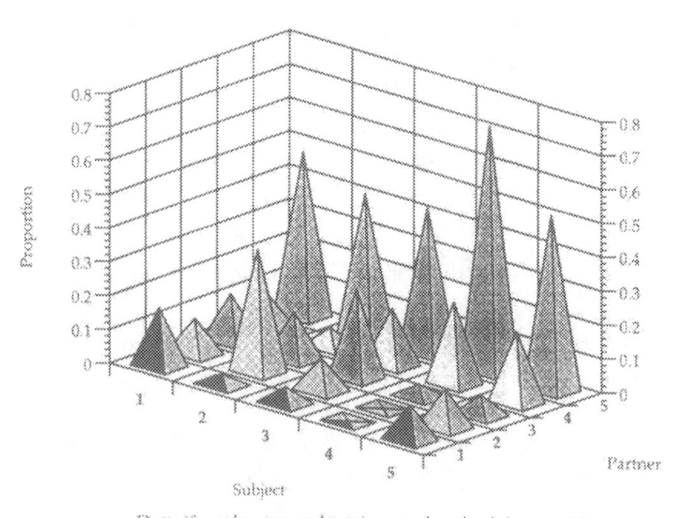

Figure 18-2. Real mixing matrices for males and females.

both constants. This was done to simplify the exposition, because a consistent theory with constant pair-formation rates for both genders holds only under very particular circumstances. In general, one has that $b_i^m \equiv b_i^m(T_j^f, T_i^m)$ and $b_j^f \equiv b_j^f(T_j^f, T_i^m)$. For a further discussion of this general case, the reader is referred to Castillo-Chavez et al. (1992, 1994b), Hsu Schmitz and Castillo-Chavez (1993, 1994), and Hsu Schmitz 1994. Castillo-Chavez et al. (1994b) show explicit ways of incorporating arbitrary mixing patterns into demographic and epidemiological

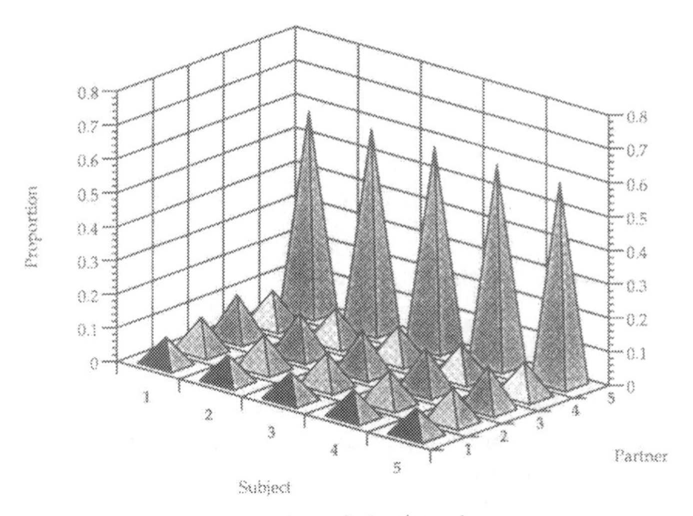

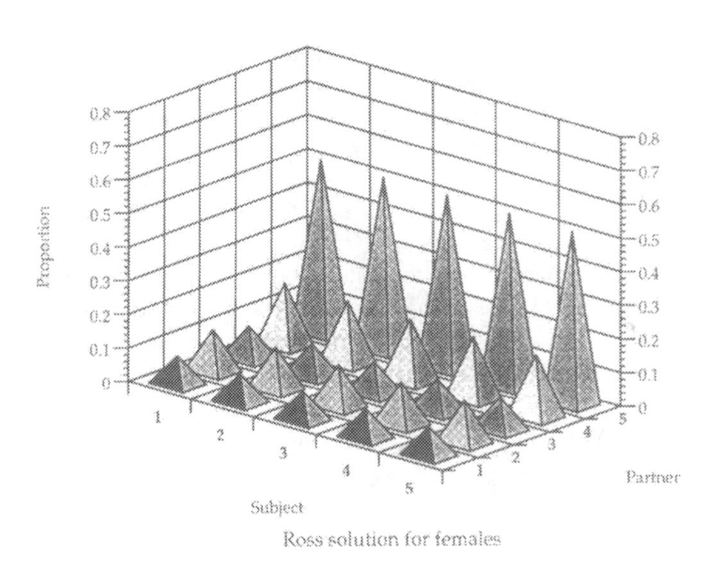

Figure 18-3. Ross solution for males and females.

models. The mathematical analysis of these general models is progressing (see Huang 1989; Huang et al. 1992).

Stochastic Models with Social Structure

In this section, we formulate a stochastic pair-formation epidemiological model by using the approach that is common to interacting particle systems (for details, see Luo and Castillo-Chavez 1991; Castillo-Chavez et al. 1994a,b). Extensions of this model to cover other scenarios should be quite evident from the following description.

We define

$$X = \{0,1, \ldots, L\} \times \{0,1\} \times \{0,1, \ldots, N\} \times \{0,1\}/\{0\} \times \{0,1\} \times \{0\} \times \{0,1\},$$

and consider the explicit stochastic process

$$\xi_t : X \rightarrow \{0, 1, 2, \ldots\}, \qquad t \geq 0.$$

Let $x = (i, u; j, v) \in X$, where i and j denote the groups of males and females and u and v denote the epidemiological statuses of males and females, respectively. If we consider an STD that does not have a long latent period, does not provide permanent immunity, and does not cause significant mortality (e.g., gonorrhea, see Hethcote and Yorke 1984), then the possible values of u and v are either 0 (susceptible) or 1 (infected). For $i > 0$ and $j > 0$, x gives the type of pair; that is, the male is from group i with epidemiological status u, and the female is from group j with epidemiological status v. If $i = 0$ but $j > 0$, then x represents a single female in group j with epidemiological status v (the value of u is not relevant), and therefore we can define $x \equiv (0; j, v) \equiv (0, 0; j, v) \equiv (0, 1; j, v)$. Similarly, if $j = 0$ but $i > 0$, then x represents a single male in group i with epidemiological status u, and again we can define $x \equiv (i, u; 0) \equiv (i, u; 0, 0) \equiv (i, u; 0, 1)$. Note that the case of $i = 0$ and $j = 0$ is not included in the domain of X. Consequently, the stochastic process $\xi_t(x)$ gives the number of pairs of type x at time t if $i > 0$ and $j > 0$; it gives the number of single males of type x at time t if $i > 0$ and $j = 0$; and it gives the number of single females of type x at time t if $i = 0$ and $j > 0$.

To complete the characterization of $\xi_t(x)$, we define $S \equiv \{0, 1, 2, \ldots\}^x$ and let $c: S \times S \rightarrow (0,\infty)$ be a real-valued function that models the flip rate. We view $\{\xi_t : t \geq 0\}$ as an S-valued Markov process with flip rate $c(.,.)$; i.e., if $\xi_t = \xi$ for some $t \geq 0$, then $c(\xi,\eta)$ denotes the instantaneous rate at which ξ_t may change to state η. Explicitly,

$$\text{Prob}(\xi_{t+h} = \eta | \xi_t = \xi) = c(\xi, \eta)h + 0(h), \qquad \forall t \geq 0.$$

The more specific definition of flip rates is as follows: for

$$\xi \in S, A \subset X, B \subset X, \text{ and } A \cap B = \varnothing, \text{ we define } \xi_B^A(x) \in S \text{ as}$$

$$\xi_B^A(x) = \begin{cases} \xi(x)+1 & \text{if } x \in A \\ \xi(x)-1 & \text{if } x \in B \\ \xi(x) & \text{otherwise.} \end{cases}$$

Thus, the system $\{\xi_t\}$ consists of a series of changing elements in the set S, which is the set of all functions on X. The dynamics of the system is described by the rates $\{c(\xi, \eta): \xi \neq \eta,\ \xi, \eta \in S\}$ at which the system changes.

We assume the existence of an underlying mixing/pair-formation matrix $(p_{ij}(\xi_t), q_{ji}(\xi_t))$ as described in the preceding section. Since ξ_t is a function of t, the mixing matrix is also a function of t. We further assume that paired individuals do not look for other partners before they separate. As the time t changes, singles may form pairs, pairs may dissolve, the disease may be transmitted within pairs from an infective to a susceptible, the infectives may be cured, and so on.

We use the indices m and f to identify the parameters associated with males and females regardless of their epidemiological status, and we use M and F to characterize those parameters associated only with infected males and females, respectively. Then the flip rate $c(.,.)$ is calculated as follows:

(a) Pair formation

For $I > 0$, $j > 0$,

$$c(\xi, \xi^{(i,u;j,v)}_{(i,u;0),(0;j,v)}) = b_j^f \xi(0;j,v) q_{j,i}(\xi) \frac{\xi(i,u;0)}{\xi(i,u;0) + \xi(i,1-u;0)}$$
$$= b_i^m \xi(i,u;0) p_{ij}(\xi) \frac{\xi(0;j,v)}{\xi(0;j,v) + \xi(0;j,1-v)};$$

(b) Pair dissolution (σ denotes the constant pair dissolution rate)

For $i > 0$, $j > 0$

$$c(\xi, \xi^{(i,u;0),(0;j,v)}_{(i,u;j,v)}) = \sigma_{ij} \xi(i,u;j,v);$$

(c) Transmission (δ denotes the constant transmission rate)

For $i > 0$, $j > 0$

$$c(\xi, \xi^{(i,1;j,1)}_{(i,0;j,1)}) = \delta_F \xi(i,0;j,1), \qquad c(\xi, \xi^{(i,1;j,1)}_{(i,1;j,0)}) = \delta_M \xi(i,1;j,0);$$

(d) Recovery (γ denotes the constant recovery rate)

For $i > 0$, $j > 0$, the recovery flip rates for one paired individual are

$$c(\xi, \xi^{(i,0;j,0)}_{(i,0;j,1)}) = \gamma_F \xi(i,0;j,1), \qquad c(\xi, \xi^{(i,0;j,0)}_{(i,1;j,0)}) = \gamma_M \xi(i,1;j,0);$$
$$c(\xi, \xi^{(i,1;j,0)}_{(i,1;j,1)}) = \gamma_F \xi(i,1;j,1), \qquad c(\xi, \xi^{(i,0;j,1)}_{(i,1;j,1)}) = \gamma_M \xi(i,1;j,1);$$

and the flip rate for both individuals is

$$c(\xi, \xi^{(i,0;j,0)}_{(i,1;j,1)}) = \gamma_{FM} \xi(i,1;j,1);$$

whereas for single infected individuals ($j = 0$ or $i = 0$), we have

$$c(\xi,\xi^{(i,0;0)}_{(i,1;0)}) = \gamma_M\xi(i,1;0), \qquad c(\xi,\xi^{(0;j,0)}_{(0;j,1)}) = \gamma_F\xi(0;j,1);$$

(e) Removal (μ denotes the constant removal rate from sexual activity)
For $i > 0$, $j > 0$,

$$c(\xi,\xi^{(i,u;0)}_{(i,u;j,v)}) = \mu_f\xi(i,u;j,v), \qquad c(\xi,\xi^{(0;j,v)}_{(i,u;j,v)}) = \mu_m\xi(i,u;j,v);$$

whereas for single individuals ($j = 0$ or $i = 0$), we have

$$c(\xi,\xi_{(i,u;0)}) = \mu_m\xi(i,u;0), \qquad c(\xi,\xi_{(0;j,v)}) = \mu_f\xi(0;j,v);$$

(f) Recruitment (Λ denotes the constant recruitment rate for susceptible singles)
For $i > 0$, $j > 0$,

$$c(\xi,\xi^{(i,0;0)} = \Lambda_i^m \text{ and } c(\xi,\xi^{(0;j,0)}) = \Lambda_j^f;$$

(g) Other
For any other $\eta \neq \xi$, we assume $c(\xi,\eta) = 0$.

(h) $c(\xi,\xi) = \sum_{\eta\neq\xi} c(\xi,\eta)$.

This concludes the characterization of our stochastic epidemiological model with pairs. Next we outline the simulation procedure of a general stochastic process that includes the above.

From the construction of the flip rates, we know that

$$c(\xi) = \sum_{\eta\neq\xi} c(\xi,\eta) < \infty.$$

If we let the sequence $0 = \rho_0 < \rho_1 < \ldots$ denote the jump times of the process, then $\tau_n = \rho_n - \rho_{n-1}$ has an exponential distribution with rate $c(\xi_{\rho_{n-1}})$. Thus, the process can be simulated as follows:

(1) First, set the initial state ξ_0 and assume that a sequence of n jump times $0 = \rho_0 < \rho_1 < \ldots < \rho_n$ and their corresponding states ξ_{ρ_i}, $1 \leq i \leq n$, have been determined.

(2) Get τ_{n+1} from $\exp\{\xi_{\rho_n}\}$ and let $\rho_{n+1} \equiv \rho_n + \tau_{n+1}$.

(3) Set $\xi_{\rho_{n+1}} + \eta$ with probability $c(\xi_{\rho_n},\eta)/c(\xi_{\rho_n})$.

(4) Define $\xi_t = \xi_{\rho_n}$ for $\rho_n < t < \rho_{n+1}$.

In Castillo-Chavez et al. (1994a), extensive simulations are carried out for pair-formation demographic models where individuals have distinct preferences. We found (and it was later justified analytically in some simple cases in Hsu Schmitz 1994) that the parameters with a bigger impact on the determination of the stable pair distribution are the dissolution rates. If these rates are identical for all groups, then the system stabilizes asymptotically to a pair-distribution that is in agreement with individual preferences. If the dissolution rates differ, however, then there is no clear connection between the "observed" mating preferences/affinities and the observed mixing patterns (stable pair-distribution). This is quite analogous to what is observed in population genetics, where there is no unique correspondence between genotype and phenotype. In fact, higher variance on the dissolution rates may reduce our ability to identify individual affinities from observed mixing patterns.

Conclusions

In this chapter, we have provided an overview on the use, formulation, and analysis of epidemiological models, including a brief introduction to the ways in which models can help us understand the mechanisms behind the (underlying) dynamics of observed time series and a concise introduction to a large class of simple and useful epidemiological models. A complete mathematical analysis is not provided, but we have taken the time to illustrate the role that the basic reproductive number plays in epidemiology.

The HIV/AIDS epidemic has forced us to look at models that incorporate long and variable periods of infectiousness; hence, we must deal with systems of difference-differential and integral equations. These models are more difficult to handle, but their analysis has been carried out in many situations. In fact, there has been a great deal of progress since the first articles on the transmission dynamics of HIV (Anderson et al. 1986; Anderson and May 1987; Pickering et al. 1986) appeared almost a decade ago.

If one wishes to construct more realistic models for the transmission dynamics of HIV/AIDS, then one must incorporate other factors, including the social structure of a population. Here we have outlined ways in which this can be accomplished and some of the difficulties encountered in the process. To limit the level of complexity, we have abandoned epidemics to concentrate on the development of methods for modeling the (dynamic) contact/pairing structure of a population. The usefulness of our approach has yet to be determined, notwithstanding the fact that it has already been applied with some degree of success to data on mixing patterns of college undergraduates (Blythe et al. 1992; Castillo-Chavez et al. 1992; Rubin et al. 1992; Hsu Schmitz and Castillo-Chavez 1994; Hsu Schmitz et al. 1993; Hsu Schmitz 1994). We have also discussed the formulation of stochastic pair-formation epidemic models. Although the analysis of stochastic models is difficult, their simulation can be easily carried out, and we have provided an outline of the simulation method. References to work being carried out in the very active field of stochastic epidemics can be found in Gabriel et al. (1990), Castillo-Chavez et al. (1994a,b), and Luo and Castillo-Chavez (1991).

This chapter has focused on modeling rather than mathematics, but we have not completely avoided all technicalities. We have tackled some and have provided extensive references for those interested in learning the basic tricks. The time devoted to studying this chapter will not be well spent, however, unless it helps the reader formulate models (preferably simple ones) that are capable of addressing important biological questions. The references provide ample documentation of the successes of mathematics when it is driven by biological research.

Acknowledgments

This research was partially supported by NSF grant DEB-925370 (Presidential Faculty Fellowship Award) to Carlos Castillo-Chavez and by the U.S. Army Research Office through the Mathematical Sciences Institute of Cornell University (contract DAAL03-91-C-0027). We thank Sam Fridman for his help with the figures, Kety Esquivel for performing many thankless tasks, and John Steele for his infinite patience. (1991 Mathematics Subject Classification: 92D30.)

References

Allen, J. C. 1989a. Are natural enemy populations chaotic? *Estimation and Analysis of Insect Populations,* Lecture Notes in Statistics **55:**190–205.

———. 1989b. Patch efficient parasitoids, chaos, and natural selection. *Florida Entomologist* **79:**52–64.

———. 1990a. Factors contributing to chaos in population feedback systems. *Ecological Modelling* **51:**281–298.

———. 1990b. Chaos in phase-locking in predator-prey models in relation to the functional response. *Florida Entomologist* **73:**100–110.

———. 1991. Chaos and coevolutionary warfare in a chaotic predator-prey system. *Florida Entomologist* **74:**50–59.

Altenberg, L. 1991. Chaos from linear frequency-dependent selection. *American Naturalist* **138:**51–68.

Anderson, R. M. (ed.). 1982. *Population Dynamics of Infectious Diseases: Theory and Applications.* Chapman & Hall, London and New York.

Anderson, R. M., S. P. Blythe, S. Gupta, and E. Konings. 1989. The transmission dynamics of the Human Immunodeficiency Virus Type I in the male homosexual community in the United Kingdom: The influence of changes in sexual behavior. *Philosophical Transactions of the Royal Society of London B* **325:**145–198.

Anderson, R. M., and R. M. May. 1987. Transmission dynamics of HIV infection. *Nature* **326:**137–142.

———. 1991. *Infectious Diseases of Humans.* Oxford Science Publications, Great Britain.

Anderson, R. M., R. M. May, G. F. Medley, and A. Johnson. 1986. A preliminary study of the transmission dynamics of the human immunodeficiency virus (HIV), the causative agent of AIDS. *IMA Journal of Mathematics Applied in Medicine and Biology* **3**:229–263.

Bailey, N. T. J. 1975. *The Mathematical Theory of Infectious Diseases and Its Applications*. Griffin, London.

Beddington, J. R., C. A. Free, and J. H. Lawton. 1975. Dynamic complexity in predator-prey models framed in difference equations. *Nature* **255**:58–60.

Bellows, T. S., and M. P. Hassell. 1988. The dynamics of age-structured host-parasitoid interactions. *Journal of Animal Ecology* **57**:259–268.

Blythe, S. P., and R. M. Anderson. 1988a. Distributed incubation and infectious periods in models of the transmission dynamics of the human immunodeficiency virus (HIV). *IMA Journal of Mathematics Applied in Medicine and Biology* **5**:1–19.

———. 1988b. Variable infectiousness in HIV transmission models. *IMA Journal of Mathematics Applied in Medicine and Biology* **5**:181–200.

Blythe, S. P., and C. Castillo-Chavez. 1989. Like-with-like preference and sexual mixing models. *Mathematical Biosciences* **96**:221–238.

———. 1990. Scaling law of sexual activity. *Nature* **344**:202.

Blythe, S. P., C. Castillo-Chavez, J. Palmer, and M. Cheng. 1991. Towards a unified theory of mixing and pair formation. *Mathematical Biosciences* **107**:379–405.

Blythe, S. P., C. Castillo-Chavez, and G. Casella. 1992. Empirical models for the estimation of the mixing probabilities for socially-structured populations from a single survey sample. *Mathematical Population Studies* 3(3):199–225.

Brauer, F. 1990. Models for the spread of universally fatal diseases. *Journal of Mathematical Biology* **28**:451–462.

———. 1991. "Models for the Spread of Universally Fatal Diseases II." In S. Busenberg and M. Martelli (eds.), *Proceedings of the International Conference on Differential Equations and Applications to Biology and Population Dynamics*. Lecture Notes in Biomathematics 92. Springer-Verlag, New York, pp. 57–69.

Busenberg, S., and C. Castillo-Chavez. 1989. "Interaction, Pair Formation and Force of Infection Terms in Sexually Transmitted Diseases." In C. Castillo-Chavez (ed.), *Mathematical and Statistical Approaches to AIDS Epidemiology*. Lecture Notes in Biomathematics 83. Springer-Verlag, New York, pp. 289–300.

———. 1991. A general solution of the problem of mixing subpopulations, and its application to risk- and age-structured epidemic models for the spread of AIDS. *IMA Journal of Mathematics Applied in Medicine and Biology* **8**:1–29.

Busenberg, S., and K. Cooke. 1993. *Vertically Transmitted Diseases: Models and Dynamics*. Biomathematics 23, Springer-Verlag, New York.

Capasso, V. 1993. *Mathematical Structures of Epidemic Systems*. Lecture Notes in Biomathematics 97. Springer-Verlag, New York.

Castillo-Chavez, C. (ed.). 1989. *Mathematical and Statistical Approaches to AIDS Epidemiology*. Lecture Notes in Biomathematics 83. Springer-Verlag, New York.

Castillo-Chavez, C., and S. Busenberg. 1991. "On the Solution of the Two-Sex Mixing Problem." In S. Busenberg and M. Martelli (eds.), *Proceedings of the International Conference on Differential Equations and Applications to Biology and Population Dynamics*. Lecture Notes in Biomathematics 92. Springer-Verlag, New York, pp. 80–98.

Castillo-Chavez, C., S. Busenberg, and K. Gerow. 1991. "Pair Formation in Structured Populations." In J. Goldstein, F. Kappel, and W. Schappacher (eds.), *Differential Equations with Applications in Biology, Physics and Engineering*. Marcel Dekker, New York, pp. 47–65.

Castillo-Chavez, C., K. Cooke, W. Huang, and S. A. Levin. 1989a. The role of long incubation periods in the dynamics of HIV/AIDS. Part I: Single population models. *Journal of Mathematical Biology* **27**:373–398.

———. 1989b. "On the Role of Long Incubation Periods in the Dynamics of HIV/AIDS. Part 2: Multiple Group Models." In C. Castillo-Chavez (ed.), *Mathematical and Statistical Approaches to AIDS Epidemiology*. Lecture Notes iln Biomathematics 83. Springer-Verlag, New York.

———. 1989c. Results on the dynamics for models for the sexual transmission of the human immunodeficiency virus. *Applied Mathematics Letters* **2(4)**:327–331.

Castillo-Chavez, C., S. Fridman, and X. Luo. 1994a. "Stochastic and Deterministic Models in Epidemiology." In *Proceedings of the First World Congress of Nonlinear Analysts*, Tampa, FL, August 19–26, 1992. (In press.)

Castillo-Chavez, C., H. Hethcote, V. Andreasen, S. A. Levin, and W.-M. Liu. 1988. "Cross-Immunity in the Dynamics of Homogeneous and Heterogeneous Populations." In T. G. Hallam, L. G. Gross, and S. A. Levin (eds.), *Mathematical Ecology*. World Scientific Publishing, Singapore, pp. 303–316.

———. 1989d. Epidemiological models with age structure, proportionate mixing, and cross-immunity. *Journal of Mathematical Biology* **27(3)**:233–258.

Castillo-Chavez, C., S.-F. Shyu, G. Rubin, and D. Umbauch. 1992. "On the Estimation Problem of Mixing/Pair Formation Matrices with Applications to Models for Sexually-Transmitted Diseases." In K. Dietz, V. T. Farewell, and N. P. Jewell (eds.), *AIDS Epidemiology: Methodology Issues*. Birkhäuser, Boston, pp. 384–402.

Castillo-Chavez, C., J. X. Velasco-Hernandez, and S. Fridman. 1994b. "Modeling Contact Structures in Biology." In S. A. Levin (ed), *Frontiers of Theoretical Biology*. Lecture Notes in Biomathematics 100. Springer-Verlag, New York. (In press).

Caswell, H., and D. E. Weeks. 1986. Two-sex models: Chaos, extinction, and other dynamic consequences of sex. *American Naturalist* **128**:707–735.

Cressman, R. 1988. Complex dynamical behaviour of frequency dependent variability selection: An example. *Journal of Theoretical Biology* **130**:167–173.

Diekmann, O., J. A. P. Heesterbeek, and J. A. J. Metz. 1990. On the definition of R_0 in models for infectious diseases in heterogeneous populations. *Journal of Mathematical Biology* **28**:365–382.

Dietz, K. 1988. On the transmission dynamics of HIV. *Mathematical Biosciences* **90**:397–414.

Dietz, K., and K. P. Hadeler. 1988. Epidemiological models for sexually transmitted diseases. *Journal of Mathematical Biology* **26:**1–25.

Ebenman, B. 1987. Niche differences between age classes and intraspecific competition in age structured populations. *Journal of Theoretical Biology* **124:**25–33.

Ellner, S. 1989. "Inferring the Causes of Population Fluctuations." In C. Castillo-Chavez, S. A. Levin, and C. A. Shoemaker (eds.), *Mathematical Approaches to Problems in Resource Management and Epidemiology*. Lecture Notes in Biomathematics 81, Springer-Verlag, New York.

Evans, A. S. 1982. *Viral Infections of Humans*. Second edition. Plenum Medical Book Company, New York.

Fredrickson, A. G. 1971. A mathematical theory of age structure in sexual populations: Random mating and monogamous marriage models. *Mathematical Biosciences* **20:**117–143.

Gabriel, J. P., C. Lefèvre, and P. Picard (eds.). 1990. *Stochastic Processes in Epidemic Theory*. Lecture Notes in Biomathematics 86. Springer-Verlag, New York.

Gardini, L., R. Lupini, C. Mammana, and M. G. Messia. 1987. Bifurcation and transition to chaos in the three dimensional Lotka Volterra map. *SIAM Journal of Applied Mathematics* **47:**455–482.

Gilpin, M. E. 1992. Spiral chaos in a predator prey model. *American Naturalist* **113:**306–308.

Guckenheimer, J., G. Oster, and A. Ipaktchi. 1977. The dynamics of density-dependent population models. *Journal of Mathematical Biology* **4:**101–147.

Gupta, S., R. M. Anderson, and R. M. May. 1989. Network of sexual contacts: Implications for the pattern of spread of HIV. *AIDS* **3:**1–11.

Hadeler, K. P. 1989a. Pair formation in age-structured populations. *Acta Applicandae Mathematicae* **14:**91–102.

———. 1989b. "Modeling AIDS in Structured Populations." In *Proceedings of the 47th Session of the International Statistical Institute,* Paris, August/September, pp. 83–99.

Hadeler, K. P., and K. Nagoma. 1990. Homogeneous models for sexually transmitted diseases. *Rocky Mountain Journal of Mathematics* **20:**967–986.

Hastings, A., C. Hom, S. Ellner, P. Turchin, and H. C. J. Godfray. 1993. Chaos in ecology: Is Mother Nature a strange attractor? *Annual Review of Ecological Systems* **24:**1–33.

Hastings, A., and T. Powell. 1991. Chaos in a three species food chain. *Ecology* **72:**896–903.

Hethcote, H. W. 1976. Qualitative analysis for communicable disease models. *Mathematical Biosciences* **28:**335–356.

———. 1978. An immunization model for a heterogeneous population. *Theoretical Population Biology* **14:**338–349.

———. 1989. "Three Basic Epidemiological Models." In S. A. Levin, T. G. Hallam, and J. Gross (eds.), *Applied Mathematical Ecology*. Biomathematics 18, Springer-Verlag, New York, pp. 119–144.

Hethcote, H. W., and S. A. Levin. 1989. "Periodicity in Epidemiological Models." In: S. A. Levin, T. G. Hallam, and L. J. Gross (eds.), *Applied Mathematical Ecology*. Biomathematics 18. Springer-Verlag, New York.

Hethcote, H. W., and J. W. van Ark. 1992. *Modeling HIV Transmission and AIDS in the United States*. Lecture Notes in Biomathematics 95, Springer-Verlag, New York.

Hethcote, H. W., and J. A. Yorke. 1984. *Gonorrhea Transmission Dynamics and Control*. Lecture Notes in Biomathematics 56, Springer-Verlag, New York.

Hochberg, M. E., M. P. Hassell, and R. M. May. 1990. The dynamics of host-parasitoid-pathogen interactions. *American Naturalist* **135:**74–94.

Hsu Schmitz, S.-F. 1994. Some theories, estimation methods, and applications of marriage and mixing functions to demography and epidemiology. Ph.D. dissertation, Cornell University, Ithaca, NY.

Hsu Schmitz, S.-F., S. Busenberg, and C. Castillo-Chavez. 1993. On the evolution of marriage functions: It Takes Two to Tango. Biometrics Unit Technical Report BU-1210-M, Cornell University, Ithaca, NY.

Hsu Schmitz, S.-F., and C. Castillo-Chavez. 1993. "Completion of Mixing Matrices for Non-Closed Social Networks." In *Proceedings of First World Congress of Nonlinear Analysts,* Tampa, FL, August 19–26, 1992. (In press.)

———. 1994. "Parameter Estimation in Non-Closed Social Networks Related to the Dynamics of Sexually-Transmitted Diseases." In E. Kaplan and M. Brandeau (eds.), *Modeling the AIDS Epidemic*. Raven, New York. (In press.)

Huang, W. 1989. Studies in differential equations and applications. Ph.D. dissertation, The Claremont Graduate School, Claremont, CA.

Huang, W., K. Cook, and C. Castillo-Chavez. 1992. Stability and bifurcation for a multiple group model for the dynamics of HIV/AIDS transmission. *SIAM Journal of Applied Mathematics*. **52(3):**835–854.

Hyman, J. M., and E. A. Stanley. 1988. Using mathematical models to understand the AIDS epidemic. *Mathematical Biosciences* **90:**415–473.

———. 1989. "The Effect of Social Mixing Patterns on the Spread of AIDS." In C. Castillo-Chavez, S. A. Levin, and C. Shoemaker (eds.), *Mathematical Approaches to Problems in Resource Management and Epidemiology*. Lecture Notes in Biomathematics 81. Springer-Verlag, New York, pp. 190–219.

Jacquez, J. A., C. P. Simon, and J. Koopman. 1989. "Structured Mixing: Heterogeneous Mixing by the Definition of Mixing Groups." In C. Castillo-Chavez (ed.), *Mathematical and Statistical Approaches to AIDS Epidemiology*. Lecture Notes in Biomathematics. Springer-Verlag, New York, pp. 301–315.

Jacquez, J. A., C. P. Simon, J. Koopman, L. Sattenspiel, and T. Perry. 1988. Modelling and analyzing HIV transmission: The effect of contact patterns. *Mathematical Biosciences* **92:**119–199.

Jewell, N. P., K. Dietz, and V. T. Farewell. 1991. *AIDS Epidemiology: Methodology Issues*. Birkhäuser. Boston.

Kaplan, E., and M. Brandeau. (eds.). 1994. *Modeling AIDS and the AIDS Epidemic*. Raven, New York. (In press.)

Kendall, D. G. 1949. Stochastic processes and population growth. *Royal Statistical Society, Series B* **2:**230–264.

Kermack, W. O., and A. G. McKendrick. 1927. A contribution to the mathematical theory of epidemics. *Proceedings of the Royal Society of London, Series A* **115:**700–721.

———. 1932. A contribution to the mathematical theory of epidemics. *Proceedings of the Royal Society of London, Series A* **138:**55–83.

Keyfitz, N. 1949. The mathematics of sex and marriage. *Proceedings of the Sixth Berkeley Symposium on Mathematical and Statistical Problems* **4:**89–108.

Kot, M., W. M. Schaffer, G. L. Truty, D. J. Graser, and L. F. Olsen. 1988. Changing criteria for imposing order. *Ecological Modelling* **43:**75–110.

Leslie, P. H. 1945. On the use of matrices in certain population mathematics. *Biometrika* **33:**183–212.

Levin, S. A. 1981. Age structure and stability in multiple-age populations. *Renewable Resources Management* **40:**21–45.

———. 1983a. "Coevolution." In H. I. Freedman and C. Strobeck (eds.), *Population Biology*. Lecture Notes in Biomathematics 52. Springer-Verlag, New York, pp. 328–334.

———. 1983b. "Some Approaches in the Modeling of Coevolutionary Interactions." In M. Nitecki (ed.), *Coevolution*. University of Chicago Press, Chicago, pp. 21–65.

Levin, S. A., and C. Castillo-Chavez. 1990. "Topics in Evolutionary Biology." In S. Lessard (ed), *Mathematical and Statistical Developments of Evolutionary Theory*. NATO ASI Series. Kluwer, Boston, pp. 327–358.

Levin, S. A., and C. Goodyear. 1980. Analysis of an age-structured fishery model. *Journal of Mathematical Biology* **9:**245–274.

Levin, S. A., and D. Pimentel. 1981. Selection of intermediate rates of increase in parasite-host systems. *American Naturalist* **117:**308–315.

Lotka, A. J. 1922. The stability of the normal age distribution. *Proceedings of the National Academy of Sciences* **8:**339–345.

———. 1923. Contributions to the analysis of malaria epidemiology. *American Journal of Hygiene,* 3 January Supplement.

Lubkin, S., and C. Castillo-Chavez. 1994. "A Pair-Formation Approach to Modeling Inheritance of Social Traits." In *Proceedings of First World Congress of Nonlinear Analysts,* Tampa, FL, August 19–26, 1992. (In press.)

Luo, X., and C. Castillo-Chavez. 1991. Limit behavior of pair formation for a large dissolution rate. *Journal of Mathematical Systems Estimation and Control* **3:**247–264.

May, R. M. 1974. Biological populations with non-overlapping generations: Stable points, stable cycles, and chaos. *Journal of Theoretical Biology* **5:**511–524.

———. 1976. Simple mathematical models with very complicated dynamics. *Nature* **261:**459–467.

———. 1985. Regulation of populations with nonoverlapping generations by microparasites: A purely chaotic system. *American Naturalist* **125:**573–584.

———. 1987. Chaos and the dynamics of biological populations. *Proceedings of the Royal Society of London, Series A* **413:**27–44.

May, R. M., and R. M. Anderson. 1983a. Epidemiology and genetics in the coevolution of parasites and hosts. *Proceedings of the Royal Society of London, Series B* **219:**281–313.

———. 1983b. "Parasite-Host Coevolution." In D. Futuyama and M. Slatkin (eds.), *Coevolution*. Sinauer, Sunderland, MA.

———. 1989. The transmission dynamics of human immunodeficiency virus (HIV). *Philosophical Transactions of the Royal Society, Series B* **321:**565–607.

May, R. M., and G. F. Oster. 1976. Bifurcations and dynamics complexity in simple ecological models. *American Naturalist* **110:**573–599.

McFarland, D. D. 1972. "Comparison of Alternative Marriage Models." In T. N. E. Greville (ed.), *Population Dynamics*. Academic Press, New York, pp. 89–106.

McKendrick, A. G. 1926. Applications of mathematics to medical problems. *Proceedings of the Edinburgh Mathematics Society* **44:**98–130.

Neubert, M. G., and M. Kot. 1992. The subcritical collapse of predator-prey models. *Mathematical Biosciences* **110:**45–66.

Nold, A. 1980. Heterogeneity in disease-transmission modeling. *Mathematical Biosciences* **52:**227–240.

Olsen, L. F., and W. M. Schaffer. 1990. Chaos versus noisy periodicity: Alternative hypotheses for childhood epidemics. *Science* **249:**499–504.

Parlett, B. 1972. "Can There Be a Marriage Function?" In T. N. E. Greville (ed.), *Population Dynamics*. Academic Press, New York, pp. 107–135.

Pickering, J. J., A. Wiley, N. S. Padian, L. E. Lieb, D. F. Echenberg, and J. Walker. 1986. Modelling the incidence of acquired immunodeficiency syndrome (AIDS) in San Francisco, Los Angeles and New York. *Mathematical Modelling* **7:**661–698.

Pollard, J. H. 1973. *Models for the Growth of Human Populations*. Cambridge University Press, London.

Pugliese, A. 1990a. Population models for disease with no recovery. *Journal of Mathematical Biology* **28:**65–82.

———. 1990b. "An S→E→I Epidemic Model with Varying Population Size." In S. Busenberg and M. Martelli (eds.), *Proceedings of the International Conference on Differential Equations and Applications to Biology and Population Dynamics*. Lecture Notes in Biomathematics 93. Springer-Verlag, New York, pp. 121–138.

Ross, R. 1911. *The Prevention of Malaria*. Second edition, with addendeum. John Murray, London.

Rubin, G., D. Umbauch, S.-F. Shyu, and C. Castillo-Chavez. 1992. Application of capture-recapture methodology to estimation of size of population at risk of AIDS and/or other sexually transmitted diseases. *Statistics in Medicine* **11:**1533–1549.

Sattenspiel, L. 1987a. Population structure and the spread of disease. *Human Biology* **59:**411–438.

———. 1987b. Epidemics in nonrandomly mixing populations: A simulation. *American Journal of Physical Anthropology* **73:**251–265.

Sattenspiel, L., and C. Castillo-Chavez. 1990. Environmental context, social interactions, and the spread of HIV. *American Journal of Human Biology* **2:**397–417.

Sattenspiel, L., and C. P. Simon. 1988. The spread and persistence of infectious diseases in structured populations. *Mathematical Biosciences* **90:**341–366.

Schaffer, W. M. 1985a. Can nonlinear dynamics elucidate mechanisms in ecology and epidemiology? *IMA Journal of Mathematics Applied in Medicine and Biology* **2:**221–252.

———. 1985b. Order and chaos in ecological systems. *Ecology* **66:**93–106.

Schaffer, W. M., and M. Kot. 1985. Nearly one dimensional dynamics in an epidemic. *Journal of Theoretical Biology* **112:**403–427.

Shaffer, W. M., L. F. Olsen, G. L. Truty, and S. L. Fulmer. 1990. "The Case of Chaos in Childhood Epidemics." In S. Krasner (ed.), *The Ubiquity of Chaos*. American Association for the Advancement of Science, Washington, D.C., pp. 138–166.

Soper, H. E. 1929. Interpretation of periodicity in disease prevalence. *Journal of the Royal Statistical Society B* **92:**34–73.

Takeuchi, Y., and N. Adachi. 1983. Existence and bifurcation of stable equilibrium in two-prey, one-predator communities. *Bulletin of Mathematical Biology* **45:**877–900.

Thieme, H. R., and C. Castillo-Chavez. 1989. "On the Role of Variable Infectivity in the Dynamics of the Human Immunodeficiency Virus Epidemic." In C. Castillo-Chavez (ed.), *Mathematical and Statistical Approaches to AIDS Epidemiology*. Lecture Notes in Biomathematics 83. Springer-Verlag, New York, pp. 157–176.

———. 1994. How may infection-age dependent infectivity affect the dynamics of HIV/AIDS? *SIAM Journal of Applied Mathematics*. (In press.)

Velasco-Hernandez, J. X., and C. Castillo-Chavez. 1994. "Modeling Vector-Host Disease Transmission and Food Web Dynamics Through the Mixing/Pair-Formation Approach." In *Proceedings in the First World Congress of Nonlinear Analysts*, Tampa, FL, August 19–26, 1992. (In press.)

Waldstätter, R. 1989. "Pair Formation in Sexually Transmitted Diseases." In C. Castillo-Chavez (ed.), *Mathematical and Statistical Approaches to AIDS Epidemiology*. Lecture Notes in Biomathematics 83, Springer Verlag, New York, pp. 260–274.

19

The Invisible Present

John J. Magnuson

All of us can sense change—the reddening sky with dawn's new light, the rising strength of lake waves during a thunderstorm, the changing seasons of plant flowering as temperature and rain have their effects on the landscapes around us. Some of us see longer-term events and remember that there was less snow last winter or that fishing was better a couple of years ago. It is the unusual person who senses with any precision changes occurring over decades. At this time scale we are inclined to think of the world as static and we typically underestimate the degree of change that does occur. Because we are unable to sense slow changes directly, and because we are even more limited in our abilities to interpret cause-and-effect relations for these slow changes, processes acting over decades are hidden and reside in what I call "the invisible present" (Magnuson et al. 1983).

The invisible present is the time scale within which our responsibilities for planet earth are most evident. Within this time scale, ecosystems change during our lifetimes and the lifetimes of our children and our grandchildren. This is the time scale of acid deposition, the invasion of nonnative plants and animals, the introduction of synthetic chemicals, CO_2-induced climate warming, and deforestation. In the absence of the temporal context provided by long-term research, serious misjudgments can occur, not only in our attempts to understand and predict change in the world around us, but also in our attempts to manage our environment. Although serious accidents in an instant of human misjudgment may cause the end of Spaceship Earth (*sensu* Fuller 1970), destruction is even more likely to occur at a ponderous pace in the secrecy of the invisible present.

Revealing the Invisible Present

Long-term or sustained research can open for view the events of the invisible present, much as time-lapse photography opens for view the blooming of a flower or the movement of a snail. A single year's observation of any structure or event, such as the duration of ice cover on Lake Mendota in the winter of 1982–1983 (Figure 19-1A), from a long-term perspective is relatively uninteresting in that it provides, in itself, no insight into the long-term behavior of a natural system. Yet when a time series of annual values is opened to 10 years, to 50 years, or to the length of record—132 years—the invisible present is put into context and can be better understood.

With 10 years of record, it is apparent that the duration of ice cover in 1983 was 40 days or so shorter than in any of the other nine years and far exceeded the typical range of variation. Also, we see that the duration of ice cover varies considerably from year to year. With 50 years of record, it becomes apparent that 1983 and other El Niño years have tended to have shorter durations of ice cover (Robertson 1989). Now the phenology of ice cover is linked to a major feature of global climate, the Southern Ocean oscillation index (Mysak 1986; Quinn et al. 1978). With 132 years of record (Robertson 1989), a general warming trend becomes visible that was invisible with the 10- and 50-year records.

A Little Ice Age ended in about 1890 (Lamb 1977; Wahl and Lawson 1970), as reflected by the decrease in the duration of ice cover. In the most recent years, there is a hint that another warming has begun, perhaps signaling CO_2-induced global climate warming (Liss and Crane 1983). The time series shows that the 1983 ice cover was of the shortest duration observed in the entire 132 years. Thus, each increase in the period of record revealed new insights about the invisible present and made the condition in 1983 more understandable and more interesting.

As with observational studies, field experiments also can be susceptible to serious misinterpretation if they are not conducted in the context of long-term ecological research (Tilman 1989). Nitrogen addition to plots in an old-field environment at the Cedar Creek Natural History Area LTER Site in Minnesota illustrates this point (Figure 19-2). If the influence of nitrogen were judged in the first year after fertilization, ecologists might have concluded that fertilization increased perennial ragweed (*Ambrosia coronopifolia)* and bluegrass (*Poa pratensis)* and had no influence on blackberry (*Rubus* sp.). The five-year time series is more revealing; perennial ragweed responded immediately and then dramatically declined to control levels, bluegrass responded most positively in the second year and by the fifth year was significantly less abundant than in control plots, and blackberry showed no change in the first year but had a highly significant increase in abundance by the fifth year (Tilman 1988). Therefore, nitrogen fertilization is time dependent. The response after one, three, or five years is each statistically significant and repeatable, but each differs significantly

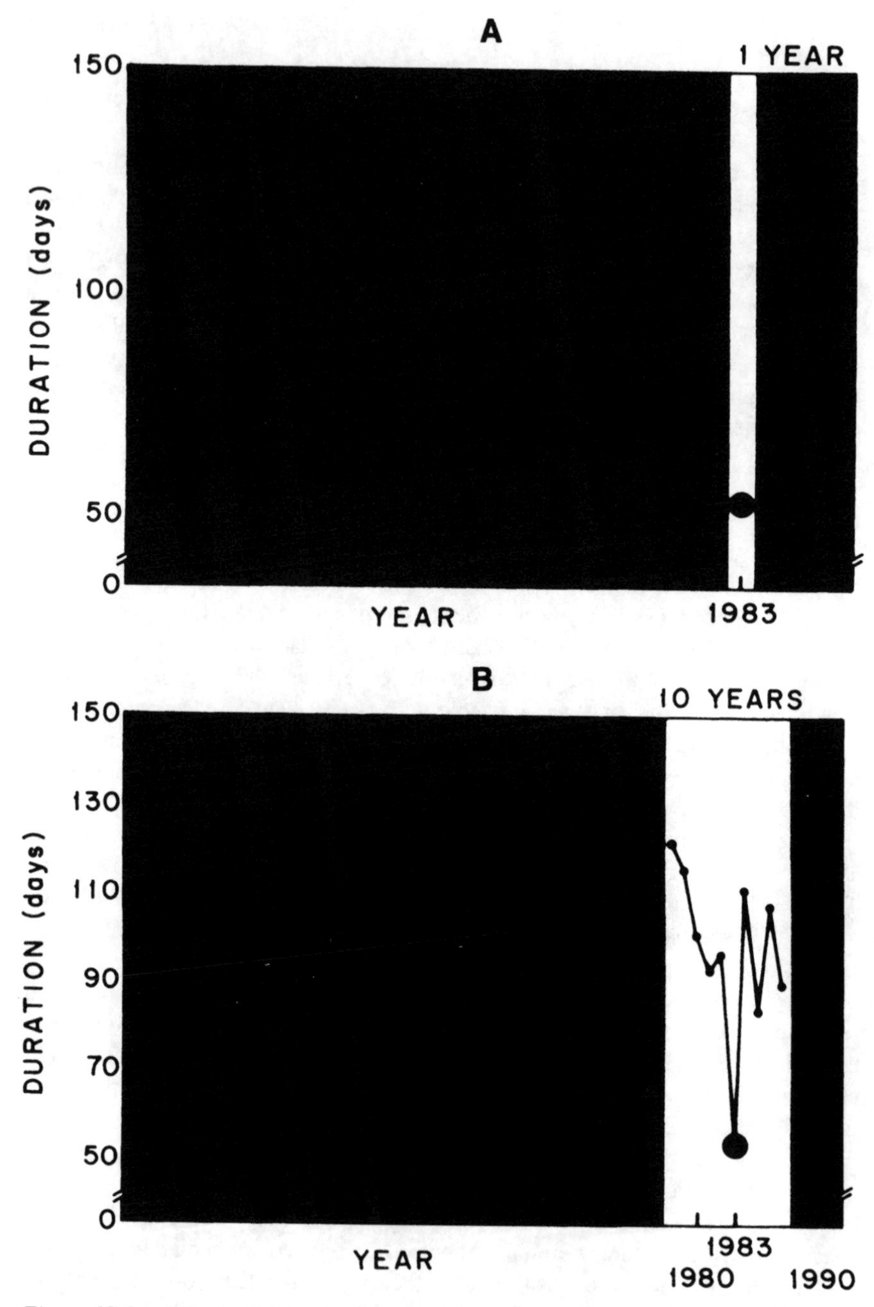

Figure 19-1. When a time series is opened up, new phenomena become apparent and the present is put into a context that makes it more understandable and more interesting. The duration of ice cover on Lake Mendota, Wisconsin, at Madison, has been recorded, originally by interested citizens beginning in 1855 and now by the state's climatologist, providing the longest limnological and climatological record in the state. The record was analyzed by Dale M. Robertson (1989).

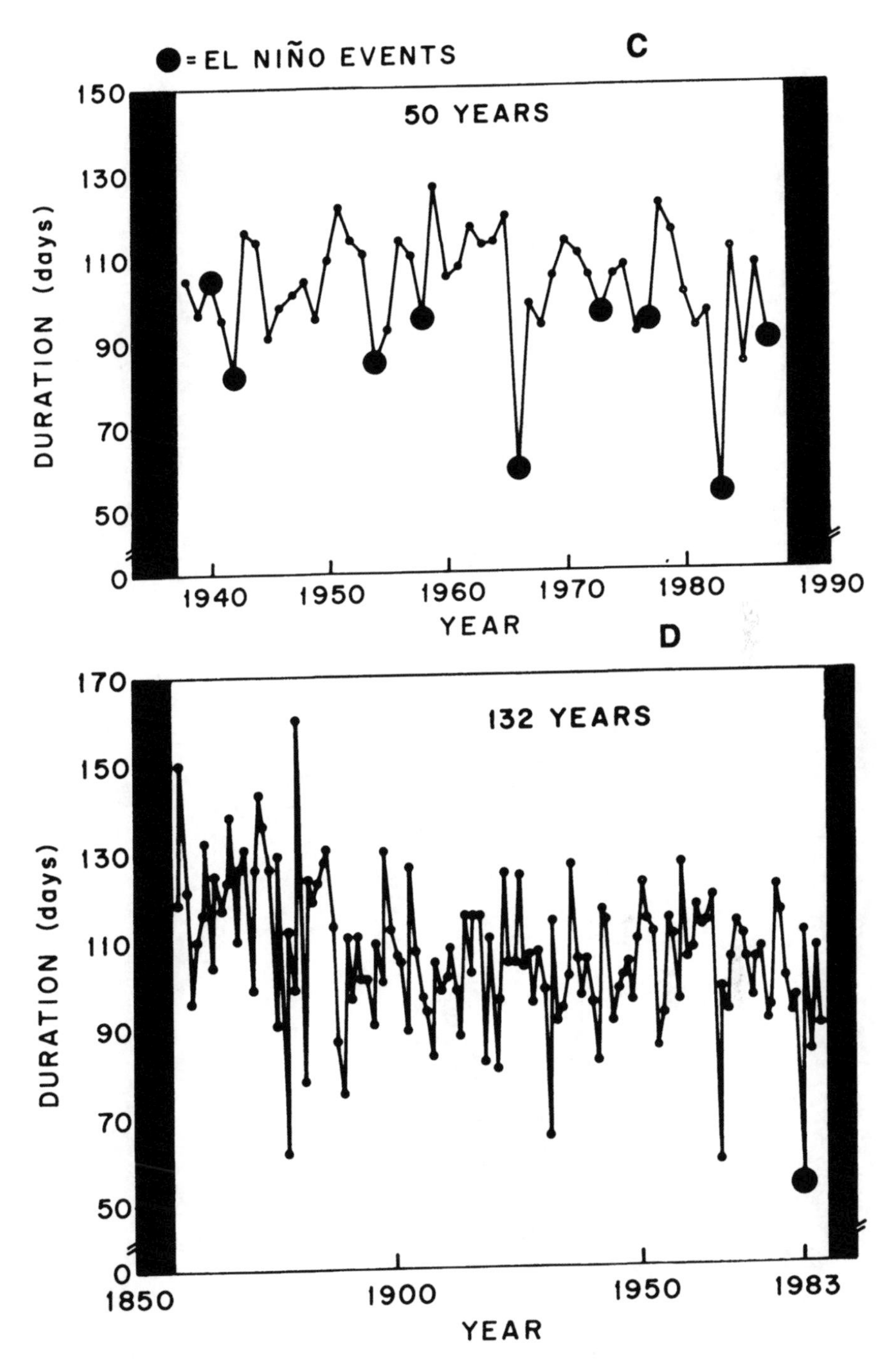

Figure 19-1. (*Continued*).

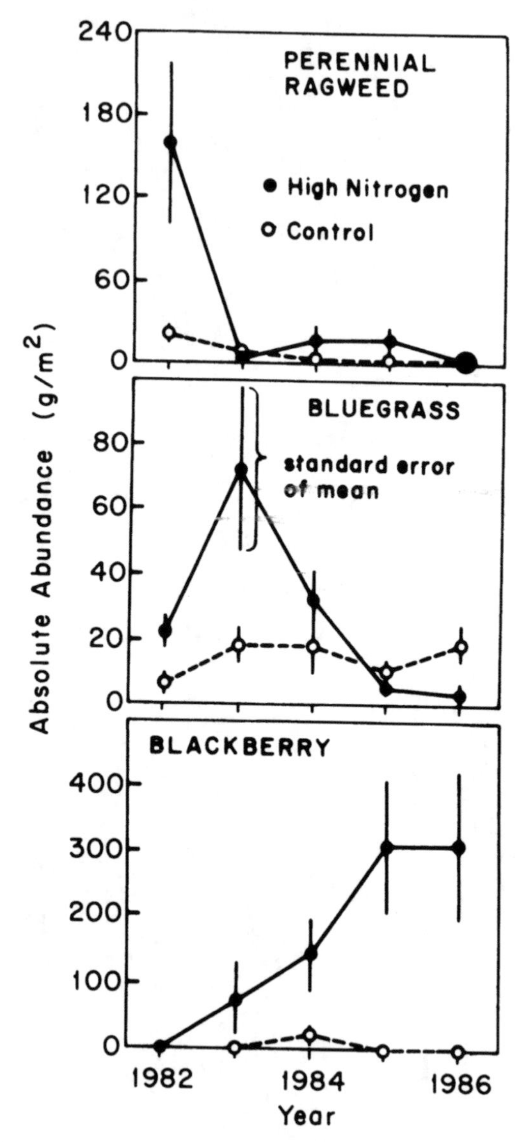

Figure 19-2. The response of a natural ecological system to a treatment or a disturbance is a time series, much like a successional sequence, rather than a single observation. Nitrogen was added to plots in an old field at the Cedar Creek Natural History Area LTER Site in Minnesota, and other plots were left as controls (Tilman 1987). The response in absolute abundance (means and standard errors) of perennial ragweed (*Ambrosia coronopifolia*), bluegrass (*Poa pratensis*), and blackberry (*Rubus* sp.) are shown from 1982 to 1986 (figure adapted from Tilman 1989).

from that in other years. This time series displays features invisible from one- or two-year experiments. Clearly, a one-year experiment, even though its results would be repeatable and statistically significant, does not reflect the change induced by the fertilization; instead of perennial ragweed and bluegrass as suggested in year one, blackberry dominates by year five.

Long-Term Ecological Research

Long-Term Ecological Research (LTER) is a program at the National Science Foundation (Brenneman 1989; Callahan 1984; Franklin et al. 1990; Magnuson and Bowser 1990; Swanson and Franklin 1988) that concentrates on unveiling the invisible present. Researchers at a network of 17 sites across the United States and one in Antarctica focus their research on time scales of years, to decades, to a century, and examine ecological processes over longer periods than possible in most previous ecological studies (Figure 19-3). The researchers also address a wide range of spatial scales—meters to kilometers to cross-continent intersite comparisons (Swanson and Sparks 1990).

Funded by the National Science Foundation's Division of Biotic Systems and Resources, the LTER sites include temperate and tropical forests, prairie, desert,

TIME SCALES

YEARS		RESEARCH SCALES	PHYSICAL RESET EVENTS	BIOLOGICAL PHENOMENA
				• Evolution of Species
10^5	100 MILLENNIA		• Continental Glaciation	
10^4	10 MILLENNIA	PALEO ECOLOGY & LIMNOLOGY		• Bog Succession • Forest Community Migration
10^3	MILLENNIUM		• Climate Change	• Species Invasion • Forest Succession
10^2	CENTURY	LTER	• Forest Fires • CO_2 Climate Warming	• Cultural Eutrophication
10^1	DECADE		• Sun Spot Cycle • El Niño	• Hare Population • Prairie Succession
10^0	YEAR		• Prairie Fires • Lake Turnover	• Annual Plants • Plankton Succession
10^{-1}	"MONTH"		• Ocean Upwelling	
10^{-2}	"DAY"	MOST ECOLOGY	• Storms • Diel Light Cycle • Tides	• Algal Bloom • Diel Migration
10^{-3}	"HOUR"			

Figure 19-3. The Long-Term Ecological Research program supported by the National Science Foundation focuses on time scales of years to decades to a century in a major effort to understand ecological structure and function hidden in the invisible present. Time scales relevant to various physical events and biological phenomena are taken from sources including Delcourt et al. (1983), Haury et al. (1978), and Stommel (1963, 1965).

alpine and arctic tundra, agricultural fields, lakes, rivers, coastal wetlands, and an estuary. At each site, major questions are addressed related to patterns and controls of primary production, food webs, population abundance and distribution, organic matter accumulation, and biogeochemical cycling, as well as questions related to disturbance frequency and effect. A base of common measurements and questions also lays the groundwork for new analyses and generalizations across different ecosystems relatively unencumbered by the unique character of each ecosystem. In addition, the accumulating record and experience provide a temporal and spatial context for individual researchers needing such a base for their experiments.

LTER sites are regional and national research facilities for long-term or sustained ecological research. Institutions operating LTER sites encourage collaborative research by scientists at other institutions and the use of the sites by visiting investigators. Paleoecology and paleolimnology complement long-term observational study because they provide an even longer-term context for interpreting the present. Because such studies have lower temporal, spatial, and biotic resolution than is possible with biological observations and experiments, they in turn are complemented by long-term research. Many important ecological changes and processes are played out over a long time scale, and they are often ones with human relevance (Likens 1983, 1989).

Long-Term Lags Between Cause and Effect

Time lags longer than a year can exist between cause and effect or until ecological responses to a disturbance permeate natural systems. These time lags occur for many reasons: certain biological and physical processes simply take time, biological relicts persist even after conditions change, movements across the landscape take time, the simultaneous occurrence of two or more necessary conditions for an event or process to occur can be rare, and a chain of events accumulates the lags between cause-and-effect events. Each of these reasons for time lags will be illustrated with examples from LTER sites.

Processes Taking Time

The accumulation of biomass is a good example of a process that simply takes time. The example for a cohort of trees is the same as for a year-class of fishes or for any population of organisms that live a number of years and whose biomass does not die back each year. After fish hatch or trees germinate, annual mortalities reduce the total number left in the cohort, but individuals grow larger. The accumulated biomass of the cohort reaches a maximum in the year when the product of the number of organisms still living and their mean body mass is greatest (see Beverton and Holt 1957; Pitcher and Hart 1982 for fish). Before the maximum, there are many trees but they are small; after the maximum, the trees are large but are few in number. The time lag between the year when a cohort is formed and when it reaches maximum biomass depends on the mortality and the growth rates of organisms in that particular population. For many fish in freshwater lakes, the time lag is two to four years, but for long-lived trees it may be 200 years or longer.

Biological Relicts

A period of successful tree reproduction can produce a forest that may persist into a future that changing conditions have made unsuitable for reproduction. The treeline at the Niwot Ridge/Green Valley Lakes LTER Site in the Rocky Mountains of Colorado has been considered to be a relict treeline, established in the more favorable climate of the altithermal period about 4000 years ago (Ives 1978; Ives and Hansen-Bristow 1983; Nichols 1982). Some evidence for this hypothesis is provided by a 1905 burn at the Niwot treeline, where, more than 80 years later, the forest shows few signs of recovery. This interpretation is controversial, and Shankman (1984) has projected from seedling and sapling establishment that recovery will yet occur given sufficient time in the absence of disturbance.

Relict biota also persist into the future after conditions have changed over time scales of only a few years. The extensive meadows dominated by the sedge *Kobresia* on Niwot are thought, on the basis of their soil characteristics, to have been in place for centuries. *Kobresia* prefers sites free of snow in winter; it rarely reproduces sexually on Niwot Ridge but instead maintains itself through vegetative tillering. Pat Webber and his students (Emerick 1976; Keigley 1987; Webber et al. 1976) assessed the resistance of the *Kobresia* community to environmental change by increasing snowpack with snow fences. In the first two years after greater snowpack, the plant produced more leaves and appeared to be vigorous. The plant apparently reallocated its resources to stature rather than to vegetative reproduction when the snowpack was present. But the snowpack prevented the generation of new shoots, and within five years *Kobresia* was sparse and by 10 years had completely disappeared. A time lag, linked causally to biology of this particular species, is evident in the disappearance of *Kobresia*; and again, as with Tilman's fertilization experiments, a conclusion made in the first two years after the manipulation would have been misleading.

Movement Across Landscapes

Landscapes in which ecosystems occur have spatial dimensions that influence the temporal ones. Movement of water, materials, and organisms takes time and introduces lags into ecosystem changes. Classic examples are the time for dispersal of exotic plants and animals across the landscape. The European starling (*Sturnus vulgaris*) dispersed from the New York City area in 1896 to western New Jersey by 1908 and to western Ohio by 1926 (Elton 1958). The sea lamprey (*Petromyzon marinus*) spread from Lake Ontario after the Welland Canal was built in 1829 into Lake Erie by 1932 (Pearce et al. 1980) and Lakes Huron,

Michigan, and Superior by the late 1930s (Smith and Tibbles 1980). The Asiatic clam (*Corbicula fluminea* Philippi) was first observed in the Columbia River in 1938; by 1960 it had reached southern Illinois, and, by 1975, southern Wisconsin via dispersal up the Mississippi River (Thompson and Sparks 1977). The Asiatic clam is thought to be inducing ecological changes in the rivers after a time lag of 50 years (Sparks pers. comm.)

Groundwater movement is slow at the North Temperate Lake LTER Site in Wisconsin, not at all like the rushing streams in more erodible landscapes. The study lakes are arranged along a topographic gradient, and groundwater takes approximately three years to flow from the uppermost, rainwater-dominated lake to the next lower lake only 100 m downslope. A surface stream would take only minutes. The slow movement of water through the sandy till increases the dissolution of ions, and the alkalinity of the water increases dramatically before entering the lower lake (Kenoyer 1986; Figure 19-4). Owing to this slow movement of groundwater down the landscape from lake to lake, each lake successively lower in the flow system has higher alkalinities and concentrations of solutes. Some of the groundwater now entering the lakes low in the landscape may have been in the flow system 100 years or more and could be, in part, a product of conditions present before the region was logged and burned near the turn of the century.

Rare Coincidence

Time lags can occur because a process requires the coincidence of two or more low-probability events; in some cases, many years may pass before the events

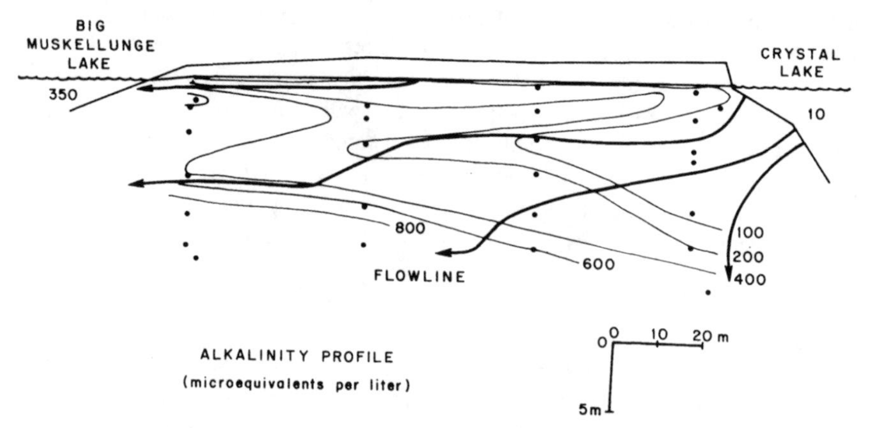

Figure 19-4. The slow movement of groundwater across the landscape allows time for chemical reactions to increase the alkalinity of the water entering the lower lake at the North Temperate Lake LTER Site in Wisconsin (adapted from Kenoyer 1986). Far more acid buffering capacity is acquired from the till, which is low in carbonates, than would have been expected if flows were faster.

occur together. In the shortgrass steppe at the Central Plains Experimental Range LTER Site in Colorado, plant communities are dominated by a perennial grass, blue grama (*Bouteloua gracilis*). Disturbances in the shortgrass steppe occur over many spatial and temporal scales, but a particular pair of conditions must occur before blue grama can recover.

Establishment of seedlings requires that viable seeds and adequate soil moisture be present. Both conditions are rare in these semiarid grasslands; viable seeds are produced approximately every seven years and appropriate soil moisture approximately every eight years. The two conditions are independent. Debra P. Coffin and William K. Lauenroth of Colorado State University devised a model to simulate the long-term changes of blue grama on small patches over 500 years (Coffin and Lauenroth 1990; Figure 19-5).

During the 500-year simulation, the plot was disturbed four times by cattle fecal pats that killed all the blue grama. Fecal pats are the most frequent small-scale disturbance in shortgrass plant communities grazed by domestic cattle and

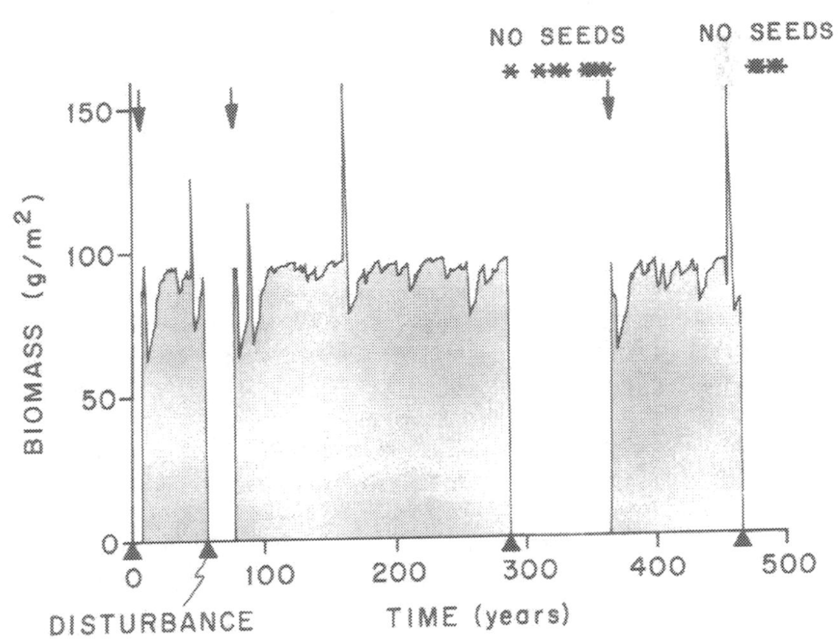

Figure 19-5. Time lags after a disturbance on the shortgrass steppe occur because the simultaneous occurrence of the events required for reestablishment of the dominant grass (i.e., viable seeds and adequate soil water) is rare. In this simulation by Coffin and Lauenroth (1990), the disturbances that eliminate the blue grama (*Bouteloua gracilis*) are indicated by four triangles on the *x*-axis, the asterisks indicate the years after a disturbance when suitable microenvironmental conditions were present but viable seeds were not, and the arrows indicate the first year after a disturbance when both seeds and adequate soil water were present.

serve as a model of larger-scale disturbances such as plowing. The time lags that resulted were, in chronological sequence, approximately 10, 20, 80, and, at the end of the record, at least 35 years in length. This model provides some insight as to why larger sections of the shortgrass steppe that were disturbed by plowing 50 to 100 years ago have not yet recovered.

Chains of Events

Time lags are also generated because a cause-and-effect chain of events accumulates the lags from each link in the chain. Changes in water clarity provide a good example (Figure 19-6). In Crystal Lake, at the North Temperate Lake LTER Site operated by the University of Wisconsin-Madison, we observed a decrease in water clarity over a three-year period from approximately 11 m to 6 m in 1984 (Magnuson et al. 1990).

This change was initiated by the formation of strong year-classes of yellow perch (*Perca flavescens*), two and three years before the lowest water clarity was observed. When the perch reached maximum biomass and moved into the open water, they preyed on and reduced the abundance of the microscopic herbivores that eat the planktonic algae. The perch-induced reduction in herbivores lessened grazing pressure on the midwater algae, and these algae increased in abundance. With more algae in the water column, light penetration was reduced and water clarity declined dramatically. As the perch year-classes aged, their biomass and numbers declined and the changes reversed.

Most of this two- to three-year lag resulted from the time it takes for a year-class of fish to obtain maximal biomass. It is at the age of maximal biomass that the perch exerted the greatest grazing pressure on their prey. These changes were not unexpected (Brooks and Dodson 1965; Carpenter 1988; Carpenter et al. 1985; Henrikson et al. 1980; Hrbacek 1958; Hrbacek et al. 1961; Northcote 1988), but we were surprised by their magnitude.

Earlier, Joan Baker and I (Baker and Magnuson 1976) had observed a change in water clarity of Crystal Lake from the early measurements by E. A. Birge and C. Juday in the 1930s and by Baker in 1973. We had concluded that a gradual change over 40 years had occurred owing to changes in human use. It now appears that a much more rapid oscillation in water clarity occurs due to the formation of strong year-classes of perch every few years. Again, Baker's short-term research, with only one recent year for comparison, did not penetrate the realm of the invisible present.

Conclusions

The natural world is dynamic, not static, and the older each of us gets, the more apparent is this truth. Time lags in ecological systems are the rule; they separate

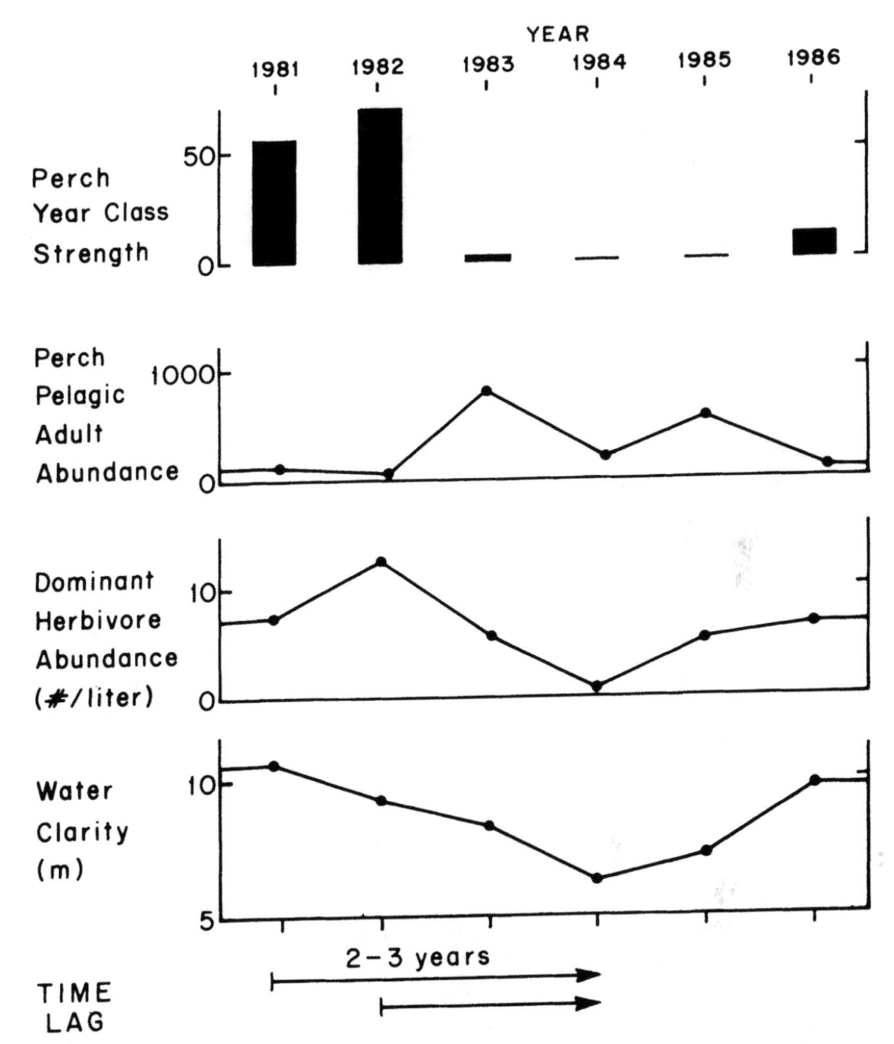

Figure 19-6. Time lags from a chain reaction in a food web occur because a fish year-class takes several years to reach maximum biomass and thus exert its maximal influence on its prey populations. The abundant perch reduce the abundance of herbivorous zooplankton, causing the algae to thrive, and in turn decreasing water clarity.

cause and effect to confuse our interpretation of the natural world, which can appear fickle and unsettled. Increasingly long records of ecosystem structure and process and long-term experiments expose new phenomena and allow understanding. Long-Term Ecological Research (LTER) formalizes this human experience. Such knowledge is sure to help us in facing the problems ahead. Our lack of understanding of events and processes in the LTER time scale of decades has been and will in the future be costly to human society. Operating Spaceship Earth blind to the invisible present in unacceptable.

The invisible present is beginning to be penetrated by researchers and agencies with an increasing awareness of the importance of serious and coordinated research at the time scales of years to decades. LTER is the most extensive of these.

Epilogue

Three years after Long-term ecological research and the invisible present appeared in *BioScience*, three developments seem worth reporting.

First, the ice cover of Lake Mendota continues to play out its dance of year-to-year variability in climate, and we continue to learn from it. Documentation of these dynamics can be found in Robertson et al. (1992), Assel and Robertson (ms.), and Anderson et al. (ms). The warming trend in late winter weather is clearly apparent during the 1970s and 1980s. Another El Niño has occurred and is signaled in the break-up dates; the length of the record now runs 139 years, from 1855 to 1993. The pattern observed for El Niño and recent warming also is apparent in other Wisconsin lakes: these are regional phenomena.

We have popularized the use of this record by preparing a more general publication (Magnuson and Drury 1991) and developing teaching materials for pre-college students (Anderson and Trewyn ms.). The popular article was reprinted in a collection of selected current articles on the environment. The teaching materials are designed to help students discover the nature of science by using our long-term ice records. The lesson for all of us is that long-term data on the world around us provide rich and useful experiences from which to learn and ask new questions.

The second and third points relate to two articles that appeared with my paper on the invisible present in the August 1990 issue of *BioScience*.

Swanson and Sparks (1990) write about a concept analogous to the invisible present: the invisible place. They point out that "the significance of research results is difficult to interpret if a site's context in space is not understood." Most ecology has been conducted at sample points or with plots up to 100 m on a side. There is a clear need for increased understanding of ecological systems at multiple scales, including the landscape, regional, and global scales. The larger spatial scales seem even more appropriate when dealing with longer-term processes. An important challenge is to learn how to scale up from small to larger spatial scales (Dale et al. 1989). Equally challenging is the task of determining how larger-scale patterns and processes influence structure and function at small spatial scales. Thus, spatial context as well as temporal context is important if one wishes to interpret meaningfully the ecological structure and function of this place, right now.

The paper by Franklin et al. (1990) describes the history, character, and contribution of a research program that is charged with replacing the ignorance hidden in those invisible places in the invisible present with a new understanding of how ecological systems work. The LTER program began with five sites in 1981. By 1993 it had grown to 18 sties ranging from Antarctic ice to Arctic tundra; 14 of the sites are in the 48 contiguous United States. Included are forest, desert, prairie, alpine, agricultural, coastal, lake, and small-stream ecosystems. Understanding requires more than simple long-term observation across broad spatial regions. The LTER studies embody and attract a rich array of research approaches and paradigms ranging broadly across the ecological and environmental sciences. It is becoming increasingly apparent from the LTER program and other research groups with similar insights that long-term ecological research at larger spatial scales will provide a macroscope in space and time that can help us understand the world around us and guide our interactions within it.

Acknowledgments

I thank the National Science Foundation for having the leadership and the insight to establish a program for a network of Long-Term Ecological Research Sites. I thank Patrick J. Webber (Niwot Ridge/Green Valley Lakes LTER Site), who made unpublished materials available for this paper. I thank Donald E. Chandler, Cheryl M. Hughes, and Mary L. Smith of the University of Wisconsin in Madison for preparation of the figures and text. This article was based on a talk presented at a 6 November 1987 briefing at the National Science Foundation in Washington, D.C., and at the LTER Symposium at the Ecological Society of America in August 1988 in Davis, California.

References

Anderson, W. L., D. M. Robertson, and J. J. Magnuson. (Ms.) Similarities in the phenology of ice thaw dates on Wisconsin lakes.

Anderson, W. L., and B. Trewyn. (Ms.) Ice out: Discovering science through ice out dates. (Lesson plans for middle and high school science.)

Assel, R. A., and D. M. Robertson. (Ms.) Changes in the winter air temperatures near Lake Michigan over the past 142 years determined from nearby ice records.

Baker, J., and J. J. Magnuson. 1976. Limnological responses of Crystal Lake (Vilas County, Wisconsin) to intensive recreational use, 1924–1973. *Transactions of the Wisconsin Academy of Sciences, Arts and Letters* **64**:47–61.

Beverton, R. H. J., and S. J. Holt. 1957. *On the Dynamics of Exploited Fish Populations*. Fishery Investigations, Series II, Volume XIX, Ministry of Agriculture, Fisheries and Food, Her Majesty's Stationery Office, London.

Brenneman, J. (ed.). 1989. *Long Term Ecological Research in the United States, a*

Network of Sites 1987 (Fifth edition). Long-Term Ecological Research Network, Forest Science Department, Oregon State University, Corvallis, OR.

Brooks, J. L., and S. I. Dodson. 1965. Predation, body size, and composition of plankton. *Science* **150:**28–35.

Callahan, J. T. 1984. Long-term ecological research. *BioScience* **34:**363–367.

Carpenter, S. R. 1988. "Transmission of Variance Through Lake Food Webs." In S. R. Carpenter (ed.), *Complex Interactions in Lake Communities*. Springer-Verlag, New York, pp. 119–135.

Carpenter, S. R., J. F. Kitchell, and J. R. Hodgson. 1985. Cascading trophic interactions and lake productivity. *BioScience* **35:**634–639.

Coffin, D. P., and W. K. Lauenroth. 1990. A gap dynamics simulation model of succession in a semi arid grassland. *Ecological Modelling* **49:**229–266.

Dale, V. H., R. H. Gardner, and M. G. Turner (guest eds.). 1989. Predicting across scales: Theory development and testing. *Landscape Ecology* **3:**145–252.

Delcourt, H. R., P. A. Delcourt, and T. Webb III. 1983. Dynamic plant ecology: The spectrum of vegetational change in space and time. *Quaternary Science Reviews* **1:**153–175.

Elton, C. S. 1958. *The Ecology of Invasions by Animals and Plants*. Methuen, London.

Emerick, J. C. 1976. Effects of artificially increased winter snow cover on plant canopy architecture and primary production in selected areas of the Colorado alpine tundra. Ph.D. dissertation. University of Colorado at Boulder, CO.

Franklin, J. F., C. S. Bledsoe, and J. T. Callahan. 1990. Contributions to ecological science: The Long-Term Ecological Research program. *BioScience* **40:**509–523.

Fuller, R. B. 1970. *Operating Manual for Spaceship Earth*. Touchstone, New York.

Haury, L. R., J. A. McGowan, and P. H. Weibe. 1978. "Patterns and Processes in the Time-Space Scales of Plankton Distributions." In J. H. Steele (ed.), *Spatial Pattern in Plankton Communities*. Plenum, New York, pp. 277–328.

Henrikson, L. H., G. Nyman, H. G. Oscarson, and J. E. Stenson. 1980. Trophic changes without changes in external nutrient loading. *Hydrobiologia* **68:**257–263.

Hrbacek, J. 1958. Density of the fish population as a factor influencing the distribution and speciation of the species in the genus *Daphnia*. *Proceedings of the XVth International Congress of Zoology*, Section X. Paper 27, pp. 794–795.

Hrbacek, J., M. Dvorakova, V. Korinek, and L. Prochazkova. 1961. Demonstration of the effect of the fish stock on the species composition of zooplankton and the intensity of metabolism of the whole plankton association. *Verhandlungen der Internationalen Vereingung fur Theoretische und Angewandeten Limnologie* **14:**192–195.

Ives, J. D. 1978. "Remarks on the Stability of Timberline." In C. Troll and W. Lauer, (eds.), *Geoecological Relations Between the Southern Temperate Zone and the Tropical High Mountains*. Erdwissenschaftliche Forschung, Franz Steiner, Wiesbaden, Germany, pp. 313–317.

Ives, J. D., and K. J. Hanson-Bristow, 1983. Stability and instability of natural and modified upper timberline landscapes in the Colorado Rocky Mountains, U.S.A. *Mountain Research and Development* **3**:149–155.

Keigley, R. B. 1987. Effect of experimental treatments on *Kobresia myosuroides* with implications for the potential effect of acid deposition. Ph.D. dissertation. University of Colorado at Boulder, CO.

Kenoyer, G. J. 1986. Evolution of groundwater chemistry and flow in a sandy aquifer in northern Wisconsin. Ph.D. dissertation. University of Wisconsin-Madison.

Lamb, H. H. 1977. *Climate: Present, Past, and Future*. Vol. II. *Climatic History and the Future*. Methuen, New York.

Likens, G. E. 1983. A priority for ecologists. *Bulletin of the Ecological Society of America* **64(4)**:234–293.

Likens, G. E. (ed.). 1989. *Long-Term Studies in Ecology*. Springer-Verlag, New York.

Liss, P. S., and A. J. Crane. 1983. *Man-Made Carbon Dioxide and Climatic Change—A Review of Scientific Problems*. Geo Books, Norwich, England.

Magnuson, J. J., M. Anderson, D. Armstrong, C. Bowser, T. Frost, and T. Kratz. 1990. Influences of a strong year class of fish and variation in hydrologic inputs on the interannual dynamics of an oligotrophic lake. Abstract of a paper delivered at V International Congress of Ecology (INTECOL 1990), Yokohama, Japan, August 23–30, 1990.

Magnuson, J. J., and C. J. Bowser. 1990. A network for long-term ecological research in the United States. *Freshwater Biology* **23**:137–143.

Magnuson, J. J., C. J. Bowser, and A. L. Beckel. 1983. The invisible present: Long-term ecological research on lakes. *L & S Magazine*, University of Wisconsin-Madison, Fall **1983**:3–6.

Magnuson, J. J., and J. A. Drury. 1991. Global change ecology. *The World and I* 6(4):304–311. [Reprinted in J. L. Allen (ed.), 1992, *Annual Editions: Environment 92/93*. Dushkin Publishing Group, Gilford, CT, pp. 39–45.]

Magnuson, J. J., T. K. Kratz, T. M. Frost, B. J. Benson, C. Bowser, and R. Nero. 1991. "Expanding the Temporal and Spatial Scales of Ecological Research and Comparison of Divergent Ecosystems: Roles for the LTER in the United States." In P. G. Risser (ed.), *Long Term Ecological Research*. SCOPE Series, Wiley & Sons, Sussex, England, pp. 45–70.

Mysak, L. A. 1986. El Niño, interannual variability and fisheries in the Northeast Pacific Ocean. *Canadian Journal of Fisheries and Aquatic Sciences* **43**:464–497.

Nichols, H. 1982. "Review of the Late Quaternary History of Vegetation and Climate in the Mountains of Colorado." In J. C. Halfpenny (ed.), *Ecological Studies in the Colorado Alpine: A Festschrift for John W. Marr*, Occasional Paper 37, University of Colorado, Institute of Arctic and Alpine Research, pp. 27–33.

Northcote, T. G. 1988. Fish in the structure and function of freshwater ecosystems: A "top-down" view. *Canadian Journal of Fisheries and Aquatic Sciences* **45**:361–379.

Pearce, W. A., R. A. Braem, S. M. Dustin, and J. J. Tibbles. 1980. Sea lamprey (*Petromyzon marinus*) in the lower Great Lakes. *Canadian Journal of Fisheries and Aquatic Sciences* **37**:1802–1810.

Pitcher, T. J., and P. B. Hart. 1982. *Fisheries Ecology*. AVI Publishing Company, Westport, CT.

Quinn, W. H., D. O. Zopf, K. S. Short, and R. T. W. Kuo Yang. 1978. Historical trends and the statistics of the southern oscillation, El Niño, and Indonesian droughts. *Fishery Bulletin* **76(3)**:663–677.

Robertson, D. M. 1989. The use of lake water temperature and ice cover as climatic indicators. Ph.D. dissertation. University of Wisconsin-Madison.

Robertson, D. M., R. A. Ragotzkie, and J. J. Magnuson. 1992. Lake ice records used to detect historical and future climatic changes. *Climatic Change* **21**:407–427.

Shankman, D. 1984. Tree generation following fire as evidence of timberline stability in the Colorado Front Range, U.S.A. *Arctic and Alpine Research* **16(4)**:413–417.

Smith, B. R., and J. J. Tibbles. 1980. Sea lamprey (*Petromyzon marinus*) in Lakes Huron, Michigan and Superior: History of invasion and control, 1936–78. *Canadian Journal of Fisheries and Acquatic Sciences* **37(11)**:1780–1801.

Sparks, R. Personal Communication. 1989. Forbes Biological Station, Havanna, Illinois, of the Illinois Natural History Survey.

Stommel, H. 1963. Varieties of oceanographic experience. *Science* **139**:572–576.

———. 1965. "Some Thoughts About Planning the Kuroshio Survey." In *Proceedings of a Symposium on the Kuroshio*, Tokyo, Japan, October 29, 1963. Oceanographic Society of Japan and UNESCO.

Swanson, F. J., and J. F. Franklin. 1988. The Long-Term Ecological Research Program. *EOS* **69(3)**:34, 36, 46.

Swanson, F. J., and R. E. Sparks. 1990. Long-term ecological research and the invisible place. *BioScience* **40**:502–508.

Thompson, C. M., and R. E. Sparks. 1977. The Asiatic clam, *Corbicula manilensis* in the Illinois River. *Nautilus* **91**:34–36.

Tilman, D. 1987. Secondary succession and the pattern of plant dominance along experimental nitrogen gradients. *Ecological Monographs* **57**:189–214.

———. 1988. *Plant Strategies and the Dynamics and Structure of Plant Communities*. Princeton University Press, Princeton, NJ.

———. 1989. "Ecological Experimentation: Strengths and Perceptual Problems." In G. Likens (ed.), *Long-Term Studies in Ecology*. Springer-Verlag, New York, pp. 136–157.

Wahl, E. W., and T. L. Lawson. 1970. The climate of the midnineteenth century United States compared to the current normals. *Monthly Weather Review* **98(4)**:259–265.

Webber, P. J., J. C. Emerick, D. C. Ebert May, and V. Komarkova. 1976. "The Impact of Increased Snowfall on Alpine Vegetation." In H. W. Steinhoff and J. D. Ives (eds.), *Ecological Impacts of Snowpack Augmentation in the San Juan Mountains, Colorado*. Final Report, San Juan Ecology Project, Colorado State University Publishers, Fort Collins.

20

Interpreting Explanatory Processes for Time Series Patterns: Lessons from Three Time Series

Mitchel P. McClaran, Put O. Ang, Jr., Angel Capurro, Douglas H. Deutschman, David J. Shafer, and Jean-Marc Guarini

What processes are responsible for time series patterns? This is the seminal question for ecologists and a critical question for society's response to changing environments. Through an historical review of three time series and the processes proposed as explanations for them, we illustrate that process interpretation begins with descriptive analysis of the time series and is clarified by greater time series length, additional time series from other areas, correlation analysis with ancillary data, simulation models, and experimentation.

Which explanatory processes are proposed is dependent on the analytical techniques used to describe the time series (Stockwell et al. Chapter 7 of this volume). Because these descriptions are inferentially based, proposed processes may be rejected when the descriptions become more accurate and precise. Greater time series length can change process interpretation by revealing less frequent patterns and by failing to replicate previous patterns (Magnuson 1990 and Chapter 19 of this volume). Similarly, additional time series from other areas describe the geographic fidelity among series and prompt reassessment of proposed processes in light of the observed spatial variation. Correlation analysis between the dependent time series variable and ancillary information is used to evaluate the strength and synchrony of potential processes. Simulation models provide theoretical evaluation of proposed processes; controlled observation and experimentation provide empirical assessment.

Our three time series examples are Canada lynx (*Lynx canadensis*) abundance estimated from pelts sold to the Hudson's Bay Company; Dungeness crab (*Cancer magister*) abundance estimated from landings in the northeastern Pacific; and global temperature changes estimated from (terrestrial and marine) surface instruments. We chose these examples because they have a rich history of proposed, amended, and rejected explanatory processes, and because they represent classic problems in ecological theory and contemporary conundrums for the management of natural resources.

Canada Lynx

Fluctuations in Canada lynx pelt sales to the Hudson's Bay Company for the years 1821–1913 stimulated significant ecological and analytical developments. These data have a long history of time series analysis: they were used in some of the earliest analyses and continue to be revisited with more sophisticated techniques. Similarly, the data have ties to the development of predator-prey theory and cycling in animal populations. The scope of the proposed explanatory processes has narrowed over time, but some processes that were initially proposed continue to find favor in results from correlation analysis, simulation modeling, and experimentation. Termination of the dataset presents the greatest limitation to interpreting the responsible processes.

Finerty (1980) and Akcakaya (1992) suggested that, because prices of pelts offered by the Hudson's Bay Company were fairly constant over time, changes in pelt sales probably reflect the true fluctuation in lynx population rather than any change in the harvest effort. There has been no documentation of a constant hunting effort, however.

Elton (1924, 1942) and Elton and Nicholson (1942) proposed a 9- to 10-year cycle of lynx abundance from visual assessment of a time series that combined information from six regions (Figure 20-1), although the display of values from 10 regions appeared to be synchronous. Cole (1951, 1954) suggested, however, that random numbers could give the impression of cycling if a time series of finite length were expressed as a sum of sine and cosine terms. Moran (1953) applied a formal time series analytical technique, a second-order autoregressive (or linear stochastic difference) model, to describe the lynx time series pattern. Bulmer (1974) suggested that Moran's model produced only irregular oscillations and proposed instead a mixed model that included a sine function to produce regular cycle amplitude and frequency. Subsequently, Bulmer (1976) used a periodogram to describe a 9.5-year cycle. Autocorrelation and periodogram analyses rejected Cole's suggestion of artificial cyclicity, because an oscillatory autocorrelation function will produce a spectral peak at the frequency of oscillation only if the amplitude of the sine wave is significantly large (Jenkins and Watts 1968; Finerty 1980).

In the 1970s, it was suggested that nonlinear time series analysis would be necessary to describe the lynx fluctuation pattern (Campbell and Walker 1977; Tong 1977). One nonlinear model described a limit cycle with a 6-year ascent and a 3-year descent, using a self-exiting autoregressive function (SETAR) and assuming a birth curve with an Allee effect (Tong 1990). This assumption has not yet been evaluated (Tong 1990). Other nonlinear models have indicated that the lynx data have a chaotic behavior (Tong 1990). These results are consistent with Schaffer's (1984, 1985) earlier findings using a Poincaré map. The short length of the series, however, weakens the power of the Poincaré map analysis. Turchin and Taylor (1992) developed a method called response surface analysis

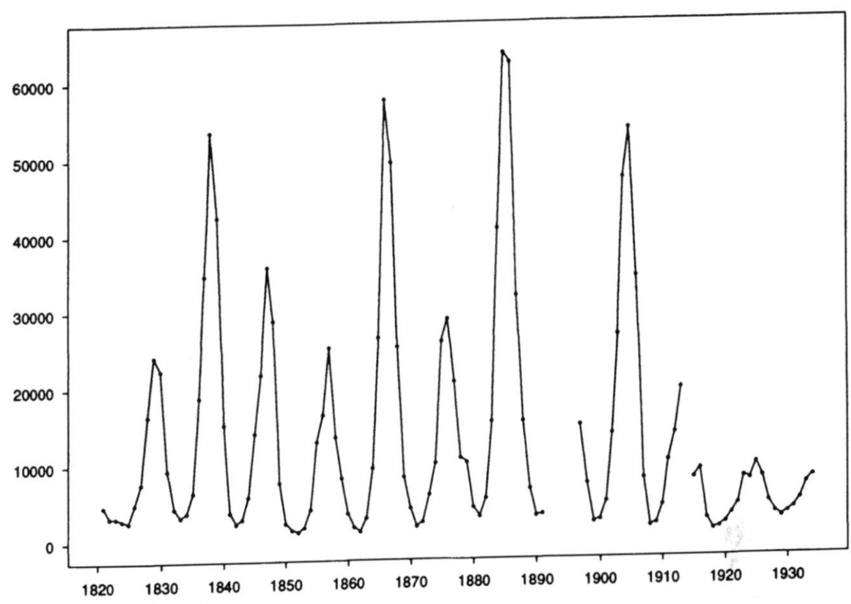

Figure 20-1. Lynx fur returns of the Northern Department, Hudson's Bay Company, 1821–1913, and of equivalent area 1915–34. (Adapted from C. S. Elton and M. Nicholson, 1942, The ten-year cycle in numbers of lynx in Canada, *Journal of Animal Ecology* 11:215–244. By permission of the British Ecological Society.)

to study dynamical systems with very few data points. They found the lynx records to have a quasi-periodic oscillation, which is consistent with the SETAR results.

Proposed processes for the lynx cycle have included sunspots, weather, and prey fluctuations, which have been evaluated using correlations with ancillary data, simulation models, and experimentation.

Sunspots were rejected as a possible explanatory process because their 11-year cycle was longer than the 9- to 10-year lynx cycle (MacLulich 1937; Elton 1942; Elton and Nicholson 1942). Conversely, Arditi (1979) used multiple regression to relate monthly temperature and precipitation to lynx population fluctuations and suggested that population size was most sensitive to temperature variations at the beginning and end of summer.

Most process interpretations have focused on fluctuations in snowshoe hare (*Lepus americanus*) populations, because snowshoe hare is the primary prey for Canada lynx. Moran (1953) suggested that predator-prey dynamics was an important process responsible for the lynx cycle. Unfortunately, the absence of a hare time series that is coincident with the lynx pelt series prevents a cross-correlation or cross-spectral analysis (Bulmer 1978). Therefore, correlation analysis has focused on processes that produce snowshoe hare cycles. These efforts suggest two possible mechanisms: hare cycles are a function of forage availability (Lack 1954; Keith 1974, 1983), or they are controlled by predators (Trostel et al. 1987, Akcakaya 1992). Fox's (1978) cross-correlation and cross-spectral analysis of 19 years of population, snowfall, and fire data suggested that fire and heavy snowfall can reduce available forage, thereby limiting hare populations and, subsequently, lynx populations. Akcakaya (1992) argued that a 19-year series was not long enough to interpret a process that creates a 9- to 10-year cycle.

Correlation analysis was also used to suggest that predation by lynx was the largest source of hare mortality, because it revealed only a 1-year lag in hare response to winter predation versus a 2-year lag in response to forage and weather factors (Trostel et al. 1987). The utility of this study, too, is limited by its length (to 6 years). Recently, Sinclair et al. (1993) used a phase analysis of the cross-correlations among annual net snow accumulation and tree-scarring ratio, hare pelt records before 1895, and sunspot number, to suggest that solar activity may have amplified some climatic cycles, which in turn affected the food supply of hares. That is, solar activity does not directly force the hare cycle, but it may provide a phase-locking mechanism that entrains the hare population cycle.

Simulation models have frequently been used to assess the predator-prey relationship as a process responsible for the lynx cycle. For example, Trostel et al. (1987) produced an 8- to 11-year hare cycle using a density-dependent prey model with a Type II functional response for the predating lynx. A ratio-dependent model also produced a realistic 9-year cycle of lynx and hare (Akcakaya 1992). This model was based on the concept that the response of the populations is a function of the ratio between predator and prey. Akcakaya (1992) used biological parameters estimated from data independent of the original time series. He found a time series that presents a 9-year cycle with a 6-year ascent and a 3-year descent, which is consistent with the response surface analysis (Turchin and Taylor 1992) and with the SETAR model (Tong 1990) results that used the original time series data.

Experimental analysis was used to evaluate the relative importance of starvation and predation to snowshoe hare mortality. When unlimited food was provided to hares, the amplitude of the hare population cycle increased because of greater rates of growth, survival, and reproduction; but there was no change in the frequency of the hare cycle (Krebs 1986a, 1986b; Smith et al. 1988). Although these results have appeal as support for the importance of lynx predation in controlling hare cycles, they have limited interpretive value for the lynx cycle, because lynx responses were not measured.

The interpretation of processes responsible for the lynx cycle has changed because the cycle was described with greater accuracy and precision using autocorrelation, nonlinear autocorrelation, and dynamical systems analyses. Greater clarification of the responsible processes has been hindered, however, by several factors: the time series was terminated; time series for possible processes were not coincident with the lynx series; experimental analysis has been continued for only short periods and has not included a lynx response; and, finally, it is not possible to assess the accuracy of pelt sales as a surrogate for lynx population dynamics.

If the time series had continued, it would now be over 170 years long, and the results of dynamical systems analysis would be more conclusive after 17 cycles than after only 10 cycles. Correlation analysis would have provided a more insightful contribution to predator-prey theory if coincident time series for snowshoe hare, precipitation, fire, and vegetation had been acquired.

The most lasting contributions to ecology from the analysis and interpretation of the Canada lynx time series have been the introduction of autocorrelation, spectral, and nonlinear time series analyses as objective measures of cycling; and the identification of areas for experimentation that may reveal processes responsible for the lynx cycle.

Dungeness Crab

The Dungeness crab (*Cancer magister*) is the main crustacean harvested on the west coast of North America (Jamieson and Armstrong 1991), where it generates nearly $50 million in annual revenues (Methot 1986; Pacific Marine Fisheries Commission 1990). Fluctuations in crab harvests (or landings) have been dramatic and relatively synchronous off the coasts of northern California, Oregon, and Washington, but there has been less pronounced cycling and synchrony in the northernmost populations off British Columbia and in Alaskan coastal waters (Figure 20-2). The fluctuations in crab landings have fostered 40 years of effort to describe these patterns and to identify responsible processes, with the eventual goal of establishing sustainable harvest strategies (see reviews by Hankin 1985; Botsford 1986; Botsford et al. 1989; Jamieson and Armstrong 1991). The range of the proposed explanatory processes has grown with each waning of the landing cycle.

The life history of the Dungeness crab and commercial harvest policies are integral to interpreting the biological and physical processes that are proposed to drive the landing fluctuations. Dungeness crabs spawn in open coastal waters of the western United States and Canada. Larvae hatch in the winter and progress through five stages before molting to the final pelagic stage, the megalopa (Jamieson and Armstrong 1991). Megalopae are initially found far offshore but eventually settle in more shallow coastal waters, which is apparently critical for survival (Botsford et al. 1989). Adults typically mate in their second or third year, depending on size and gender. In general, male crabs mate with numerous females that are all considerably smaller than themselves (Smith and Jamieson 1991). Harvest is restricted to male crabs of larger than 160-mm carapace width (Botsford et al. 1989).

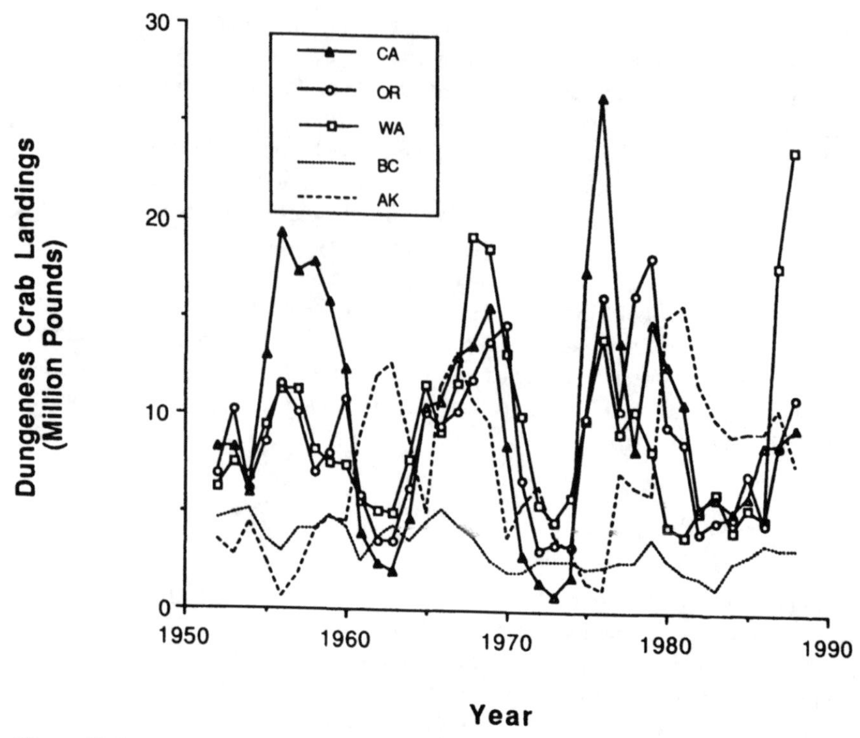

Figure 20-2. Estimates of commercial landings (in million pounds) of Dungeness crab by state and province. California, Oregon, and Washington landings oscillate coherently, unlike those for Alaska and British Columbia. (Data source: Pacific Marine Fisheries Commission 1990).

A 10-year periodicity and spatial synchrony for population fluctuations was proposed in the mid-1960s, when the Dungeness crab entered its first well-documented period of low harvest. It was generally believed that similar large-scale and synchronous fluctuations in abundance had occurred earlier and were characteristic of the fishery (Pacific Marine Fisheries Commission 1965; Poole 1968). Autocorrelation analysis was first used to describe the pattern about 10 years later (Peterson 1973; Botsford and Wickham 1975). In the late 1970s and early 1980s, it became apparent that landings were an imprecise and potentially biased estimate of adult population size. Estimates of effort-corrected harvest rates ranged from as low as 40% to nearly 90% of all adult males (Pacific Marine Fisheries Commission 1978; Botsford et al. 1983). Consequently, landings were adjusted for variable harvest effort and reanalyzed using autocorrelation, cross-correlation, and coherence techniques (Botsford et al. 1983; Johnson et al. 1986; Berryman 1991). The consensus emerging from these studies was a 10.5-year frequency for crab population fluctuations. The fluctuations are roughly synchronous among the northern California, Oregon, and Washington populations, but the Alaskan and Canadian populations fluctuate differently (Figure 20-2).

Most of the processes proposed to explain the fluctuations in crab landings are predicated on the assumption that recruitment into the first-year class is most critical. However, population estimates are based on adult males that are generally 3 to 5 years of age. Consequently, most hypotheses have assumed a 3- to 5-year lag between an event and its appearance in the landing time series (Cleaver 1949; Pacific Marine Fisheries Commission 1965; Botsford and Wickham 1975). Therefore, most research has assumed that a 3- to 5-year lag exists between changes in recruitment and observed changes in adult harvest. The processes suggested to limit recruitment have represented either fluctuations in the physical environment or biological interactions. The former group includes ocean temperatures, currents, and wind stress; the latter includes unfertilized females, intraspecies competition, cannibalism, egg parasitism, and excessive harvest levels. Correlation analysis and simulation modeling have been used to evaluate these processes, but conclusions have been limited by bias from the landing-based population estimates, the shortness of the time series, and the absence of life history information.

In 1968, it was assumed that nearly all legal-sized males were harvested, and thus that landings reflected adult male abundance. At the time, the lack of correlation with ancillary information was used to reject unfertilized females as an explanatory process. Surveys showing a high frequency of fertilized females were interpreted as supporting the contention that there was adequate egg fertilization but poor recruitment into the juvenile classes (Poole 1968). As a consequence, most of the research and discussion focused on the environmental determinants of recruitment. In particular, larval survival was linked with ocean temperature and currents (Pacific Marine Fisheries Commission 1965; Poole and Gotshall 1965; Poole 1968).

The early 1970s, the second well-documented period of low crab abundance, witnessed renewed effort to use correlations with ancillary information to interpret the role of environmental processes in the crab landing fluctuations. Peterson (1973) reported that landings followed changes in ocean upwelling, lagging by two years in California and one year in Oregon and Washington. A reanalysis of the data using time series methods suggested that the correlation may have been spurious (Botsford and Wickham 1975): the crab landings showed a distinct cycle with positive autocorrelations and a 10-year lag, but the index of upwelling was not cyclical. The lack of periodicity in upwelling and the absence of a biological mechanism that enhances adult survival cast serious doubt on this type of environmental forcing.

Ancillary information stimulated the use of simulation models to evaluate biological processes in the late 1970s and early 1980s. Fortuitous but sketchy ancillary findings of immature crabs in adult crab stomachs and of parasites in eggs led to the development of density-dependence models (Gotshall 1977; Botsford and Wickham 1978; Pacific Marine Fisheries Commission 1978; McKelvey et al. 1980; Botsford 1981; McKelvey and Hankin 1981). Experimental investigation of these proposed processes has not occurred because the processes are difficult to control and quantify over long periods and large areas of coastal waters (Jamieson and Armstrong 1991). Furthermore, variable fishing effort and its consequences for population estimates were revealed during this period. The large fluctuations in harvest rate confound the ability to compare landings with adult densities predicted from the population models. Consequently, the inaccuracy of the time series description limited progress in the evaluation of proposed explanatory processes.

Over the last 10 years, debate over the environmental or density-dependent mechanisms responsible for the cycles has continued (see reviews in Hankin 1985; Botsford et al. 1989; and Jamieson and Armstrong 1991). For example, simulation analysis suggests that fishing effort itself may influence the cycle period or amplitude (Botsford et al. 1983; Berryman 1991). Environmental forcing, such as southward wind stress in spring, was correlated with lagged harvest, and the wind stress appeared to be periodic (Johnson et al. 1986). The supporting evidence for a number of proposed processes has led to interpretations that multiple, interacting processes might be responsible for population fluctuations (Jamieson 1986; Jamieson and Armstrong 1991).

Changes in the interpretation of the Dungeness crab cycle have occurred because proposed processes were not correlated with the time series, the time series was corrected for fishing effort, and a multiple process paradigm gained support. The presence of fertilized females and the lack of correlation between the upwelling pattern and crab landings eliminated these proposed processes from further consideration. Reassessment of the landing fluctuations in light of variable fishing effort led to a clarification of the time series pattern and to suggestions that fishing effort may influence the crab cycle. Some authors have proposed that a multiple-process interpretation may be more realistic than a single-process interpretation, as there is some support for southward wind stress, fishing effort, and a number of density-dependent processes.

Further progress in interpreting the crab cycle will occur with additional life-history information describing the age-size relationship, continuation of the effort-corrected landing time series, comparison of the time series among different fisheries, and experimental assessment of multiple processes. Until the age-size relationship is described, any interpretation that is based on the assumption of a 3- to 5-year lag with landing fluctuations will remain suspect. Time series continuation is critical because it will increase the power of cross-correlation and cross-spectral analyses. Similarly, comparing longer time series among the fisheries should reveal the scale of processes responsible for fluctuations in crab landings.

Increasing Global Temperature

Increasing global surface temperature between 1861 and 1989 (Figure 20-3) has been the subject of significant efforts to identify the responsible processes. These efforts have focused on the role of anthropogenic emission of CO_2 and SO_4-aerosols in, respectively, increasing or decreasing radiative forcing and the amount of incoming solar radiation (Houghton et al. 1990). Increasingly sophisticated simulation models have been used to predict climate sensitivity to increasing CO_2 and SO_4-aerosols, and correlation of model predictions with empirical trends has been applied. The relative emphasis placed on these processes has varied in response to short-term cooling trends and, more recently, lower-than-expected warming from increased CO_2 levels. Variability in the times series and uncertainty in model predictions have prevented the identification of clear and necessary responses to anthropogenic emission of CO_2 and SO_4-aerosols.

The global temperature time series is relatively long, but there is some inherent error: records are available since 1861, but Folland et al. (1990) suggest that urbanization and incomplete spatial coverage may introduce an error of 0.1° C. Remarkably similar estimates for the rate of warming have been produced by simple and sophisticated analytical methods. Using simple linear trends analysis, Folland et al. (1990) estimated rates of 0.0045° C/yr since 1861 and 0.0053° C/yr since 1880, and all rates were other than zero. Bloomfield (1992) suggested warming rates of 0.0038° C/yr from 1861 to 1989 and 0.0057° C/yr from 1880 to 1989 using autoregressive (AR1 and AR4) and autoregressive integrated moving average (ARIMA 0,d,0 and ARIMA 0,d,1) techniques. The rate estimates were all different than zero and they were nearly identical for all the techniques, though the confidence intervals were different.

The basis for the propositions that increasing atmospheric CO_2 and SO_4-aerosols will increase or decrease global surface temperatures lies in the empirically derived radiative properties of CO_2 and the reflective properties of SO_4-aerosols. However, describing the realized influence of CO_2 and SO_4-aerosols has been complicated by limited knowledge of cloud, water vapor, and ocean-atmosphere parameters, and by the geographic extent of the increase in SO_4-aerosols. Simulation models have been evolving to improve atmospheric parameterization with a goal of predicting climate sensitivity to a doubling of the atmospheric components.

Since 1795, atmospheric CO_2 has increased from 279 ppm to 356 ppm (Shine et al. 1990). Using this increase in concentration, and the radiative properties of CO_2, it is estimated that radiative forcing at the top of the troposphere has increased 1.5 W/m^2 since 1795 (Shine et al. 1990). Evidence from Greenland ice cores suggests that the Northern Hemisphere SO_4-aerosol concentration nearly tripled between 1895 and 1978 (Neftel et al. 1985). Unfortunately, there has been no continuous, direct recording of SO_4-aerosol concentration outside urban and industrial areas in the Northern Hemisphere (Charlson 1988). Theoretically, SO_4-aerosols can reduce incoming radiation directly, and, by increasing cloud albedo, Wigley (1989) estimated that, in the Northern Hemisphere, incoming radiation could have been reduced by 0.5–1.5 W/m^2 since 1900.

Early estimates of climate sensitivity to CO_2 doubling of 2.1° C (Callender

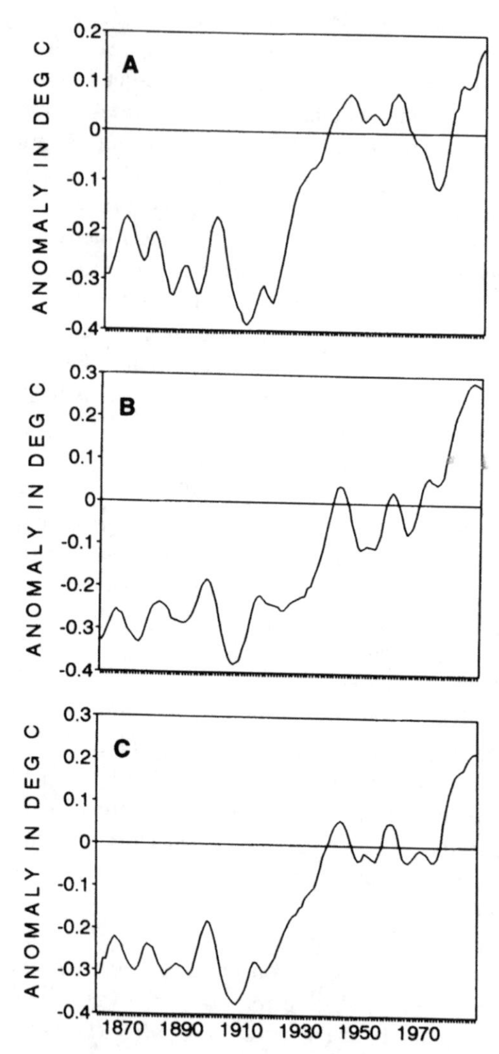

Figure 20-3. Combined land, air, and sea surface temperatures, 1861–1989, relative to 1951–1980. (A) Northern Hemisphere; (B) Southern Hemisphere; (C) global. (Adapted from C. K. Folland, T. R. Karl, and K. Y. A. Vinnikov, 1990, "Observed Climate Variations and Change," in J. T. Houghton, G. J. Jenkins, and J. J. Ephraums (eds.), *Climate Change: The IPCC Scientific Assessment*, pp. 194–238. Cambridge University Press, U.K. By permission of the Intergovernmental Panel on Climate Change.)

1949) and 3.6° C (Plass 1956) were based on simple dry-body calculations. Moller's (1963) one-dimensional atmospheric model estimated a 1.5° C sensitivity when the reflective properties of clouds was included. The addition of convection into a one-dimensional model by Manabe and Wetherald (1967) estimated a 2.4° C sensitivity. General circulation models (GCMs) led to more sophisticated parameterization and feedback mechanisms, and resulted in theoretically sound predictions of temperature increases in the troposphere and decreases in the stratosphere (Henderson-Sellers 1990). The first GCM suggested a 2.93° C sensitivity (Manabe and Wetherald 1975). Over the past ten years, simulation models have advanced in the parameterization of clouds, water vapor and ocean-atmosphere temperature feedback, and in finer-scale spatial resolution (Cubash and Cess 1990). Sensitivity estimates from these GCMs range from 1.9° to 5.2° C, with most estimates between 3.5° and 4.5° C.

Correlation analysis has been used to compare the simulation model projections with the empirical 0.3°–0.6° C warming and 25% increase in atmospheric CO_2 since 1861. Using an upwelling-diffusion climate model, Wigley and Raper (1991) suggested that the observed temperature series is most similar to a sensitivity of 1.43° C. They estimated the 95% confidence limit at 0.73°–4.25° C using variability from model parameters of ocean mixing, observed climate variability since 1881, and error associated with measurement of the temperature record. Bloomfield (1992) used ARIMA models to estimate a maximum likelihood of 1.39° C with a 0.69°–2.19° C 95% confidence interval for climate sensitivity from the observed time series. However, this estimate changed when he included the 0.1° C measurement error in the temperature time series proposed by Folland et al. (1990).

These evaluations suggest that observed changes in global temperature are at the low end of the predictions from model estimates of climate sensitivity. Because the confidence interval overlaps with modeled estimates of climate sensitivity, the CO_2 forcing hypothesis cannot be rejected, though Wigley and Raper (1991) suggest that even if global temperatures increase 0.6° C in the next 10 years, there is too much variation in the observed record and modeled estimates to provide conclusive evidence for the proposed CO_2 process.

The low coherence between modeled predictions and empirical sensitivity estimates has resulted in two basic responses: build better models that pay particular attention to a lag in responses (Bretherton et al. 1990), and reevaluate other processes such as the effect of SO_4-aerosols that may be dampening the expected radiative influence of CO_2 (e.g., Wigley and Raper 1991).

The recent search for a cooling process is not the first in the many efforts to interpret processes responsible for global temperature changes. In the early 1970s, projections of significant SO_4-aerosol cooling coincided with a global temperature cooling between 1940 and 1970 (Rasool and Schneider 1971; Study of Man's Impact on Climate 1971). These projections were largely dismissed on technical grounds, and the warming trend after the late 1970s eroded their popularity.

More recent attention to SO_4-aerosol cooling processes has incorporated more sophisticated analysis than previously, but the impetus again has been the need to explain cooler-than-expected conditions (Bloomfield 1992; Wigley and Barnett 1990; Wigley and Raper 1991). In fact, Wigley (1991) suggested that reduced CO_2 emissions associated with restrictions on fossil fuel burning could lead to increased global temperatures if less burning also resulted in reduced SO_2 emissions. This suggestion is based on the different half-lives of the pollutants; SO_4-aerosols remain in the atmosphere about one year and CO_2 concentrations persist for many decades. Because the empirical forcing characteristics of SO_4-aerosols are largely unknown, however, the expected cooling effects can only be a matter of speculation (Shine et al. 1990; Wigley 1991).

Interpretations of the CO_2 and SO_4-aerosol influences on global temperature since 1861 have changed because of a longer time series and correlation analysis. The longer time series was instrumental in putting the cooling trend of 1940–1970 into perspective. If the series had stopped in 1974, propositions concerning significant SO_4-aerosol cooling would have found greater favor than was the case when the warming trend continued and increased after the mid-1970s. The SO_4-aerosol cooling hypothesis was resurrected when correlations between empirical and predicted sensitivity turned out to be lower than expected.

Progress will be made in interpreting the relative importance of CO_2 and SO_4-aerosols to changing global temperatures with the continuation of the temperature and CO_2 time series, expansion and continuation of the SO_4-aerosol series, improved ancillary information describing ocean-atmosphere temperature and CO_2 interactions, and simulation models that slowly increase CO_2 and have finer geographic resolution. Continuation of all the time series will increase the power of correlation analyses and provide temporal perspective. Understanding the role of deep-ocean water in CO_2 cycling and heat absorption will be critical to realistic simulation models (Bretherton et al. 1990). Lagging CO_2 increases and finer geographic resolution will also improve model realism. The irony of simulation model improvement, however, is the similarity between climate sensitivity estimates produced by the early, most simple models and those produced by the recent, more sophisticated models.

It is not surprising to observe an ebb and flow in the popularity of seemingly diametrically opposed processes when there is variation in the time series pattern. In this situation, there is substantial theoretical and empirical evidence that two such opposed processes both play important roles in global temperature fluctuations. However, because their relative roles and the importance of anthropogenic activities have eluded description, there is some uncertainty as to the appropriate societal response.

Summary

Changes in the interpretation of the processes responsible for the three time series reviewed in this chapter illustrate that several factors have been important in the development of realistic and accurate interpretations. These factors include time series analytical techniques; the availability of time series that are long, continuous, and geographically distinct; the availability of ancillary information for correlation analysis; simulation models; and experimentation.

Time series analytical techniques were critical to descriptions of the lynx and crab cycles. Effort-adjusted estimates of crab landings were important in changing process interpretations; by contrast, the absence of data on harvest effort in the case of the Canada lynx casts doubt on the lynx cycle description and proposed explanatory processes.

The importance of long and continuous time series was evident in all three examples. As cyclic crab landings waned, research effort was renewed and new processes were proposed. Similarly, process interpretation has changed in response to short-term temperature fluctuations. The point appears most relevant for the lynx time series: although this record is the longest of the three, its termination over 80 years ago has prevented conclusive determination of chaotic or quasi-periodic behavior. The apparent synchrony of lynx pelt sales for 11 geographic regions was a major stimulus for the investigation of population cycles. Attention has only recently been paid to the geographic variation in crab landings, but it shows significant promise for improving interpretations.

The utility of correlation analysis was revealed in all three cases. The absence of a coincident snowshoe hare series severely limits the evaluation of a predator-prey process. Correlation analysis was used to reject several proposed processes in the crab landing cycle, but the paucity of long and extensive ancillary data prevents further use of these techniques. CO_2 and SO_4-aerosol interpretation has been refined with correlations between model predictions and empirical patterns, although correlation analysis has not led to conclusive descriptions.

Simulation modeling has been aggressively pursued in all three cases, but its influence has been the greatest in the case of interpretations of the global temperature series. Interestingly, the disparity between empirical climate sensitivity and model predictions has prompted significant reevaluations of proposed explanatory processes.

Experimentation has been used only for the lynx cycle, and unfortunately the response of only the snowshoe hare was recorded. The many difficulties of performing experiments in oceans has prevented the use of experimentation for the crab series. The global temperature series does not have the benefit of a control planet for use in experimentation, though one might say that civilization is performing a significant atmospheric experiment by emitting large amounts of CO_2 and SO_4-aerosols.

How do these lessons apply to classic problems in ecology and contemporary conundrums for our management of natural resources? The simplest and most obvious lessons are the importance of long-term data and the utility of time series patterns to define fruitful areas for experimental efforts. But these lessons are of little use for urgent problems that demand a response. In such situations, incorporating probability estimates into time series descriptions and proposed processes is a very valuable lesson. For example, the recent efforts to evaluate climate sensitivity have enlarged the role of variability in estimates by describing confidence intervals based on model variability and measurement errors. The utility of disclosing the level of uncertainty may be the most significant lesson of all to be drawn from a review of these time series.

References

Akcakaya, H. R. 1992. Population cycles of mammals: Evidence for a ratio-dependent predation hypothesis. *Ecological Monographs* **62:**142–199.

Arditi, R. 1979. Avoiding significance tests in stepwise regression: A Monte Carlo method applied to meteorological theory for the Canadian lynx cycle. *International Journal of Biometeorology* **33:**24–26.

Berryman, A. A. 1991. Can economic forces cause ecological chaos? The case of the northern California Dungeness crab fishery. *Oikos* **62:**106–109.

Bloomfield, P. 1992. Trends in global temperature. *Climatic Change* **21:**1–16.

Botsford, L. W. 1981. Comment on cycles in the northern California Dungeness crab population. *Canadian Journal of Fisheries and Aquatic Sciences* **38:**1295–1296.

———. 1986. "Population Dynamics of the Dungeness Crab (*Cancer magister*)." In G. S. Jamieson and N. Bourne (eds.), *North Pacific Workshop on Stock Assessment and Management of Invertebrates*. Canadian Special Publication in Fisheries and Aquatic Science 92, pp. 140–153.

Botsford, L. W., D. A. Armstrong, and J. M. Shenker. 1989. "Oceanograpnic Influences on the Dynamics of Commercially Fished Populations." In M. R. Landry and B. M. Hickey (eds.), *Coastal Oceanography of Washington and Oregon*. Elsevier Science Publishers, Amsterdam, pp. 511–565.

Botsford, L. W., R. D. Methot, Jr., and W. E. Johnston. 1983. Effort dynamics of the northern California Dungeness crab (*Cancer magister*) fishery. *Canadian Journal of Fisheries and Aquatic Sciences* **40:**337–346.

Botsford, L. W., and D. E. Wickham. 1975. Correlation of upwelling index and dungeness crab catch. *U.S. Fish and Wildlife Service Fisheries Bulletin* **73:**901–907.

———. 1978. Behavior of age-specific, density-dependent models and the northern California dungeness crab (*Cancer magister*) fishery. *Journal of the Fisheries Resources Board of Canada* **35:**833–843.

Bretherton, F. P., K. Bryan, and J. D. Woods. 1990. "Time-Dependent Greenhouse-Gas-Induced Climate Change." In J. T. Houghton, G. J. Jenkins, and J. J. Ephraums (eds.), *Climate Change: The IPCC Scientific Assessment*. Cambridge University Press, London, pp. 173–193.

Bulmer, M, G. 1974. A statistical analysis of 10 year cycle in Canada. *Journal of Animal Ecology* **43:**701–718.

———. 1976. The theory of predator prey oscillations. *Theoretical Population Biology* **9:**137–150.

———. 1978. "The Statistical Analysis of the Ten Year Cycle." In H. H. Shugart (ed.), *Time Series and Ecological Processes*. Society for Industrial and Applied Mathematics, Philadelphia, pp. 141–153.

Callender, G. S. 1949. Can carbon dioxide influence weather? *Weather* **4:**310–314.

Campbell, M. J., and A. M. Walker. 1977. A survey of statistical work on the Mackenzie River series of annual Canadian lynx trappings for the years 1821–1934 and a new analysis. *Journal of the Royal Statistical Society* A 140, Part **4:**411–431.

Charlson, R. J. 1988. "Have the Concentrations of Tropospheric Aerosols Changed?" In F. S. Rowland and I. S. A. Isaksen (eds.), *The Changing Atmosphere*. J. Wiley & Sons, Chichester, U.K., pp. 70–89.

Cleaver, F. C. 1949. Preliminary results of the coastal crab (*Cancer magister*) investigation. Washington Department of Fisheries Biology, Rep. 49A:47–82.

Cole, L. C. 1951. Population cycles and random oscillations. *Journal of Wildlife Management* **15:**233–252.

———. 1954. Some features of random population cycles. *Journal of Wildlife Management* **19:**2–24.

Cubash, U., and R. D. Cess. 1990. "Processes and Modelling." In J. T. Houghton, G. J. Jenkins, and J. J. Ephraums (eds.), *Climate Change: The IPCC Scientific Assessment*. Cambridge University Press, London, pp. 69–91.

Elton, C. S. 1924. Periodic fluctuations in the numbers of animals: Their causes and effects. *Journal of Experimental Biology* **2:**119–163.

———. 1942. *Voles, Mice and Lemmings*. Clarendon Press, Oxford, U.K.

Elton, C. S., and M. Nicholson. 1942. The ten-year cycle in numbers of lynx in Canada. *Journal of Animal Ecology* **11:**215–244.

Finerty, J. P. 1980. *The Population Ecology of Cycles in Small Mammals*. Yale University Press, New Haven, CT.

Folland, C. K., T. R. Karl, and K. Y. A. Vinnikov. 1990. "Observed Climate Variations and Change." In J. T. Houghton, G. J. Jenkins, and J. J. Ephraums (eds.), *Climate Change: The IPCC Scientific Assessment*. Cambridge University Press, London, pp. 194–238.

Fox, J. R. 1978. Forest fires and the snowshoe hare-Canada lynx cycle. *Oecologia* **31:**349–374.

Gotshall, D. W. 1977. Stomach contents of northern California dungeness crab, *Cancer magister*. *California Fish and Game* **63:**43–51.

Hankin, D. G. 1985. "Proposed Explanations for the Fluctuations in Abundance of Dungeness Crabs: A Review and Critique." In *Proceedings of the Symposium on Dungeness Crab Biology and Management*. University of Alaska Sea Grant Rep. **85-3:**305–326.

Henderson-Sellers, A. 1990. Modelling and monitoring 'greenhouse' warming. *Trends in Ecology and Evolution* **5:**270–275.

Houghton, J. T., G. J. Jenkins, and J. J. Emphraums (eds.), 1990. *Climate Change: The IPCC Scientific Assessment*. Cambridge University Press, London.

Jamieson, G. S. 1986. Implications of fluctuations in recruitment in selected crab populations. *Canadian Journal of Fisheries and Aquatic Science* **43:**2085–2098.

Jamieson, G. S., and D. A. Armstrong. 1991. Spatial and temporal recruitment patterns of dungeness crab in the northeast Pacific. *Memoirs of the Queensland Museum* **31:**365–381.

Jenkins, G. M., and D. G. Watts. 1968. *Spectral Analysis and its Applications*. Holden-Day, San Francisco.

Johnson, D. F., L. W. Botsford, R. D. Methot, Jr., and T. C. Wainwright. 1986. Wind stress and cycles in dungeness crab (*Cancer magister*) catch off California, Oregon, and Washington. *Canadian Journal of Fisheries and Aquatic Science* **43:**838–845.

Keith, L. B. 1974. Some features of population dynamics in mammals. *Proceedings of the International Congress of Game Biologists* **11:**17–58.

———. 1983. Role of food in hare population cycles. *Oikos* **40:**385–395.

Krebs, C. J., S. Boutin, and B. S. Gilbert. 1986a. A natural feeding experiment on a declining snowshoe hare population. *Oecologia* **70:**194–197.

Krebs, C. J., B. S. Gilbert, S. Boutin, and J. N. M. Smith. 1986b. Population biology of snowshoe hares. I. Demography of food supplemented populations in southern Yukon, 1976–84. *Journal of Animal Ecology* **55:**963–982.

Lack, D. 1954. *The Natural Regulation of Animal Numbers*. Oxford University Press, London.

MacLulich, D. A. 1937. Fluctuations in the numbers of varying hare. *University of Toronto Studies in Biology Series*, No. 43.

Magnuson, J. J. 1990. The invisible present. *BioScience* **40(7):**495–501.

Manabe, S., and R. T. Wetherald. 1967. Thermal equilibrium of the atmosphere with a given distribution of relative humidity. *Journal of the Atmospheric Sciences* **24:**241–259.

———. 1975. The effects of doubling the CO_2 concentration of the climate of a general circulation model. *Journal of the Atmospheric Sciences* **3:**3–15.

McKelvey, R., and D. Hankin. 1981. Reply to comment on cycles in the northern California dungeness crab population. *Canadian Journal of Fisheries and Aquatic Science* **38:**1296–1297.

McKelvey, R., D. Hankin, K. Yanosko, and C. Snygg. 1980. Stable cycles in multistage recruitment models: An application to the northern California dungeness crab (*Cancer magister*) fishery. *Canadian Journal of Fisheries and Aquatic Science* **37:**2323–2345.

Methot, R. D., Jr. 1986. "Management of Dungeness Crab Fisheries." In: G. S. Jamieson and N. Bourne (eds.), North Pacific Workshop on stock assessment and management of invertebrates. *Canadian Special Publication on Fisheries and Aquatic Science* 92, pp. 326–334.

———. 1989. "Management of a Cyclic Resource: The Dungeness Crab Fisheries of the Pacific Coast of North America." In J. F. Caddy (ed.), *Marine Invertebrate Fisheries: Their Assessment and Management*. John Wiley & Sons, New York, pp. 205–223.

Moller, F. 1963. On the influence of changes in the CO_2 concentration in air on the radiation of the Earth's surface and on climate. *Journal of Geophysics Research* **68:**3877–3886.

Moran, P. A. 1953. The statistical analysis of the Canadian lynx cycle: I. Structure and prediction. *Australian Journal of Zoology* **1:**163–173.

Neftel, A., J. Beer, H. Oeschger, F. Zurcher, and R. C. Finkel. 1985. Sulphate and nitrate concentrations in snow from south Greenland 1895–1978. *Nature* **314:**611–613.

Pacific Marine Fisheries Commission. 1965. Discussion following the report on dungeness crabs. Pacific Marine Fisheries Commission, Annual Rep. **16/17:**38–39.

———. 1978. Dungeness crab project of the state-federal fisheries management program. Pacific Marine Fisheries Commission, 200 pp. with appendices.

———. 1990. Pacific coast fishery review reports: Dungeness crab fishery in 1989–1990. Pacific Marine Fisheries Commission, Annual Rep. 43:10.

Peterson, W. T. 1973. Upwelling indices and annual catches of dungeness crab, *Cancer magister*, along the west coast of the United States. *U.S. Fish and Wildlife Service Fisheries Bulletin* **71:**902–910.

Plass, G. N. 1956. Carbon dioxide theory of climate. *Tellus* **8:**140–153.

Poole, R. 1968. The market crab—will it come back? *Outdoor California* **29(1):**1–3.

Poole, R., and D. Gotshall. 1965. Market crab fishery: Facts on the fishery and information on new regulations. *Outdoor California* **26(9):**7–8.

Rasool, S. I., and S. H. Schneider. 1971. Atmospheric carbon dioxide and aerosols: Effects of large increases on global climate. *Science* **173:**138–141.

Schaffer, W. M. 1984. Stretching and folding in lynx fur returns: Evidence for a strange attractor? *American Naturalist* **124:**798–820.

———. 1985. Order and chaos in ecological systems. *Ecology* **66:**93–106.

Shine, K. P., R. G. Derwent, D. J. Wuebbles, and J. J. Morcette. 1990. "Radiative Forcing of Climate." In J. T. Houghton, G. J. Jenkins, and J. J. Ephraums (eds.), *Climate Change: The IPCC Scientific Assessment*. Cambridge University Press, London, pp. 41–68.

Sinclair, A. R. E., J. M. Gosline, G. Holdsworth, C. J. Krebs, S. Boutin, J. N. M. Smith, R. Boonstra, and M. Dale. 1993. Can the solar cycle and climate synchronize the snowshoe hare in Canada? Evidence from tree rings and ice cores. *American Naturalist* **141:**173–198.

Smith, B. D., and G. S. Jamieson. 1991. Possible consequences of intensive fishing for mating opportunities of dungeness crabs. *Transactions of the American Fisheries Society* **120:**650–653.

Smith, J. N. M., C. J. Krebs, A. R. E. Sinclair, and R. Boonstra. 1988. Population biology of snowshoe hares. II. Interactions with winter food plants. *Journal of Animal Ecology* **57:**269–286.

Study of Man's Impact on Climate. 1971. *Inadvertent Climate Modification*. MIT Press, Cambridge, MA.

Tong, H. 1977. Some comments on the Canadian lynx data. *Journal of the Royal Statistical Society* A 140, Part **4:**432–436.

———. 1990. *Non-linear Time Series: A Dynamical Approach*. Oxford University Press, London.

Trostel, K., A. R. E. Sinclair, C. J. Walters, and C. J. Krebs. 1987. Can predation cause the 10 year hare cycle? *Oecologia* **74:**185–192.

Turchin, P., and A. Taylor. 1992. Complex dynamics in ecological time series. *Ecology* **73:**289–305.

Wigley, T. M. L. 1989. Possible climate change due to SO_2-derived cloud condensation nuclei. *Nature* **339:**365–367.

———. 1991. Could reduced fossil-fuel emissions cause global warming? *Nature* **349:**503–506.

Wigley, T. M. L., and T. P. Barnett. 1990. "Detection of the Greenhouse Effect in the Observations." In J. T. Houghton, G. J. Jenkins, and J. J. Ephraums (eds.), *Climate Change: The IPCC Scientific Assessment*. Cambridge University Press, London, pp. 237–255.

Wigley, T. M. L., and S. C. B. Raper. 1991. "Detection of the Enhanced Greenhouse Effect on Climate." In J. Jager and H. L. Ferguson (eds.), *Climate Change: Science, Impacts and Policy*. Cambridge University Press, London, pp. 231–242.

Wild, P. W., and R. N. Tasto (eds.). 1983. Life history, environment and maraculture studies of the dungeness crab, *Cancer magister*, with emphasis on the central California fishery resource. *California Department of Fish and Game Fisheries Bulletin* **172:**1–352.

Index